Gas Phase Ion Chemistry

Volume 3
IONS AND LIGHT

CONTRIBUTORS

J. L. BEAUCHAMP
J. H. BEYNON
VERONICA M. BIERBAUM
JOHN I. BRAUMAN
ALAN CARRINGTON
PAUL S. DRZAIC
ROBERT C. DUNBAR
G. BARNEY ELLISON
F. M. HARRIS
RICHARD A. KENNEDY
STEPHEN R. LEONE
J. J. LEVENTHAL
W. C. LINEBERGER
JOHN P. MAIER
JEFFREY MARKS
ROY D. MEAD
CH. OTTINGER
AMY E. STEVENS
FRITZ THOMMEN
L. R. THORNE

Gas Phase Ion Chemistry

Edited by *MICHAEL T. BOWERS*

DEPARTMENT OF CHEMISTRY
UNIVERSITY OF CALIFORNIA
SANTA BARBARA, CALIFORNIA

Volume 3

IONS AND LIGHT

1984 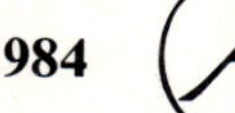

ACADEMIC PRESS, INC.
(Harcourt Brace Jovanovich, Publishers)

Orlando San Diego San Francisco New York London
Toronto Montreal Sydney Tokyo São Paulo

ACADEMIC PRESS, INC.
Orlando, Florida 32887

United Kingdom Edition published by
ACADEMIC PRESS, INC. (LONDON) LTD.
24/28 Oval Road, London NW1 7DX

Library of Congress Cataloging in Publication Data
(Revised for volume 3)
Main entry under title:

Gas phase ion chemistry.

Includes bibliographies and indexes.
Vol. 3 has special title: Ions and light.
1. Chemical reaction, Conditions and laws of--
Addresses, essays, lectures. 2. Gases, Ionized.
I. Bowers, Michael T.
QD501.G324 541.3'9 78-19666
ISBN 0-12-120803-6 (V. 3)

PRINTED IN THE UNITED STATES OF AMERICA

84 85 86 87 9 8 7 6 5 4 3 2 1

CONTENTS

Chapter 19 Photodissociation in Beams: Organic Ions

F. M. HARRIS AND J. H. BEYNON

Chapter 20 Photodissociation in the ICR Ion Trap

ROBERT C. DUNBAR

Chapter 21 Electron Photodetachment from Gas Phase Molecular Anions

PAUL S. DRZAIC, JEFFREY MARKS, AND JOHN I. BRAUMAN

CONTRIBUTORS

Numbers in parentheses indicate the pages on which the authors' contributions begin.

J. L. BEAUCHAMP (41), Arthur Amos Noyes Laboratory of Chemical Physics, California Institute of Technology, Pasadena, California 91125

J. H. BEYNON (99), Royal Society Research Unit, University College of Swansea, Swansea SA2 8PP, United Kingdom

VERONICA M. BIERBAUM (1), Department of Chemistry, University of Colorado, and Joint Institute for Laboratory Astrophysics, University of Colorado and National Bureau of Standards, Boulder, Colorado 80309

JOHN I. BRAUMAN (167), Department of Chemistry, Stanford University, Stanford, California 94305

ALAN CARRINGTON (393), Department of Chemistry, University of Southampton, Southampton, England

PAUL S. DRZAIC* (167), Department of Chemistry, Stanford University, Stanford, California 94305

ROBERT C. DUNBAR (129), Department of Chemistry, Case Western Reserve University, Cleveland, Ohio 44106

G. BARNEY ELLISON (1), Department of Chemistry, University of Colorado, and Joint Institute for Laboratory Astrophysics, University of Colorado and National Bureau of Standards, Boulder, Colorado 80309

F. M. HARRIS (99), Royal Society Research Unit, University College of Swansea, Swansea SA2 8PP, United Kingdom

* Present address: Raychem Corporation, Menlo Park, California 94025.

RICHARD A. KENNEDY (393), Department of Chemistry, University of Southampton, Southampton, England

STEPHEN R. LEONE (1), Department of Chemistry, University of Colorado, and Joint Institute for Laboratory Astrophysics, University of Colorado and National Bureau of Standards, Boulder, Colorado 80309

J. J. LEVENTHAL (309), Department of Physics, University of Missouri —St. Louis, St. Louis, Missouri 63121

W. C. LINEBERGER* (213), Department of Chemistry, University of Colorado, and Joint Institute for Laboratory Astrophysics, University of Colorado and National Bureau of Standards, Boulder, Colorado 80309

JOHN P. MAIER (357), Physikalisch-Chemisches Institut, Universität Basel, Basel, Switzerland

JEFFREY MARKS (167), Department of Chemistry, Stanford University, Stanford, California 94305

ROY D. MEAD (213), Department of Chemistry, University of Colorado, and Joint Institute for Laboratory Astrophysics, University of Colorado and National Bureau of Standards, Boulder, Colorado 80309

CH. OTTINGER (249), Max-Planck-Institut für Strömungsforschung, D 3400 Göttingen, Federal Republic of Germany

AMY E. STEVENS (213), Department of Chemistry, University of Colorado, and Joint Institute for Laboratory Astrophysics, University of Colorado and National Bureau of Standards, Boulder, Colorado 80309

FRITZ THOMMEN (357), Physikalisch-Chemisches Institut, Universität Basel, Basel, Switzerland

L. R. THORNE† (41), Arthur Amos Noyes Laboratory of Chemical Physics, California Institute of Technology, Pasadena, California 91125

* Present address: Department of Chemistry, University of Colorado, Boulder, Colorado 80309.

† Present address: Sandia National Laboratories, Livermore, California 94550.

PREFACE

The field of gas phase ion chemistry is about twenty years old, although early work began nearly 30 years ago. The growth of this field over the past ten years has been explosive. There are now at least three journals completely dedicated to gas phase ion chemistry of one sort or another, and ion chemistry papers are making major contributions to most main line chemical journals. The roots of this growth are firmly set in technological development. Modern gas phase work requires a sophisticated combination of high vacuum, electronic, and computer technology. With the present high level of development of these technologies, a wide assortment of experiments are possible and, indeed, have been developed.

In this book the emphasis is on one aspect of gas phase ion chemistry—ions and light. Within this subfield, a variety of techniques have emerged, and more are being developed each day. This subfield is really only about ten years old, although a few scattered papers appeared earlier. There are two broad subdivisions. In the first, ions are formed by electron impact, electrical discharge, or ion–molecule reactions. The chemistry or structure of these ions is then probed by subjecting the ion to radiation—usually supplied by a laser. Of the chapters in this book, six fall into this category. The second broad subdivision is the use of light emitted by excited molecules to characterize either the chemistry that formed the excited ion, the structure of the excited ion, or both. Four of the ten chapters fall into this category.

There are some absences in the book. There is nothing on multiphoton ionization. At the time this book was organized, it was not clear that this technique would lead to interesting information on *ion* chemistry. Hence, I did not include it. In the past year some clever uses have been made of multiphoton ionization for preparing state-selected ions that subsequently undergo unimolecular reactions. This promises to be a very powerful technique in the future. Another area that was omitted was photoion–

photoelectron coincidence studies. This also was intentional since an excellent review of the field had been included in Volume 1. Several other areas were omitted, notably, spectroscopy of ions in the solid state and microwave spectroscopy of ions of interstellar interest. These were omitted because of space limitations and my own current inclination toward experiments with dynamical applications.

The group of ten chapters that is included is evenly split between the chemistry of "largish" polyatomic molecules (five chapters) and small systems (five chapters). The chapters by Drzaic, Marks, and Brauman, and Mead, Stevens, and Lineberger between them contain a complete up-to-date compendium of electron affinities.

Each set of authors was chosen because they are world leaders in their subdisciplines. Their choice of topics and method of organization are their own. I am grateful for their willingness to contribute to this volume, for their observance of deadlines, and for the uniform high quality of their contributions.

CONTENTS OF PREVIOUS VOLUMES

Chapter 17

Flowing afterglow studies of ion reaction dynamics using infrared chemiluminescence and laser-induced fluorescence

VERONICA M. BIERBAUM

*G. BARNEY ELLISON**

STEPHEN R. LEONE†

DEPARTMENT OF CHEMISTRY
UNIVERSITY OF COLORADO
AND
JOINT INSTITUTE FOR LABORATORY ASTROPHYSICS
UNIVERSITY OF COLORADO AND NATIONAL BUREAU OF STANDARDS
BOULDER, COLORADO

* Alfred P. Sloan Fellow.

† Staff member, Quantum Physics Division, National Bureau of Standards, Boulder, Colorado.

ISBN 0-12-120803-6

I. INTRODUCTION

An understanding of energy disposal in ion–molecule reactions has been a major research goal since the earliest days of mass spectrometry. For many ion reactions, the formation of translationally and electronically excited products and their state distributions has been determined. However, due to experimental constraints, very little is known about the channeling of reaction exothermicity into vibrational and rotational levels of the ground electronic state of the product, e.g.,

$$A^- + BC \rightarrow AB(v, J) + C^-. \tag{1}$$

There are many thermal energy (25-meV) ion–molecule processes which can energetically access these levels and whose dynamics remain largely unexplored. In this chapter we describe new experiments to determine the initial vibrational energy distribution of ion reaction products in their ground electronic states; in the most recent work, information on nascent rotational populations and on energy transfer processes of ions is also obtained. These studies are made possible by monitoring optically the excited products of ion–molecule reactions carried out in a flowing afterglow apparatus. Two complementary optical techniques are used: (1) direct observation of wavelength-dispersed infrared chemiluminescence from $AB(v, J)$ and (2) laser-induced fluorescence detection, i.e., laser excitation of product $AB(v, J)$ molecules to bound electronic states and detection of the resulting visible fluorescence.

Methods to probe the final product states resulting from neutral–neutral reactions (A + BC) are well established. A similar experimental program concerned with ions is a much more difficult enterprise for several reasons. The number density of reacting ions is much lower (by several orders of magnitude) than for comparable neutral reactants. Reactant ions are difficult to prepare under known, controllable conditions. Moreover, in many cases, thermochemical and spectroscopic information is not available for ions. To overcome some of these difficulties we employ the flowing afterglow technique, which provides high densities of thermal ions and serves as a well-characterized medium for the study of ion reaction kinetics and dynamics. This method, which offers a variety of valuable features, also imposes a serious limitation; due to the presence of helium carrier gas and other molecules, energy transfer and collisional deactivation processes of the excited ion and neutral products must be fully understood.

In the following sections we will survey the optical techniques and the flowing afterglow method; we will then present an overview of the vibrational state distributions and their dynamical interpretations for the wide variety of ion–molecule processes studied in the past several years.

II. SURVEY OF OPTICAL TECHNIQUES

Modern explorations of reaction dynamics make extensive use of optical techniques such as visible and infrared chemiluminescence and laser-induced fluorescence. These methods are reviewed by Levy (1979) for neutral reactions and by Tiernan and Lifshitz (1981) for ion–molecule reactions. Several other chapters in this book consider the use of visible or ultraviolet chemiluminescence to obtain product state distributions of ion–molecule reactions carried out at high collision energies (Ottinger, Leventhal). A few thermal energy ion–molecule reactions are sufficiently exothermic to produce visible or ultraviolet chemiluminescence (Ferguson *et al.*, 1969; Tsuji *et al.*, 1982; Bowers and Laudenslager, 1976; Laudenslager, 1979). In this section we briefly survey the techniques of infrared chemiluminescence and laser-induced fluorescence which are used in our studies of ion–molecule interactions.

A. *Infrared Chemiluminescence*

Spectrally resolved infrared emission is a powerful tool for studying reaction dynamics (Smith, 1980). The infrared region provides information on excited vibrational state distributions, and under high resolution one can obtain rotational state populations as well. Polanyi and co-workers have exploited this method to obtain detailed product state information on numerous $A + BC \rightarrow AB(v, J) + C$ reactions (Anlauf *et al.*, 1967, 1972; Polanyi, 1972). In the typical experiment, infrared chemiluminescence is observed from excited diatomic products of chemical reactions carried out in a low-pressure chamber. The spectral data are deconvoluted with known transition strengths to obtain relative vibrational and rotational populations. Although the first experiments were performed with monochromators, addition of a Fourier transform infrared spectrometer provides a powerful tool for rapid acquisition of high-resolution data (Charters *et al.*, 1971; Moehlmann *et al.*, 1974; Agrawalla *et al.*, 1981).

Typically, infrared radiative lifetimes are long, ranging from 10^{-3} sec to 1 sec, and infrared emission signals are obscured by blackbody photons emitted by room temperature objects. Thus, in order to obtain an adequate signal-to-noise ratio, the concentration of emitting species must be large. Since maximum achievable ion densities are typically many orders of magnitude smaller than neutral radical concentrations, infrared emission signals in a comparable "best case" ion reaction will be much weaker than for a neutral reaction. Improvements in infrared detection skills (Keyes, 1977; Wolfe and Zissis, 1978) now make it possible to detect the weak signals from ion–molecule infrared chemiluminescence.

Ding (1982) has used a high-resolution Fourier transform infrared emission apparatus to investigate the rotationally resolved spectrum of H_3^+ and its isotopic variations in a large volume, low-pressure reactor. In our work, we employ a low-resolution device [0.05–0.10 μm, full width at half maximum (FWHM)] in order to obtain a higher signal-to-noise ratio. High-resolution detection is generally not necessary for obtaining vibrational distributions in the flowing afterglow environment since rotational populations are completely relaxed by numerous collisions with the helium buffer

TABLE I

Some Diatomic Ions and Neutral Molecules Amenable to Study by Infrared Chemiluminescence

Molecule	Radiative lifetime, $v = 1$–0 (msec)	Frequency[a], 1–0 (cm^{-1})	Anharmonicity[a], $\omega_e x_e$ (cm^{-1})
HF	5.3[b]	3962	89.9
DF	18[b]	2907	45.8
HCl	29[c]	2886	52.8
DCl	105[c]	2091	27.2
HBr	140[d]	2559	45.2
CO	33[e]	2144	13.3
NeH^+	1.3[f]	2678[g]	113[g]
ArH^+	2.3[h]	2589[i]	61.6[i]
NO^+	92[j]	2344	16.3

[a] From Huber and Herzberg (1979), except as noted.
[b] Sileo and Cool (1976).
[c] Herbelin and Emanuel (1974).
[d] Malins and Setser (1980).
[e] Benedict *et al.* (1962).
[f] Rosmus and Reinsch (1980).
[g] Wong *et al.* (1982).
[h] Rosmus (1979).
[i] Brault and Davis (1982).
[j] Werner and Rosmus (1982).

gas. However, a low-resolution device is best suited to molecules whose emission bands are spaced out by large anharmonicities.

A number of diatomic ion and neutral molecules which are amenable to study by low-resolution infrared emission are listed in Table I. Also given are their vibrational frequencies, calculated or measured radiative lifetimes, and anharmonic constants. It is interesting to note that the calculations of Rosmus on NeH^+ and ArH^+ (Rosmus and Reinsch, 1980; Rosmus, 1979) indicate that the radiative lifetimes of positive molecular ions are an order of magnitude shorter than those of their isoelectronic neutral counterparts. Thus, they should make excellent candidates for both spectroscopic studies and state dynamics measurements by infrared emission. In addition to the diatomic molecules in Table I, a number of polyatomic molecules may also be detected, including CO_2, HCN, OCS, N_2O, NO_2, H_2O, and CH_4. However, vibrational relaxation by the helium buffer gas is considerably more rapid for polyatomic species, making it difficult to measure nascent vibrational distributions.

B. Laser-Induced Fluorescence

There are many ion–molecule reactions which cannot be conveniently characterized by infrared chemiluminescence techniques. Laser-induced fluorescence is an independent and complementary technique for studying a whole host of ion and neutral species, often with greater sensitivity and resolution than by infrared emission methods. In addition, the population in $v'' = 0$ can be probed, something which is impossible by infrared chemiluminescence detection. Zare and co-workers were the first to apply the laser-induced fluorescence technique to probe product states of neutral chemical reactions (Schultz *et al.*, 1972). The method has been reviewed in detail (Zare and Dagdigian, 1974; Kinsey, 1977). The laser is used to excite transitions from specific vibrational states to electronic states which then fluoresce in the visible or ultraviolet regions. Because the state selection is accomplished by tuning the narrow-band laser excitation source, rotational resolution is possible. The laser excitation spectrum is deconvoluted with the proper Franck–Condon factors and rotational degeneracies to obtain vibrational population distributions. In the collision-dominated flow tube environment, rotational information is unfortunately lost, but the thermalized rotational distribution makes the measurement of vibrational populations easier. In a single-collision ion reaction apparatus described in Section IV.D, laser-induced fluorescence detection provides complete vibrational and rotational state information. The method can also be used to probe selected states for rates of reaction or deactivation.

Engelking and Smith (1975) were the first to detect an ion, N_2^+, by laser-induced fluorescence. Miller and Bondybey (1980) have reviewed the spectral determinations of many diatomic and polyatomic ions studied by laser-induced fluorescence techniques. Laser-induced fluorescence of ions has been used to investigate a wide variety of physical phenomena. Allison *et al.* (1979) measured the state distribution of N_2^+ produced by electron impact ionization of N_2. Ding *et al.* (1981) detected N_2^+ in an ion beam extracted from a plasma source. Heaven *et al.* (1982) demonstrated the first jet-cooled expansion of ions detected by laser-induced fluorescence. The jet-cooled expansion has been exploited by Lineberger (1983) to study reactive events of ions at low temperatures. Laser-induced fluorescence has also been used to study the spectroscopy (Grieman *et al.*, 1980; Danon *et al.*, 1982) and reactive dynamics (Danon and Marx, 1982; Mahan *et al.*, 1982) of ions trapped in an ion cyclotron resonance device.

TABLE II

Some Diatomic Ions and Neutral Molecules Amenable to Study by Laser-Induced Fluorescence[a]

Molecule	Transition	Wavelength (nm)	τ_{rad} (nsec)	ω_e (cm^{-1}) of ground state
Br_2^+	$A\,^2\Pi_u - X\,^2\Pi_g$	500	—	376
C_2^-	$B\,^2\Sigma_u^+ - X\,^2\Sigma_g^+$	544	—	1781
Cl_2^+	$A\,^2\Pi_u - X\,^2\Pi_g$	500	—	646
Cl_2^+	$B\,^2\Sigma_g^+ - X\,^2\Pi_g$	290	—	646
CN	$B\,^2\Sigma^+ - X\,^2\Sigma^+$	390	60[b]	2069
CO^+	$A\,^2\Pi - X\,^2\Sigma^+$	480	3000[c]	2214
HBr^+	$A\,^2\Sigma^+ - X\,^2\Pi$	350	4500[d]	2442
HCl^+	$A\,^2\Sigma^+ - X\,^2\Pi$	350	2600[e]	2674
N_2^+	$B\,^2\Sigma_u^+ - X\,^2\Sigma_g^+$	393	60[f]	2207
NH	$A\,^3\Pi - X\,^3\Sigma^-$	335	400[g]	3282
NH^+	$A\,^2\Sigma^- - X\,^2\Pi$	450	1000[h]	2922
O_2^+	$A\,^2\Pi_u - X\,^2\Pi_g$	246	700[i]	1905
OH	$A\,^2\Sigma^+ - X\,^2\Pi$	306	700[j]	3738

[a] From Huber and Herzberg (1979), except as noted.
[b] Luk and Bersohn (1973).
[c] Holland and Maier (1972).
[d] Möhlmann and de Heer (1976).
[e] Möhlmann *et al.* (1977).
[f] Smith *et al.* (1975).
[g] Smith *et al.* (1976).
[h] Brzozowski *et al.* (1974).
[i] Jeunehomme (1966).
[j] German (1975).

Table II lists some of the diatomic neutral and ion species amenable to study with laser-induced fluorescence. There are also several triatomic species which can be studied, including NH_2, $CO_2{}^+$, $CS_2{}^+$, N_2O^+, and OCS^+. The method holds promise for at least qualitative study of excitation in more complex species as well.

III. SURVEY OF THE FLOWING AFTERGLOW METHOD

Studies of vibrational product state distributions of gas phase ion reactions are possible by coupling the flowing afterglow technique with the optical detection methods of infrared chemiluminescence and laser-induced fluorescence. This section will describe the conventional flowing afterglow apparatus with its features and constraints, the modifications required for optical studies, as well as experimental details of operation, data acquisition, and analysis.

A. *The Flowing Afterglow*

The flowing afterglow technique was developed in 1963 by Ferguson, Fehsenfeld, and Schmeltekopf of the National Oceanic and Atmospheric Administration Laboratories in Boulder, Colorado. These investigators have used the technique extensively to study the ion chemistry of the earth's atmosphere. In their laboratories and many others, the technique has provided a wealth of data in areas as diverse as fundamental atomic and molecular physics as well as the chemistry of gas phase organic ions. In the 20 years since its inception, the capabilities of the flowing afterglow method have been continuously refined and extended so that today it constitutes a very versatile and productive approach to the study of ion–molecule interactions in the gas phase.

1. *Apparatus Design*

The conventional flowing afterglow technique has been extensively reviewed (Ferguson *et al.*, 1969; Fehsenfeld, 1975a); therefore, the description here will concentrate on aspects important to studies of product state distributions. The apparatus, shown in Fig. 1, consists of a flow tube in which reactant ions are produced in a flow of buffer gas maintained by a large pumping system. The ion–molecule reactions are carried out by the addition of a neutral reactant. Optical emissions either in the infrared or by laser-induced fluorescence are detected at various ports just after the neutral addition. The ion population is continuously monitored downstream by a quadrupole mass filter coupled with a particle multiplier.

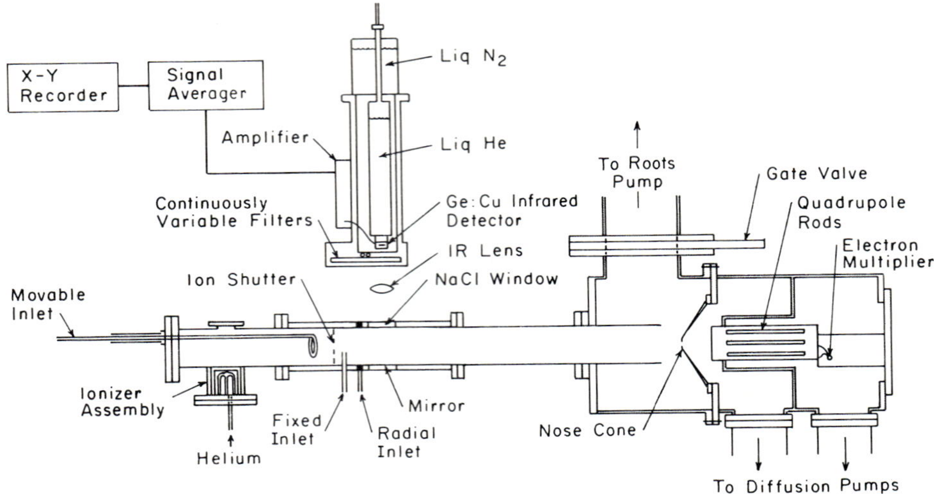

FIG. 1 Schematic drawing of the flowing afterglow infrared chemiluminescence apparatus.

The stainless steel flow tube is 7.3-cm i.d. and is constructed of several 30-cm detachable sections which incorporate a variety of inlets, ports, and guard rings to allow a wide range of experiments in reactive and collision dynamics. The total flow tube length is about 1 m. Helium buffer gas pressures and velocities are typically 100 Pa and 80 m sec^{-1}, respectively. Small flows of other gases are added near the electron impact ionizer to generate the desired reactant ion. For example, addition of F_2, CCl_4, and BrCN serve as efficient sources of F^-, Cl^-, and CN^-, respectively; O^- is formed by addition of O_2 while N^+ and $N_2{}^+$ result from addition of N_2. Regulated emission currents of 0.25–100 mA are employed to maintain stable ion densities of 10^8–10^9 cm^{-3} in the reaction region for several hours of data collection. The plasma of reactant ions in helium is allowed to flow about 50 cm before addition of the neutral reactant; this ensures that diffusion and flow properties are well defined and that the ions are collisionally relaxed.

The neutral reactant is added through one of several inlets: through a movable ring inlet, through a short length of unconstricted Teflon tubing, or radially through a series of small holes in the circumference of the flow tube. Most of the gas mixture is exhausted by the large-capacity Roots pumping system. However, a small fraction is sampled through an orifice in a molybdenum disk, analyzed with a quadrupole mass filter, and detected with a continuous channel electron multiplier. Reaction kinetics are studied by monitoring reactant and product ion densities as a function of reaction distance or of neutral reactant flow rate. Ion counts are monitored with a frequency meter for a single mass, or they are accumulated in a signal averager as a function of mass to produce a mass spectrum. For very high ion densities required in the optical studies, the mass-selected ion currents are monitored with a Faraday plate coupled to an electrometer.

2. *Features of the Flowing Afterglow Technique for the Study of Reaction Dynamics*

The flowing afterglow technique offers several valuable features for the study of product state distributions in ion–molecule reactions:

(a) The measurement of kinetics is reliable and straightforward. Albritton (1978) has recently tabulated data for approximately 1600 ion–neutral reactions measured by the flowing afterglow technique. Thus, the reaction of interest can be fully characterized by conventional methods.

(b) Both the ionic and neutral reactants have well-defined thermal energy distributions due to the spatial separation of ion production and reaction regions. The ions are collisionally relaxed before initiation of the ion–molecule reaction, and addition of the neutral reactant downstream of the ionization region prevents unwanted excitation or emissions from the neutral reagents.

(c) The flowing afterglow technique is unparalleled in its chemical versatility. Reactant ions can be generated directly by electron impact or synthesized from a precursor ion by chemical reaction in the flow tube. Due to the flow nature of the technique, highly reactive neutral reagents such as H atoms may be generated (Section IV.B).

(d) Energy variability is possible. Reactions have been examined at well-defined temperatures between 80 and 900 K (Dunkin *et al.*, 1968). By replacing the conventional flow tube with a drift tube, ion kinetic energies between 0.04 eV and 10 eV in the center of mass reference frame have been attained (McFarland *et al.*, 1973). The resulting internal excitation of the ions has been probed directly by optical means (Section IV.E).

(e) The flowing afterglow technique uniquely provides high densities and large volumes of ions. This feature is crucial to the optical detection experiments described here and has allowed the innovations of neutral product collection from ion–molecule reactions (Smith *et al.*, 1980) and the use of mass-selected ion injection techniques (Adams and Smith, 1976).

(f) The flowing afterglow is inherently a stable, steady-state environment, allowing for easy coupling of various ion modulation techniques to enhance the signal-to-noise ratio in the optical detection methods.

3. *Constraints of the Flowing Afterglow Technique for the Study of Reaction Dynamics*

The flowing afterglow technique has several complications and imposes several constraints for the study of state-resolved ion reaction dynamics.

(a) A variety of ions, metastable species, and reactive radicals may be formed in the initial ionization process; therefore, a reaction product has several potential precursors.

(b) A variety of neutral reagents are also commonly present, which can cause subsequent reactions of the product of interest.

(c) The flowing afterglow is inherently a collision-dominated technique. The high-collision frequency ensures that the reactants possess well-defined energies; however, the study of initial vibrational and rotational populations of products can be complicated by collisional redistribution of energy in these species. In many cases, all rotational state information is lost but vibrational states can be observed before relaxation.

(d) There is considerable background light from the ion source and the subsequent plasma relaxation processes. Indeed the term *flowing afterglow* arises from the characteristic emission of the carrier gas after its excitation. This can obscure optical detection.

(e) Because of the large flow rates of the buffer gas and neutral reagents, there is high consumption of chemicals, precluding the routine use of exotic compounds.

B. *The Flowing Afterglow Coupled with Infrared Chemiluminescence Detection*

1. *Apparatus Design*

Incorporation of infrared chemiluminescence detection requires three changes to the conventional flowing afterglow (Fig. 1) (Zwier *et al.*, 1980a): (1) modification of the flow tube to allow collection of light, (2) use of a sensitive infrared detector and associated electronics, and (3) incorporation of a shutter to provide modulation of the ions. The infrared port consists of a 2.5-cm-diameter NaCl window inserted tangent to the inside wall of the flow tube. This recessed design allows close placement of the infrared detector to the flow tube for viewing a volume of up to 120 cm^3. Since the wavelength transmission of interference filters depends on the angle of incidence, larger port-to-detector separations are usually used to minimize collection of off-normal light. A 10-cm focal length aluminum mirror is placed opposite the window, and a 5.1-cm-diameter f/1 CaF_2 lens is used above the window to collect and image the infrared emission onto the detector.

Two infrared detector assemblies are used in the infrared chemiluminescence experiments: (1) an indium antimonide detector for the reactions of chloride and cyanide ions and (2) a copper-doped germanium detector for the studies involving fluoride ion and N^+. The photovoltaic InSb detector is a liquid-nitrogen-cooled, 1.27-cm-diameter element with a wavelength response of 1.0–5.5 μm and a detectivity, $D^* = 1 \times 10^{11}$ cm $\mathrm{Hz}^{1/2}$ W^{-1} at 5 μm. The large area of the detector element, the high D^* value, and the ease of liquid-nitrogen cooling make this detector best suited for initial studies. Wavelength resolution of the transmitted light is provided by a series of room temperature fixed frequency interference filters mounted below the detector. Cold gas filter cells containing HCl, HF, or HCN are also employed to selectively absorb more than 99% of the emission from $v = 1 \rightarrow 0$. The output of the InSb detector is amplified, and modulated signals are accumulated in a signal averager.

The photoconductive Ge:Cu detector is a liquid-helium-cooled, 3 mm × 10 mm element with a wavelength response of 2–30 μm and a detectivity, $D^* \sim 1 \times 10^{12}$ cm $\mathrm{Hz}^{1/2}$ W^{-1} at 3 μm, under background limited conditions. Liquid-helium-cooled magnesium fluoride and/or sapphire windows are mounted directly on the detector heat sink to provide a long-wavelength blackbody cutoff. The detector views emission through a wavelength-selective, low-resolution rotatable circular variable interference filter assembly (CVF) and a 1.5 mm × 12 mm slit, both cooled to 77 K (Zwier *et al.*, 1981). The output of the Ge:Cu detector is amplified by current feedback circuitry and then accumulated in the signal averager.

The circular variable filters consist of three 90° annular sections of a transparent substrate coated with continuously varying thicknesses of multiple dielectrics. Each radial slice through the annulus is a narrow bandpass interference filter whose peak transmission wavelength varies linearly around the circumference of the annulus. The three filter segments span the range 2.5–14.5 μm and are calibrated to ± 0.05 μm using a broadband infrared source and a monochromator. The resolution is slit-width-limited and varies from 0.055 μm at 2.5 μm to 0.11 μm at 8 μm; signal collection is maximized by this relatively low-resolution configuration. The sensitivity of the detector and filter combination as a function of wavelength is measured by chopping and imaging the output of a calibrated blackbody source onto the detector; the output versus CVF setting is normalized to the calculated blackbody emission curve.

The ion source generates a variety of reactive radical and metastable species in addition to the reactant ion of interest. In order to distinguish the desired infrared emission of the ion–molecule reaction from all other interfering signals, the ions must be modulated. Ion modulation is achieved by repetitively applying a repulsive potential to a 95% transmission tungsten mesh stretched across the flow tube about 40 cm downstream of the ionizer. For negative ions the grid is typically pulsed to -10 V for 0.3 msec and returned to ground for the remainder of the 5- or 10-msec period, as shown in Fig. 2 for chloride ion. This causes a decrease in the chloride ion density as detected at the quadrupole mass spectrometer 6 msec later, but does not

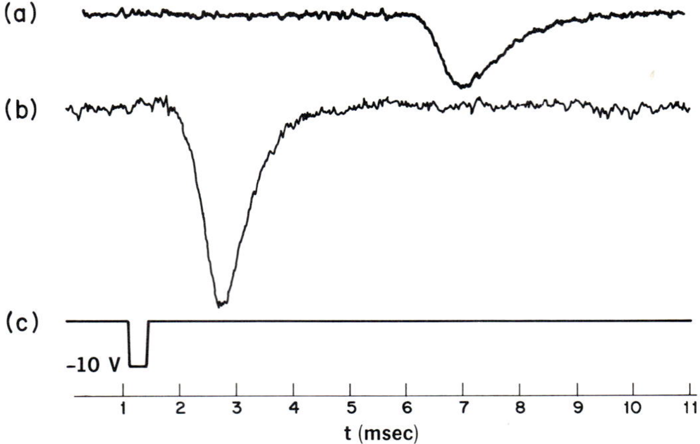

FIG. 2 Ion modulation for infrared chemiluminescence detection. (a) Chloride ion counts observed downstream with the quadrupole mass spectrometer when no reactant is added. The spreading of the modulated ion packet is due to axial diffusion. (b) HCl(v) infrared emission, showing a decrease in emission when the ions are stopped; the delay is the flow time between the shutter and IR detector. (c) Minus 10-V pulse repetitively applied to the ion shutter. [From Zwier *et al.* (1980a).]

affect the density of uncharged species. If hydrogen iodide is added to this modulated plasma, vibrationally hot HCl product is formed. The decrease in the ion density appears as a decrease in the infrared emission intensity due to the interruption of the ion–molecule reaction; therefore, the area of the emission curve is directly proportional to the amount of HCl(v) arising from the ion–molecule reaction. These signals are monitored as a function of transmitted wavelength to construct a spectrum.

2. *Experimental Method for Population Measurements by Infrared Chemiluminescence*

Reactant ions are formed in the ion source, thermalized in a 40-cm section of flow tube, and modulated with the shutter. The addition of neutral reagent 10 cm downstream of the mesh initiates the ion–molecule reaction, and wavelength-resolved infrared emission is detected in phase with the ion modulation by the infrared detector positioned 2–5 cm downstream. The signal-to-noise ratio before averaging is about 0.1–1.0, in good agreement with estimates based on ion densities, radiative lifetimes, collection efficiency, and detector characteristics. Typically, 10^4–10^5 modulation periods are accumulated and averaged at each wavelength, corresponding to 1–10 min of data collection. The raw infrared emission intensities are corrected for detector and filter sensitivity, and a complete spectrum is constructed (Fig. 3). For

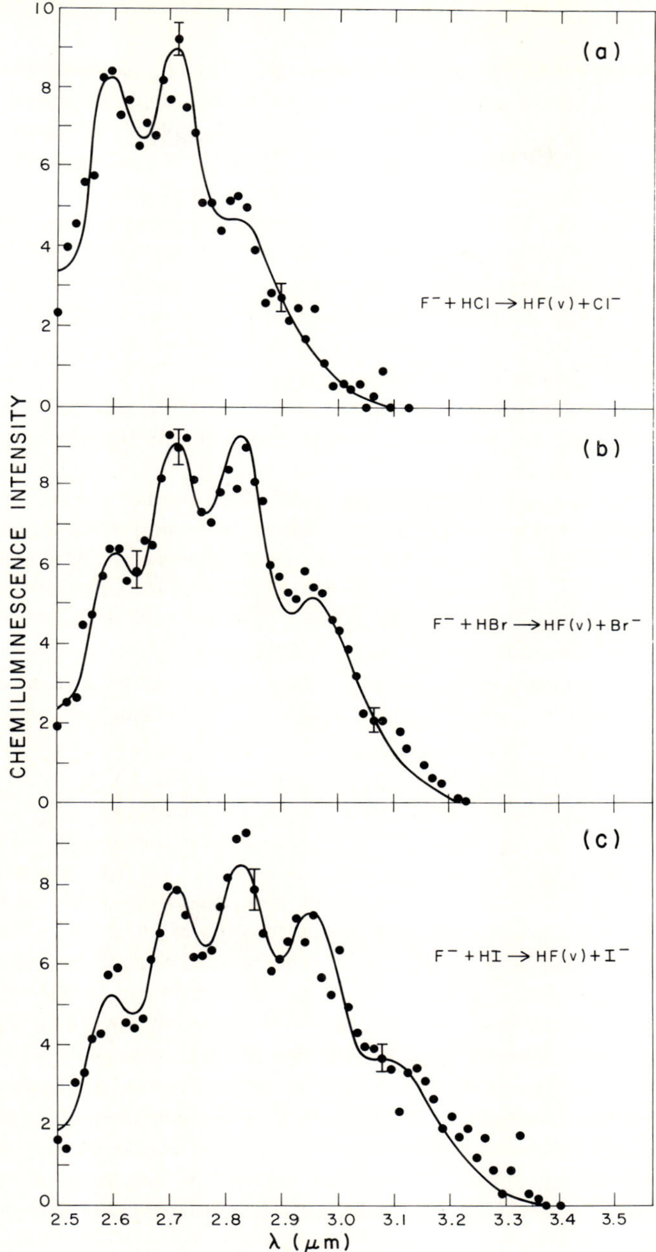

FIG. 3 Low-resolution infrared chemiluminescence spectra of HF(v) produced from the ion reactions F^- + (a) HCl, (b) HBr, and (c) HI. The anharmonicity of the vibrational levels creates an overlap between P and R branches of adjacent bands, producing the multiple-peaked spectrum. The exothermicities are sufficient to produce HF up to $v = 3$, 4, and 5 for the HCl, HBr, and HI reactants, respectively. [From Weisshaar *et al.* (1981).]

fixed interference filter experiments, populations are deduced from the transmission through several filters and the known values of the Einstein coefficients A_v (Maricq *et al.*, 1981). For the CVF data the chemiluminescence spectra are computer analyzed by a linear least squares procedure in which relative populations N_v are extracted (Zwier *et al.*, 1981). In those calculations, the molecular line positions, the transition strength of each line, and the spectral resolution are specified. Small corrections to the resulting vibrational populations are then applied to account for radiative and collisional cascading between the time of formation and detection of the vibrationally excited product. In most cases a thermal distribution of rotational states is observed and used in the computer fitting routine. When possible, vibrational surprisal plots are extrapolated to obtain estimates of the $v = 0$ population and of the average fraction of available energy deposited into product vibration (Zwier *et al.*, 1981).

A variety of experimental checks are performed to ensure that the infrared emission arises solely from the ion–molecule reaction of interest and that the vibrational distributions are nascent. Ion modulation with phase-sensitive detection of chemiluminescence eliminates contributions from neutral reactions. Ion–molecule reactions are monitored with the mass spectrometer before and after addition of the neutral reagent. In a valid experiment, the reactant species is depleted, the product species is generated, and all other ion signals remain constant. Other evidence that the detected emission arises from the ion reaction is the proper correlation of chemiluminescence intensity with ion density and neutral reactant flow and the correspondence of the time evolution of the infrared emission to that of the ion arrival waveform at the mass spectrometer. In some test experiments, a second mesh is inserted into the flow tube upstream of the ion shutter and held at a constant, highly repulsive potential. This effectively stops all ions, and the infrared emission intensity decreases by greater than 98%. Finally, for some reactions, e.g., Cl^- or $F^- + HI$, the neutral reaction counterpart is known to populate vibrational levels higher than allowed by the ion process exothermicity. Absence of emission in this region confirms that the neutral process does not contribute to the detected infrared signals.

Gas phase deactivation, wall deactivation, and radiative cascading must be considered for quantitative assignment of initial vibrational distributions. Gas phase deactivation is addressed by monitoring infrared emission over a wide range of flow rates of precursor gas and neutral reactant. Data are obtained under experimental conditions where vibrational deactivation is completely negligible or very small. In these latter cases the data are corrected using published collisional relaxation rates. Although vibrational relaxation of many diatomics by helium is extremely inefficient (Leone, 1982), rotational relaxation is rapid on the time scales of our experiment (Sung and

Setser, 1978). This is experimentally checked in several ways. For example, in the formation of HCl from Cl^- + HBr, where only $v = 1$ is formed, emission is totally blocked by a gas filter, indicating the absence of hot rotational levels. In other experiments, emission to red or blue wavelengths, which would reflect the presence of non-Boltzmann rotational distributions, is not observed. Moreover, in the computer simulation of emission intensity versus wavelength, a rotational temperature of 300 K gives excellent agreement to the data; assumed temperatures above 450 K result in considerably degraded fits. Therefore, vibrational excitation is largely preserved but rotational relaxation is complete.

Vibrational deactivation at the walls of the flow tube is negligible. Calculations indicate that only 5% of the product species experience a collision with the wall before detection of chemiluminescence. This is supported by the consistency of data obtained with different distances between the neutral inlet and the infrared detector.

Finally, radiative cascading, i.e., depletion in the population of level v and the resulting increase in the population of level $v - 1$ by radiative processes, must be carefully evaluated. The raw populations are corrected for the known radiative lifetimes by a simple kinetic model to deduce the initial vibrational populations. Corrections never exceed 20% and are generally considerably smaller.

C. *The Flowing Afterglow Coupled with Laser-Induced Fluorescence (LIF) Detection*

1. *Apparatus Design*

Use of laser-induced fluorescence detection requires several extensions of the flowing afterglow, as illustrated in Fig. 4 (Hamilton *et al.*, 1983b): incorporation of the laser; flow tube modifications to allow entry of laser light, collection of fluorescence, and reduction of background light; a photomultiplier tube, lenses, and filters to detect the fluorescence; and signal collection and processing electronics.

A Nd:YAG pumped dye laser is used, which is capable of producing tunable pulsed radiation at 0–20 Hz with energies of 5–50 mJ/pulse in the 380–700 nm range. Use of nonlinear optical techniques extends the output range to 217 nm through harmonic generation and mixing in angle phase-matched crystals. Most experiments are performed with 0.1–1 mJ/cm^2 to avoid saturation of the molecular transition. The short pulse duration of 5 nsec allows gating out of the scattered laser light. Careful control of amplified spontaneous emission is essential to avoid broadband background

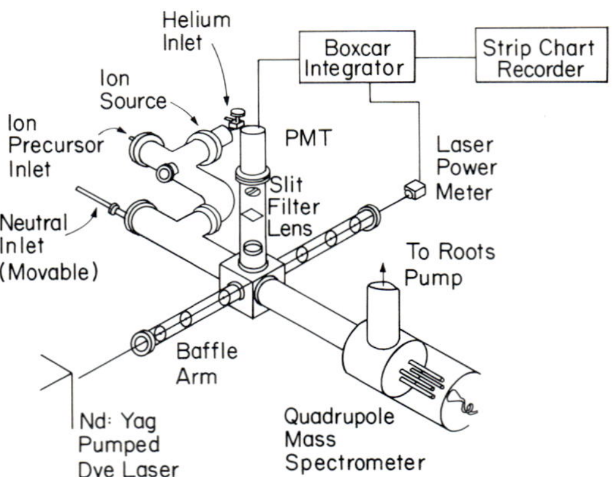

FIG. 4 Flowing afterglow apparatus incorporating laser-induced fluorescence detection. [From Hamilton *et al.* (1983b).]

signals. The full width at half maximum of 0.02 nm provides rotational resolution in most cases. Three species have been studied using laser-induced fluorescence: N_2^+, CO^+, and OH. Their transitions are listed in Table II.

A section of flow tube incorporates two sidearms perpendicular to the direction of flow to allow the entrance and exit of laser light and a third port above the flow tube for the collection of fluorescence. Windows are made of ultraviolet grade quartz. The baffle arms contain several apertures to reduce scattered laser light. A blackened right-angle bend is added to the flow tube between the ionizer and LIF regions to reduce background light from the ion source. The photomultiplier tube is positioned directly above the flow tube in the region of laser probing to detect the resulting fluorescence. The output of the photomultiplier is amplified and fed into a boxcar integrator.

To reduce the background signals, the photomultiplier field of view is restricted to the small volume of gas intersected by the laser beam by a combination lens and slit arrangement. The fluorescence signal is collected by a 5.1-cm-diameter f/1 CaF_2 lens and focused onto a 5-mm slit positioned directly in front of the phototube, resulting in a factor of 30 improvement in the signal-to-noise ratio. The afterglow background is also reduced by chemical quenching by addition of trace gases and by the use of filters in the photomultiplier housing; short-wavelength cutoff filters are effectively used in combination with bandpass filters. Nevertheless, the ultimate signal-to-noise ratio is still limited by the continuous afterglow background light.

A boxcar signal averager provides a variable width, variable position window for gated collection of fluorescence. The boxcar is triggered by the Q switch of the Nd:YAG laser, the gate is normally delayed by about one molecular radiative lifetime to minimize scattered light, and it is then opened for several lifetimes for signal acquisition. The current from the photomultiplier is transformed by the boxcar into a digitally held voltage, which is a measurement of the average signal over the gate width. The laser power as measured by a thermopile is simultaneously monitored in the second channel of the boxcar. The fluorescence signal normalized for laser power is then taken directly as an output.

2. *Experimental Method for Population Measurements by Laser-Induced Fluorescence*

The reactant ion is generated in the ion source and reacts downstream with a neutral reagent to form vibrationally excited products just before the laser excitation port. The tunable dye laser induces a transition from specific vibrational and rotational levels of the products to a bound electronic state; the total visible or ultraviolet fluorescence from this state is monitored with the photomultiplier. The boxcar signal averager provides the gated detection of the fluorescence and the normalization for laser power while the dye laser wavelength is slowly scanned through a vibronic band of the species of interest. The resulting LIF excitation spectrum is recorded by coupling the boxcar output with a strip chart recorder or with a multichannel analyzer connected to a minicomputer.

A laser-induced fluorescence spectrum for

$$OH[A\,^2\Sigma^+(v' = 1) - X\,^2\Pi(v'' = 0)]$$

formed in the reaction of O^- with HF is shown in Fig. 5. The excellent signal-to-noise ratio possible with LIF is clearly evident; signal levels of 2000

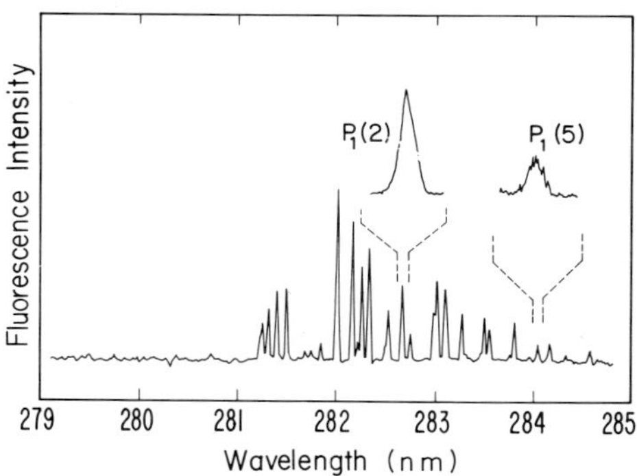

FIG. 5 Laser-induced fluorescence spectrum of the (1, 0) band of OH ($A\,^2\Sigma^+ - X\,^2\Pi$) produced in the reaction $O^- + HF \rightarrow OH(v = 0, 1) + F^-$. The well-isolated lines $P_1(2)$ and $P_1(5)$ are shown for a typical slow scan. The areas of these peaks are used to determine the population in the $v'' = 0$ state. [From Hamilton *et al.* (1983b).]

single potential energy surface, a fact which greatly simplifies the interpretative tasks. All reagents and products are closed-shell species whose first excited electronic states are high in energy so that, at thermal energies, surface hopping trajectories are expected to be unimportant (Preston and Tully, 1971; Krenos *et al.*, 1974).

In light of the enormous amount of research on the dynamics of simple neutral reactive encounters, $Y + HX \rightarrow HY + X$ (Kuntz, 1976), it is illuminating to compare the vibrational state distributions for corresponding neutral and ion reactions. The bottom of Table III provides data for the neutral reactions. It should be noted that the halogen atom reactions sometimes involve several interacting electronic states which lead to a number of coupled potential surfaces. Another important difference between neutral and ion reactions is the intermolecular potential. The potential between neutrals is only weakly attractive and generally includes a barrier which must be surmounted for reaction to take place. In contrast, the potentials of ion–molecule reactions are dominated by the attractive ion–dipole and ion-induced dipole terms and, consequently, pass through potential minima. Figure 6 sketches possible potential energy curves for collinear collisions of F with HCl and F^- with HCl. In both cases, the product HF can be excited up to $v = 3$.

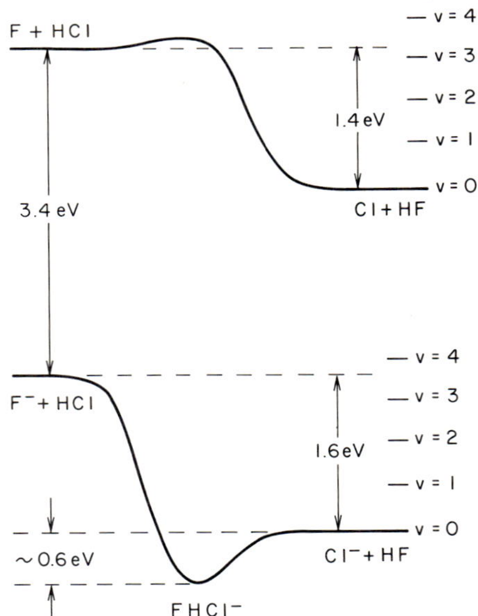

FIG. 6 Schematic drawing showing the difference between typical potential energy surfaces for a neutral reaction, $F + HCl \rightarrow HF(v = 0\text{–}3) + Cl$, and an ion reaction $F^- + HCl \rightarrow HF(v = 0\text{–}3) + Cl^-$. [From Weisshaar *et al.* (1981).]

As indicated in Table III, the exothermicities of the ion reactions are comparable to their neutral analogs but the rate constants are roughly two or three orders of magnitude larger. The fast rates of ion–molecule reactions are generally attributed to the presence of the large attractive well. In general, both the ion and neutral reactions have a tendency to dispose of a large fraction of the exothermicity into product vibration.

The details of the reaction dynamics of several proton transfer reactions have been studied by infrared chemiluminescence (Zwier *et al.*, 1980a; Weisshaar *et al.*, 1981):

$$Cl^- + HBr, HI \rightarrow HCl(v) + Br^-, I^- \tag{2}$$

and

$$F^- + HCl, HBr, HI \rightarrow HF(v) + Cl^-, Br^-, I^-. \tag{3}$$

For the reaction of Cl^- with HBr, there is only enough energy (0.43 eV) to produce $HCl(v = 0, 1)$. Infrared emission from $HCl(v = 1)$ at 3.47 μm is observed, but there are no direct means of monitoring the population born in $v = 0$. In the reaction of Cl^- with HI, production of $HCl(v = 0, 1, 2)$ is possible and the emission of $HCl(v = 1)$ at 3.47 μm is separated from that of $HCl(v = 2)$ at 3.60 μm by use of an HCl gas filter. From these measured chemiluminescence intensities and the known Einstein coefficients, the relative populations, corrected for radiative cascading and collisional relaxation, are found to be $(v = 2)/(v = 1) = 0.85$. Comparison of the Cl^- + HBr and HI emission intensities and further comparison of Cl^- + HI with the Cl^- + H associative detachment reaction allows approximate populations in $v = 0$ to be assigned (Zwier *et al.*, 1980a). This comparison is based on the fact that the associative detachment reaction forms very little $v = 0$ product. The numbers in Table III are based on the surprisal analysis for Cl^- + HI, and they differ only slightly from the experimental comparison.

The reactions of F^- with HCl, HBr, and HI provide a much broader dynamical range; excitation of HF up to $v = 5$ is possible with HI. Figure 3 shows the emission spectrum obtained from the product $HF(v)$ formed in each case. These spectra are composed of multiple features which arise from the superposition of the *P* and *R* branches of the anharmonic bands. Figure 3 shows that successively higher vibrational bands are produced on going through the series F^- + HCl, HBr, and HI. Indeed the highest vibrational level observed in emission corresponds in all cases to the highest vibrational state allowed by the exothermicity. About 40–60 points define each spectrum in Fig. 3, and these spectra are inverted as described earlier to obtain vibrational populations. In this calculation, a room temperature Boltzmann distribution is assumed, and the positions, line strengths, and spectral resolution are specified. The average results of the relative vibrational populations after correction for radiative cascading are displayed in Table III (Weisshaar *et al.*, 1981).

The relative populations can be used to generate the complete product distribution if the $v = 0$ population is known. These populations are estimated from a surprisal analysis (Levine, 1978) and are included in Table III in parentheses. Therefore, one can approximately determine $\langle f_v \rangle$, the average fraction of the available energy deposited in product vibration. As evident from Table III, these fractions are substantial, ranging from 0.30 to 0.46 for the proton transfer reactions of Cl^- and F^-.

Comparison of the ion and neutral populations in Table III shows many similarities. Both types of reactions lead to highly vibrationally excited HY product molecules but with the neutral reactions producing distinctly higher excitation. It is important to note that the kinematic effects for both ion and neutral reactions may be similar since both have identical mass combinations of two heavy particles (H and H′) and one light particle (L): $H + LH' \rightarrow H' + LH$. In both cases the mass-weighted coordinate system (which diagonalizes the kinetic energy) is sharply skewed (Kuntz, 1976). This means that the entrance and exit valleys are rotated from each other by about 15°. Although the kinematics for the ion and neutral are similar, the detailed mechanisms of their interactions are expected to be very different. The possible presence of a long-lived intermediate in the ion reaction may allow for extensive redistribution of vibrational energy relative to the neutral process. The similarity of vibrational populations for ion and neutral reactions suggests that these simple proton transfer reactions, $Y^- + HX$, proceed through direct collisions with their dynamics being dictated largely by the kinematics rather than by the shape of the intermolecular potential. However, the lower $\langle f_v \rangle$ in most cases for the ion reactions suggests that the long-range attractive forces do influence the final vibrational distribution and cause it to be somewhat more statistical.

Another interesting proton transfer process is

$$O^- + HF \rightarrow OH(v = 0, 1) + F^-. \tag{4}$$

The corresponding neutral reaction is endothermic by 1.48 eV and does not occur at room temperature. In this case, LIF is used to detect the product OH molecules; a typical fluorescence spectrum is shown in Fig. 5 (Hamilton *et al.*, 1983b). The LIF detection allows direct measurement of the populations of both vibrational states, and no surprisal analysis is necessary. The absolute vibrational populations of OH products from this reaction are $N_0 = 0.82$ and $N_1 = 0.18$. The observed fraction in $v = 1$ is significantly greater than the value (0.07) expected by a statistical "prior" distribution (Levine, 1978), even though the available energy is barely able to populate the $v = 1$ state. A detailed interpretation of this reaction may be complicated by the multiple states of $O^-(^2P_{3/2}$ and $^2P_{1/2})$ and $OH(^2\Pi_{3/2}$ and $^2\Pi_{1/2})$. The two spin–orbit states of O^- are split by only 22 meV (Hotop and Lineberger, 1975) so that a mixture of reactant states is present in the flowing afterglow. However, the reactivity and detailed dynamics of these states may differ; e.g., Tully (1974) finds a substantial difference in reactivity between the two fine structure states in the reaction of $F(^2P_{3/2}$ and $^2P_{1/2})$ with H_2. The splitting between the two spin–orbit states of OH is 16 meV (Evenson *et al.*, 1970); the numerous collisions with helium preclude detection of any preferential excitation in the $^2\Pi_{3/2}$ and $^2\Pi_{1/2}$ product states of OH.

The reactions discussed so far have all been those of an atomic ion with a diatomic molecule to produce a vibrationally excited hydrogen halide. We next consider the reactions of CN^- with HX to produce vibrationally excited HCN (Maricq *et al.*, 1981):

$$CN^- + HCl, HBr, HI \rightarrow HCN(v) + Cl^-, Br^-, I^-. \qquad (5)$$

The study of IR emission from a triatomic molecule is considerably more difficult than from a diatomic species. It is now necessary to distinguish chemiluminescence from the C≡N stretch (ν_1), a doubly degenerate H—C≡N bend (ν_2), and the C—H stretch (ν_3). With so many modes, the spectroscopy becomes more complicated, and the separation and assignment of emission lines become more difficult. Worse yet, the energy may scramble via intramolecular pathways so that the resulting emission bears little resemblance to the initial product state distribution. In an extreme case, the internal energy may rapidly leak out by collisional deactivation of a low-frequency vibrational mode, resulting in no emission at all.

Because of the favorable spectroscopic features (Allen *et al.*, 1956) and relaxation rates of HCN (Hariri *et al.*, 1976; Petersen and Smith, 1979; Arnold *et al.*, 1980), it is possible to extract the relative populations in the ν_3 mode produced by the reactions CN^- with HCl, HBr, and HI. These are summarized in Table III; no dynamical information is available for the corresponding CN reactions (Roden, 1975). Since population inversions are observed, it is apparent that the excess vibrational energy does not become completely scrambled among the available vibrational modes as the reactants pass through the deep potential well separating reactants and products. Unfortunately, very little can be said about the energy distribution in other modes except that a considerable number of bending quanta must be excited (ν_2).

A second important and unexpected result of these studies is the detection of hydroisocyanic acid (HNC), produced in the reaction of CN^- with HI. In fact, HNC represents roughly 20% of the excited products detected in this reaction. HNC is identifiable by its characteristic N—H stretch at 2.74 μm. Figure 7 shows the results of a point-by-point scan of the Ge:Cu detector

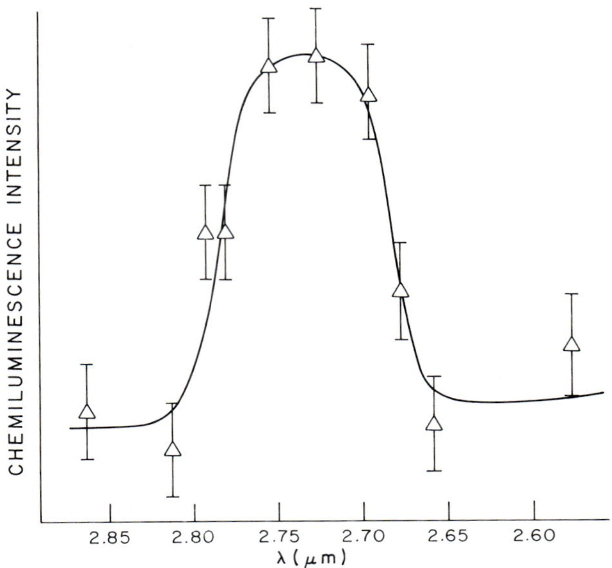

FIG. 7 Emission band from the NH stretch of HNC produced in the ion reaction of CN^- with HI. [From Maricq *et al.* (1981).]

equipped with the circular variable filter through the high-frequency N—H stretch region for HNC resulting from the $CN^- + HI$ reaction. This is a nice example of the ability of the infrared method to sense a fragile reaction product not easily detected in any other way.

The thermochemistry of HNC is of some interest, and from our data we can make some quantitative statements. The difference in enthalpy between HCN and HNC is the isomerization energy ΔH_{iso}, and the height of the barrier to isomerization of HNC($v = 0$) to HCN is the isomerization barrier ΔE_{iso}. By considering which CN^- reactions produce excited HNC, the enthalpy difference between the two species can be bracketed (Maricq *et al.*, 1981)

$$1.14(\pm 0.11)\ \text{eV} \geqslant \Delta H_{iso} \geqslant 0.74(\pm 0.11)\ \text{eV}.$$

The uncertainty in the heats of reaction is traceable to the lack of precision in the heat of formation of HCN (Berkowitz *et al.*, 1969). Two recent experimental studies have independently measured ΔH_{iso} in a direct manner. A study of microwave absorption as a function of temperature reports a value of 0.44 ± 0.05 eV (Maki and Sams, 1981) while a deprotonation experiment in an ion cyclotron resonance spectrometer finds a value of 0.64 ± 0.09 eV (Pau and Hehre, 1982). Both our results and recent computational work (Redmon *et al.*, 1980; Dykstra and Secrest, 1981) favor the latter value. Using

Pau and Hehre's value for ΔH_{iso} and the fact that none of the HCN(v) product of the CN^- + H reaction isomerizes to HNC, we can likewise place a lower bound on the isomerization barrier energy:

$$\Delta E_{iso} \geqslant 0.92\ (\pm 0.11)\ \text{eV}.$$

This result is consistent with a calculated barrier of 1.3 eV (Redmon *et al.*, 1980).

B. *Associative Detachment Reactions*

Associative detachment reactions, $A^- + B \rightarrow AB(v, J) + e^-$, offer a novel opportunity to probe the transition of a negative ion potential energy surface into a neutral plus electron continuum. The flowing afterglow infrared method provides detailed information on these processes by investigation of state distributions in the neutral product molecule. There is tremendous interest in several processes which proceed through negative molecular ion states, including electron scattering, dissociative attachment, and associative detachment (Shimamura and Matsuzawa, 1979). Thus far, there is no unifying theoretical description of these interrelated processes. Associative detachment is unique among these in that it samples the negative ion surface from long range. As the internuclear separation decreases, the neutral surface becomes unstable with respect to autodetachment. [For representative HF and HF^- surfaces, see Segal and Wolf (1981).] The vibration–rotation state distribution can provide a tremendous amount of information concerning the geometry and dynamics in the autodetachment region. There are numerous early discussions of the associative detachment process and of the important theoretical criteria (Fehsenfeld, 1975b; Herzenberg, 1967).

The flowing afterglow infrared chemiluminescence technique has provided almost all of the recent data on vibrational product states in associative detachment reactions. The technique has been applied to five systems:

$O^- + CO \rightarrow CO_2(v) + e^-$ (Bierbaum *et al.*, 1977), (6)

$Cl^- + H \rightarrow HCl(v = 0\text{–}2) + e^-$ (Zwier *et al.*, 1980b), (7)

$CN^- + H \rightarrow HCN(v) + e^-$ (Maricq *et al.*, 1981), (8)

$F^- + H \rightarrow HF(v = 0\text{–}5) + e^-$ (Zwier *et al.*, 1981; Smith and Leone, 1983), (9)

$F^- + D \rightarrow DF(v = 0\text{–}7) + e^-$ (Smith and Leone, 1983). (10)

The O^- + CO reaction was the first ion–molecule reaction studied by infrared chemiluminescence in the flowing afterglow. Direct excitation of the product CO_2 antisymmetric stretch mode is observed. When a CO_2 gas filter cell is used to block the (001 → 000) emission, 30% of the fluorescence is transmitted, compared to about 10% which would be transmitted for a sample thermalized in the bending modes (0n0). This clearly indicates excitation in states higher than (001), which might also include excitation of other modes.

In the $Cl^- + H$ reaction, the HCl vibrational distribution of $(v = 2)/(v = 1)$ is measured to be 0.6. This is less than the $(v = 2)/(v = 1)$ ratio of 0.85 for the $Cl^- + HI$ reaction. However, comparison of the total emission intensity of these two reactions, which have similar exothermicities, suggests that the formation of HCl $(v = 0)$ in the associative detachment reaction is zero or small. This is corroborated by the application of microscopic reversibility to data obtained by Allan and Wong (1981) on the reverse dissociative attachment process (Zwier *et al.*, 1980b). It is found that the associative detachment process strongly favors population of the highest vibrational states and that the reverse, dissociative attachment cross sections are greatly enhanced by vibrational excitation. Figure 8 shows this propensity for high vibrational

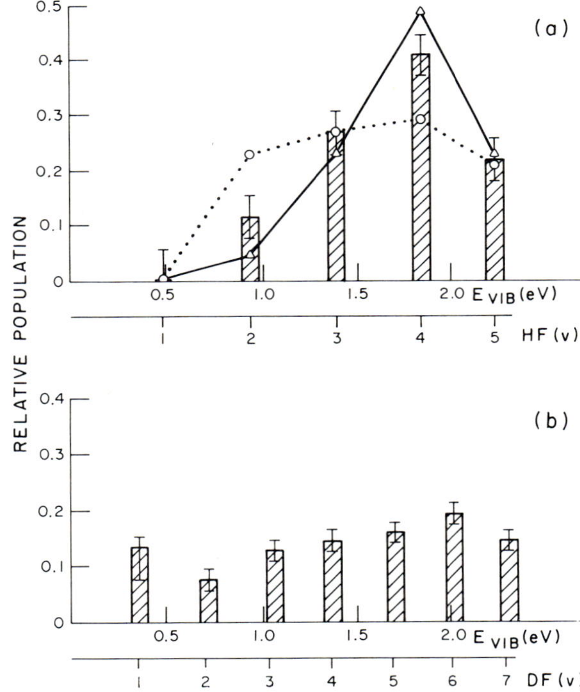

FIG. 8 Vibrational population distributions for the (a) $F^- + H \rightarrow HF(v = 1-5) + e^-$ and (b) $F^- + D \rightarrow DF(v = 1-7) + e^-$ associative detachment reactions (Smith and Leone, 1983). Earlier data (○) of Zwier *et al.* (1981) were found to be partially relaxed. Theoretical calculations (△) of Gauyacq (1982) (calculated at 300 K) show good agreement with the present results for HF. The results in this figure incorporate a small correction as reported by Hamilton *et al.* (1983a).

excitation even more dramatically for the $F^- + H$ and D systems, which produce HF up to $v = 5$ and DF up to $v = 7$. There is a strong peaking of the vibrational excitation into the highest states accessible, which is a good indication that the electron detachment occurs at large separations in the initial approach of the $F^- + H$. The colliding atom and ion sample the highest vibrational states first and have a substantial probability of detaching into those high levels. Indeed, this physical description is supported by detailed theoretical calculations (Gauyacq, 1982; see also Zwier *et al.*, 1981). There is a significantly greater population of the lower vibrational states in the $F^- + D$ case. This is attributed to a reduction in the probability for detachment at large internuclear separations, increasing the fraction of collisions which survive to closer separations and sample the lower vibrational levels (Smith and Leone, 1983). This result has not yet been satisfactorily explained by theory.

The release of the extremely light electron in the associative detachment process provides an interesting dynamical constraint due to conservation of energy and angular momentum. A substantial fraction of the orbital angular momentum of the incoming reactants is converted into rotational excitation of the neutral product molecule. For molecules with large rotational constants, this has the effect of reducing the range of allowed impact parameters, and hence the cross section, that can lead to the highest vibrational levels. This truncation of population in the highest vibrational states is readily apparent in Fig. 8. Simple model arguments can predict the near one-to-one mapping of impact parameter into rotational state (Zwier *et al.*, 1980b; 1981). In the more complete theoretical description when both vibrational and rotational states are taken into account, each vibrational state spans predominantly a small range of J values, and the cross sections into each successive vibrational level turn on and off at a certain J (Gauyacq, 1982). The production of high rotational states has recently been observed directly by carrying out the $F^- + D$ reaction in an argon buffer gas, which is much less efficient than helium at rotational relaxation (Smith and Leone, 1983).

The population distribution in the C—H stretch of HCN is obtained for the $CN^- + H$ associative detachment reaction and found to be

$$(0.47)_{v=1}:(0.39)_{v=2}:(0.09)_{v=3}:(0.05)_{v=4}$$

(Maricq *et al.*, 1981). Although there is substantial excitation, this distribution decreases with increasing vibrational level in contrast to the diatomic cases. This is attributed to the large number of vibrational states accessible in the polyatomic product, which reduces the probability of depositing all of the reaction exothermicity into the C—H mode. In the reaction of CN^- with HI, the HNC product is detected. Even though the $CN^- + H$ reaction has a nearly identical exothermicity, the vibrationally excited HNC product is not observed in this case. It is conjectured that perhaps the approach of the H atom to the nitrogen end of the CN^- leads to a significant barrier in the potential surface, preventing the HNC^- species from ever reaching the autodetaching region (Maricq *et al.*, 1981).

C. *Heavy Atom Transfer*

With the flowing afterglow infrared chemiluminescence apparatus, the first quantitative measurements of infrared emission from an ionic product of an ion–molecule reaction have been made, specifically from vibrationally excited NO^+ formed in the reaction of N^+ with O_2 (Smith *et al.*, 1983):

$$N^+ + O_2 \rightarrow NO^+ + O \quad \sim 45\%, \tag{11}$$

$$\rightarrow O_2^+ + N \quad \sim 50\%, \tag{12}$$

$$\rightarrow O^+ + NO \quad \sim 5\%. \tag{13}$$

The branching fractions have been summarized by Albritton *et al.* (1979). This reaction serves as a major loss of N^+ in the earth's atmosphere. The reaction is rapid ($k = 6.1 \times 10^{-10}$ cm^3 sec^{-1}) (McFarland *et al.*, 1973) and the channel forming NO^+ is extremely exoergic, releasing 6.6-eV energy.

The disposal of this large exothermicity is an intriguing question. Tully *et al.* (1971) found low product translational energies in their crossed-beam experiments. Albritton *et al.* (1979) found less than 2% formation of the first electronically excited state, NO^+ $(a\,^3\Sigma^+)$, in flowing afterglow studies. Therefore, vibrational excitation is highly probable, and levels up to $v = 28$ are energetically accessible. This would produce infrared emission between 4 and 7 μm, a region where there are unidentified emissions in the earth's atmosphere. With oxygen pressures where the $N^+ + O_2$ reaction goes to approximately 50% completion, infrared emission is observed between 4.1 and 5.2 μm, peaking near 4.5 μm (Fig. 9). There is no emission at longer wavelengths. If the oxygen pressure is increased tenfold, an intense new emission feature appears, centered at 5.7 μm and spanning 5.2–6.4 μm; the original emission remains essentially unchanged.

These results suggest a more complex reaction scheme. The reactions of N^+ with O_2 are sufficiently exothermic so that several electronically excited states of the product atoms are accessible:

$$\Delta H$$

$$N^+ + O_2 \rightarrow NO^+ + O(^3P) \quad -6.6 \text{ eV}, \tag{14}$$

$$\rightarrow NO^+ + O(^1D) \quad -4.6 \text{ eV}, \tag{15}$$

$$\rightarrow NO^+ + O(^1S) \quad -2.4 \text{ eV}, \tag{16}$$

$$\rightarrow O_2^+ + N(^4S) \quad -2.5 \text{ eV}, \tag{17}$$

$$\rightarrow O_2^+ + N(^2D) \quad -0.1 \text{ eV}. \tag{18}$$

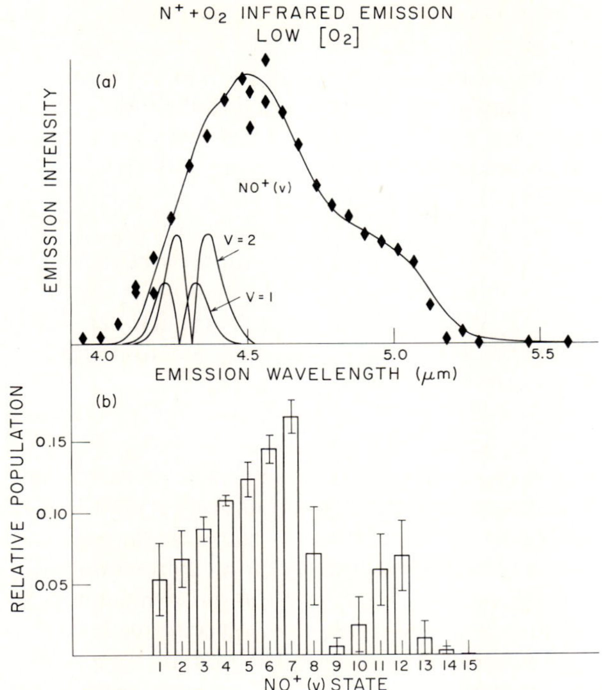

FIG. 9 (a) Infrared chemiluminescence spectrum of $NO^+(v)$ from the $N^+ + O_2 \rightarrow NO^+(v) +$ O reaction. The individual contributions from the $v = 2$–1 and $v = 1$–0 bands are shown underneath. The solid line is the best fit given by the bar graph populations (b). [From Smith *et al.* (1983).]

Although $NO^+(v = 28)$ can be populated in the reaction forming the $O(^3P)$ product, only levels up to $v = 18$ and $v = 8$ are allowed for the $O(^1D)$ and $O(^1S)$ channels, respectively. Likewise, in addition to formation of O_2^+ and $N(^4S)$, there is a nearly energy resonant channel that can generate $N(^2D)$. $N(^2D)$ can undergo a well-known reaction at the higher pressures of oxygen (Lin and Kaufman, 1971):

$$N(^2D) + O_2 \rightarrow NO(v) + O. \tag{19}$$

This reaction proceeds with a rate constant of 6×10^{-12} cm^3 sec^{-1} and is sufficiently exothermic to populate $NO(v \leqslant 18)$, which appears to be the source of the additional emission. The corresponding reaction of $N(^4S)$ with O_2 would not occur under our experimental conditions ($k = 9 \times 10^{-17}$ cm^3 sec^{-1}) (Baulch *et al.*, 1973). Using radiative lifetimes for NO^+ (Werner and Rosmus, 1982) and for NO (Billingsley, 1976), we have carried out a computer fitting of the infrared data and obtained relative vibrational populations. Those for $NO^+(v)$ are shown in Fig. 9. The $NO^+(v)$ distribution is bimodal, peaking at $v = 7$ and 12, indicating partitioning of the corresponding O-atom product into at least two of the three accessible states, ^3P, ^1D, and ^1S. The absence of any population in levels greater than $v = 14$ suggests that the $O(^3P)$ channel does not occur significantly. With the observation of low translational (Tully *et al.*, 1971) and low vibrational excitation, occurrence of the $O(^3P)$ pathway would require the unlikely channeling of about 3 eV into rotation. The rapid cutoff in emission near $v = 8$ corresponds to the exothermicity of the product channel forming $O(^1S)$. The sum of $NO^+(v = 1–8)$ and $NO^+(v = 9–14)$ populations is in an approximate ratio of 5:1. While it is tempting to assign these emissions to the $O(^1S)$ and $O(^1D)$ channels respectively, other interpretations of the data are possible. In light of these results the $N^+ + O_2$ reaction is being reconsidered as the source of the auroral green line (Rees, 1983).

The longer wavelength emission which arises at high oxygen pressure is identified as arising from $NO(v)$ formed by further reaction of the $N(^2D)$ product. Our best fit population for $NO(v)$ is in excellent agreement with the experimental results of Kennealy and co-workers (1978). Comparison of the relative intensities of NO^+ and NO emission suggests that the $N(^2D)$ is the major charge transfer product channel, representing 0.7 ± 0.2 of the total charge transfer process. The formation of $NO(v)$, therefore, serves as a convenient probe of otherwise "dark" ion–molecule reaction channels. These results present an interesting contrast with those of HCl and HF, where very high vibrational excitation is observed. When electronically excited atomic products are possible, it appears that these channels are preferred, leaving less energy for vibrational excitation of the diatomic product.

D. *Rotationally Resolved Charge Transfer*

In the flowing afterglow, product molecules typically suffer thousands of collisions with the helium buffer gas before detection. This has the unfortunate consequence that rotational state distributions in all but a few cases are completely thermalized. In many ion–molecule reactions the rotational state populations will provide an important determination of the dynamical characteristics. A new apparatus has been developed which preserves the advantageous features of the high ion density and thermalized ions available in the flowing afterglow, but in which the ion reaction is carried out in a low-pressure chamber under nearly single-collision conditions (Guyer *et al.*, 1983).

Figure 10 shows the source and interaction region of this apparatus. A flowing afterglow source is coupled to a low-pressure chamber by a large diameter (3–5 mm) orifice. With roughly 130-Pa pressure behind the nozzle orifice, the helium-entrained ions form a novel, low Mach number, free jet

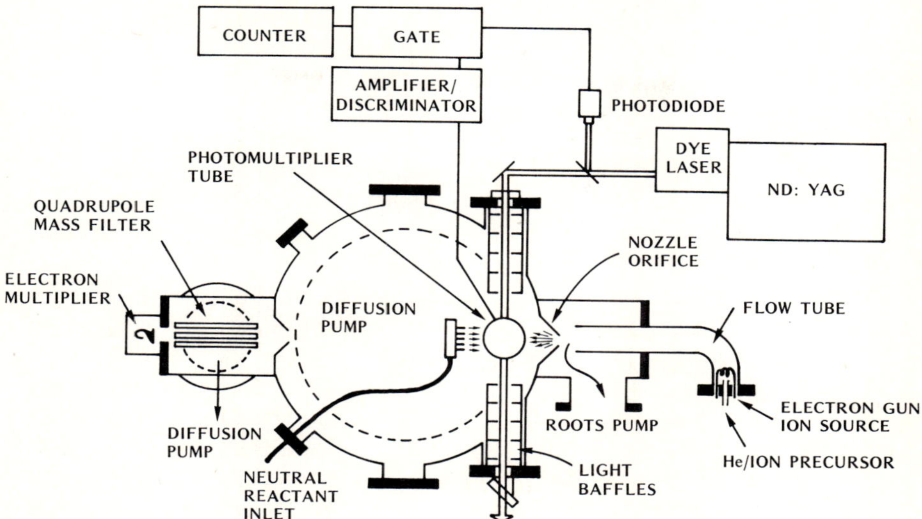

FIG. 10 Schematic drawing of the apparatus used to study thermal energy ion–molecule reactions under single-collision conditions. [From Guyer *et al.* (1983).]

expansion into the high-vacuum region. Densities of thermalized ions of approximately 10^6 cm^{-3} are achieved in the reaction zone with a background pressure of 0.04 Pa. Only about 10% of the ions undergo reaction with the crossed spray of reagent molecules, so that nearly single-collision conditions are achieved. The product state distribution is probed by laser-induced fluorescence. In order to obtain a sufficient signal-to-noise ratio, the pulsed excitation laser is operated in the saturating regime. The signal count rates are typically one photon per laser pulse on a single line in the laser-induced fluorescence spectrum, with background count rates of 0.1 per laser pulse.

The nozzle expansion produces ions with well-characterized translational and internal energies. Laser-induced fluorescence measurements of diatomic ions such as CO^+ delivered through the expansion show rotational distributions that are typically non-Boltzmann. The low rotational states are about 60 K and the higher states about 160 K. This can be explained by the varying rates of rotational relaxation, which decrease with increasing energy spacing between rotational levels. The translational energy is typically 0.1 eV, only slightly above thermal.

Results have been obtained for the charge transfer reaction

$$N^+ + CO \rightarrow N + CO^+, \tag{20}$$

which is sufficiently exothermic to populate CO^+ up to $v = 2$. Population is predominantly observed in the $v = 0$ state of CO^+, with only a small

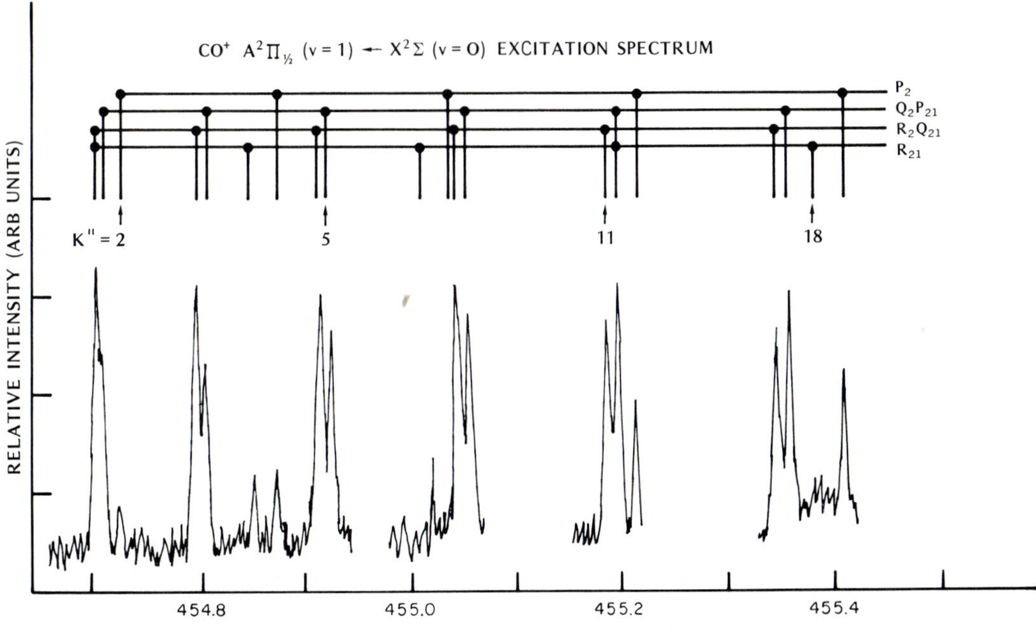

FIG. 11 Laser-induced fluorescence spectrum of a small portion of the rotational states in the CO^+ (A–X) system used to probe the rotational state populations in the X, $v'' = 0$ state produced by the N^+ + CO charge transfer reaction. [From Guyer *et al.* (1983).]

amount in the resonant state $v = 2$ (Guyer *et al.*, 1983). A portion of the LIF spectrum from $CO^+(v = 0)$ produced in the reaction is shown in Fig. 11. The rotational distribution in the $v = 0$ state is characterized closely by a Boltzmann distribution with a temperature of 410 K. Therefore, the reaction energy is almost completely released into translation. A physical picture of this reaction is one in which the electron transfer occurs at large internuclear distances, delivering a near-Franck–Condon population in the CO^+ products. Such a process is called an electron jump with momentum transfer (Marx, 1979). It is characterized by an electron transfer at a distance where the difference potential between the $N^+ + CO$ system and the $CO^+ + N$ system goes to zero. For $N^+ + CO$, this crossing occurs at long range. The low degree of rotational excitation suggests that the receding products do not exert much torque on the CO^+ rotor in the exit channel. This may be the case if, once the electron is transferred, the energy is released relatively slowly. This is consistent with the fact that the ion-induced forces are comparable in magnitude to the overall exothermicity. There are many other possibilities where the capability for rotationally resolved product state distributions can be used to test important dynamical concepts in near-thermal energy ion–molecule reactions.

E. Collisional Excitation and Relaxation of Ions

The flowing afterglow coupled with laser-induced fluorescence allows studies of collisional energy transfer of ions. This is a challenging and important new area for which little experimental data exist. The kinetic energy dependence of ion–molecule reaction rate constants and branching ratios has been addressed in drift tubes and flow-drift systems, and the translational energy distribution of atomic ions in atomic buffer gases has been characterized. However, the extent of internal excitation of diatomic or polyatomic ions in a drift field is unknown. Lindinger *et al.* (1975) first observed a number of unusual changes in the reactivity of ions in a flow-drift system, which they attributed to vibrational excitation. Recently, large enhancements of reaction rate constants by vibrational excitation have been reported for a variety of charge transfer reactions in flow-drift tube experiments (Durup-Ferguson *et al.*, 1983).

Strikingly little information exists on vibrational energy transfer processes for ions. Smith and Adams (1981) find essentially no relaxation of $N_2^+(v = 1)$ by helium, and Bien (1978) initially reported relatively slow deactivation rates of $NO^+(v)$. However, there is evidence that symmetric charge transfer processes rapidly deactivate vibrationally excited ions (Mahan *et al.*, 1982). In preliminary studies we have found that relaxation of $CO^+(v = 1)$ is rapid by both CO and N_2, occurring with a probability between 0.1 and 1.0 per collision (Hamilton *et ai.*, 1983b). Deactivation by both CO and N_2 can occur by near resonant vibrational energy transfer and, with CO, charge transfer deactivation is also possible. Danon and Marx (1982) have also reported rapid vibrational relaxation of CO^+ in their LIF/ion trap studies of the reaction of Ar^+ with CO. Recently Böhringer *et al.* (1983) have reported rate constants for collisional relaxation of vibrationally excited $O_2{}^+$ ions by a variety of quenchers; many of these rates are very rapid. Similar results have now been obtained for $NO^+(v)$ (Dobler *et al.*,1983).

To study collisional excitation and relaxation of ions, we have incorporated a drift section immediately before and throughout the region of laser interrogation, as shown in Fig. 12. This region is composed of about 40 thin cylindrical sections of flow tube, separated by Mylar spacers and connected by precision resistors. In the region of LIF detection, a small opening is covered with wire mesh to allow collection of fluorescence by the photomultiplier. For these experiments the pulsed dye laser is introduced through a window in the end flange of the flow tube and directed along the flow tube axis. Therefore, the baffle arms and laser ports in the region of laser probing are eliminated and field perturbations are minimized. With applied potentials of up to 5 V/cm and helium buffer gas pressures of about 65 Pa, E/N values of up to 32 Td are obtained (1 Townsend (Td) = 10^{-17} V cm^2).The field is estimated to be 99% uniform to within 78% of the flow tube radius (Albritton, 1967). For proper operation of a drift tube, ionic species of only one polarity can be present. Efficient separation of positive and negative ions requires

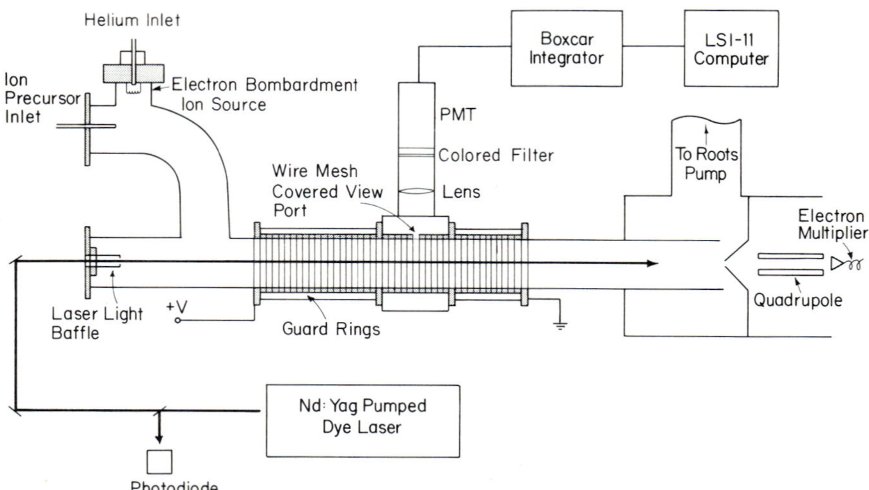

FIG. 12 Schematic drawing of the flow-drift tube used to study the internal excitation of ions in a drift field by laser-induced fluorescence.

densities of 10^7 cm^{-3} or less; therefore, in comparison with other studies, these drift experiments are performed with considerably lower ion concentrations and a reduced signal-to-noise ratio.

The N_2^+ ion is an excellent candidate for the first drift studies due to its short radiative lifetime and straightforward spectroscopy (Engelking and Smith, 1975). Figure 13a shows a portion of the LIF spectrum for the $B\,^2\Sigma_u^+$ $(v' = 0)$–$X\,^2\Sigma_g^+$ $(v'' = 0)$ transition in N_2^+ in the absence of a drift field (Duncan *et al.*, 1983). A least squares analysis of the relative peak intensities in the *R* branch yields a rotational temperature of ~300 K, indicating a relaxed thermal Boltzmann distribution of ground-state rotational levels. When a field of 30.9 Td is applied to N_2^+ in helium buffer gas, the LIF spectrum in Fig. 13b results. The maximum intensity has shifted from $K'' = 6$ to $K'' = 10$, and a rotational temperature of ~700 K is obtained. Varying either the length of drift section before laser interrogation or the time delay between application of the field and laser probing does not change the rotational temperature for a given E/N. This suggests that rotational excitation and relaxation processes in helium are very rapid, and equilibrium is

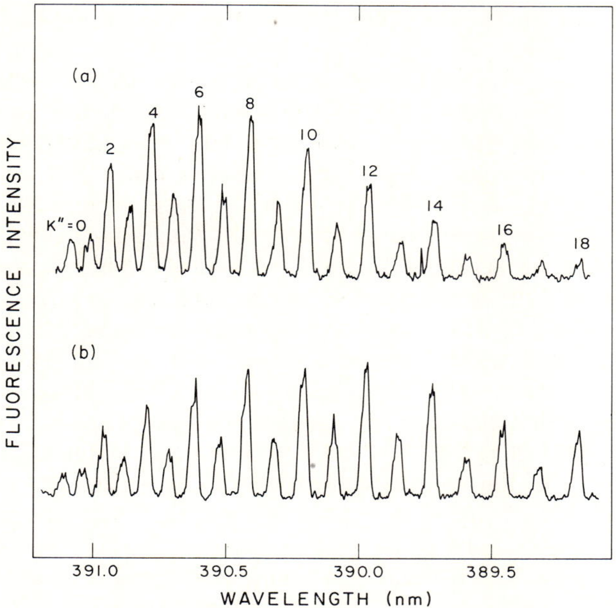

FIG. 13 Laser-induced fluorescence spectra of N_2^+ $(B\,^2\Sigma_u - X\,^2\Sigma_g)$ obtained by scanning the (0, 0) transition while collecting the (0, 1) emission through a bandpass filter. N_2^+ ions at (a) room temperature (~300 K, zero field) and (b) drifted at 30.9 Td in helium showing the effect of rotational heating (~700 K, 5V/cm). The spectra shown are not corrected for the transmission of the bandpass filter used.

achieved in less than ten collisions. Rapid translation-to-rotation energy transfer has also been reported by Mahan *et al.* (1982) in N_2^+-Ar collisions in an rf trap.

Viehland *et al.* (1981) have developed a theoretical description of collisional excitation of ions in a drift field; in this treatment the internal temperature of an ion is simply given by the collision energy of the ions with the neutral buffer gas. Our initial results for the dependence of rotational temperature on field strength are in good agreement with this theory.

V. CONCLUSION

Extension of the classical flowing afterglow technique to include infrared chemiluminescence and laser-induced fluorescence detection has provided a powerful new method for studying ion reaction dynamics. The diversity of processes outlined in Section IV is a testimony to the broad nature and versatility of this approach. Initial vibrational distributions for a variety of thermal energy ion–molecule reactions are determined for the first time; for some processes, detailed rotational distributions are also obtained; collisional energy transfer processes are studied both at thermal energy and as a function of kinetic energy.

However, major difficulties remain to be contended with. Although detailed spectroscopic constants are readily available for most neutral diatomic molecules and some polyatomic molecules, data are much more scarce for positive and negative ions. This state of affairs seems certain to improve in the next decade. A more serious problem, for those experiments which do not employ a low-pressure reaction chamber, is that of molecular relaxation. In order to convert measured distributions to initial distributions, all relaxation processes of the species must be understood. This is a demanding task for even simple triatomic molecules and a nearly intractable issue for complicated polyatomic species. Moreover, signal-to-noise limitations remain a serious consideration for detection of species with long radiative lifetimes.

Nevertheless, investigations in a variety of new areas are immediately feasible with these techniques. The study of ion reactions producing several diatomic products, ion–ion recombination processes, and the reactions of translationally excited reagents represent important new directions. Additional information on energy transfer processes in ions can be obtained by studies at higher drift velocities and with other buffer gases. Moreover, vibrationally excited ions can be formed by chemical reaction or discharge methods; relaxation or selective reactivity upon addition of neutral reagents can be probed. Major extensions, such as incorporation of selected ion flow tube methods will essentially eliminate background radiation and substantially reduce relaxation processes by removal of the neutral precursors of the ionic species. It is clear that the present results only skim the surface of a field that is rich and fertile and holds great promise of future development.

ACKNOWLEDGMENTS

We gratefully acknowledge support of this research by the Air Force Office of Scientific Research, the National Science Foundation, and the National Bureau of Standards. This work would not have been possible without the excellent collaboration of our co-workers. We are grateful for the expert assistance of the technical and secretarial staffs of the Joint Institute for Laboratory Astrophysics.

REFERENCES

Adams, N. G., and Smith, D. (1976). *Int. J. Mass Spectrom. Ion Phys.* **21**, 349.

Agrawalla, B. S., Manocha, A. S., and Setser, D. W. (1981). *J. Phys. Chem.* **85**, 2873.

Albritton, D. L. (1967). Ph.D. thesis, Georgia Institute of Technology, Atlanta, Georgia.

Albritton, D. L. (1978). *At. Data Nucl. Data Tables* **22**, 1.

Albritton, D. L., Viggiano, A. A., Dotan, I., and Fehsenfeld, F. C. (1979). *J. Chem. Phys.* **71**, 3295.

Allan, M., and Wong, S. F. (1981). *J. Chem. Phys.* **74**, 1687.

Allen, Jr., H. C., Tidwell, E. D., and Plyler, E. K. (1956). *J. Chem. Phys.* **25**, 302.

Allison, J., Kondow, T., and Zare, R. N. (1979). *Chem. Phys. Lett.* **64**, 202.

Anlauf, K. G., Kuntz, P. J., Maylotte, D. H., Pacey, P. D., and Polanyi, J. C. (1967). *Disc. Faraday Soc.* **44**, 183.

Anlauf, K. G., Horne, D. S., Macdonald, R. G., Polanyi, J. C., and Woodall, K. B. (1972). *J. Chem. Phys.* **57**, 1561.

Arnold, G. S., Fernando, R. P., and Smith, I. W. M. (1980). *J. Chem. Phys.* **73**, 2773.

Baulch, D. L., Drysdale, D. D., Horne, D. G., and Lloyd, A. C. (1973). "Evaluated Kinetic Data for High Temperature Reactions. II. Homogeneous Gas Phase Reactions of the H_2-N_2-O_2 System." CRC Press, Cleveland, Ohio.

Benedict, W. S., Herman, R., Moore, G. E., and Silverman, S. (1962). *Astrophys. J.* **135**, 277.

Berkowitz, J., Chupka, W. A., and Walter, T. A. (1969). *J. Chem. Phys.* **50**, 1497.

Bien, F. (1978). *J. Chem. Phys.* **69**, 2631.

Bierbaum, V. M., Ellison, G. B., Futrell, J. H., and Leone, S. R. (1977). *J. Chem. Phys.* **67**, 2375.

Billingsley, F. P. II (1976). *J. Mol. Spectrosc.* **61**, 53.

Böhringer, H., Durup- Ferguson, M., Fahey, D. W., Fehsenfeld, F. C., and Ferguson, E. E. (1983). *J. Chem. Phys.*, in press.

Bowers, M. T., and Laudenslager, J. B. (1976). *In* "Principles of Laser Plasmas" (G. Bekefi, ed.), p. 89. Wiley, New York.

Brault, J. W., and Davis, S. P. (1982). *Phys. Scr.* **25**, 268.

Brzozowski, J., Elander, N., Erman, P., and Lyyra, M. (1974). *Phys. Scr.* **10**, 241.

Charters, P. E., Macdonald, R. G., and Polanyi, J. C. (1971). *Appl. Opt.* **10**, 1747.

Danon, J., and Marx, R. (1982). *Chem. Phys.* **68**, 255.

Danon, J., Mauclaire, G., Govers, T. R., and Marx, R. (1982). *J. Chem. Phys.* **76**, 1255.

Ding, A. (1982). Private communication.

Ding, A., Richter, K., and Menzinger, M. (1981). *Chem. Phys. Lett.* **77**, 523.

Dobler, W., Federer, W., Howorka, F., Lindinger, W., Durup-Ferguson, M., and Ferguson, E. E. (1983). *J. Chem. Phys.* **79**, 1543.

Duncan, M. A., Bierbaum, V. M., Ellison, G. B., and Leone, S. R. (1983). *J. Chem. Phys.*, in press.

Dunkin, D. B., Fehsenfeld, F. C., Schmeltekopf, A. L., and Ferguson, E. E. (1968). *J. Chem. Phys.* **49**, 1365.
Durup-Ferguson, M., Böhringer, H., Fahey, D. W., and Ferguson, E. E. (1983). *J. Chem. Phys.* **79**, 265.
Dykstra, C. E., and Secrest, D. (1981). *J. Chem. Phys.* **75**, 3967.
Engelking, P. C., and Smith, A. L. (1975). *Chem. Phys. Lett.* **36**, 21.
Evenson, K. M., Wells, J. S., and Radford, H. E. (1970). *Phys. Rev. Lett.* **25**, 199.
Fehsenfeld, F. C. (1975a). *Int. J. Mass Spectrom. Ion Phys.* **16**, 151.
Fehsenfeld, F. C. (1975b). *In* "Interactions Between Ions and Molecules" (P. Ausloos, ed.), p. 387. Plenum Press, New York.
Ferguson, E. E., Fehsenfeld, F. C., and Schmeltekopf, A. L. (1969). *Adv. At. Mol. Phys.* **5**, 1.
Gauyacq, J. P. (1982). *J. Phys. B.* **15**, 2721.
German, K. R. (1975). *J. Chem. Phys.* **62**, 2584; **63**, 5252.
Grieman, F. J., Mahan, B. H., and O'Keefe, A. (1980). *J. Chem. Phys.* **72**, 4246.
Guyer, D. R., Hüwel, L., and Leone, S. R. (1983). *J. Chem. Phys.* **79**, 1259.
Hamilton, C. E., Bierbaum, V. M., and Leone, S. R. (1983a). *J. Chem. Phys.*, submitted.
Hamilton, C. E., Duncan, M. A., Zwier, T. S., Weisshaar, J. C., Ellison, G. B., Bierbaum, V. M., and Leone, S. R. (1983b). *Chem. Phys. Lett.* **94**, 4.
Hariri, A., Petersen, A. B., and Wittig, C. (1976). *J. Chem. Phys.* **65**, 1872.
Heaven, M., Miller, T. A., and Bondybey, V. E. (1982). *J. Chem. Phys.* **76**, 3831.
Herbelin, J. M., and Emanuel, G. (1974). *J. Chem. Phys.* **60**, 689.
Herzenberg, A. (1967). *Phys. Rev.* **160**, 80.
Holland, R. F., and Maier, W. B. II (1972). *J. Chem. Phys.* **56**, 5229.
Hotop, H. and Lineberger, W. C. (1975). *J. Phys. Chem. Ref. Data* **4**, 539.
Huber, K. P., and Herzberg, G. (1979). "Molecular Spectra and Molecular Structure. IV. Constants of Diatomic Molecules." Van Nostrand-Reinhold, New York.
Jeunehomme, M. (1966). *J. Chem. Phys.* **44**, 4253.
Katayama, D. H., Miller, T. A., and Bondybey, V. E. (1980). *J. Chem. Phys.* **72**, 5469.
Kennealy, J. P., Del Greco, F. P., Caledonia, G. E., and Green, B. D. (1978). *J. Chem. Phys.* **69**, 1574.
Keyes, R. J. (ed.) (1977). "Optical and Infrared Detectors." Springer-Verlag, Berlin and New York.
Kinsey, J. L. (1977). *Ann. Rev. Phys. Chem.* **28**, 349.
Krenos, J. R., Preston, R. K., Wolfgang, R., and Tully, J. C. (1974). *J. Chem. Phys.* **60**, 1634.
Kuntz, P. J. (1976). *In* "Dynamics of Molecular Collisions, Part B" (W. H. Miller, ed.), p. 53. Plenum Press, New York.
Laudenslager, J. B. (1979). *In* "Kinetics of Ion-Molecule Reactions" (P. Ausloos, ed.), p. 405. Plenum Press, New York.
Leone, S. R. (1982). *J. Phys. Chem. Ref. Data.* **11**, 953.
Levine, R. D. (1978). *Ann. Rev. Phys. Chem.* **29**, 59.
Levy, M. R. (1979). *Prog. Reaction Kinet.* **10**, 1.
Lin, C., and Kaufman, F. (1971). *J. Chem. Phys.* **55**, 3760.
Lindinger, W., McFarland, M., Fehsenfeld, F. C., Albritton, D. L., Schmeltekopf, A. L., and Ferguson, E. E. (1975). *J. Chem. Phys.* **63**, 2175.
Lineberger, W. C. (1983). Private communication.
Luk, C. K., and Bersohn, R. (1973). *J. Chem. Phys.* **58**, 2153.
Mahan, B. H., Martner, C., and O'Keefe, A. (1982). *J. Chem. Phys.* **76**, 4433.
Maki, A. G., and Sams, R. L. (1981). *J. Chem. Phys.* **75**, 4178.
Malins, R. J., and Setser, D. W. (1980). *J. Chem. Phys.* **73**, 5666.
Maricq, M. M., Smith, M. A., Simpson, C. J. S. M., and Ellison, G. B. (1981). *J. Chem. Phys.* **74**, 6154.

Marx, R. (1979). *In* "Kinetics of Ion-Molecule Reactions" (P. Ausloos, ed.), p. 103. Plenum Press, New York.

Maylotte, D. H., Polanyi, J. C., and Woodall, K. B. (1972). *J. Chem. Phys.* **57**, 1547.

McFarland, M., Albritton, D. L., Fehsenfeld, F. C., Ferguson, E. E., and Schmeltekopf, A. L. (1973). *J. Chem. Phys.* **59**, 6610.

Miller, T. A., and Bondybey, V. E. (1977). *Chem. Phys. Lett.* **50**, 275.

Miller, T. A., and Bondybey, V. E. (1980). *J. Chim. Phys. Phys-Chim. Biol.* **77**, 695.

Möhlmann, G. R., and de Heer, F. J. (1976). *Chem. Phys.* **17**, 147.

Moehlmann, J. G., Gleaves, J. T., Hudgens, J. W., and McDonald, J. D. (1974). *J. Chem. Phys.* **60**, 4790.

Möhlmann, G. R., Bhutani, K. K., and de Heer, F. J. (1977). *Chem. Phys.* **21**, 127.

Pau, C. F., and Hehre, W. J. (1982). *J. Phys. Chem.* **86**, 321.

Petersen, A. B., and Smith, I. W. M. (1979). *J. Chem. Phys.* **71**, 3346.

Polanyi, J. C. (1972), *Acc. Chem. Res.* **5**, 161.

Preston, R. K., and Tully, J. C. (1971). *J. Chem. Phys.* **54**, 4297.

Redmon, L. T., Purvis, G. D. III, and Bartlett, R. J. (1980). *J. Chem. Phys.* **72**, 986.

Rees, M. H. (1983). *Planet. Space Sci.*, submitted.

Roden, G. (1975). Ph.D. thesis, Univ. of Göttingen.

Rosmus, P. (1979). *Theoret. Chim. Acta (Berlin)*, **51**, 359.

Rosmus, P., and Reinsch, E. A. (1980). *Z. Naturforsch.* **A35**, 1066.

Schultz, A., Cruse, H. W., and Zare, R. N. (1972). *J. Chem. Phys.* **57**, 1354.

Segal, G. A., and Wolf, K. (1981). *J. Phys. B.* **14**, 2291.

Shimamura, I., and Matsuzawa, M. eds. (1979). *Symp. Electron-Mol. Collisions.* Univ. of Tokyo, Tokyo, Japan.

Sileo, R. N., and Cool, T. A. (1976). *J. Chem. Phys.* **65**, 117.

Smith, A. J., Read, F. H., and Imhof, R. E. (1975). *J. Phys. B.* **8**, 2869.

Smith, D. and Adams, N. G. (1981). *Phys. Rev.* **A23**, 2327.

Smith, I. W. M. (1980). "Kinetics and Dynamics of Elementary Gas Reactions." Butterworths, London.

Smith, M. A., and Leone, S. R. (1983). *J. Chem. Phys.* **78**, 1325.

Smith, M. A., Barkley, R. M. and Ellison, G. B. (1980). *J. Am. Chem. Soc.* **102**, 6851.

Smith, M. A., Bierbaum, V. M., and Leone, S. R. (1983). *Chem. Phys. Lett.* **94**, 398.

Smith, W. H., Brzozowski, J., and Erman, P. (1976). *J. Chem. Phys.* **64**, 4628.

Sung, J. P., and Setser, D. W. (1978). *J. Chem. Phys.* **69**, 3868.

Tamagake, K., Setser, D. W., and Sung, J. P. (1980). *J. Chem. Phys.* **73**, 2203.

Tiernan, T. O., and Lifshitz, C. (1981). *Adv. Chem. Phys.* **45**, 81.

Tsuji, M., Susuki, T., Endoh, M., and Nishimura, Y. (1982), *Chem. Phys. Lett.* **86**, 411.

Tully, J. C., Herman, Z., and Wolfgang, R. (1971). *J. Chem. Phys.* **54**, 1730.

Tully, J. C. (1974). *J. Chem. Phys.* **60**, 3042.

Viehland, L. A., Lin, S. L., and Mason, E. A. (1981). *Chem. Phys.* **54**, 341.

Weisshaar, J. C., Zwier, T. S., and Leone, S. R. (1981). *J. Chem. Phys.* **75**, 4873.

Werner, H.-J., and Rosmus, P. (1982). *J. Mol. Spectrosc.* **96**, 362.

Wolfe, W. L., and Zissis, G. J. (eds.) (1978). "The Infrared Handbook." Environmental Research Institute of Michigan.

Wong, M., Bernath, P., and Amano, T. (1982). *J. Chem. Phys.* **77**, 693.

Zare, R. N., and Dagdigian, P. J. (1974). *Science* **185**, 739.

Zwier, T. S., Bierbaum, V. M., Ellison, G. B., and Leone, S. R. (1980a) *J. Chem. Phys.* **72**, 5426.

Zwier, T. S., Maricq, M. M., Simpson, C. J. S. M., Bierbaum, V. M., Ellison, G. B., and Leone, S. R. (1980b). *Phys. Rev. Lett.* **44**, 1050.

Zwier, T. S., Weisshaar, J. C., and Leone, S. R. (1981). *J. Chem. Phys.* **75**, 4885.

Chapter 18
Infrared photochemistry of gas phase ions

*L. R. THORNE**

J. L. BEAUCHAMP

ARTHUR AMOS NOYES LABORATORY
OF CHEMICAL PHYSICS
CALIFORNIA INSTITUTE OF TECHNOLOGY
PASADENA, CALIFORNIA

* Present address: Sandia National Laboratories, Livermore, California 94550.

ISBN 0-12-120803-6

I. INTRODUCTION

Experiments involving molecules which are isolated from collisions for periods exceeding several milliseconds require special particle storage techniques. In the case of ions, crossed electric and magnetic fields can be used to confine particle motion for time periods exceeding several hours, establishing virtually collision-free conditions. For charged particles stored in such fields, the phenomenon of ion cyclotron resonance (ICR) provides a sensitive and selective means of detection. At pressures below 10^{-5} torr, stored ions are forced to maintain energy equilibrium with their environment by absorption and emission of infrared radiation rather than by collisions with other molecules. Under these conditions, infrared lasers provide a means of upsetting the equilibrium by exposing the ions to enormous narrowband photon fluxes and thereby driving photochemical processes.

What fraction of the ion population will absorb infrared radiation at a specific wavelength? Can more than one photon be absorbed? When the ions can be photodissociated, how does the dependence of dissociation yield on laser wavelength (photodissociation spectrum) compare to the small signal absorption spectrum? Is the dissociation spectrum homogeneously broadened? (That is, does each molecule absorb at all wavelengths within the photodissociation band?) Or is it heterogeneously broadened? (That is, does each molecule absorb at slightly different wavelengths within the photodissociation band?)

Recently, these questions have been answered, in part, by using the techniques of ion cyclotron resonance spectroscopy to monitor ion populations irradiated with either low-power cw or high-power pulsed CO_2 lasers. These experiments represent a unique application of ICR spectroscopy and lay the groundwork for several applications described in this chapter.

The important aspects of infrared multiphoton absorption are illustrated in Fig. 1. M represents a molecular species (neutral or ion) in its ground state. Infrared absorption involving bound molecular states will occur only at discrete wavelengths if intramolecular vibrational relaxation (IVR) is slow compared to the photon absorption rate. Generally, IVR rates increase rapidly with increasing internal energy due to the rapidly increasing density of vibrational energy levels and increasing anharmonic coupling between vibrational modes. If sufficiently energized, a molecule can be in what is termed the *vibrational quasi-continuum.* Onset of the quasi-continuum can be considered to occur when the density of vibrational levels of a molecule exceeds the inverse of the transition bandwidth, or for an homogeneously broadened transition, when the vibrational state density exceeds the inverse of the laser bandwidth. Under these circumstances, it is highly probable that at least one other vibrational energy level matches the energy level pumped by the laser. This "resonance" enhances IVR rates by transferring energy rapidly out of the pumped mode into other vibrational modes of the molecule. Perturbations which occur during collisions can also increase IVR rates.

Exciting a molecule into the quasi-continuum may be accomplished several ways, as shown in Fig. 1. At room temperature a sufficiently large molecule may be in the quasi-continuum solely as a result of its internal thermal energy content. However, smaller molecules must be suitably activated. This can be accomplished by collisional processes, by highly exothermic chemical reactions which yield vibrationally excited products, or by electronic excitation followed by radiationless relaxation to the ground electronic state (internal conversion). Alternatively, molecules can be "forced" into the vibrational quasi-continuum, up a ladder of discrete vibrational energy levels belonging to the same vibrational mode, by the resonant or near-resonant absorption of several infrared photons. In this case, IVR initially involves highly excited energy levels of the pumped mode. Such forced multiphoton excitation gives photodissociation spectra which are often shifted to longer wavelengths relative to the fundamental absorption. This is due to anharmonicity, which causes adjacent vibrational levels high on the ladder to be more closely spaced than those nearer the bottom. Although few such photodissociation spectra have been reported, shift of the photodissociation spectra to longer wavelengths relative to the small signal absorption spectrum appears to be characteristic of small molecules dissociated by high-power pulsed infrared lasers.

The total number of photons to which a molecular system is exposed during excitation is measured by laser fluence, which is usually expressed in joules per centimeter squared. Fluences in excess of 1 J cm^{-2} (e.g., 1 MW cm^{-2} exposure for 1 μsec with a pulsed laser) are often sufficient to effect

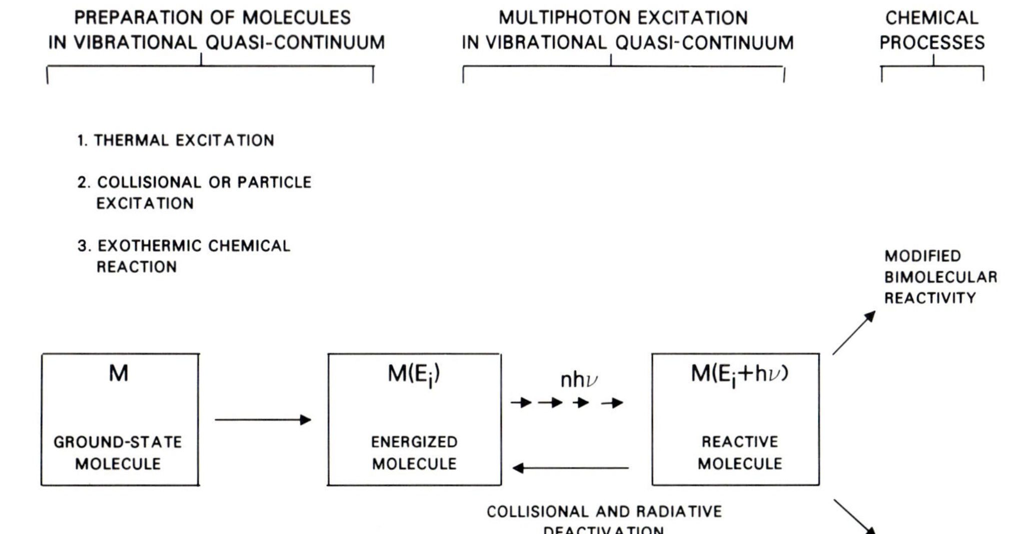

FIG. 1 Schematic representation of laser-driven photochemistry indicating processes which promote molecules into the vibrational continuum, cause deactivation, and lead to chemical reactivity.

dissociation of an absorbing species. Although nonlinear effects are now well recognized, early experiments with SF_6 (Black *et al.*, 1977, 1979) suggested that, at low conversion, the probability of dissociation is independent of peak power if the fluence is constant. This led us to consider the use of continuous wave (cw) infrared lasers for dissociation experiments, since 50 W cm^{-2}, which is easily obtained from commercial line-tunable cw lasers, delivers 1 J cm^{-2} in only 20 msec. Clearly, the success of such experiments depends upon radiative and collisional relaxation rates being low because of the comparatively long time required for irradiation. Estimates of infrared cooling rates of ions suggested that the radiative relaxation rates would not be a problem and that collisional relaxation rates could be slowed by lowering the cell pressure and using trapped ion techniques. The success of our early experiments using cw infrared lasers showed that infrared multiphoton dissociation is indeed a facile process. This has been discussed in two previous reviews: Bomse *et al.* (1978) and Woodin *et al.* (1979b). Since that time, we have observed multiphoton dissociation of (1) "large" molecules, which are in the vibrational quasi-continuum at room temperature; (2) "medium"-size molecules, for which collisions enhance IVR rates; and (3) "activated" molecules, energized by one of the processes indicated in Fig. 1.

The majority of our observations suggest that in order to effect infrared multiphoton dissociation using cw lasers, the molecule must be in the vibrational quasi-continuum before irradiation; i.e., there must be some mechanism to populate the quasi-continuum other than the forced excitation which characterizes the dissociation of small molecules with high-power pulsed lasers. This is because, with low-power cw radiation, the competing processes of radiative and collisional deactivation are more rapid than sequential photon absorption within a single vibrational mode.

As illustrated in Fig. 1, once in the quasi-continuum, further multiphoton excitation can lead to modified bimolecular reaction rates or unimolecular dissociation. Both processes produce a change in the ion population, which can be detected using ICR techniques.

In this chapter we summarize our observations of multiphoton excitation and dissociation processes using cw and pulsed lasers to excite ions which are formed, stored, and detected using the techniques of ion cyclotron resonance spectroscopy.

II. EXPERIMENTAL METHODS

Beauchamp (1971) and Lehman and Bursey (1976) have presented reviews of the theory, techniques, and instrumentation of trapped ion ICR spectrometry. Figure 2 shows a schematic view of an ICR cell modified for photochemical studies (Freiser and Beauchamp, 1974). The cell is operated in the

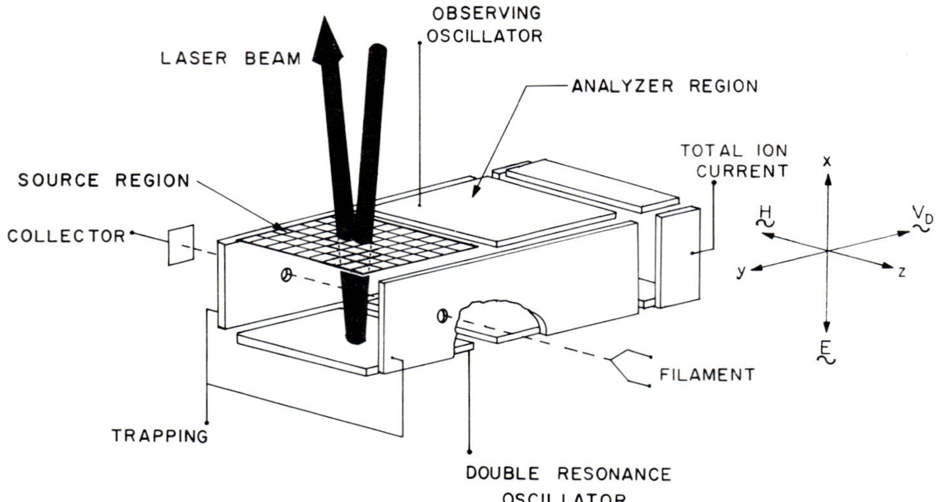

FIG. 2 Cutaway view of cyclotron resonance cell. The electron beam is collinear with the magnetic field. The laser beam enters through an open mesh grid comprising the top plate of the source region and is reflected by the lower plate. [From Bomse *et al.* (1979).]

pressure range of 10^{-7}–10^{-5} torr. Depending on the collision partners, this corresponds to a mean ion–molecule collision frequency of between 1 and 100 sec^{-1}. Ions are generated from the neutral gas in the cell by electron impact with a pulsed electron beam. Typically, a beam pulse of 10 msec forms a cloud of 10^5 ions along its path, parallel to the magnetic field lines. The cloud collapses within microseconds to the center of the source region due to the crossed electric and magnetic fields. Ions thus trapped may be stored without appreciable loss for several seconds.

Since the trapped ions are localized in a small volume, overlap with unfocused laser output may be made very efficient. While trapped, the ions may participate in ion–molecule or photochemical reactions or be selectively ejected using ICR double-resonance techniques. The ability to eject ions of a given mass permits positive identification of reaction pathways which occur during the trapping period, and provides a means of suppressing ion–molecule reactions which may interfere with the measurements. After a predetermined trapping period, ions are swept into the analyzer region by applying the appropriate drift potentials to the plate in the trapping and analyzer regions. The ions are then detected by using the upper and lower plates in the analyzer region as a capacitor in a marginal oscillator circuit and bringing the ions into resonance by adjusting the magnetic field strength. At resonance, ions of a particular mass-to-charge ratio absorb energy from the marginal oscillator in proportion to their total number. Energy loss of the marginal oscillator is detected by a boxcar averager, which is used to average repetitive cycles of ion formation, trapping, and detection.

The timing sequence for a typical photochemical experiment is illustrated in Fig. 3. Electronic control of an optical shutter permits the infrared beam to irradiate ions during any portion of the trapping sequence. Photodissociation rate constants can be determined by irradiating the ions on alternate trapping cycles and monitoring the ion signal intensities using a two-channel boxcar integrator. As the detect pulse is swept through the irradiation period, the logarithm of the ratio of the two signals produces a line whose slope is proportional to the photodissociation rate. Alternatively, the magnetic field may be swept for a fixed detect pulse delay to trace out a mass spectrum in the presence and absence of irradiation.

As shown in Fig. 2, replacement of one of the source plates of the ICR cell with a screen grid permits irradiation of trapped ions. The laser beam is directed through the grid into the cell and reflected back out of the cell by a mirror finish on the back source plate. The angle of incidence is made to be nearly normal, so the ions are exposed to effectively twice the irradiance impinging on the cell. Brauman and co-workers have made use of a similar cell design, but their laser beam is directed lengthwise down the cell, coincident with the drift path of the ions. This configuration provides improved overlap of the laser beam with the ions if the cell is operated in drift mode.

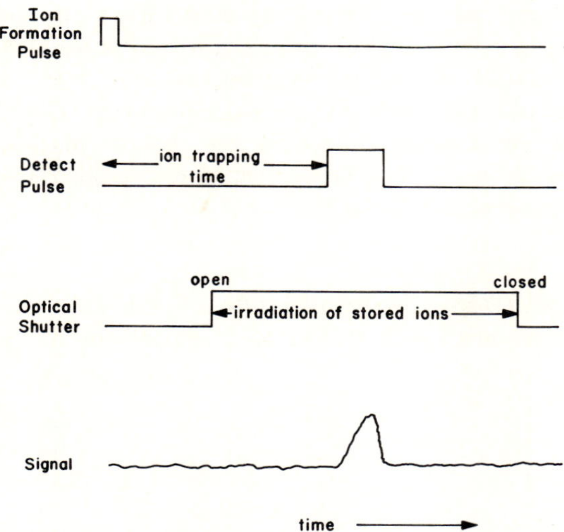

FIG. 3 Timing sequence for trapped ion ICR photodissociation experiments. Ions may be irradiated during any portion of the trapping period. [From Bomse *et al.* (1979).]

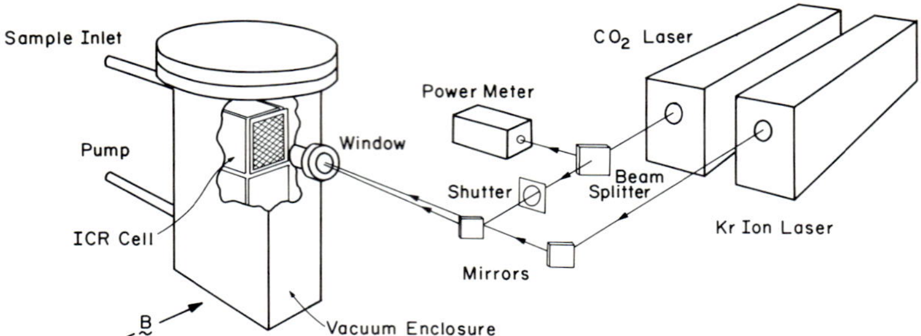

FIG. 4 Schematic view of the ICR cell and optical configuration. The vector **B** indicates the orientation of the magnetic field. [From Wight and Beauchamp (1981).]

Figure 4 shows a schematic view of the optical configuration used in our studies. Several experiments utilized an additional visible wavelength laser. Its position is also shown in Fig. 4. Either a line-tunable cw CO_2 laser or a line-tunable CO_2 TEA laser was used as the infrared radiation source. Both were operated in TEM_{00} and have beam-intensity profiles which are nearly Gaussian [full width at half maximum (FWHM) = 6 mm]. A cw krypton ion laser having a 2-mm beam diameter was used for the experiments requiring visible laser radiation. Laser beams were used unfocused to maximize uniform irradiation of the ions. With this arrangement, ions could be exposed with up to 120 W cm^{-2} (cw) or fluences up to 1.6 J cm^{-2} (pulsed) in the infrared or 5 W cm^{-2} in the visible. Irradiation of the ions with the cw laser was controlled by a mechanical shutter with a 5-msec opening time. The frequency of the laser radiation was measured with a spectrum analyzer. Additional experimental details are given by Bomse *et al.* (1979).

Typical ICR operating conditions in our experiments include marginal oscillator frequencies of 70–300 kHz, magnetic fields of 7–21 kG, and electron impact energies of 8–70 eV.

III. MULTIPHOTON DISSOCIATION OF IONS WITH LOW-INTENSITY CW INFRARED RADIATION

Multiphoton dissociation using cw infrared laser radiation has been observed for over 30 ions. In general, the multiphoton dissociation process for a particular ion may be characterized by the ion chemistry with and without laser radiation, the thermochemistry of the dissociation, and the dependence of the photodissociation yield on the laser wavelength, laser irradiance, and neutral gas pressure. In this section we present a summary of these observations. First, a general summary of the characteristics common to most of the ions studied is given; then results for specific ions are used to illustrate various aspects of the multiphoton dissociation process. Finally, a phenomenological model is presented which correlates many of the observations.

A. Summary of Multiphoton Dissociation Results

Table I lists the ions studied in our laboratory using ICR techniques and cw laser irradiance in the range of 1–100 W cm^{-2} or pulsed laser energies up to 1.6 J cm^{-2}. In each case the decomposition pathway was confirmed by direct observation of the product ion or by double-resonance techniques. The identity of the neutral product was inferred from mass balance considerations. Enthalpy changes given in Table I were calculated from heats of formation. Activation energies, except where noted, were not determined, and they may be in excess of the reaction endothermicity. The minimum number of photons needed to reach the thermodynamic threshold for dissociation n is given by the enthalpy change divided by the photon energy.

All observed dissociation reactions are identical to the known lowest energy decomposition pathway. Indeed, there is strong experimental evidence (see the results for diethyl ether in Section III.A.1) that with cw infrared multiphoton excitation, dissociation occurs at energies close to the thermodynamic threshold for the lowest energy dissociation channel. For proton-bound dimers of some alcohols, there are multiple products, but the lowest energy process is always observed (see Section IV.A). In this case, multiple channels are open due to a common transition state.

For all ions studied, the observed decomposition pathway is not dependent on laser wavelength. However, the photodissociation yield P_D generally is dependent on laser wavelength. P_D is defined as the fraction of the initial ion population dissociated during a specified exposure time at a specified laser irradiance:

$$P_D = 1 - I/I_0. \tag{1}$$

Here, I is the signal intensity of the dissociating ion measured at the end of the exposure period. I_0 is the signal without irradiation at the end of the same period. In every case studied (except for CF_3I^+ and $C_3F_6{}^+$ at high irradiance), the photodissociation process is characterized by the first-order kinetics given by Eq. (2),

$$I/I_0 = \exp(-k_D t), \tag{2}$$

where k_D is the dissociation rate constant.

For all ions studied, the dissociation rate constant, typically 2–30 sec^{-1}, is dependent on laser irradiance. If the dependence is linear, k_D can be

TABLE I

Observed Low-Intensity IR Multiphoton Dissociation Reactions

Reactant	Products	Δh (kcal/mole)	n^a
$[(C_2H_5)_2O]_2H^{+b,c}$	$(C_2H_5)_2OH^+ + (C_2H_5)_2O$	31	12
$(C_2H_5)_2OH^{+b,c}$	$C_2H_5OH_2^+ + C_2H_4$	27	10
$(C_2D_5)_2OD^{+b,c}$	$C_2D_5OD_2^+ + C_2D_4$	27	10
$(C_2D_5)(C_2H_5)OH^{+b,c}$	$C_2D_5OH_2^+ + C_2D_4$	27	10
$CH_3CHOC_2H_5^{+b}$	$CH_3CHOH^+ + C_2H_4$	34	13
$(CH_2CH_2CH_2CH_2O)_2H^{+b}$	$CH_2CH_2CH_2CH_2OH^+ +$ $CH_2CH_2CH_2CH_2O$	31	12
$C_3F_6^{+d,e}$	$C_2F_4^+ + CF_2$	56^f	20
$C_6H_{12}^{+d}$	$C_5H_9^+ + CH_3$	30	11
$CpRh(CO)_2H^{+b,g}$	$CpRhH^+ + 2CO$	—[h]	—[h]
CF_3I^{+i}	$CF_3^+ + I$	10	3
$(CH_3OH)_2H^{+j}$	$(CH_3)_2OH^+ + H_2O$	17	7
$(C_2H_5OH)_2H^{+j}$	$(CH_2H_5)_2OH^+ + H_2O$	15	6
$(n\text{-}C_3H_7OH)_2H^{+j}$	$n\text{-}C_3H_7OH_2^+ + C_3H_7OH(71\%)$	33	12
	$(n\text{-}C_3H_7)_2OH^+ + H_2O(17\%)$	17	6
	$(n\text{-}C_3H_7OH)H^+(H_2O) + C_3H_6(12\%)$	19	7
$(i\text{-}C_3H_7OH)_2H^{+j}$	$(i\text{-}C_3H_7)_2OH^+ + H_2O$	19	7
$(s\text{-}C_4H_9OH)_2H^{+j}$	$(s\text{-}C_4H_9)_2OH^+ + H_2O(38\%)$	15	6
	$(s\text{-}C_4H_9)H^+(H_2O) + C_4H_8(62\%)$	20	7
$(i\text{-}C_3H_7OH)H^+(s\text{-}C_4H_9OH)^j$	$(i\text{-}C_3H_7)(C_4H_9)OH^+ + H_2O(37\%)$	19	7
	$(i\text{-}C_3H_7OH)H^+(OH_2) + C_4H_8(63\%)$	20	7
$(t\text{-}C_4H_9OH)_2H^{+j}$	$(t\text{-}C_4H_9)OH^+ + H_2O$	20	7
$(t\text{-}C_5H_{11}OH)_2H^{+j}$	$(t\text{-}C_5H_{11}OH)H^+(H_2O) + C_5H_{10}$	22	8
$(i\text{-}C_3H_7OH)H^+(t\text{-}C_4H_9OH)^j$	$(i\text{-}C_3H_7)(t\text{-}C_4H_9)OH^+ + H_2O$	13	5
$(CD^{35}_{32})Cl^{+k}$	$CD_2{}^{35}Cl^+ + CD_4$	18	7
$(CD_3)_2^{37}Cl^{+k}$	$CD_2{}^{37}Cl^+ + CD_4$	18	7
$C_6H_5CN^{+l}$	$C_6H_4^+ + HCN$	—	—
$i\text{-}C_3H_7ClLi^+$	$C_3H_6Li^+ + HCl$	—	—
$(p\text{-methoxybenzaldehyde})Li^+$	$p\text{-methoxybenzaldehyde} + Li^+$	—	—

[a] Calculated for 944 cm^{-1}, except for $C_3F_6^+$ which is given for 1047 cm^{-1} (maximum of photodissociation probability curve).

[b] Ion–molecule reaction product.

[c] Bomse *et al.* (1978, 1979) and Woodin *et al.* (1978).

[d] Formed by electron impact ionization.

[e] Bomse *et al.* (1978) and Woodin *et al.* (1979a,b).

[f] Golden *et al.* (1981).

[g] $C_p = (\eta^2\text{-}C_5H_5)$.

[h] Value not known.

[i] Thorne and Beauchamp (1981).

[j] Bomse and Beauchamp (1981a).

[k] Bomse and Beauchamp (1981b).

[l] Wight and Beauchamp (1981b).

expressed in terms of phenomenological dissociation cross section σ_D,

$$k_D = \sigma_D \Phi, \tag{3}$$

where Φ is the photon flux.

At a given laser wavelength the photon flux is proportional to the irradiance (1 W cm^{-2} = 5.34×10^{19} photons cm^{-2} sec^{-1} for 947 cm^{-1} photons). Typical values of σ_D are in the range of $0.5 - 2 \times 10^{-20}$ cm^2.

Our studies of infrared multiphoton dissociation have been limited primarily to cations. Brauman and co-workers, however, have studied a number of anionic species, for example, CF_3O^- (Jasinski and Brauman, 1980) and CH_3OHF^- (Rosenfeld *et al.*, 1980, 1982). In addition, they have obtained results for $[(CH_3CH_2)_2O]_2H^+$ and $C_3F_6^+$ (Jasinski *et al.*, 1982), using pulsed laser radiation.

1. *Diethyl Ether*

The first infrared multiphoton dissociation studies of ions produced by electron impact on diethyl ether were reported by Beauchamp and co-workers (Woodin *et al.*, 1978; Bomse *et al.*, 1979). A subsequent study comparing the results for both pulsed and cw IR irradiation was reported by Jasinski *et al.* (1982).

At long trapping times and low pressures the major ions in the electron impact mass spectrum of diethyl ether are $(C_2H_5)_2OH^+$ and $CH_3CHOC_2H_5^+$. At higher diethyl ether pressures (5×10^{-7} torr) and long trapping times (>500 msec), appreciable amounts of proton-bound dimer are formed, characterized by a bimolecular reaction rate constant of $k = 1.9(2) \times 10^{-11}$ cm^3 $molecule^{-1}$ sec^{-1}

$$(C_2H_5)_2OH^+ + (C_2H_5)_2O \rightarrow [(C_2H_5)_2O]_2H^+. \tag{4}$$

Further details of the diethyl ether ion–molecule chemistry are given by Bomse *et al.* (1979).

A typical experiment monitoring the proton-bound dimer

$$[(C_2H_5)_2O]_2H^+,$$

is shown in Fig. 5. Ions are produced by a 10-msec electron beam pulse and are stored for up to 2 sec. At 1 sec of trapping time the remaining $(C_2H_5)_2OH^+$ is rapidly ejected by double resonance; this prevents further formation of the dimer. This is evidenced by the constant abundance of $[(C_2H_5)_2O]_2H^+$ after 1 sec (Fig. 5, upper trace). The laser, tuned to 944 cm^{-1} and 14 W cm^{-2} irradiance, is gated on at 1 sec of trapping time, coincident with ejection of $(C_2H_5)_2OH^+$. This causes an exponential decay of the dimer (Fig. 5, lower trace). At this irradiance, no appreciable photodissociation of $(C_2H_5)_2OH^+$ is observed, and the increase in abundance of this species

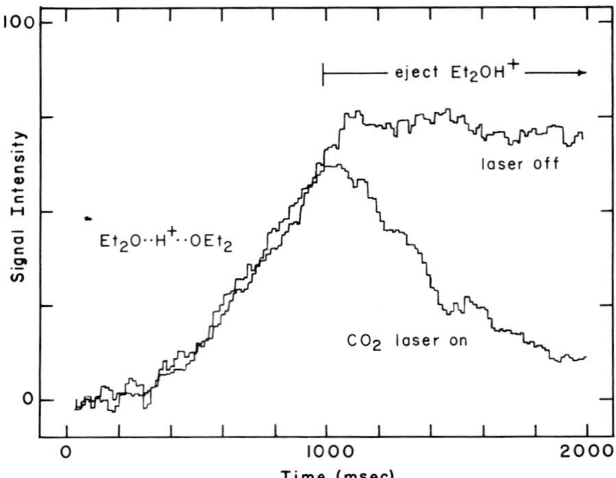

FIG. 5 Ion intensity versus trapping time for a typical multiphoton dissociation experiment (Bomse *et al.*, 1979). At a diethyl ether pressure of 5.5×10^{-7} torr, ions are formed by a 10-msec 70-eV electron beam pulse. The upper trace is the proton-bound dimer signal with the laser off. Ejection of $(C_2H_5)_2OH^+$ beginning at 1-sec trapping time halts further dimer formation. CW irradiation by the infrared laser (14 W cm^{-2} at 944 cm^{-1}) coincident with ejection of $(C_2H_5)_2OH^+$ (lower trace) results in photodissociation of the dimer. At this pressure the time between collisions is approximately 50 msec. [From Bomse *et al.* (1979).]

exactly matches the decrease in abundance of the proton-bound dimer, verifying that there is only one decay channel open

$$[(C_2H_5)_2O]_2H^+ + nh\nu \rightarrow (C_2H_5)_2OH^+ + (C_2H_5)_2O. \tag{5}$$

At low pressures, where proton-bound dimer formation is not significant, multiphoton dissociation of the protonated ether $(C_2H_5)_2OH^+$ can be studied. The laser-induced process and postulated four-center intermediate are shown in reaction (6)

$$(C_2H_5)_2OH^+ + nh\nu \rightarrow \left[\begin{array}{ccc} & H^+ & \\ & | & \\ C_2H_5 & O\text{—} & CH_2 \\ & \vdots & \vdots \\ & H\cdots & CH_2 \end{array}\right]^* \rightarrow \begin{array}{l} C_2H_5OH_2^+ + CH_2CH_2, \\ \Delta H = 27 \quad \text{kcal/mole.} \end{array} \tag{6}$$

The other major ion present at long times, $CH_3CHOC_2H_5{}^+$, also undergoes infrared multiphoton dissociation:

$$CH_3CHOC_2H_5{}^+ + nh\nu \rightarrow CH_3CHOH^+ + CH_2CH_2. \tag{7}$$

Ions derived from the electron impact ionization of $(C_2D_5)_2O$ undergo the same reactions as do corresponding unlabeled species. However, with partially deuterated ether, $C_2H_5OC_2D_5$, an interesting isotope effect is observed in the decomposition of the protonated molecular ion. Chemical ionization of $C_2D_5OC_2H_5$ at low (12-eV) electron energies, using cyclohexane as the protonating agent, permits selective formation of $(C_2H_5)(C_2D_5)OH^+$ with only trace amounts of $(C_2H_5)(C_2D_5)OD^+$. Reaction (6) suggests that two product ions could result from decomposition of $(C_2H_5)(C_2D_5)OH^+$:

$$C_2H_5\overset{H}{\underset{+}{O}}C_2D_5 + nh\nu \rightarrow C_2D_5OH_2{}^+ + CD_2CD_2 \quad (8a)$$

$$\rightarrow C_2H_5OHD^+ + CH_2CH_2. \quad (8b)$$

Yet, during laser irradiation, $C_2D_5OH_2{}^+$ is the only product ion [reaction (8a)]. Thus, hydrogen transfer in the four-center intermediate is more facile than deuterium transfer. Consideration of ion detection limits in this experiment provides a lower limit of $\geqslant 6$ for the combined primary and secondary isotope effects (defined as the ratio of rates of product ion formation). The observed specificity arises from the energetics of decomposition and cannot be attributed to selective laser pumping of only one-half of the ion. By comparison, the observed isotope effect is ~ 2 when $(C_2D_5)(C_2D_5)OH^+$ is formed by highly exothermic proton transfer, where the protonated ether internal energy greatly exceeds the threshold for decomposition [in accordance with reaction (8)].

These results imply that multiphoton dissociation occurs at an energy only slightly in excess of the thermodynamic threshold. Large primary isotope effects have also been reported for metastable ion decompositions at threshold energies (Williams and Hvistendahl, 1974). Thus, the excitation rate must be slow compared to the unimolecular decomposition rate.

2. *Perfluoropropylene Cation*

Woodin *et al.* (1978) and Bomse *et al.* (1979) have reported the infrared photodissociation of $C_3F_6{}^+$ using cw laser radiation. Recently, Jasinski *et al.* (1982) have reported a similar investigation using pulsed laser radiation. Electron impact ionization (14–70 eV) of perfluoropropylene $CF_3CF{=}CF_2$ produces only four major ions: $C_3F_6{}^+$, $C_3F_5{}^+$, $C_2F_4{}^+$, and $CF_3{}^+$. Fluoride abstraction by $CF_3{}^+$ occurs, leaving three ions at long trapping times

$$CF_3{}^+ + C_3F_6 \rightarrow C_3F_5{}^+ + CF_4. \quad (9)$$

Dissociation of the parent cation $C_3F_6{}^+$ according to reaction (10) is the only observed infrared laser-induced process

$$C_3F_6{}^+ + nh\nu \rightarrow C_2F_4{}^+ + CF_2, \qquad \Delta H \cong 56 \quad \text{kcal/mole}. \quad (10)$$

The product is both stable to laser irradiation and chemically unreactive.

$C_3F_5^+$ is totally unaffected by laser irradiation despite the availability of a decay channel, reaction (11), which has an energy requirement comparable to that of the observed photodissociation reaction (Bomse *et al.*, 1979)

$$C_3F_5^+ + nh\nu \rightarrow CF_3^+ + C_2F_2, \qquad \Delta H \cong 53 \quad \text{kcal/mole.} \tag{11}$$

That $C_3F_5^+$ is not dissociated by the laser field proves that the observed photodissociation reactions result from absorption by the ions and are not due to nonspecific heating of the ICR cell contents.

3. *Dimethyl Chloronium Cation*

The ion–molecule chemistry of CH_3Cl is described in detail elsewhere (Beauchamp *et al.*, 1972). Briefly, the dimethyl chloronium ion $(CH_3)_2Cl^+$ is formed in the two-step sequence [reactions (12) and (13)]. $(CH_3)_2Cl^+$ does not react further with methyl chloride.

$$CH_3Cl^+ + CH_3Cl \rightarrow CH_3ClH^+ + CH_2Cl. \tag{12}$$

$$CH_3ClH^+ + CH_3Cl \rightarrow (CH_3)_2Cl^+ + HCl. \tag{13}$$

In a mixture of CD_3Cl and CH_3Cl, the isotopically mixed species $(CH_3)_2Cl^+$, $(CD_3)_2Cl^+$, and $CH_3CD_3Cl^+$ are formed as indicated in Fig. 6. Only $(CD_3)_2Cl^+$ can be photodissociated with laser irradiance $\leqslant 100$ W cm^{-2} [reaction (14)] (Bomse and Beauchamp, 1981b)

$$(CD_3)_2Cl^+ + nh\nu \rightarrow CD_2Cl^+ + CD_4, \qquad \Delta H = 19 \quad \text{kcal/mole.} \tag{14}$$

No chlorine isotope effect is observed.

4. *Trifluoroiodomethane*

The infrared photodissociation of CF_3I^+ has been studied by Coggiola *et al.* (1980), Thorne and Beauchamp (1981), and Jarrold *et al.* (1983). The ion–molecule chemistry of CF_3I^+ has been studied by Berman *et al.* (1981). Briefly, the major ions formed by 70-eV electron impact on CF_3I are CF_3^+, I^+, CF_2I^+, and CF_3I^+ (Berman *et al.*, 1981). Only the concentrations of CF_3I^+, CF_2I^+, and CF_3^+ are affected by CO_2 laser radiation at CF_3I pressures below 10^{-6} torr; CF_3I^+ decreases, with a concomitant increase in CF_2I^+ and CF_3^+ (Thorne and Beauchamp, 1981). Ion cyclotron double-resonance results show that the increase in CF_2I^+ is due entirely to the increase in CF_3^+ via its reaction with neutral CF_3I. This establishes CF_3^+ as

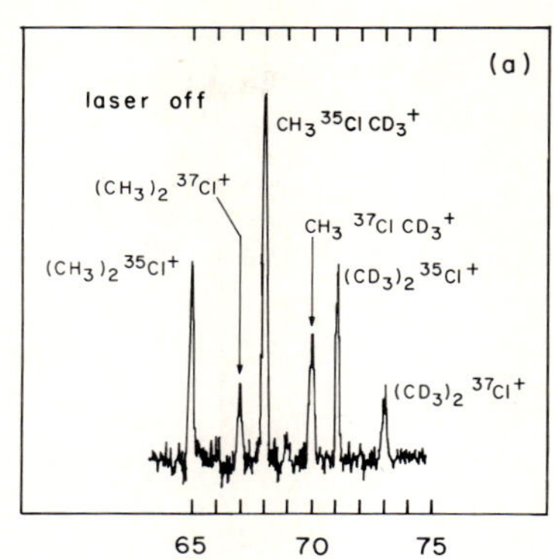

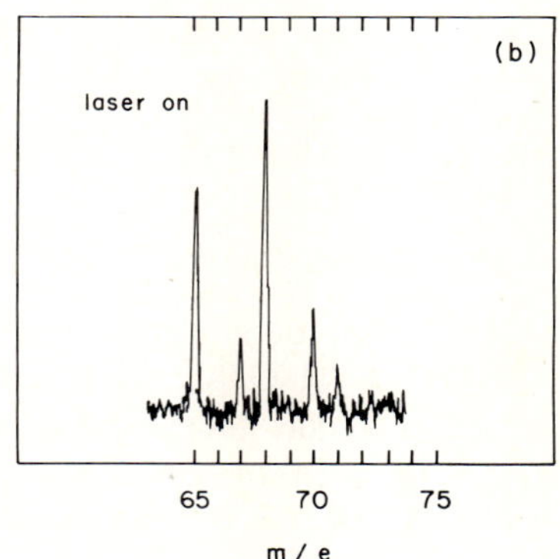

FIG. 6 (a) Trace is 2-sec time delay mass spectrum of 1:1 CH_3Cl:CD_3Cl. Ionization is by 50-msec pulse of 14-eV electrons. Total pressure is 1.9×10^{-7} torr. (b) Trace shows the same mass region for experiment in which ions undergo 2-sec continuous infrared laser irradiation (23 W cm^{-2} at 967 cm^{-1}). [From Bomse and Beauchamp (1981b).]

the exclusive photoproduct

$$CF_3I^+ + nh\nu \rightarrow CF_3{}^+ + I, \qquad \Delta H = 24 \quad \text{kcal/mole}. \tag{15}$$

Unlike the photodissociation kinetics of the ions mentioned previously, those of CF_3I^+ cannot be characterized by a simple exponential decay, as indicated in Fig. 7; a single exponential decay would give a straight line. Rather, the CF_3I^+ photodissociation kinetics show evidence of at least two populations of CF_3I^+: one which dissociates rapidly and another which dissociates slowly or not at all. These two populations are made in roughly equal amounts by electron impact ionization of CF_3I. Experimental evidence (Thorne and Beauchamp, 1981) supports the conclusion that the rapidly dissociating fraction is CF_3I^+ in the higher energy level of the ground-state spin-orbit doublet $\tilde{X}^2E_{1/2}$. The slowly dissociating fraction is in the lower doublet level, $\tilde{X}^2E_{3/2}$. Kinetics of the rapidly dissociating fraction follow approximately a single exponential decay which is proportional to laser irradiance, $k_D = 0.80(6)$ sec^{-1} W^{-1} cm^2 or $\sigma_D = 1.5 \times 10^{-20}$ cm^2.

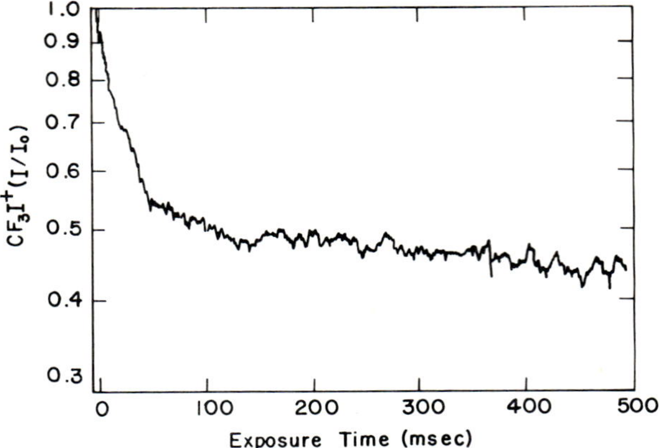

FIG. 7 Photodissociation kinetics of CF_3I^+ at 70-eV electron impact energy, 1.5×10^{-7} torr CF_3I pressure, and 50 W cm^{-2} laser irradiance at 954.6 cm^{-1}. The slope of the line at a given point gives the apparent photodissociation rate constant. [From Thorne and Beauchamp (1981).]

5. *Alcohol Dimers*

The infrared photodissociation of the proton-bound alcohol dimers of CH_3OH, C_2H_5OH, i-C_3H_7OH, n-C_3H_7OH, s-C_4H_9OH, t-C_4H_9OH, and t-$C_5H_{11}OH$ has been investigated by Bomse and Beauchamp (1981a). Full details of alcohol ion–molecule chemistry, including reaction sequences leading to $(ROH)_2H^+$ formation and pertinent thermochemistry, are given in the references cited by Bomse and Beauchamp (1981a). Infrared laser irradiation causes the multiphoton dissociation of each of the proton-bound dimers studied. The three lowest energy routes to decomposition are given by reactions (16)–(18). Temporal variation of the ion population during laser irradiation is first order in all cases, and dissociation of 100% of the dimers is possible

$$ROH_2^+ \; ROH \rightleftharpoons [(ROH)_2H^+]^* \begin{cases} \rightarrow ROH_2^+ + ROH & (16) \\ \rightarrow R_2OH^+ + H_2O & (17) \\ \rightarrow (ROH)H^+(H_2O) + \text{olefin.} & (18) \end{cases}$$

6. *Cyanobenzene Molecular Ion*

Wight and Beauchamp (1981b) have reported the infrared photochemistry of vibrationally excited cyanobenzene cation prepared by internal conversion following electronic excitation. The cyanobenzene cation undergoes a sequential 2-photon dissociation in the presence of visible laser radiation between 450 and 600 nm (2.1–2.8 eV) (Orlowski and Freiser, 1976)

$$C_6H_5CN^+ \xrightarrow{h\nu} [C_6H_5CN^+]^* \xrightarrow{h\nu} C_6H_4^+ + HCN. \tag{19}$$

The overall endothermicity of the reaction is 3.2 eV. Kinetic modeling of the dissociation reveals that the intermediate state is long lived and that the principal mode of deactivation is charge transfer to neutral cyanobenzene. Based on a comparison of photodissociation and photoelectron spectra (Neijzen and de Lange, 1978; Rabalais and Colton, 1972), the initial excitation appears to involve promotion of an electron on the nitrile group to the highest occupied ring orbital of cyanobenzene. Internal conversion to excited vibrational levels of the ground state is followed by absorption of a second photon, which raises the energy of the molecule above the threshold for the lowest energy dissociation channel.

When the ions are irradiated with up to 60 W cm^{-2} from a cw CO_2 laser for 1 sec, no dissociation is observed at any of the available laser wavelengths. Irradiation by both visible and infrared lasers, however, results in a significant increase in the overall dissociation rate compared with visible irradiation alone under the same conditions. This enhancement of the dissociation rate is due to infrared multiphoton dissociation of vibrationally excited cyanobenzene cations which have undergone internal conversion. Since the infrared laser is effective in dissociating only those molecules which have been "activated" by absorption of a visible photon, this technique represents a convenient photochemical probe of the molecules which have a high degree of internal excitation.

B. *Variation of Photodissociation Yield with Laser Wavelength*

For all of the ions studied, the decomposition pathway does not depend on laser wavelength, but the photodissociation yield does. This dependency is the basis of isotopic and isomeric selectivity reported in Sections IV.D. and IV.E. Of the ions listed in Table I, only the larger proton-bound alcohol dimers show little wavelength dependence.

Analysis of photodissociation spectra is not always straightforward, partly because our studies have been limited to the tuning range of the CO_2 laser, 925–1090 cm^{-1}. However, comparisons with gas phase spectra of related neutrals and consideration of changes in bonding due to ionization allow a reasonable interpretation of the features present in many infrared dissociation spectra. The analysis rests on the assumption that multiphoton dissociation spectra mimic the small signal absorption spectra of closely related neutral molecules (see Section III.E). The relationship between small signal absorption spectra and the wavelength dependence of dissociation yield using pulsed megawatt lasers is well established (Ambartzumian and Letokhov, 1977; Hartford, 1978). The primary difference between small signal absorption spectra of neutrals and the photodissociation spectra of the corresponding ions is that the latter can be broadened by up to 50% and shifted by up to 10 cm^{-1}. Further, photodissociation bands are generally featureless, with no P(Q)R-type band contours. Despite these differences, multiphoton dissociation represents one of the few techniques available for obtaining spectral information on gas phase ions in the infrared region.

1. *Perfluoropropylene*

Figure 8 shows the variation of $C_3F_6^+$ photodissociation yield P_D with laser wavelength. P_D is defined, in this case, as the fraction of ions dissociated during 2 sec of irradiation at 34 W cm^{-2} in 4.8×10^{-7} torr C_3F_6. The two sets of data are for electron impact ionization at 70 and 20 eV. Also shown in Fig. 8 is the small signal gas-phase absorption spectrum of perfluoropropylene over the same wavelength range. The infrared absorption band of the neutral molecule ($\nu_{max} = 1037$ cm^{-1}) has been assigned to a C–F stretching mode of A′ symmetry (Nielsen *et al.*, 1952). Comparison with other fluorinated species suggests that this vibrational mode involves primarily stretching motion of the CF_3 group. The absorption frequency of this particular mode in the ion is expected to be similar to that in the neutral,

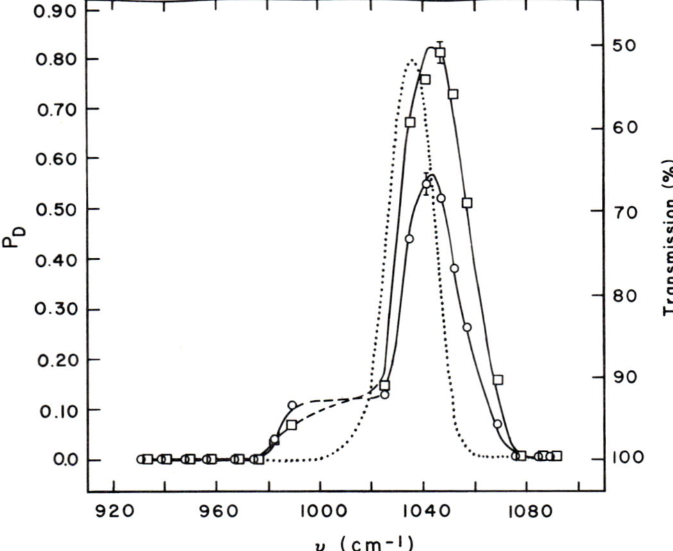

FIG. 8 Photodissociation spectrum of $C_3F_6^+$ over the CO_2 laser spectral range (Bomse *et al.*, 1978). Left ordinate is fraction of $C_3F_6^+$ dissociated after 2 sec of irradiation at 34 W cm^{-2}. The two solid curves are for ionization energies of 70 eV (□) and 20 eV (○). Perfluoropropylene pressure is 4.8×10^{-7} torr. Dotted line is infrared absorption spectrum of perfluoropropylene at 0.8 torr in a 10-cm length cell [C-F stretch (neutral)]. [From Woodin *et al.* (1979b).]

since the lowest energy ionization process in C_3F_6 entails removing an electron from the C–C double bond, which should not appreciably affect the bonding or the vibrational frequencies of the CF_3 group. Therefore, the photodissociation spectrum and the neutral absorption band are nearly superimposed. Although the infrared absorption spectrum of the neutral shows a combination band at 978 cm^{-1}, its intensity relative to the major peak at 1037 cm^{-1} is considerably smaller than the feature at 985 cm^{-1} in the $C_3F_6{}^+$ photodissociation spectrum. The small peak in the photodissociation spectrum occurs at an energy which is too low to attribute it to a $v = 1 \rightarrow v = 2$ transition. Thus, it was tentatively assigned as a combination band. Attempts to probe in detail the characteristics of $C_3F_6{}^+$ photodissociation at 985 cm^{-1} were thwarted by a lack of laser intensity in this spectral region.

Energy deposition into the internal degrees of freedom of ions formed by electron impact increases with increasing electron energy. Thus, the population of vibrationally excited $C_3F_6{}^+$ will be greater when formed with 70-eV electrons than with 20-eV electrons. The increase in photodissociation yield, at 70 eV compared to 20 eV (Fig. 8), is therefore attributed to the presence of "hot" ions formed at 70 eV.

The pulsed laser results for $C_3F_6{}^+$ reported by Jasinski *et al.* (1982) give a peak in the photodissociation yield at 1052 cm^{-1}, about 6–10 cm^{-1} higher than for cw laser dissociation. The cross section determined using pulsed laser radiation is about 30 times larger than that found using cw radiation. Differences between the results for the two modes of excitation are attributed to the greater influence of collisional and radiative relaxation in the cw experiments (Jasinski *et al.*, 1982).

2. *Diethyl Ether*

The wavelength dependences for multiphoton dissociation of $(C_2H_5)_2OH^+$ and $[(C_2H_5)_2O]_2H^+$ are shown in Fig. 9a; data for $(C_2D_5)_2OD^+$ are shown in Fig. 9b. Also shown are the gas phase absorption spectra of the corresponding neutrals over the range of CO_2 laser wavelengths. (Note the change of scale in the axes for percent transmission for the spectra of the neutral species.) For both $(C_2H_5)_2OH^+$ and $(C_2D_5)_2OD^+$, experimental conditions were nearly identical; therefore, differences in P_D values for the monomer species are a direct measure of differences in cross sections for multiphoton dissociation at each wavelength. No such direct comparison regarding dissociation cross sections can be made between the protonated ether and the proton-bound dimer, owing to differences in laser power and ether pressure in the two experiments.

The pulsed laser results for $(C_2H_5)_2OH^+$ (Jasinski *et al.*, 1982) again show a higher dissociation cross section by an order of magnitude than for the cw results, but the wavelength dependence is similar.

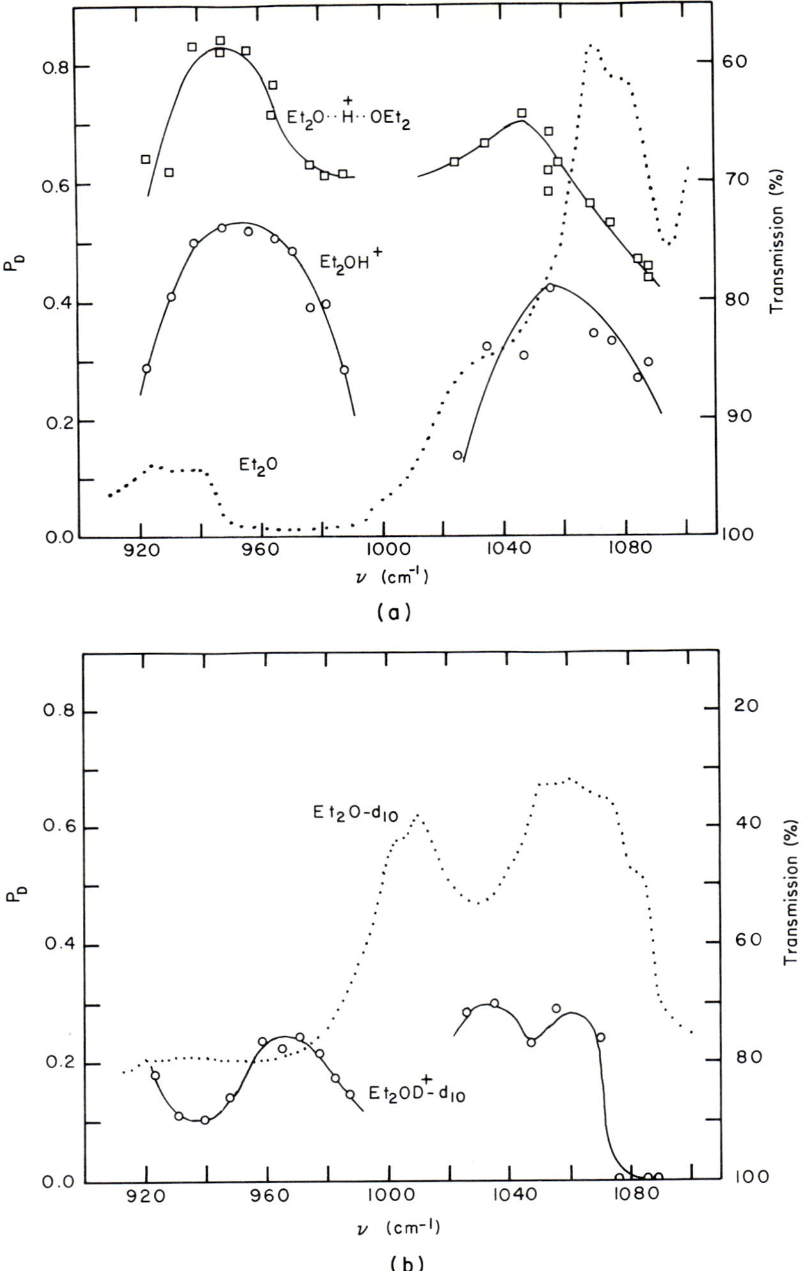

FIG. 9 (a) Photodissociation spectra of $(C_2H_5)_2OH^+$ (○) and $[(C_2H_5)_2O]_2H^+$ (□) over the CO_2 laser spectral range. For $(C_2H_5)_2OH^+$, the curve indicates the fraction of ions dissociated after 1.9 sec of irradiation at 48 W cm^{-2}; $(C_2H_5)_2O$ pressure is 8.8×10^{-8} torr. P_D for $[(C_2H_5)_2O]_2H^+$ is defined as the fraction of ions dissociated after 2.0 sec of irradiation at 10 W cm^{-2}; $(C_2H_5)_2O$ pressure is 4.7×10^{-7} torr. Ionization energy for both experiments is 14 eV. Dotted line is the infrared absorption spectrum of diethyl ether at 20 torr in a 10-cm length cell. (b) Photodissociation spectrum of $(C_2D_5)_2OD^+$ over the CO_2 laser spectral range. Experimental conditions are the same as for photodissociation of $(C_2H_5)_2OH^+$ in (a). Dotted line is the infrared absorption spectrum of $(C_2D_5)_2O$ at 16 torr in a 20-cm length cell. [From Bomse *et al.* (1979).]

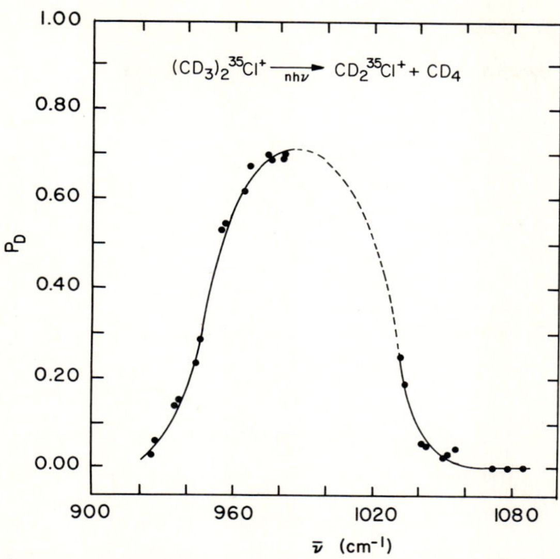

FIG. 10 Infrared photodissociation spectrum of $(CD_3)_2\,{}^{35}Cl^+$. CD_3Cl pressure is 1.3×10^{-7} torr. Ionization same as in Fig. 6. Laser power is 54 W cm^{-2}. P_D denotes the fraction of ions dissociated during 2-sec continuous laser irradiation. [From Bomse and Beauchamp (1981b).]

3. *Dimethyl Chloronium*

Figure 10 shows the variation with laser wavelength of $(CD_3)_2\,{}^{35}Cl^+$ photodissociation yield P_D. In this case, P_D is defined as the fractional change in ion intensity resulting from 2-sec continuous laser irradiation at 51 W cm^{-2}. The band maximum is at 980 cm^{-1}; its width is 70 cm^{-1}. No chlorine isotope effect is observed; the $(CD_3)_2\,{}^{37}Cl^+$ photodissociation spectrum is identical to that shown in Fig. 10.

4. *Trifluoroiodomethane*

The photodissociation spectrum of CF_3I^+ is shown in Fig. 11. The band (solid circles) peaks at 960 cm^{-1}, within the frequency region expected for the v_1 absorption of CF_3I^+ ($\tilde{X}\,{}^2E_{1/2}$), and is about three times wider than for the absorption band of neutral CF_3I. This band is also 1.2 times wider and at 13 cm^{-1} higher energy than the same band previously reported for CF_3I^+ photodissociation in a fast ion beam (Coggiola *et al.*, 1980). There is no detectable photodissociation in the 1030–1085 cm^{-1} region where the absorption of $CF_3I^+(\tilde{X}\,{}^2E_{3/2})$ is expected to occur. The limits of detectability for CF_3I^+ loss in these experiments place an upper limit of 2% on the extent of dissociation in this frequency region. This is in agreement with earlier results (Coggiola *et al.*, 1980) for CF_3I^+, which showed that photodissociation is 40 times more efficient near 940 cm^{-1} than near 1080 cm^{-1}. In the

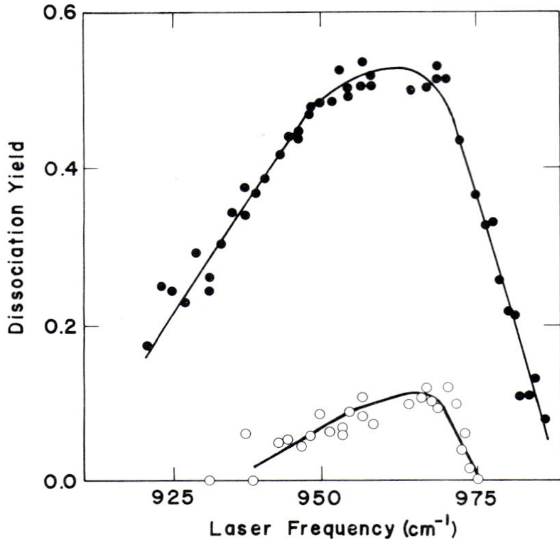

FIG. 11 Dependence of CF_3I^+ photodissociation on laser frequency using cw radiation. The extent of dissociation is monitored by CF_3I^+ signal intensity. Experimental conditions are 5×10^{-6} torr CF_3I pressure, 70-eV electron impact energy, 50 W cm^{-2} irradiance, 500 msec exposure time. Data are shown for no delay (●) and 200-msec delay (○) between ion formation and irradiation. [From Thorne and Beauchamp (1981).]

present case, a delay of 200 msec between ion formation and irradiation greatly reduces the extent of photodissociation, implying loss of internal energy; but the laser frequency dependence is unchanged, as shown by the open circles in Fig. 11.

5. *Cyanobenzene*

Photodissociation of $C_6H_5CN^+$ by visible radiation is enhanced by cw infrared radiation. Dependence of the enhancement on laser frequency has a pronounced peak at 970 cm^{-1}, as shown in Fig. 12. Here, the enhancement is given by $(I - I')/I$, where I is the ion population with visible irradiation only and I' is the population with irradiation by both lasers. Since only excited state molecules are sensitive to the infrared laser, the enhancement spectrum in Fig. 12 is a photodissociation spectrum of a molecule with 2 eV of internal energy. The density of vibrational states at this level of excitation is estimated to be 2×10^{11} states per cm^{-1}, so the excitation is truly in a continuum of overlapping states. The distinct band in the photodissociation spectrum, however, indicates that, although transitions are energetically possible at every wavelength, transition oscillator strengths remain spectrally localized.

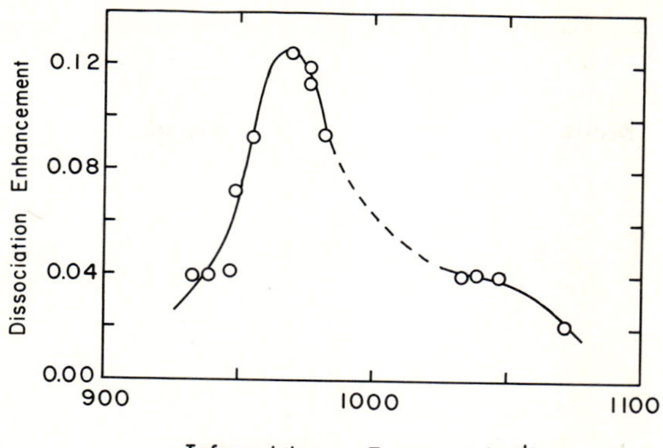

FIG. 12 Plot of dissociation enhancement versus infrared laser frequency for cyanobenzene cations. Laser conditions: Kr^+, 568 nm and 5 W cm^{-2}; CO_2, 50 W cm^{-2}. No CO_2 laser lines capable of 50 W cm^{-2} were available in the region of the dotted curve. [From Wight and Beauchamp (1981b).]

C. *Effects of Collisions on Photodissociation Yield*

The role of collisions in infrared multiphoton dissociation of gas phase ions is not clearly understood. Both enhancement and reduction in photodissociation yield occur with increasing neutral gas pressure. However, for most ions studied, the yield decreases with increasing pressure. In this case, the dominant mechanism involves transfer of internal energy from the excited ion to its collision partner, thus reducing the energy ultimately available to the ion for dissociation. These collisional deactivation processes can be particularly efficient when the ion M^+ collides with its corresponding neutral M. For these collisions, symmetric charge transfer can occur, and the close match between the vibrational frequencies of the two collision partners can enhance intermolecular vibrational energy transfer. In the case of increasing yield with increasing pressure, collisions may serve to repopulate the specific rotational levels which are depopulated by laser radiation, thus making absorption of subsequent photons more efficient. The observed pressure-dependence of the photodissociation yield probably reflects the competition between these two effects.

1. *Proton-Bound Dimer of Diethyl Ether*

Collisional effects both prior to and during irradiation have been studied in the photodissociation of $[(C_2H_5)_2O]_2H^+$ (Bomse *et al.*, 1979). Since proton-bound dimer formation, reaction (4), is bimolecular and exothermic,

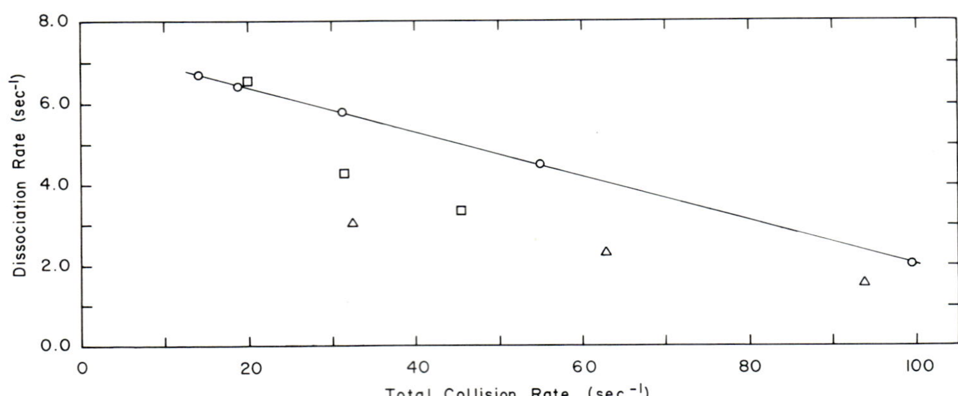

FIG. 13 $[(C_2H_5)_2O]_2H^+$ (○) multiphoton dissociation rate as a function of added buffer gases: SF_6 (△) and *i*-C_4H_{10} (□). Dissociation rate is plotted as a function of total collision rate [$(C_2H_5)_2O$ plus buffer gas] to allow for direct comparison of collision efficiencies. SF_6 or *i*-C_4H_{10} is added to 3.7×10^{-7} torr of diethyl ether. Ionization energy is 14 eV. Laser intensity $= 8.1 \pm 0.5$ W cm^{-2} and $\bar{\nu} = 939$ cm^{-1}. [From Bomse *et al.* (1979).]

a question arose as to whether or not $[(C_2H_5)]_2OH^+$ is in a highly vibrationally excited state prior to irradiation. An experiment similar to the one depicted in Fig. 5 was carried out, modified so that the laser was delayed for up to 900 msec following onset of $(C_2H_5)_2OH^+$ ejection. At various diethyl ether and buffer gas (SF_6 and *i*-C_4H_{10}) pressures, no change in $[(C_2H_5)_2O]_2H^+$ photodissociation rate was observed with increasing laser delay. Invariance of dissociation rates with laser delay (at constant pressure) implies that $[(C_2H_5)_2O]_2H^+$ is vibrationally relaxed prior to irradiation.

The data in Fig. 13 indicate that $[(C_2H_5)_2O]_2H^+$ multiphoton dissociation rates decrease with increasing pressure at constant laser intensity. In addition to diethyl ether, results obtained by adding either of two buffer gases, SF_6 and *i*-C_4H_{10}, to a small amount of diethyl ether are included. To allow for a direct comparison of deactivation efficiencies of the three gases, dissociation rates are plotted as a function of ion–molecule collision frequency (calculated from average dipole orientation (ADO) theory (Bass *et al.*, 1975; Su and Bowers, 1979). Both SF_6 and *i*-C_4H_{10} appear to be more effective than diethyl ether at quenching photodissociation.

2. *Perfluoropropylene*

In sharp contrast to the collisional effects demonstrated in Fig. 13 for $[(C_2H_5)_2O]_2H^+$, $C_3F_6{}^+$ multiphoton dissociation rates increase with increasing pressure up to 5×10^{-6} torr, as shown in Fig. 14 (Woodin *et al.*, 1979b). These results were obtained for C_3F_6, Ar, N_2, and SF_6 as buffer

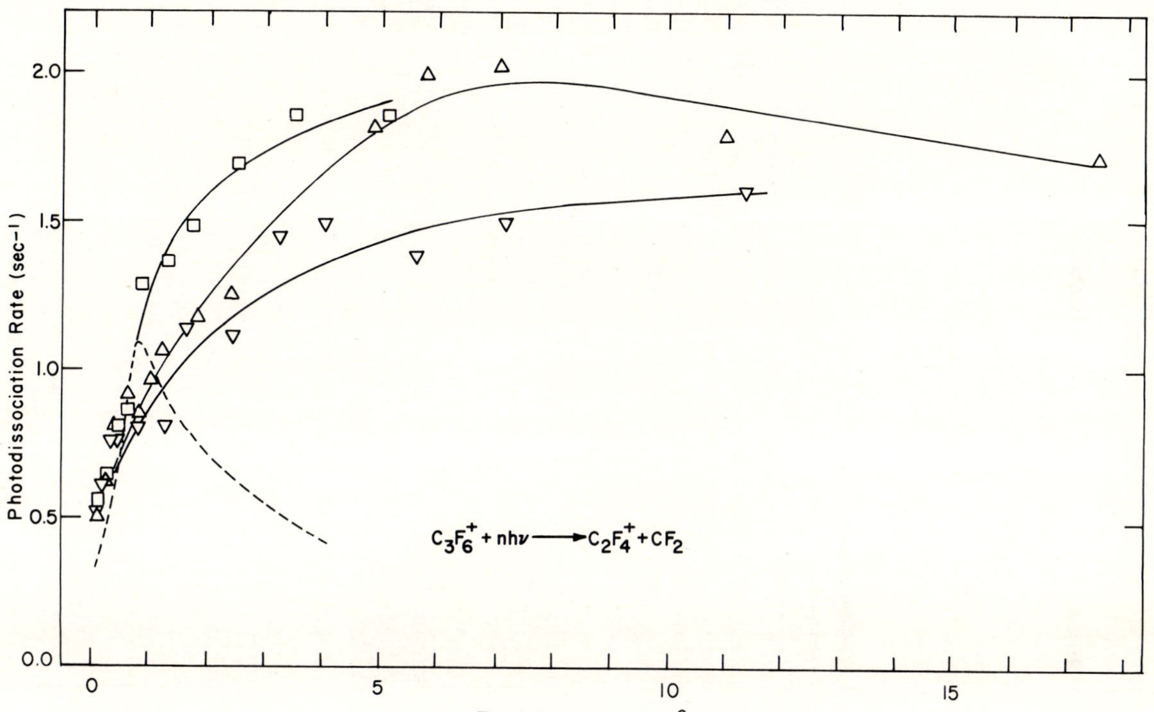

FIG. 14 $C_3F_6^+$ multiphoton dissociation rate as a function of added buffer gases Ar(△), N_2(▽), and SF_6(□). The dashed line indicates $C_3F_6^+$ multiphoton dissociation rate with C_3F_6 as the added buffer gas. Ar, N_2, or SF_6 is added to 1.2×10^{-7} torr C_3F_6. The abscissa indicates total pressure. Laser intensity = 48 W cm^{-2} and $\bar{\nu} = 1047$ cm^{-1}. [From Woodin *et al.* (1979b).]

gases. In order to minimize possible hot ion effects (as seen in Fig. 8) and to avoid ionization of buffer gases other than C_3F_6, an electron impact ionization energy of 14 eV was used. Only C_3F_6 (dashed line in Fig. 14) shows a marked tendency to quench dissociation, beginning at pressures above 1.5×10^{-6} torr. This is attributable to deactivation by symmetric charge transfer processes. In Fig. 14, photodissociation rates are plotted against buffer gas pressure rather than collision frequency because, for all four gases used, rate constants for collision with $C_3F_6^+$ are nearly identical.

D. *Variation of Photodissociation Yield with Laser Irradiance*

Variation of multiphoton dissociation yields with laser irradiance provides a useful probe of the excitation mechanism and offers a means of comparison with megawatt pulsed laser results. In studies of SF_6 at high laser irradiance, the dissociation probability per pulse is proportional to I_{las}^n, where I_{las} is the laser irradiance. Experimentally derived values (Ambartzumian and Letokov, 1977) of n vary from 1.5 to 14. The use of focused laser beams leads to large uncertainties in measurements of power density and irradiated volumes, possibly accounting for the lack of agreement in reported values of n. The effects of laser power and sample pressure are intimately related since, at constant pressure, increased laser power reduces the time scale for excitation. Thus, fewer collisions are experienced during excitation. Therefore, the ideal experiment for determining the variation of photodissociation yield with I_{las} would utilize an unfocused laser beam and allow extrapolation to a low-pressure limit. Such conditions are readily obtained in the ICR studies of slow multiphoton dissociation.

1. *Proton-Bound Dimer of Diethyl Ether*

Careful examination of multiphoton dissociation of $[(C_2H_5)_2O]_2H^+$ (Bomse *et al.*, 1979) reveals an induction period prior to the onset of decomposition. An induction period is defined as the time delay between the opening of the shutter and the first observable photodissociation. The variation of induction period τ_{ind} with laser irradiation is illustrated in Fig. 15 for $[(C_2H_5)_2O]_2H^+$ at constant ether pressure. When corrected for a shutter rise time of 5 msec, the data in Fig. 15 closely fit Eq. (20) (solid line in Fig. 15), indicating an energy fluence threshold of 0.3 J cm^{-2}

$$\tau_{ind} I_{las} = 0.3 \quad \text{J cm}^{-2}. \tag{20}$$

That an energy fluence threshold was observed is in agreement with pulsed laser multiphoton dissociation results for SF_6 (Kolodner *et al.*, 1977).

Assuming that decomposition is rapid compared to excitation in the ICR experiments (Section III.E), the induction period represents the time required

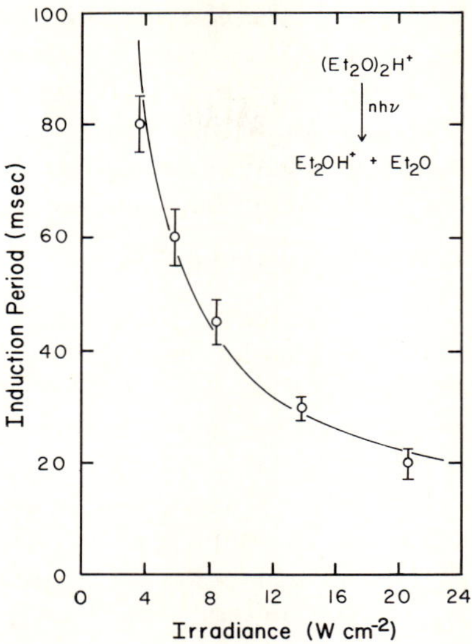

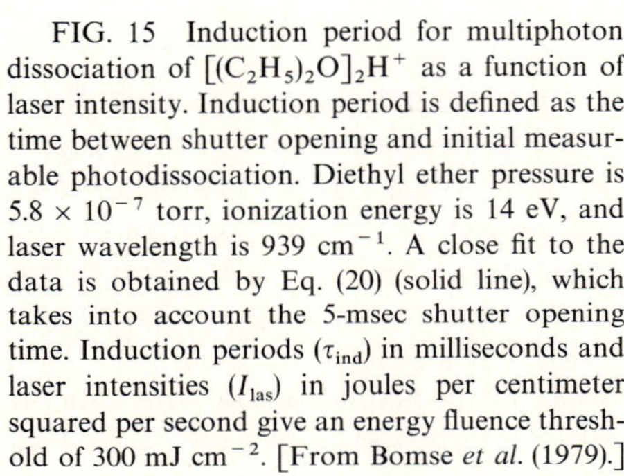

FIG. 15 Induction period for multiphoton dissociation of $[(C_2H_5)_2O]_2H^+$ as a function of laser intensity. Induction period is defined as the time between shutter opening and initial measurable photodissociation. Diethyl ether pressure is 5.8×10^{-7} torr, ionization energy is 14 eV, and laser wavelength is 939 cm^{-1}. A close fit to the data is obtained by Eq. (20) (solid line), which takes into account the 5-msec shutter opening time. Induction periods (τ_{ind}) in milliseconds and laser intensities (I_{las}) in joules per centimeter squared per second give an energy fluence threshold of 300 mJ cm^{-2}. [From Bomse *et al.* (1979).]

to pump ions to energies sufficient for dissociation. This argument predicts that photodissociation rates are pressure-independent for pressures at which the time between collisions is long compared to the observed induction period. This criterion defines the low-pressure limit. Extrapolation of data such as that given in Fig. 13 to the low-pressure limit yields the collision-free photodissociation rate. A plot of the logarithm of the collision-free dissociation rate as a function of $\ln(I_{las})$ (Fig. 16) is linear with slope

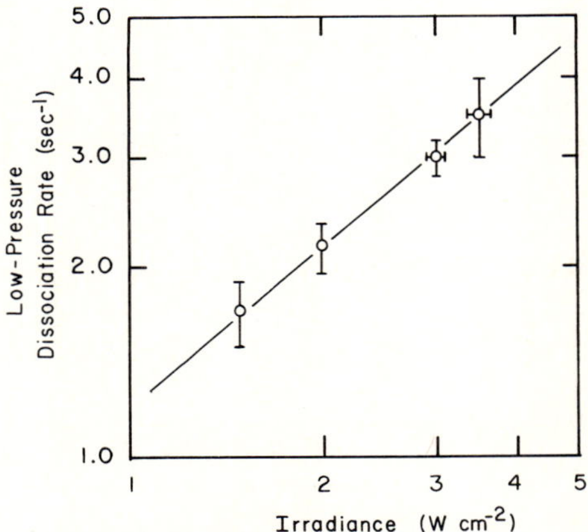

FIG. 16 Log–log plot indicating the dependence of the photodissociation rate of $[(C_2H_5)_2O]_2H^+$ on laser intensity. Slope = 0.84 ± 0.22 and $\bar{\nu} = 944$ cm^{-1}. [From Bomse *et al.* (1979).]

0.84 ± 0.22. Therefore, within experimental error, multiphoton dissociation of $[(C_2H_5)_2O]_2H^+$ is first order in laser intensity, i.e., $n = 1$. Over the range of laser irradiance (4–20 W cm^{-2}), which permits accurate measurement of dissociation rates, no saturation effects are observed. The ratio of the low-power dissociation rate to the photon flux gives $2.0 \pm 0.5 \times 10^{-20}$ cm^2 for the multiphoton dissociation cross section of $[(C_2H_5)_2O]_2H^+$.

2. Perfluoropropylene

Since collisions in $C_3F_6^+$ are seen to enhance multiphoton infrared dissociation, extrapolation of these data to a low-pressure limit is not straightforward. $C_3F_6^+$ photodissociation has been studied as a function of laser power at several pressures. For C_3F_6 pressures $\geqslant 1 \times 10^{-6}$ torr and C_3F_6 plus buffer gas (N_2) mixtures at total pressures $\geqslant 1 \times 10^{-6}$ torr, photodissociation rates increase linearly with increasing irradiance up to the highest attainable laser intensity, 100 W cm^{-2}. At C_3F_6 pressures $\leqslant 1 \times 10^{-6}$ torr, the $C_3F_6^+$ signal does not exhibit a simple exponential decay with time at higher laser powers. The observed dissociation process can be described by the decay of two separate ion populations with different rate constants. A complete quantitative analysis is not currently available, but the mechanistic implications of this result are discussed in Section III.E.

3. Trifluoroiodomethane

Unlike the ions previously mentioned, the variation of CF_3I^+ photodissociation yield with laser irradiance shows saturation effects. As indicated in

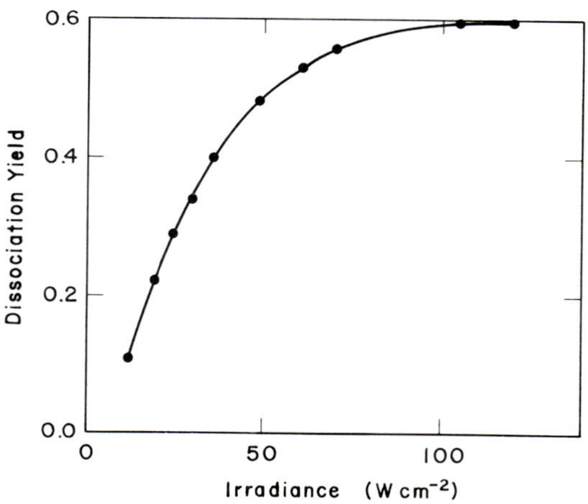

FIG. 17 Variation of the dissociation yield on laser irradiance as monitored by the CF_3I^+ signal. Experimental conditions are 70-eV electron energy, 6×10^{-8} torr CF_3I pressure, 500-msec exposure time, and 954.6 cm^{-1} laser frequency. Longer exposure time does not increase the yield. [From Thorne and Beauchamp (1981).]

Fig. 17, the photodissociation yield rises monotonically to a limiting value of 60%. This is a result of electron impact on CF_3I, producing CF_3I^+ in two different, nearly equally populated electronic states: CF_3I^+ ($\tilde{X}\,^2E_{1/2}$) and CF_3I^+ ($\tilde{X}\,^2E_{3/2}$). In the presence of laser radiation the $\tilde{X}\,^2E_{1/2}$ population rapidly dissociates while the $\tilde{X}\,^2E_{3/2}$ population photodissociates slowly or not at all. Thus, only about half of the CF_3I^+ can be photodissociated during a 1–2-sec exposure.

E. *Infrared Multiphoton Dissociation Model*

While it is tempting to apply current theories for megawatt pulsed laser multiphoton dissociation (Golden *et al.*, 1981; Schulz *et al.*, 1979; Quack, 1978, 1979; Cantrell *et al.*, 1979; Mukamel and Jortner, 1976) to the low-intensity IR photolyses of gas phase ions, there are inherent differences between the two types of experiments which necessitate modification of the existing theories. In particular, the time scale of the ICR experiments requires consideration of spontaneous emission as a viable deactivation mechanism. Equally important, the low laser intensities obviate power broadening (Mukamel and Jortner, 1976) as a mechanism for overcoming anharmonicities in hot band absorptions. In addition, any mechanistic model must allow for collisional enhancement of dissociation rates (as observed in $C_3F_6{}^+$).

1. *Excitation Processes*

Several authors have pointed out that, at high levels of vibrational excitation, the spectrum of transitions is nearly continuous (Isenor *et al.*, 1973; Akulin *et al.*, 1975). This realm of molecular excitation is generally known as the quasi-continuum of vibrational states. Onset of the quasi-continuum begins when the molecular vibrational state density is roughly equal to the laser bandwidth, typically 1000 states/cm^{-1} for cw CO_2 laser radiation. Figure 18 shows the vibrational state density as a function of vibrational energy for several of the ions mentioned above. The density of states was determined by a direct count of states for internal energies below 4.5 kcal/$mole^{-1}$ except for $(CH_3)_2Cl^+$ and CF_3I^+, where the direct count was used up to 8 kcal/mole and 20 kcal/mole, respectively. The Whitten–Rabinovitch approximation (Robinson and Holbrook, 1972) was used for higher energies. Because vibrational frequencies are not known for $(CH_3)_2Cl^+$, the frequencies of $(CH_3)_2S$ (Ypenburg and Gerding, 1971) were used to approximate the state density of $(CH_3)_2Cl^+$. Similarly, C_3F_6 (Nielsen *et al.*, 1952) frequencies were used for $C_3F_6{}^+$, CF_3I (Edgell and May, 1954) frequencies for CF_3I^+, and $(C_2H_5)_2O$ (Wieser *et al.*, 1968) frequencies for $[(C_2H_5)_2O]_2H^+$. In the latter case, all of the $(C_2H_5)_2O$ frequencies were used twice, and the nine additional frequencies of $[(C_2H_5)_2O]H^+$ were estimated for the vibrational

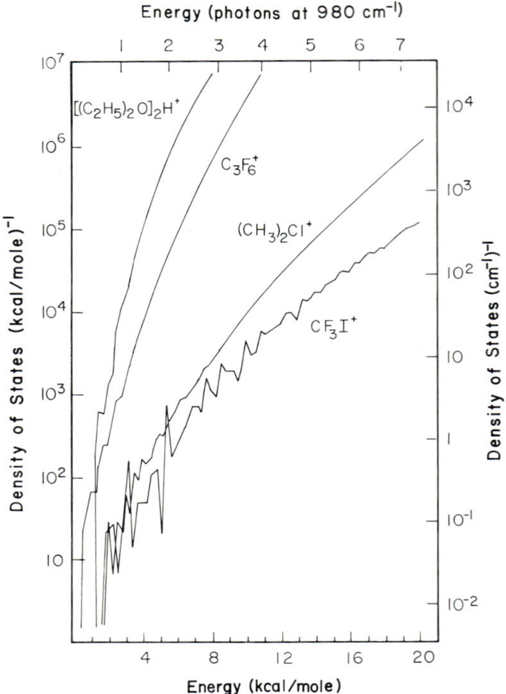

FIG. 18 Calculated vibrational state density as a function of vibrational energy content for $[(C_2H_5)_2O_2]_2H^+$, $C_3F_6{}^+$, $(CH_3)_2Cl^+$, and CF_3I^+.

modes which result from addition of H^+ and the relative motions of the two $(C_2H_5)_2O$ groups. Theories of excitation through the quasi-continuum treat the process as a sequence of incoherent single-photon events. Photoacoustic experiments indicate that, for SF_6, absorption cross sections in the quasi-continuum decrease exponentially with increasing internal excitation (Black *et al.*, 1977). Model calculations, for SF_6 using cross sections obtained from the photoacoustic results, agree well with pulsed laser multiphoton dissociation yields (Grant *et al.*, 1978). For megawatt powers and short pulse durations [Eq. (20)] predicts induction periods commensurate with the model calculations. In a complementary fashion, scaling the model calculations to low laser intensities yields results very similar to experimentally observed multiphoton dissociation reactions with cw radiation (Bomse *et al.*, 1979). Thus, absorption through the quasi-continuum is expected to be similar for both high-power pulsed laser excitation and low-power cw experiments. The only difference may arise from deactivation due to spontaneous emission in the case of low irradiance photolysis. Estimates of spontaneous emission rates in the range of 1–100 sec^{-1} are given by Dunbar (1975).

The primary difference between megawatt multiphoton activation and the ICR experiments involves the mechanism of initial excitation to the quasi-continuum. The ease of populating the quasi-continuum depends on the number of photons required to reach this region. Molecules possessing many degrees of freedom have a significant amount of internal energy at room temperature. The combination of many vibrational modes and appreciable thermal energy content serves to locate such molecules near to the quasi-continuum prior to laser excitation. It is thus natural to consider large molecule and small molecule limits. For $[(C_2H_5)_2O]_2H^+$, the densities of states at energies corresponding to absorption of one and two IR photons (1000 cm^{-1}) are 120 states/cm^{-1} and 9×10^4 states/cm^{-1}, respectively. Thus, following resonant absorption of just one IR photon, a near continuum of states (where the separation between adjacent levels and the laser bandwidth are comparable) is available for subsequent absorptions. Furthermore, at room temperature, the thermal energy content (3–4 kcal/mole) in excess of the zero point energy for $[(C_2H_5)_2O]_2H^+$ is comparable to excitation by a single photon. This puts the proton-bound dimer of diethyl ether in the large molecule category, where the density of vibrational states is large at low energies and multiphoton excitation is facile.

Density of state calculations for $C_3F_6^+$ give 5 states/cm^{-1} and 213 states/cm^{-1} at energies corresponding to absorption of one and two IR photons, respectively. The quasi-continuum of vibrational states in $C_3F_6^+$ is not reached until it has absorbed approximately four photons; therefore, $C_3F_6^+$ is classified as a small molecule. The mechanism by which these four photons are absorbed is significant in determining the $C_3F_6^+$ multiphoton dissociation rate. Even with the highest laser irradiance available, sequential excitations within one vibrational mode are not possible without energy transfer out of the excited mode. This is because the power broadening, typically 5×10^{-3} cm^{-1} or less, is insufficient to overcome the vibrational anharmonicities. Among current theoretical treatments, an excitation mechanism based on the formalism of a single resonant absorption appears best suited to these experiments (Quack, 1978; Cantrell *et al.*, 1979). This model assumes that a two-level system is in resonance with the laser field which populates the upper level; the upper level can then be depopulated by intramolecular vibrational energy transfer. When the resonant mode returns to the ground state, it is free to absorb another photon. The process is repeated until the quasi-continuum is reached. The requirement of a single resonant vibrational mode is consistent with the sharp frequency dependence observed in the $C_3F_6^+$ photodissociation spectrum and that of the smaller ions studied. The slow step in the excitation process is intramolecular *V–V* transfer since, at all laser powers, resonant transitions from the ground state are saturated (Allen and Eberly, 1975). Therefore, the simplified model predicts that photodissociation rates are independent of photon flux because intramolecular V–V transfer is a nonradiative process. Even though photolysis of $C_3F_6{}^+$ shows a change in behavior at low pressures and high laser power, it does not exhibit the predicted saturation effect, indicating that modifications to this simplified model are needed to describe low-intensity infrared multiphoton dissociation processes.

2. *Effects of Collisions on Multiphoton Excitation*

For species at the large molecule limit, excitation occurs exclusively via sequential photon absorptions through the quasi-continuum; collisions act only to depopulate excited vibrational states. This is observed in the decrease in $[(C_2H_5)_2O]_2H^+$; dissociation rates with increasing pressure (Fig. 13).

Collisions have three possible effects on the excitation of small molecules. The first, and most obvious, is simple deactivation. Second, collisions can enhance the rate of intramolecular V–V transfer, thus increasing the rate at which the molecule is pumped to the vibrational quasi-continuum. The third effect results from the narrowness of the laser linewidth. At a given laser frequency, only a few rotational states lead to a resonant transition from the ground vibrational level. These states are depleted by photoexcitation. However, collisions may serve to repopulate certain crucial rotational levels.

Either collision-induced intramolecular V–V transfer (Frankel and Manuccia, 1978; Knudtson and Flynn, 1973) or collisional repopulation of depleted rotational states would account for an increase in photodissociation rate with increasing pressure (Quigley, 1978). To date, no experimental results conclusively distinguish between these mechanisms.

The effects of collisions on the multiphoton dissociation of $C_3F_6{}^+$ have been studied by Woodin *et al.* (1979b) and Bomse *et al.* (1981). The two-population decay observed in $C_3F_6{}^+$ photodissociation at low pressure and high laser intensity (Section III.4) results from either the presence of vibrationally excited ions prior to irradiation or varying rates of collisional redistribution of rotational states. In the first case, the fraction of $C_3F_6{}^+$ with appreciable vibrational excitation decomposes rapidly because it has already passed the intramolecular V–V transfer bottleneck. At the lowest pressures used ($8 \times 10^{-8} - 2 \times 10^{-7}$ torr), there are not sufficient numbers of collisions to deactivate vibrationally hot $C_3F_6{}^+$ formed by electron impact ionization (Fig. 8). An alternative explanation for the two-population decay is derived from observations by Polanyi and Woodall (1972) and Ding and Polanyi (1975), who reported that probabilities for collisionally induced rotational transitions vary inversely with the change in rotational energy. During the course of infrared laser photolysis the involvement of species originally in rotational states energetically far removed from the resonant state leads to a slowing of repopulation rates and hence to a change in the observed dissociation rate.

The obvious complexities in low-intensity, infrared multiphoton dissociation preclude further detailing of the excitation process. It is reasonable to expect that modeling calculations similar to those performed for megawatt pulsed laser photolysis of SF_6 will provide new insight into the nature of the phenomena.

3. Unimolecular Dissociation of Activated Molecules

Observation of a large isotope effect in the photodissociation of $(C_2H_5)(C_2D_5)OH^+$ (Section III.A) establishes that dissociation occurs before the molecule can absorb an appreciable amount of energy above the dissociation threshold. The relatively slow time scale for low-intensity IR laser excitation implies that standard statistical treatments of unimolecular reactions, i.e., RRKM (Rice–Ramsperger–Kassel–Marcus) theory (Robinson and Holbrook, 1972), can be utilized to describe the decomposition step and that decomposition is generally much more rapid than sequential photon absorption.

4. Kinetic Model for Multiphoton Dissociation

One of the objectives in studying infrared laser-induced chemical reactions is to better understand the mechanism by which many photons are sequentially deposited in a molecule. Individual absorption and deactivation events are difficult to probe directly, however, and usually the only physical observable is the overall dissociation of the molecules. With this limitation in mind, it is instructive to construct simple models which can provide some insight into the excitation mechanism by reproducing various aspects of the observed dissociation kinetics.

One approach which has met with some success is the development of an energy-grained master equation following the basic ideas put forth in Fig. 1 (Wight and Beauchamp, 1983c; Jasinski *et al.*, 1982). Each molecule is assigned energy levels which are spaced by the energy of one laser photon. If n photons are required to reach the dissociation threshold, there are $n + 1$ levels and $n^2 + n$ possible rate constants. Further, it is assumed that laser excitation can involve only adjacent levels and that collisional deactivation reduces the internal energy of a molecule to the lowest level. Crude estimates for infrared radiative emission rates may be calculated from the known infrared spectra of model compounds (such as the neutral parents of molecular ions) following the basic method outlined by Dunbar (1975). (See Wight and Beauchamp, 1983c.)

Analytical solutions of the master equation are difficult for $n > 2$, so a numerical stochastic approach to the solutions was carried out using the method developed by Gillespie (1976, 1977) and applied to multiphoton dissociation kinetics by Barker (1980). The relative populations of the energy levels are determined by a series of trajectories, or random walks, through the various energy levels. Each trajectory is stopped when a dissociation event occurs or when a set time limit is exceeded. The transition probabilities are determined by the rate constants, and the results can be made arbitrarily precise merely by increasing the number of trajectories.

This method has been used successfully to account for behavior such as the exponential decay of ion population with irradiation period, the linear dependence of the dissociation rate on laser intensity, the variation of the induction period with laser intensity, and the effects of collisional deactivation on the dissociation yield.

It must be noted that the available data do not allow a unique determination of each of the rate constants in the master equation. This type of modeling, however, is useful for gaining a qualitative understanding of the importance of various rate constants in determining the overall dissociation kinetics. In the case of $[(C_2H_5)_2O]_2H^+$, for instance, it is believed that radiative emission plays an important role in the deactivation of molecules in the upper energy levels. In this particular case, the dissociation rates and induction periods are accounted for successfully using a model in which the rate constants for laser excitation between levels are independent of the energy content of the molecule.

Recently, Baldwin and Barker (1982) have presented an approximate solution to the master equation, which assumes that decomposition is rapid once a critical energy is reached. This "sudden death" approximation may be well suited for describing the photodissociation of ions using cw lasers since, in general, decomposition is rapid relative to further laser pumping (Section III.E.3).

As the infrared multiphoton dissociation kinetics of additional molecules are characterized, it is hoped that a generalized model will be developed which will be of value in predicting the behavior of untested systems.

IV. APPLICATIONS OF MULTIPHOTON DISSOCIATION

A. *Multiphoton Excitation as a Probe of Bimolecular and Unimolecular Reaction Energetics*

Proton-bound alcohol dimers provide an opportunity to investigate the reaction potential energy surface for ion–molecule bimolecular reactions, where the dimer is considered to be the intermediate (Bomse and Beauchamp,

1981a). In the general case, where two different product ions are possible from an ion–molecule reaction, the reaction will follow one of the three potential energy surfaces given in Fig. 19

$$A^{+*} + B \rightarrow [AB^{+}]^{*} \begin{cases} \rightarrow C^{+} + D, & \Delta H_1 < 0, \quad E_{a1} \qquad (21) \\ \rightarrow F^{+} + G, & \Delta H_2 < 0, \quad E_{a2}. \qquad (22) \end{cases}$$

For most ion–molecule systems, the values of activation energies E_{a1} and E_{a2} are not known. When reaction A^{+} with B yields both C^{+} and F^{+}, the most that can be inferred is that no point along paths followed on the potential energy surface from reactants to products exceeds the energy E^{*} available to the reactive intermediate AB^{+}. No information is available to distinguish among cases I, II, and III in Fig. 19. However, gas phase ions

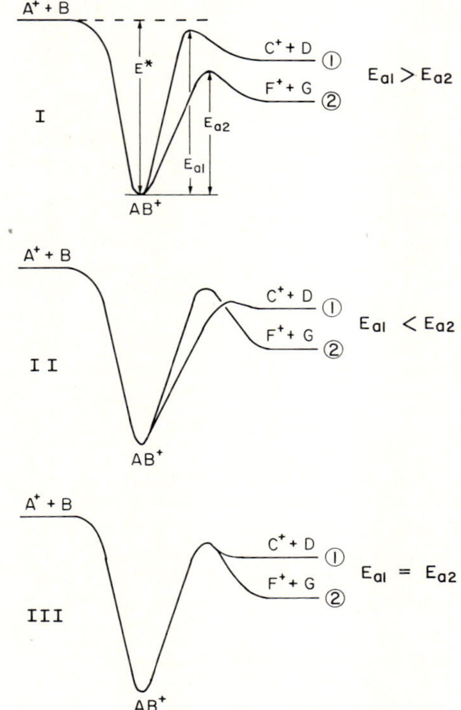

FIG. 19 Three possible potential energy surfaces for the exothermic ion–molecule reaction of $A^{+} + B$ leading to products $C^{+} + D$ and $F^{+} + G$. [From Bomse and Beauchamp (1981a).]

exposed to low-power cw infrared radiation undergo slow sequential multiphoton excitation followed by rapid dissociation (see Section III.E). Thus, the potential energy surface accessible to a decomposing ion is sampled thoroughly below (but not above) the lowest energy decomposition pathway. Consequently, in cases where AB^+ can be formed initially with little or no internal energy, as for the proton-bound alcohol dimers, laser photodissociation of AB^+ identifies the lowest energy decomposition reaction.

Table II lists the proton-bound alcohol dimers and enthalpy changes for reactions corresponding to processes (16)–(18), and the observed multiphoton dissociation pathways. Infrared activation of dimers **1**, **2**, **4**, **7**, and **9** leads to formation of the lowest possible enthalpy products, as would be expected for type I potential energy surfaces (Fig. 19). $(t\text{-}C_5H_{11}OH)_2H^+$, dimer **8** is the one example which is described by a type II potential energy surface (Ridge and Beauchamp, 1971). The remaining proton-bound alcohol dimers listed in Table II (**3**, **5**, and **6**) when photodissociated, yield two or more sets of products. Significantly, these are also the first examples of slow multiphoton dissociation following more than one reaction pathway.

TABLE II

Unimolecular Decomposition Reactions of Proton-Bound Alcohol Dimers[a]

	Dimer ion $(R_1OH)H^+(R_2OH)$	ΔH, reaction (16)	ΔH, reaction (17)	ΔH, reaction (18)	$\Delta H'$[b], reaction (18)
1	$R_1 = R_2 = CH_3$[c]	31	<u>18</u>	89	—
2	$R_1 = R_2 = C_2H_5$	32	<u>15</u>	19	—
3	$R_1 = R_2 = n\text{-}C_3H_7$	<u>33</u>(71%)	<u>17</u>(17%)	<u>19</u>(12%)	—
4	$R_1 = R_2 = i\text{-}C_3H_7$	31	<u>19</u>	22	—
5	$R_1 = R_2 = s\text{-}C_4H_9$	32	<u>15</u>(38%)	<u>20</u>(62%)	—
6	$R_1 = i\text{-}C_3H_7$; $R_2 = s\text{-}C_4H_9$	33[d]	<u>19</u>(37%)	*18*[e](63%)	24[f]
7	$R_1 = R_2 = t\text{-}C_4H_9$	31	<u>20</u>	23	—
8	$R_1 = R_2 = t\text{-}C_5H_{11}$	31	19	<u>22</u>	—
9	$R_1 = i\text{-}C_3H_7$; $R_2 = t\text{-}C_4H_9$	30[d]	<u>13</u>	22[g]	23[h]

[a] Thermochemical data, in kilocalories per mole, taken from Bomse and Beauchamp (1981c). Underlined numbers denote observed IR laser-driven reactions. Numbers in parentheses are product distributions for laser-driven reactions in which more than one product is observed.

[b] $\Delta H'$ refers to enthalpy change for the alternate reaction (18) pathway available when R_1 and R_2 are not identical.

[c] Laser-accelerated reaction rate (Bomse and Beauchamp, 1980).

[d] Assumes that more basic alcohol is protonated.

[e] Products are $(i\text{-}C_3H_7OH)H^+(OH_2)$ + *trans*-2-butene.

[f] Products are $(s\text{-}C_4H_9OH)H^+(OH_2)$ + propene.

[g] Products are $(t\text{-}C_4H_9OH)H^+(OH_2)$ + propene.

[h] Products are $(i\text{-}C_3H_7OH)H^+(OH_2)$ + isobutylene.

Low-power infrared laser activation does not open any additional reaction channels to $(ROH)_2H^+$. Products of the multiphoton dissociation of $(ROH)_2H^+$ are always a subset of the products generated by the exothermic reaction of $ROH_2{}^+$ and ROH. The only difference is that laser activation is more selective than the exothermic reaction. For example, multiphoton dissociation of $(C_2H_5OH)_2H^+$ yields exclusively $(C_2H_5)_2OH^+$ and H_2O (Table II). However, above 10^{-5} torr two other exothermic channels are open (Bomse and Beauchamp, 1981c).

Proton-bound alcohol dimers **3**, **5**, and **6** each yield more than one set of products from slow multiphoton dissociation (Table II). The measured branching ratios do not vary with changes in laser wavelength (942 – 1079 cm^{-1}) or laser power (27 to 63 W cm^{-2}). Dimers **3**, **5**, and **6** contain *n*-propanol or 2-butanol. Of the alcohols studied, only these two can isomerize to more stable alcohols. We infer that more than one set of dissociation products is formed because an isomerization occurs in the transition state, which releases sufficient energy to make two or three reaction channels accessible. This corresponds to case III in Fig. 19. Results of previous studies indicate that slow multiphoton dissociation is extremely sensitive to small differences in activation energies. Therefore, more than one set of products can arise from different transition states only if the corresponding activation energies are almost identical (Bomse *et al.*, 1979). However, it is highly unlikely that an accidental degeneracy would exist in three different proton-bound dimers. Thus, the three products must come from a common transition state.

A candidate for a transition state which allows for alkyl isomerization is composed of an alkyl cation R^+, water, and alcohol ROH, in which R^+ isomerizes to R'^+ [**1** → **2** in (23)]:

$$\left[H_2O \cdots R^+ \cdots O \begin{matrix} \diagup H \\ \diagdown R \end{matrix} \right] \longrightarrow \left[H_2O \cdots R'^+ \cdots O \begin{matrix} \diagup H \\ \diagdown R \end{matrix} \right]. \tag{23}$$

1 **2**

Loss of H_2O results in the formation of protonated ether $RR'OH^+$. The other products result from deprotonation of R'^+ to yield $(R'OH)\,H^+(OH_2)$ plus an olefin or recombination of R'^+ and H_2O to form $R'OH_2{}^+$ and ROH. In fact, all observed alcohol dimer decompositions, including those giving multiple sets of products, are consistent with the formation of a "solvated" alkyl cation as an electrostatically bound intermediate.

Proton-bound dimers **3** and **4** have identical charge-to-mass ratios and identical infrared photodissociation spectra. However, the isomers yield different characteristic sets of products from multiphoton dissociation. Thus, multiphoton dissociation reactions can be used to distinguish structural isomers of these ions (Section IV.E).

To date, most detailed information about reaction coordinate profiles in ion–molecule reactions has been obtained from studies of metastable ion fragmentations (Williams, 1977; Bowen and Williams, 1980; Williams and Howe, 1972; Levsen and Schwarz, 1976; Cooks *et al.*, 1973). In those experiments the decomposing species are ions initially formed with internal energy at, or slightly above, the threshold for unimolecular decomposition. In light of these results infrared activation provides an additional and complementary technique for the study of unimolecular decomposition of gas phase ions having internal energies slightly in excess of dissociation thresholds.

Furthermore, because of current interest in mass spectrometric techniques (McLafferty, 1980; McClusky *et al.*, 1978) to analyze complicated mixtures and to "fingerprint" ions by their secondary fragmentation patterns, infrared multiphoton excitation represents a novel method for producing ion fragmentation. Although low-power (<100 W cm^{-2}) infrared radiation leads to decomposition exclusively by the lowest energy pathway (Bomse *et al.*, 1979; Woodin *et al.*, 1979a), pulsed laser irradiation at high peak powers (>1 MW cm^{-2}) is known to make multiple reaction channels energetically accessible (Sudbø *et al.*, 1978, 1979, and references contained therein; Collussi *et al.*, 1977). The only requirement is that infrared absorption rates exceed the unimolecular decomposition rates of the irradiated molecule.

B. *Multiphoton Dissociation as a Probe of Molecular Relaxation Rates*

Multiphoton dissociation provides a sensitive probe of the internal energy content of CF_3I^+ since only those ions with sufficient energy before irradiation can be dissociated (Thorne and Beauchamp, 1981). With pulsed CO_2 laser radiation, photoproducts are produced on a submicrosecond time scale (Jarrold *et al.*, 1983) and can be detected quantitatively before they can react with neutral CF_3I. Increasing the time delay between ion formation and laser irradiation thus reduces the photodissociation yield due to relaxation of $CF_3I^+(\tilde{X}\,^2E_{1/2})$ to the ground state. The exponential decrease in photodissociation yield with increasing delay can be related to a relaxation rate, as shown in Fig. 20 for two different CF_3I pressures. Similar data obtained at various pressures indicate that the relaxation of CF_3I^+ can be described by a pressure-dependent, first-order rate constant k_r, which is proportional to the CF_3I pressure. These data, shown in Fig. 21, give $k_r = 3.1 \pm 0.4 \times 10^{-9}$ cm^3 mole^{-1} sec^{-1} with a zero-pressure intercept of 11.3 ± 1.3 sec^{-1}. This value for k_r is three times the ion–molecule collision rate constant, indicating that collisions are highly effective in causing $CF_3I^+(\tilde{X}\,^2E_{1/2})$ relaxation. However, a value for k_r of 2.4×10^{-10} cm^3 mole^{-1} sec^{-1} obtained by Jarrold

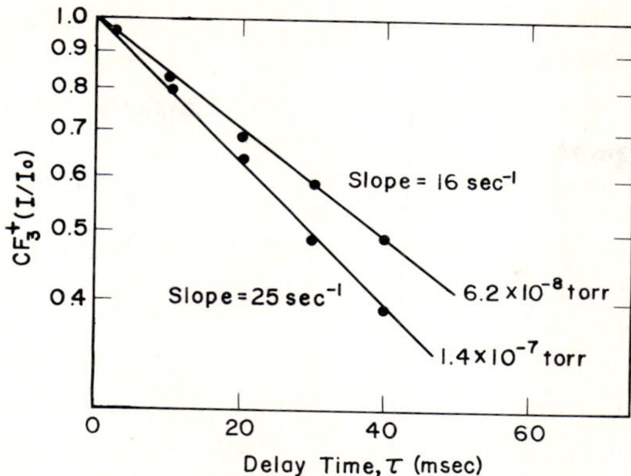

FIG. 20 Relaxation rate of CF_3I^+ at two different pressures as measured by CF_3^+ signal intensity. The delay time is the time between ion formation and irradiation. The slopes of the lines give the relaxation rates. Experimental conditions: 70-eV electron impact energy, CF_3I pressure as indicated, 1.6 J cm^{-2}, 40-nsec laser pulse at 952.9 cm^{-1}. [From Thorne and Beauchamp (1981).]

et al. (1983) in an ion beam experiment is substantially lower and indicates that only one in every three collisions removes enough energy from CF_3I^+ to preclude photodissociation. The zero-pressure intercept indicated in Fig. 21 sets an upper limit to the spin–orbit relaxation rate, which is comparable to the same rate in I (5.9 sec^{-2}) (Comes and Pionteck, 1976) and Xe^+ (18 sec^{-1}) (Armentrout *et al.*, 1978).

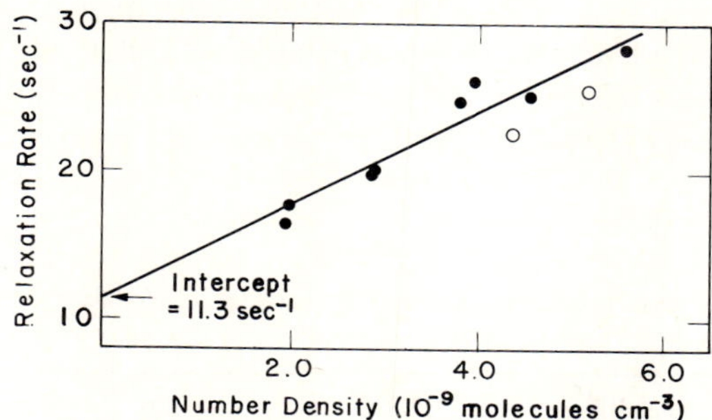

FIG. 21 Dependence of the CF_3I^+ relaxation rate on total pressure for CF_3I (●) and 6×10^{-8} torr CF_3I^+ with the balance Xe (○). The slope of the line gives a collisional relaxation rate constant of 3.1×10^{-9} cm^3 $molecule^{-1}$ sec^{-1}, and the intercept gives the zero-pressure relaxation rate constant of 11.3 sec^{-1}. Same experimental conditions as for Fig. 20. [From Thorne and Beauchamp (1981).]

C. Multiphoton Dissociation as a Probe of the Vibrational Quasi-Continuum

Cyanobenzene ions are dissociated by absorbing two visible photons. The decay of the cyanobenzene ion population with time has the functional form (Wight and Beauchamp, 1981b)

$$I/I_0 = (r_+ - r_-)^{-1}[r_+ \exp(r_- t) - r_- \exp(r_+ t)],$$

where

$$r_\pm = -\tfrac{1}{2}\{\Phi\sigma_1 + \Phi\sigma_2 + k_3 + [(\Phi\sigma_1 + \Phi\sigma_2 + k_3)^2 - 4\Phi^2\sigma_1\sigma_2]^{\frac{1}{2}}\}, \tag{24}$$

where Φ is the photon flux, σ_1 and σ_2 are the photoabsorption cross sections for the first and second photons, and k_3 is the charge-transfer (deactivation) rate at the measured pressure of cyanobenzene neutrals. The photoabsorption cross sections at several of the argon ion laser wavelengths have been reported by Orlowski and Freiser (1976). Table III gives the analogous cross sections determined at two of the krypton ion laser wavelengths. The visible photodissociation is enhanced by CO_2 laser radiation showing peak enhancement at 970 cm^{-1} (see Fig. 12). The two possible mechanisms which explain the dissociation enhancement are illustrated in Fig. 22.

Ground-state molecules may gain enough vibrational energy from the infrared laser to increase the Franck–Condon overlaps between electronic states significantly, thereby increasing the photoabsorption cross sections of the visible laser. Alternatively, molecules having absorbed 2.2 eV in the form of a visible photon may then absorb nine infrared photons to reach the dissociation limit.

The observation that no dissociation takes place with irradiation by the infrared laser alone is evidence that only excited-state molecules may absorb an appreciable number of infrared photons. This restriction may be due to a combination of factors. First, the intramolecular vibrational energy transfer rate in ground-state (cold) molecules is not fast enough to relocate the absorbed energy out of the pump modes so that laser excitation cannot

TABLE III

Photoabsorption Cross Sections for the Sequential Two-Photon Dissociation of the Cyanobenzene Cation[a]

Wavelength (nm)	σ_1 (10^{-19} cm^{-2})	σ_2 (10^{-19} cm^{-2})
568.2	0.85	6.8
530.9	3.9	120.0

[a] Wight and Beauchamp (1981b).

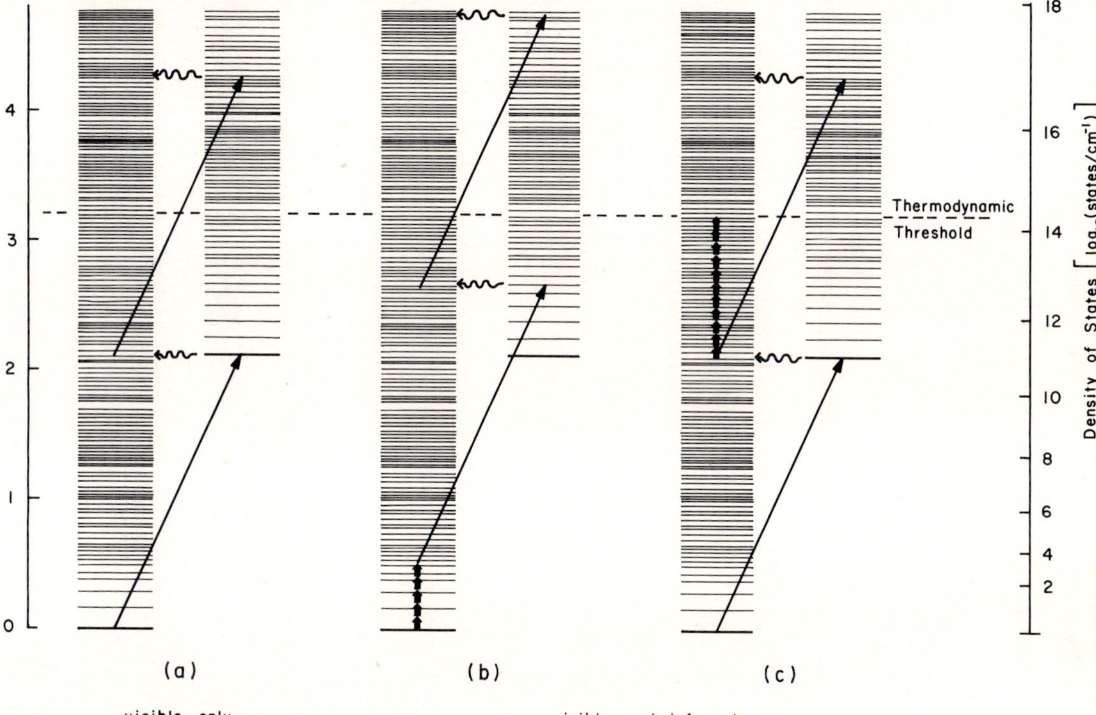

FIG. 22 Energy level diagram depicting three mechanisms for photodissociation of cyanobenzene cations. (a) Sequential two-photon absorption with visible laser only. (b) Infrared excitation from the ground state enhances the photoabsorption cross sections for the visible laser. (c) Infrared multiphoton dissociation and single-photon dissociation deplete the population of vibrationally excited intermediates produced by absorption of a visible photon. [From Wight and Beauchamp (1981b).]

compete efficiently with radiative and collisional relaxation. Second, the absorption profile in cold molecules is probably inhomogeneously broadened so that only a fraction of the molecules may absorb at a given infrared laser frequency. Both of these problems may be overcome in molecules which contain 2.2 eV of energy because the density of vibrational states is extremely high. Using the Whitten–Rabinovitch approximation (Robinson and Holbrook, 1972) and the vibrational frequencies of neutral cyanobenzene (Varsanyi, 1974), a density of 2×10^{11} states/cm^{-1} is calculated. Extremely fast intramolecular energy transfer rates at this level cause spectroscopic transitions to be dominated by homogeneous broadening, as evidenced by the Lorentzian profiles of overtone spectra in benzene (Bray and Berry, 1979). Thus, mechanism (c) of Fig. 22 most closely describes the observed results.

It is now possible to extend the kinetic model represented by reaction (24) to include dissociation of the intermediate by the infrared laser. The term $\Phi\sigma_2$ is simply replaced by the sum of dissociation rate constants $\Phi\sigma_2 + \Phi_{IR} \cdot \sigma_{IR}$, where σ_{IR} is a phenomenologically defined dissociation cross section for the infrared laser radiation, which is dependent on wavelength. The new equation is used to fit the intensity data. The result is a cross section of $\sigma_{IR} = 6 \times 10^{-21}$ cm^2 at 969 cm^{-1}, which is comparable to cross sections for low-power infrared multiphoton dissociation of other molecules. Thus, the infrared-enhanced visible dissociation of cyanobenzene provides a measure of the infrared absorption cross section in the quasi-continuum. Frequency dependence of the enhancement (Fig. 12) gives a measure of the uniformity of the density of states in the quasi-continuum. In the present case, the observed peak may be related to an overtone of the ring skeletal mode at 488 cm^{-1}.

D. *Isotopic Selectivity*

One key result obtained from low-power multiphoton dissociation is that the dissociation spectra exhibit linewidths comparable to those observed in infrared spectra of corresponding neutrals. Thus, low-power multiphoton dissociation may be selective with respect to isotopic substitutions if the spectral shifts are sufficient to separate the dissociation bands of the isotopic components.

In a 1:1 mixture of CH_3Cl and CD_3Cl, all six isotopic variants of dimethyl chloronium ion are present, as shown in Fig. 6. Only $(CD_3)_2\,^{35}Cl^+$ and $(CD_3)_2\,^{37}Cl^+$ photodissociate in response to the CO_2 laser (Bomse and Beauchamp, 1981b). In this case, deuterium-containing molecules may be selectively dissociated; however, the isotope shift induced by the central chlorine atom is not enough to selectively dissociate either ^{35}Cl or ^{37}Cl-containing ions. There are two plausible explanations for the observed selectivity. One possibility is that $(CH_3)_2Cl^+$ and $(CH_3)Cl^+(CD_3)$ are both transparent in the 925–1080 cm^{-1} region. Alternatively, the distribution and the density of vibrational states in $(CD_3)_2Cl^+$ might make it uniquely suited for multiphoton excitation. Unfortunately, there are no reported infrared absorption spectra of gas phase dimethyl chloronium to allow a test of the first hypothesis.

Assuming that deuterium substitution alters photodissociation band positions but not line shapes, a shift of $\geqslant 160$ cm^{-1} to higher energy is necessary to move the peak shown in Fig. 10 out of the CO_2 laser wavelength region. Dimethyl sulfide is isoelectronic with dimethyl chloronium. Examination of reported vibrational frequencies of $(CH_3)_2S$, $(CH_3)S(CD_3)$, and $(CD_3)_2S$ shows five fundamental modes with isotope shifts in excess of 160 cm^{-1} (Ypenburg and Gerding, 1971; Ypenburg, 1972). Furthermore, almost all $(CD_3)_2S$ vibrational frequencies of ≈ 1000 cm^{-1} exhibit large shifts to higher energy in $(CD_3)S(CH_3)$ or $(CH_3)_2S$. These changes suggest one possibility for the observed isotopic selectivity, namely, that $(CD_3)_2Cl^+$ is the only species which absorbs strongly in the CO_2 laser wavelength region. However, by the same arguments, it appears likely that all the ions have at least one infrared active vibrational mode within the tuning range of the laser. A second factor which may influence multiphoton dissociation probabilities is the intramolecular energy transfer rate. The uniqueness of $(CD_3)_2Cl^+$ may reflect its relatively large density and, with more speculation, a favorable distribution of states which leads to large statistical factors favoring facile energy transfer. Magnetic field-induced level splittings may also enhance multiphoton excitation of $(CD_3)_2Cl^+$ (Duperrex and van den Bergh, 1980).

In contrast to the spectral isotopic selectivity just described, a special case of kinetic isotopic selectivity is observed in the decomposition of protonated diethyl ether. As mentioned, photodissociation occurs by way of a four-center intermediate reaction (6). However, photodissociation of $(C_2D_5)\cdot(C_2H_5)OH^+$ yields exclusively $C_2D_5OHD^+$—not $C_2H_5OHD^+$. Since both products have the same precursor, isotopic selectivity must occur in the decomposition step as a result of differing activation energies for β-D or β-H transfer. This observation further demonstrates that decomposition competes favorably with further absorption of photons above the dissociation threshold, as such a process would significantly dilute the kinetic isotopic selectivity.

E. *Isomeric Selectivity*

Isomeric ions may be differentiated if they have different photodissociation spectra or if their photoproducts differ. For example, $C_3F_6^+$ molecular ions may be formed from either perfluoropropylene or perfluorocyclopropane, affording the possibility of observing cyclic or acyclic ions of the same mass-to-charge ratio. Ions formed by both methods have identical photodissociation spectra, however, indicating that ring opening accompanies ionization of perfluorocyclopropane (Bomse and Beauchamp, 1981c).

The proton-bound alcohol dimers $(n\text{-}C_3H_7OH)_2H^+$ and $(i\text{-}C_3H_7OH)_2H^+$ have identical charge-to-mass ratios and identical photodissociation spectra. However, the isomers yield different but characteristic sets of photoproducts (see Table II) (Bomse and Beauchamp, 1981a).

V. MULTIPHOTON ELECTRON DETACHMENT (MED) FROM NEGATIVE IONS

Molecular anions present an interesting case for the study of infrared multiphoton excitation because the lowest energy dissociation pathway is usually detachment of an electron. By this process, vibrationally hot anions become relatively cool neutrals. Thus, multiphoton electron detachment (MED) may be thought of as a type of inverse electronic relaxation. It is, therefore, conceptually distinct from multiphoton dissociation and single-photon electron detachment, since vibrational energy must be converted to electronic energy.

Although single-photon electron detachment at visible wavelengths is a well-established technique, only recently has multiphoton electron detachment been observed. Brauman and co-workers first reported infrared multiphoton electron detachment produced by high-power, pulsed CO_2 radiation (Drzaic and Brauman, 1981; Meyer *et al.*, 1982; Rosenfeld *et al.*, 1979). Subsequently, Simons (1981) presented a theoretical analysis in which propensity rules were given for vibration-induced electron detachment. Wight studied multiphoton electron detachment using low-power cw CO_2 laser radiation and observed isomeric and isotopic selectivity (Wight and Beauchamp, 1983a, 1981a). Electron detachment from vibrationally excited anions was also observed (Wight and Beauchamp, 1983b,c). Based on these results, a kinetic model for multiphoton electron detachment has been developed (Wight and Beauchamp, 1983c). Because the overall detachment kinetics are governed by laser-induced vibrational excitation, there are many similarities to the kinetics of multiphoton dissociation.

A. *Isomeric and Isotopic Selectivity*

Selectivity has been observed for two sets of isomeric anions: $C_7H_7^-$ and $C_4H_7^-$ (Wight and Beauchamp, 1983a,c). In addition, isotopic selectivity has been observed for $C_7H_7^-$.

For $C_7H_7^-$, the anions of interest are formed by proton-transfer reactions of anionic bases with the neutral C_7H_8 isomers in reactions (25–27):

$$CH_3O^- + C_6H_5CH_3 \rightarrow C_6H_5CH_2^- + CH_3OH \quad (25)$$

$$CH_3O^- + c\text{-}C_7H_8 \rightarrow c\text{-}C_7H_7^- + CH_3OH \quad (26)$$

$$NH_2^- + \text{norbornadiene} \rightarrow C_7H_7^- + NH_3. \quad (27)$$

After an initial ion-formation period, laser irradiation causes the ion population to decay exponentially in time with an induction period, which is evident at laser intensities less than 20 W cm^{-2} (Bomse *et al.*, 1979; Bien, 1978). Recycling of the detached electrons to form additional $C_7H_7^-$ is prevented by double-resonance ejection of the precursor base during the irradiation period.

Infrared spectra of the $C_7H_7^-$ anions, cycloheptatrienyl and benzyl, are obtained by monitoring the extent of photodetachment as a function of laser wavelength at a fixed intensity and irradiation period. Irradiation of $C_7H_7^-$ formed by deprotonation of norbornadiene with NH_2^-, reaction (27), results in no detectable electron detachment, indicating a third distinct $C_7H_7^-$ isomer. An alternate explanation for these observations is that $C_7H_7^-$ formed from the NH_2^- reaction does not next undergo MED because it has too little internal energy before irradiation and that $C_7H_7^-$ made by the other two reactions may have more energy and thus undergo MED. This possibility seems unlikely in the present case, and deuterium isotope exchange reactions in a flowing afterglow (Squires *et al.*, 1981) apparatus does support the existence of three distinct isomers.

For benzyl anion, a distinct isotope effect is observed in the photodetachment spectra (Wight and Beauchamp, 1983c). The photodetachment yield for $C_7H_7^-$ is greatest near 945 cm^{-1} and greatest for $C_7D_7^-$ near 932 cm^{-1}.

To generate butenyl anions, $C_4H_7^-$, ion formation is initiated by a 10-msec electron beam pulse at approximately 5 eV. Dissociative electron attachment to ammonia produces H^- and NH_2^-, both of which react with C_4H_8 isomers to form butenyl anions in reaction (28),

$$B^- + C_4H_8 \rightarrow C_4H_7^- + BH, \quad (28)$$

where $B^- = NH_2^-, H^-$. Subsequent irradiation with a cw CO_2 laser results in loss of ion signal intensity due to multiphoton electron detachment. It is expected that absorption of only three or four photons is sufficient to cause detachment since the electron affinity of $CH_3CHCHCH_2$ is 9 kcal/mole (see Wight and Beauchamp, 1983a, and references therein). The detached electrons remain in the ion trap but do not reattach to NH_3 because the electron attachment cross section is sharply peaked at 5 eV (Wight and Beauchamp, 1983a). In a typical experiment, $C_4H_7^-$ is formed via reaction (28) for 100 msec in the absence of laser irradiation. Subsequent irradiation for 100 msec at 929 cm^{-1} and 30 W cm^{-2} results in a loss of 50% or more of the ions. By varying the laser frequency while keeping all other experimental conditions the same, an infrared multiphoton electron detachment spectrum is obtained. Spectra obtained for these conditions are shown in Fig. 23 for $C_4H_7^-$ derived from *cis*- and *trans*-2-butene. The spectra are qualitatively similar to each other, with the maximum photodetachment yields occurring at the low-frequency limit of the CO_2 laser (about 927 cm^{-1}). The yields in the case of the trans-precursor are significantly higher, however, at the high-frequency limit (about 1085 cm^{-1}). Fortunately, low-power laser excitation affords an opportunity to examine the photodetachment *rate* as well as the total yield. The decay of $C_4H_7^-$ population with increasing laser irradiation time is shown in Fig. 24. Using *cis*-2-butene as the neutral precursor, the disappearance of $C_4H_7^-$ is exponential in time for irradiation at 929 and 1076 cm^{-1} (Fig. 23). It is thus concluded that all of the anions produced this way have the same phenomenological photodetachment cross section and that with sufficiently long irradiation periods, 100% of the ions may be destroyed.

If $C_4H_7^-$ is formed from *trans*-2-butene, very different behavior is observed. The disappearance of the ion signal is no longer a single exponential decay for 1076 cm^{-1} irradiation. Instead, 14% of the ions undergo rapid detachment with $\sigma = 30 \times 10^{-21}$ cm^2, while the remaining 86% have a

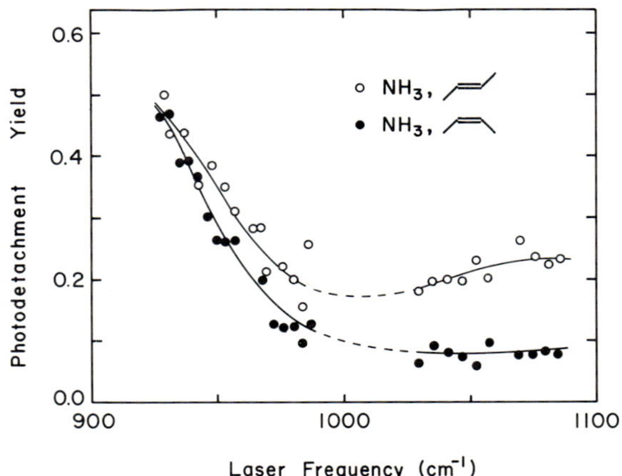

FIG. 23 Infrared multiphoton electron detachment spectra of $C_4H_7^-$ produced by ionization of mixtures of ammonia with *trans*-2-butene (○) and *cis*-2-butene (●). Neutral pressures of ammonia and butene were 4.5×10^{-6} and 1.3×10^{-6} torr in each case. A 100-msec ion formation period was followed by 100 msec of laser irradiation at 30 W cm^{-3} at each laser frequency. [From Wight and Beauchamp (1983a).]

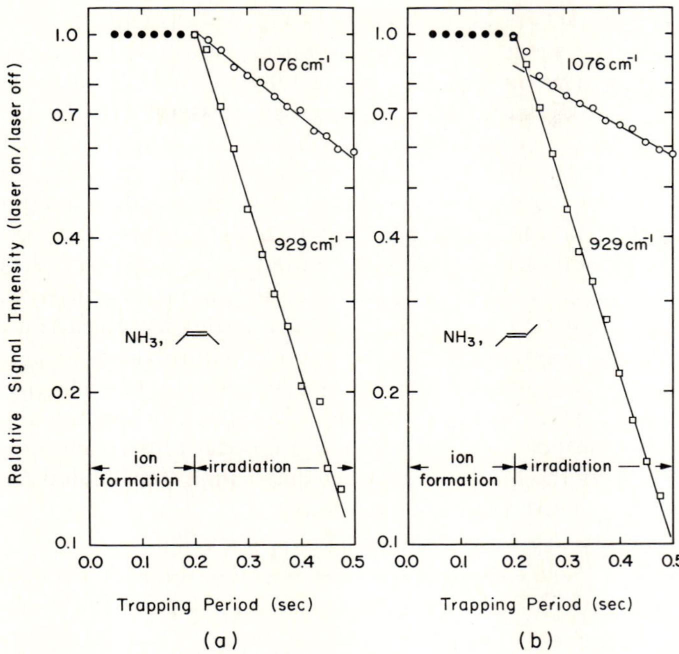

FIG. 24 Semilogarithmic plots of relative $C_4H_7^-$ intensity versus trapping period. A 200-msec ion-formation period precedes laser irradiation at 40 W cm^{-2}. (a) Decay of anions from *cis*-2-butene is shown for 929 and 1076-cm^{-1} irradiation. The respective phenomenological photodetachment cross sections (σ) are 3.9×10^{-21} and 1.0×10^{-21} cm^2. (b) The corresponding data for the *trans*-2-butene precursor are shown. Decay of $C_4H_7^-$ using 929-cm^{-1} irradiation is homogeneous, with $\sigma = 3.6 \times 10^{-21}$ cm^2. Data at 1076-cm^{-1} reveal that 14% of the ions decay with $\sigma = 30 \times 10^{-21}$ cm^2, while the remainder have $\sigma = 0.7 \times 10^{-21}$ cm^2. Neutral pressures of NH_3 and butene are 2.1×10^{-6} and 1.1×10^{-6} torr, respectively, corresponding to a total ion–molecule collision rate of approximately 90 sec^{-1}. [From Wight and Beauchamp (1983a).]

much lower cross section, comparable to $C_4H_7^-$ formed from *cis*-2-butene. The most likely explanation of these results is that proton abstraction from *cis*-2-butene forms exclusively the *cis*-butenyl anion, whereas abstraction from *trans*-2-butene results in a mixture of 86% *cis*-$C_4H_7^-$ and 14% *trans*-C_4H_7. Irradiation of the mixture at 929 cm^{-1} again results in a single exponential decay, suggesting that the *cis*- and *trans*-butenyl anions have the same photodetachment cross section at this frequency, within experimental uncertainty.

These results are in good agreement with the spectra in Fig. 23. Since the cross sections for the *cis*- and *trans*-butenyl anions are the same at 927 cm^{-1}, the photodetachment yields for $C_4H_7^-$ from the isomeric 2-butenes are the same at this frequency. At the high-frequency end of the spectrum (around 1076 cm^{-1}), only about 8% of the *cis*-butenyl anions are destroyed, under the conditions given in Fig. 23, whereas 98% of the *trans*-butenyl anions are destroyed as a result of their much larger cross section. The increase in photodetachment yield at 1076 cm^{-1} is limited to about 14% because only that fraction of the $C_4H_7^-$ population exists as the *trans*-butenyl isomer.

For comparison, spectra of butenyl anions from two additional precursors were obtained. Proton abstraction from 1-butene is expected to form either *cis*- or *trans*-butenyl anion. The spectrum in Fig. 25 is again consistent with formation of a mixture of *cis*- and *trans*-butenyl anions, although only about 10% of the mixture is trans in this case. The results are consistent with the observation that 1-butene prefers the gauche conformation (75%) to the cis conformation (25%) at room temperature (Barnes and Howells, 1973). Also shown in Fig. 25 is the spectrum of $C_4H_7^-$ resulting from proton abstraction from 2-methyl propene. As expected, this anion has a spectrum which is distinguishable from the spectra of both *cis*- and *trans*-butenyl anions and probably exists as a stable 2-methylallyl anion.

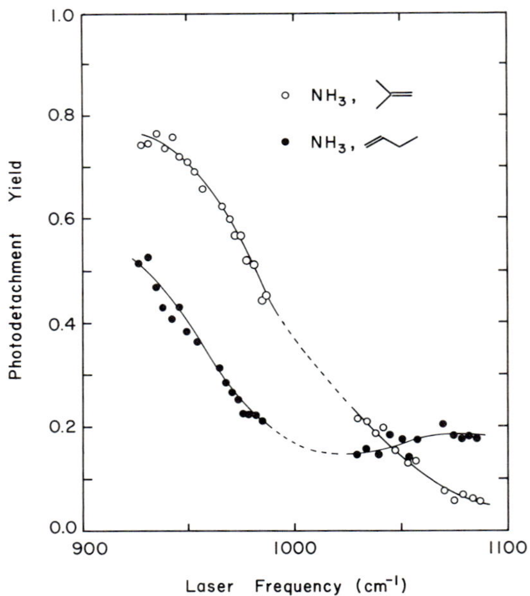

FIG. 25 Infrared multiphoton electron detachment spectra of $C_4H_7^-$ produced by ionization of mixtures of ammonia with 2-methyl propene (○) and 1-butene (●). Neutral pressures of ammonia and butene were 3.0×10^{-6} and 1.2×10^{-6} torr in each case. A 100-msec ion formation period was followed by 100 msec of laser irradiation at 30 W cm^{-2} at each laser frequency. [From Wight and Beauchamp (1983a).]

B. *Electron Detachment from Vibrationally Excited Anions*

Wight and Beauchamp (1983b) have demonstrated electron photodetachment from allyl anions, $C_3H_5^-$, having 10–13 kcal/mole of vibrational energy in excess of the zero-point energy. Either of two reactions, fluoride abstraction from allyltrimethylsilane (ATMS) or proton abstraction from propene, was used to generate $C_3H_5^-$:

$$F^- + (CH_3)_3Si(C_3H_5) \rightarrow C_3H_5^- + (CH_3)_3SiF \tag{29}$$

or

$$B^- + CH_3CHCH_2 \rightarrow C_3H_5^- + BH \tag{30}$$

where $B^- = H^-$ or NH_2^-.

A two-step reaction scheme was used to obtain values for the photodetachment cross sections for vibrationally excited and vibrationally relaxed $C_3H_5^-$:

$$F^- + ATMS \rightarrow A^- + A^{-*} + TMSF \tag{31}$$

$$A^{-*} \rightarrow A^- \tag{32a}$$

$$A^{-*} \rightarrow A + e^-. \tag{32b}$$

Here, k_r and k_d are the relaxation and detachment rate constants, respectively. The cross section $\sigma_d{}^*$ for excited allyl anions was found to be 6.5×10^{-20} cm^2 and that for the relaxed anions, σ_d, was 6.3×10^{-23} cm^2. The fraction of $C_3H_5^-$ produced in vibrationally excited states of sufficiently high energy to permit photodetachment can also be estimated using the scheme. This fraction was found to be 0.29 for $C_3H_5^-$ produced from $(CH_3)_3Si(C_3H_5)$.

Because of the weak dependence of the relaxation rate on pressure, the predominant relaxation mechanism is thought to be radiative emission and is characterized by a rate constant of 53 sec^{-1}. This is much larger than the theoretical rate obtained from the procedure outlined by Dunbar (1975), even for anions having up to 12.7 kcal/mole (the adiabatic electron affinity of C_3H_5) of vibrational energy. Because of this, it is believed that photodetachment occurs only for those anions which have no less vibrational energy than the adiabatic electron affinity of C_3H_5 less the energy of one CO_2 laser photon. That is, the electron photodetachment of $C_3H_5^-$ using cw infrared radiation is probably a one-photon process.

C. *Multiphoton Electron Detachment Model*

Wight and Beauchamp (1983c) have recently proposed a kinetic model for multiphoton electron detachment and have used it to analyze data obtained for benzyl anion. The model is essentially that given in Section III.E. Nine energy levels are used to specify the system.

Each trajectory is initiated from the lowest energy level, since the average thermal internal energy of benzyl anion at 300 K is only 1.6 kcal/mole. This value is calculated using the vibrational frequencies of toluene and assumes that no excess energy is acquired as a result of ion formation. The internal energy of cycloheptatrienyl anion is probably somewhat higher, since it is the product of an exothermic ion–molecule reaction.

The simplest way to choose transition rate constants for transitions between adjacent energy levels is to make them all the same. Doing so, however, results in calculated population decay curves which are conspicuously nonexponential (Fig. 26, dashed line) and which have induction periods that are much longer than those observed. The experimentally observed first-order decays and the linear dependence of the photodetachment rate con-

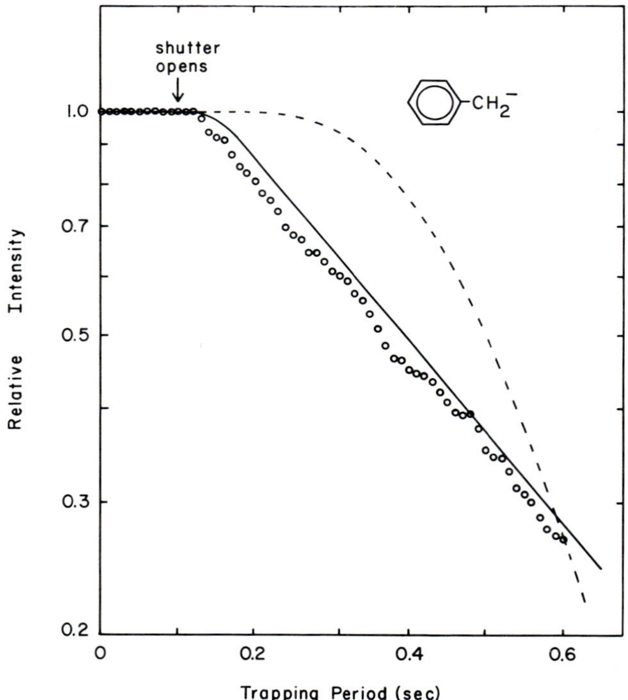

FIG. 26 Semilogarithmic plot of I/I_0 versus trapping period for benzyl anion, where I and I_0 are the ion signals in the presence and absence of laser irradiation, respectively. The mechanical shutter opens at 100 msec. Laser intensity is 50 W cm^{-2} at 948 cm^{-1}. Neutral pressures of methyl nitrite and toluene are 2×10^{-8} and 4×10^{-7} torr, respectively. The open circles represent experimental data while the solid and dashed curves are calculated using the kinetic model outlined in the text.

stant on laser intensity suggest that one of the adsorption steps is severely rate limiting. This "bottleneck" is presumably one of the first steps in the absorption process, although there is no direct evidence for this. Setting the cross section for absorption of the first photon equal to 1.0×10^{-21} cm^2 and the remainder of the absorption cross sections to 50.0×10^{-21} cm gives good agreement between the calculated and experimental decay curves in Fig. 26 (solid line).

Radiative emission rates between the various energy levels are somewhat more difficult to obtain. Crude estimates for benzyl anion are obtained using the basic method outlined by Dunbar (1975). Using toluene as a model compound, spontaneous emission rate coefficients are calculated for infrared active vibrational modes using integrated infrared absorption spectra (Table IV). Assuming a thermal distribution of internal energy over the various vibrational modes, transition rate constants between the energy levels may be estimated. Details of this procedure are outlined in the references cited by Wight and Beauchamp (1982c). Spontaneous emission rates calculated in this way are small compared to the excitation rates at moderate laser intensities. This relationship is in good qualitative agreement with the observation that photodetachment yields for benzyl anion are independent of laser intensity at constant fluence.

Although we have chosen the absorption cross sections empirically and the radiative emission rate constants calculated from infrared spectra of toluene, a check for self-consistency is possible at this point. The rate coefficient for spontaneous emission (Einstein *A* coefficient) is related to the

TABLE IV

Einstein A Coefficients for Spontaneous Emission from Infrared Active Vibrations of Toluene[a]

Band position (cm^{-1})	A (sec^{-1})
3050	58.43
2900	29.49
1610	1.97
1500	4.12
1080	0.59
1030	0.55
730	1.99
700	0.86

[a] Obtained from the gas phase infrared spectrum of toluene as outlined by Wight and Beauchamp (1982c).

absorption cross section σ by

$$A_{10} = \frac{8\pi h\nu^3}{C^2} B_{10} \frac{\Delta\nu_{\text{trans}}}{\Delta\nu_{\text{laser}}} = \frac{8\pi\nu^2\sigma\Delta\nu_{\text{trans}}}{C^2} \tag{33}$$

where $\Delta\nu_{\text{laser}}$ and $\Delta\nu_{\text{trans}}$ are the bandwidths of the laser and molecular transition, respectively. The laser-stimulated emission rate coefficient B_{10} is equal to that for absorption when nondegenerate vibrations are considered. Using the empirical absorption cross section $\sigma = 50.0 \times 10^{-21}$ cm^2 and a transition bandwidth of 35 cm^{-1} (Wight and Beauchamp, 1983c), reaction (33) gives $A_{10} = 1.2$ sec^{-1}, in good agreement with the values calculated for toluene listed in Table IV.

Using the empirically derived absorption cross sections and the spontaneous emission rate constants in Table V, the model reproduces many of the observed features of the photodetachment kinetics. Calculated curves of photodetachment yield versus irradiation period show that the decay of ion population with irradiation period is exponential (Fig. 26). The induction periods calculated using the model display the correct qualitative intensity dependence, but the values are larger than those observed experimentally.

There are many different ways to account for collisional deactivation using the model. The dashed curve in Fig. 27 illustrates the effect of "strong collisions" which remove all of the internal energy in the anion. Such a description might be appropriate to describe reactive collisions in which transfer of a proton from a cold toluene molecule to a hot benzyl anion would leave the resultant anion with very little internal energy. The disagreement with experiment is apparent. The solid curve, calculated for weak collisions which

TABLE V

Spontaneous Emission Rate Constants[a] Used for Model Calculations[b]

Initial internal energy	Final internal energy (kcal/mole)						
	0	2.71	5.42	8.13	10.83	13.54	16.25
2.71	0.15	0.03	0.00	0.01	—	—	—
5.42	—	0.29	0.07	0.06	0.02	—	—
8.13	—	—	0.46	0.12	0.13	0.04	—
10.83	—	—	—	0.63	0.17	0.22	0.06
13.54	—	—	—	—	0.80	0.22	0.33
16.25	—	—	—	—	—	0.98	0.27
18.96	—	—	—	—	—	—	1.14

[a] Rate constants have units of second^{-1}.

[b] Calculated from infrared spectra of toluene and adjusted for a model having equally spaced energy levels.

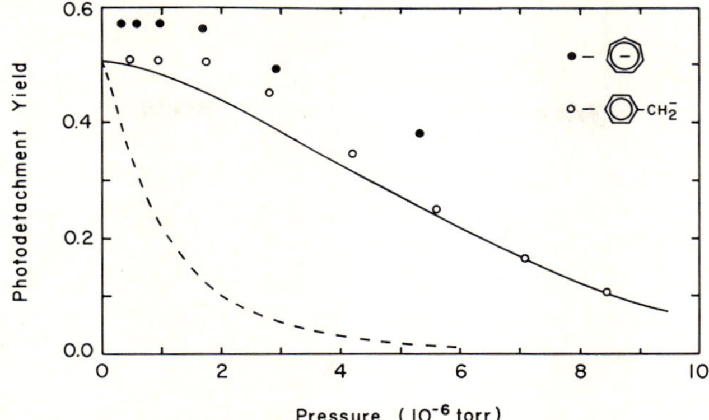

FIG. 27 Pressure dependence of the multiphoton electron detachment yield for cycloheptatrienyl anion (●) and benzyl anion (○). For all data, the pressure of methyl nitrite is 1.0×10^{-7} torr, with neutral C_7H_8 making up the balance. Results of a model calculation for benzyl anion are shown for strong collisions which remove all of the internal energy of the anion (dashed curve) and weaker collisions which remove only 25% of the internal energy (solid curve).

remove only 25% of the internal energy, gives better overall agreement with experiment but predicts too sharp a drop in yield at low pressures.

The available data do not permit a unique determination of each rate constant incorporated into the above model. This type of approach is instructive, however, for gaining a qualitative understanding of the relative importance of various rate constants in determining the overall photodetachment kinetics. Radiative emission, for instance, is believed to play a relatively unimportant role. Collisions are only moderately effective at deactivating excited intermediates, and there is some suggestion that they may enhance the net excitation rate at one or more steps. Finally, the shape of the decay curves strongly suggests that the net rate of photon absorption at one step is much slower than the rest.

VI. SELECTIVE ENHANCEMENT OF BIMOLECULAR REACTION RATES USING LOW-INTENSITY CW LASER RADIATION

Bomse and Beauchamp (1980) have used low-intensity cw laser radiation to shift the equilibrium given in reaction (34) by enhancing the reverse reaction rate:

$$(CH_3OH)H^+(OH_2) + CH_3OH \rightleftharpoons (CH_3OH)_2H^+ + H_2O \qquad (34)$$

In the absence of laser irradiation the forward rate constant k_f is 5.0×10^{-10} cm^3 mole^{-1} sec^{-1} and the reverse rate constant k_r is 8.2×10^{-15} cm^3

mole^{-1} sec^{-1}. k_f has been measured using ICR techniques, whereas k_r is calculated from k_f and the equilibrium constant K. The value $\Delta G = 6.5 \pm 1.0$ kcal/mole (Bomse and Beauchamp, 1981c) for reaction (34) gives $K = 6.1 \times 10^4$ in favor of the proton-bound methanol dimer $(CH_3OH)_2H^+$ at room temperature.

During cw laser irradiation ($\nu = 947$ cm^{-1}) at 34 W cm^{-2}, the reverse reaction rate is enhanced by more than three orders of magnitude to $k_r^{IR} = 2.6 \times 10^{-11}$ cm^3 mole^{-1} sec^{-1} (Bomse and Beauchamp, 1980). The forward reaction is unaffected by laser irradiation. Both the forward and reverse reactions in equilibrium (34) proceed through a common intermediate, $(CH_3OH)_2\text{—}H^+(OH_2)$. The competitive dissociation of this species can be evaluated using RRKM theory (Robinson and Holbrook, 1972), where the internal energy is taken as absorbed infrared energy added to a 300 K Boltzmann distribution of vibrational energy. At an added energy of 10.5 kcal/mole, the calculated ratio of H_2O to CH_3OH loss is equal to the observed value of $k_f/k_r^{IR} = 19.8$. This implies that $(CH_3OH)_2H^+$ absorbs an average of 3.9 infrared photons ($\nu = 947$ cm^{-1}) prior to bimolecular reaction with H_2O.

Selective excitation of reactants represents not only an interesting tool for experimental chemical dynamics but also offers the possibility of using measured changes in reaction rates as a spectroscopic probe. For example, the use of infrared excitation to alter bimolecular reaction rates could provide a general technique for obtaining vibrational spectra of ions and transient molecules.

VII. PROGNOSIS

Although additional experiments are in progress to further elucidate the photophysics of infrared multiphoton excitation, a reasonably clear picture of the mechanism of this process, involving both pulsed and cw lasers, has emerged over the past several years. Of the studies currently in progress, we are particularly excited about the use of multicolor excitation with continuously tunable sources to provide new insights into this phenomenon. All of the experiments reported to date have utilized only discrete infrared line sources.

The development of infrared laser technology is continuing at a rapid pace, and it should be feasible eventually to extend the experiments described above over most of the infrared spectrum. This will facilitate the identification of molecular structures, making it possible to obtain infrared spectra with sensitivity approaching that of mass spectra. By ionizing a mixture of compounds appropriately, e.g., by Li^+ attachment, it should be possible to obtain an infrared spectrum of each component without preseparation. This is an improvement over conventional infrared absorption spectroscopy, where the spectra of components are summed together. The slow accumulation of energy in increments of 2 kcal/mole allows a molecule to sample the potential energy surface for the system thoroughly at or below the activation energy for the lowest energy process. This is different than collisional activation, where energy is deposited in a comparatively short time period, normally with a distribution which allows for several competitive dissociation processes. It is not difficult to imagine many future applications of this aspect of infrared multiphoton activation. For example, a complex species could be sequentially fragmented, giving only a single product ion at each stage of dissociation. In this manner it might be possible to sequence a suitably ionized polypeptide. Combined with other, newer developments in ion cyclotron resonance spectroscopy (utilizing Fourier transform detection and high field superconducting magnets), such experiments are entirely feasible.

REFERENCES

Akulin, V. M., Alimprev, S. S., Kalov N. V., and Shelepin L. A. (1975). *Zh. Eksp. Theor. Fiz.* **69**, 836 [*English transl.: Sov. Phys. JETP* **42**, 427].

Allen, L., and Eberly, J. H. (1975). *In* "Optical Resonance and Two Level Atoms." Wiley (Interscience), New York.

Ambartzumian, R. V., and Letokhov, V. S. (1977) *In* "Chemical and Biochemical Applications of Lasers" (C. B. Moore, ed.), Vol. III. Academic Press, New York.

Armentrout, P. B., Berman, D. W., and Beauchamp (1978). *Chem. Phys. Lett.* **53**, 255.

Baldwin, A. C., and Barker, J. R. (1982). *Chem. Phys. Lett.* **86**, 55.

Barker, J. R. (1980). *J. Chem. Phys.* **72**, 3686.

Barnes, A. J., and Howells, J. D. R. (1973). *J. Chem. Soc. Faraday Trans. II* **69**, 532.

Bass, L., Su, T., Chesnavich, W. J., and Bowers, M. T. (1975). *Chem. Phys.* **34**, 119.

Beauchamp, J. L. (1971). *Ann. Rev. Phys. Chem.* **22**, 527.

Beauchamp, J. L., Holtz, D., Woodgate, S. D., and Pratt, S. L. (1972). *J. Am. Chem. Soc.* **94**, 2798.

Berman, D. W., Beauchamp, J. L., and Thorne, L. R. (1981). *Int. J. Mass Spectrom. Ion Phys.* **39**, 47.

Bien, F. (1978). *J. Chem. Phys.* **69**, 2631.

Black, J. G., Kolondner, P., Shultz, M. J., Yablonovitch, E., and Bloembergen, N. (1979). *Phys. Rev. A* **19**, 704–716.

Black, J. G., Yablonovitch, E., Bloembergen, N., and Mukamel, S. (1977). *Phys. Rev. Lett.* **38**, 1131.

Bomse, D. S., and Beauchamp, J. L. (1980). *J. Am. Chem. Soc.* **102**, 3967–3969.

Bomse, D. S., and Beauchamp, J. L. (1981a). *J. Am. Chem. Soc.* **103**, 3292.

Bomse, D. S., and Beauchamp, J. L. (1981b). *Chem. Phys. Lett.* **77**, 25.

Bomse, D. S., and Beauchamp, J. L. (1981c). *J. Phys. Chem.* **85**, 448.

Bomse, D. S., Woodin, R. L., and Beauchamp, J. L. (1978). *In* "Advances in Laser Chemistry" (A. H. Zewail, ed.), pp. 362–373. Springer-Verlag, Berlin and New York.

Bomse, D. S., Berman, D. W., and Beauchamp, J. L. (1981). *J. Am. Chem. Soc.* **103**, 3967.

Bomse, D. S., Woodin, R. L., and Beauchamp, J. L. (1979). *J. Am. Chem. Soc.* **101**, 5503.

Bowen, R. D., and Williams, D. H. (1980). *J. Am. Chem. Soc.* **102**, 2752.

Bray, R. G., and Berry, M. J. (1979). *J. Chem. Phys.* **71**, 4909.

Cantrell, C. D., Freund, S. M., and Lyman, J. L. (1979). *In* "Laser Handbook," Vol. III. North Holland Publ., Amsterdam.

Coggiola, M. J., Crosby, P. C., and Peterson, J. R. (1980). *J. Chem. Phys.* **72**, 6507.

Collussi, A. J. Benson, S. W., and Huang, R. S. (1977). *Chem. Phys. Lett.* **52**, 349.

Comes, F. J., and Pionteck, S. (1976). *Chem. Phys. Lett.* **42**, 558.

Cooks, R. G., Benyon, J. H., Caprioli, R. M., and Lester, G. R. (1973). *In* "Metastable Ions." Elsevier, Amsterdam.

Ding, A. M. G., and Polanyi, J. C. (1975). *Chem. Phys.* **10**, 39.

Druzaic, P. S., and Brauman, J. I. (1981). *Chem. Phys. Lett.* **83**, 508.

Dunbar, R. C. (1975). *Spectrochim. Acta* **31A**, 797.

Duperrex, R., and van den Bergh, H. (1980). *J. Chem. Phys.* **73**, 585.

Edgell, W. F., and May, C. E. (1954). *J. Chem. Phys.* **22**, 1808.

Frankel, D. S. Jr., and Manuccia, T. (1978). *J. Chem. Phys. Lett.* **54**, 451.

Freiser, B. S., and Beauchamp, J. L. (1974). *J. Am. Chem. Soc.* **96**, 6260.

Gillespie, D. T. (1976). *J. Comput. Phys.* **22**, 403.

Gillespie, D. T. (1977). *J. Phys. Chem.* **81**, 2340.

Golden, D. M., Rossi, M. J., Baldwin, A. C., and Barker, J. R. (1981). *Accounts Chem. Res.* **14**, 56.

Grant, E. R., Schulz, P. A., Sudbø, Aa. S., Shen, Y. R., and Lee, Y. T. (1978). *Phys. Rev. Lett.* **40**, 115.

Hartford, A. Jr. (1978). *Chem. Phys. Lett.* **3**, 503.

Isenor, N. R., Merchant, V., Hallsworth, R. S., and Richardson, M. S. (1973). *Can. J. Phys.* **51**, 1281.

Jarrold, M. F., Illies, A. J., Krochner, N. J., Bowers, M. T., Mandich, M., and Beauchamp, J. L. (1983). *J. Phys. Chem.* (in press).

Jasinski, J. M., and Brauman, J. I. (1980). *J. Chem. Phys.* **73**, 6191.

Jasinski, J. M., Rosenfeld, R. N., Meyer, F. K., and Brauman, J. I. (1982). *J. Am. Chem. Soc.* **104**, 652.

Knudtson, J. T., and Flynn, G. W. (1973). *J. Chem. Phys.* **58**, 1467.

Kolodner, P., Winterfeld, C. W., and Yablonovitch, E. (1977). *Opt. Commun.* **20**, 119.

Lehman, T. A., and Bursey, M. M. (1976). *In* "Ion Cyclotron Resonance Spectroscopy." Wiley (Interscience), New York.

Levsen, K., and Schwarz, H. (1976). *Angew. Chem.* **88**, 589.

McClusky, G. A., Kondrat, R. W., and Cooks, R. G. (1978). *J. Am. Chem. Soc.* **100**, 6045.

McLafferty, F. W. (1980). *Accounts Chem. Res.* **13**, 33.

Meyer, F. K., Jasinski, J. M., Rosenfeld, R. N., and Brauman, J. I. (1982). *J. Am. Chem. Soc.* **104**, 663.

Mukamel, S., and Jortner, J. J. (1976). *J. Chem. Phys.* **65**, 5204.

Neijzen, B. J. M., and de Lange, C. A. (1978). *J. Electron Spectrom.* **14**, 187.

Nielsen, J. R., Claassen, H. H., and Smith, D. C. (1952). *J. Chem. Phys.* **20**, 1916.

Orlowski, T. E., and Freiser, B. S. (1976). *Chem. Phys.* **16**, 439.

Polanyi, J. C., and Woodall, K. B. (1972). *J. Chem. Phys.* **56**, 1563.

Quack, M. J. (1978). *Chem. Phys.* **69**, 1294.

Quack, M. J. (1979). *Ber. Bunsenges. Physik. Chem.* **83**, 757.

Quigley, G. P. (1978). *In* "Advances in Laser Chemistry" (A. H. Zewail, ed.), Springer Series in Chemical Physics. Springer-Verlag, Berlin and New York and references contained therein.

Rabalais, J. W., and Colton, R. J. (1972). *J. Electron Spectrosc. Related Phenom.* **1**, 83–89.
Ridge, D. R., and Beauchamp, J. L. (1971). *J. Am. Chem. Soc.* **93**, 5925.
Robinson, P. J., and Holbrook, K. A. (1972). *In* "Unimolecular Reactions." Wiley, New York.
Rosenfeld, R. N., Jasinski, J. M., and Brauman, J. I. (1979). *Chem. Phys.* **71**, 1030.
Rosenfeld, R. N., Jasinski, J. M., and Brauman, J. I. (1980). *Chem. Phys. Lett.* **71**, 400.
Rosenfeld, R. N., Jasinski, J. M., and Brauman, J. I. (1982). *J. Am. Chem. Soc.* **104**, 658.
Schulz, P. A. *et al.* (1979). *Ann. Rev. Phys. Chem.* **30**, 379.
Simons, J. (1981). *J. Am. Chem. Soc.* **103**, 3971.
Squires, R. R., Wight, C. A., and DePuy, C. H., unpublished results.
Su, T., and Bowers, M. T. (1979). *In* "Gas Phase Ion Chemistry," (M. T. Bowers, ed.), Vol. I. Academic Press, New York.
Sudbø, Aa., Schulz, P. A., Grant, E. R., Shen, Y. R., and Lee, Y. T. (1979). *J. Chem. Phys.* **70**, 912 and references therein.
Sudbø, Aa., Schulz, P. A., Grant, E. R., Shen, Y. R., and Lee, Y. T. (1978). *J. Chem. Phys.* **68**, 1306.
Thorne, L. R., and Beauchamp, J. L. (1981). *J. Chem. Phys.* **74**, 5100.
Varsanyi, G. (1974). *In* "Assignments for Vibrational Spectra of Seven Hundred Benzene Derivatives," p. 73. Wiley, New York.
Wieser, H., Laidlaw, W. G., Krueger, P. J. and Fuhrer, H. (1968). *Spectrochim. Acta* **24A**, 1055.
Wight, C. A., and Beauchamp, J. L. (1981a). *J. Amer. Chem. Soc.* **103**, 6449.
Wight, C. A., and Beauchamp, J. L. (1981b). *Chem. Phys. Lett.* **77**, 30.
Wight, C. A., and Beauchamp, J. L. (1983a). *J. Amer. Chem. Soc.* (in press).
Wight, C. A., and Beauchamp, J. L. (1983b). *J. Phys. Chem.* (accepted for publication).
Wight, C. A., and Beauchamp, J. L. (1983c). *Chem. Phys.* (accepted for publication).
Williams, D. H. (1977). *Accounts Chem. Res.* **10**, 280.
Williams, D. H., and Howe, I. (1972). *In* "Principles of Organic Mass Spectrometry." McGraw Hill, New York.
Williams, D. H., and Hvistendahl, G. (1974). *J. Am. Chem. Soc.* **96**, 6753–6755.
Woodin, R. L., Bomse, D. S., and Beauchamp, J. L. (1978). *J. Am. Chem. Soc.* **100**, 3248.
Woodin, R. L., Bomse, D. S., and Beauchamp, J. L. (1979a), *In* "Chemical and Biochemical Applications of Lasers" (C. B. Moore, ed.), Vol. IV., p 355. Academic Press, New York.
Woodin, R. L., Bomse, D. S. and Beauchamp, J. L. (1979b). *Chem. Phys. Lett.* **63**, 630.
Ypenburg, J. W. (1972). *Rec. Trav. Chim.* **91**, 671.
Ypenburg, J. W., and Gerding, H. (1971). *Rec. Trav. Chim.* **90**, 885.

Chapter 19
Photodissociation in beams: organic ions

F. M. HARRIS

J. H. BEYNON

ROYAL SOCIETY RESEARCH UNIT
UNIVERSITY COLLEGE OF SWANSEA
SWANSEA, UNITED KINGDOM

ISBN 0-12-120803-6

I. INTRODUCTION

The large majority of studies to date of the dissociation of organic ions has been carried out using sectored mass spectrometers and ion beams having several kiloelectron volts of translational energy (Cooks *et al.*, 1973; Beynon and Gilbert, 1979; Cooks, 1978). Most of the earlier studies were concerned with the dissociation of metastable ions but these have been considerably added to over the past 15 years by investigations in which stable ions are induced to dissociate as the result of collisional excitation. The number of publications reporting both types of investigations is now incredibly large. On the other hand, photodissociation of ions has only recently begun to be studied extensively; the first results for organic ions were obtained using ion cyclotron resonance (ICR) spectrometers in the early 1970s. Such instruments were found to be well suited for photodissociation studies since ions have relatively long residence times within them, and sufficient photofragment-ion yields were obtained using conventional arc lamp sources and monochromators.

With the availability of commercial laser sources of radiation, it became feasible to photodissociate ions in other types of mass spectrometers. Sophisticated spectroscopic techniques were developed for the study of small ions in high-energy beams while, for the study of organic ions, triple quadrupole mass spectrometers were constructed in which low-energy beams were irradiated. In parallel with these developments, photodetachment of electrons from ions in negative-ion beams was being thoroughly investigated and irradiation of slow-moving ions in drift tubes was also being undertaken. These various developments and the results obtained from the use of the specially constructed equipment are the subject of a detailed review by Miller (1981).

Bearing in mind the wealth of information on the properties of organic ions that had already been obtained using sectored instruments, the authors' group decided in the late 1970s to introduce laser radiation into commercial mass spectrometers of this type and study the photodissociation of ions taking place within them. The number of ICR spectrometers available in laboratories is quite small in comparison to commercial and specially built sectored instruments. The demonstration at Swansea that studies of photodissociation can successfully be carried out on commercial sectored instruments has opened up this type of research to many more scientists, and already at least five other groups have modified or are about to modify their sectored instruments to introduce laser radiation into them.

In this volume, detailed descriptions can be found of studies of the photodissociation of ions undertaken for a wide variety of ions under various conditions. The contents of this chapter are concerned specifically with the photodissociation of organic ions in beams having several kiloelectron volts of translational energy.

II. EXPERIMENTAL TECHNIQUES

A Required Laser Power

It is of interest to estimate the power of the laser light required to give acceptable yields of photofragment-ion currents under normal working conditions of a sectored mass spectrometer. For maximum yields, it is preferable to introduce the radiation in such a way as to provide a long interaction path length, say, by merging the ion and laser beam over the entire length of a suitable field-free region. Other requirements, such as accurate location of the interaction region, may necessitate transmitting the laser beam at right angles across the ion beam, which will give a very small interaction length, i.e., either the height or the width of the ion beam in single-pass experiments. For crossed-beam work, intracavity operation of the laser would be ideal if the spectrometer can be suitably modified by fitting Brewster-angle windows. In this way, yields can be increased by between one and two orders of magnitude; such an irradiation method has been used on instruments specially built for the study of small ions (e.g., van Asselt *et al.*, 1974; Huber *et al.*, 1977).

In the work at Swansea, two spectrometers have been used and, in both, the ion and laser beams have been brought into axial coincidence in a field-free region. For the ZAB-2F spectrometer, the second field-free region was used giving an interaction path length l, of about 30 cm. If the photon flux is reduced from I_0 to I photons $\sec^{-1}$, owing to absorption of photons by ions having a cross section for photodissociation of σ_d, then the photofragment-ion current is given by $I_0 - I \simeq I_0 N \sigma_d l$, where N is the number density of ions. For a 5×10^{-10}-A beam of 6-keV ions of, say, 100-u mass, the value of N in the second field-free region of the ZAB-2F is estimated to be about 2×10^3 ions cm^{-3} and thus $N\sigma_d l = 3 \times 10^{-13}$ if $\sigma_d = 5 \times 10^{-18}\ \text{cm}^2$, a value approximately in the middle of the range of published data for organic ions (Dunbar, 1979). Thus, if the photofragment-ion yield is required to be 6.25×10^4 ions $\sec^{-1}$ (equivalent to a current of 10^{-14} A), the photon flux must be about $2 \times 10^{17}\ \sec^{-1}$, which corresponds to a laser power 88 mW at a wavelength of 450 nm. This requirement is immediately doubled in cw experiments when phase-sensitive detection is used and the laser is mechanically chopped giving a mark-to-space ratio of 1. Further, overlap of the beams may not be perfect and ions may be studied having values of σ_d considerably lower than $5 \times 10^{-18}\ \text{cm}^2$. Thus, in the work done so far, lasers having at least 750 mW of power at individual wavelengths have been used.

B. *Experimental Arrangements of Spectrometers and Lasers*

The first commercial mass spectrometer to be modified for studies of photodissociation of ions was the ZAB-2F spectrometer at Swansea (Mukhtar *et al.*, 1981a). It is a double-focusing spectrometer of reverse geometry having a magnetic sector of 300-mm radius, which mass selects the ions to be irradiated in the second field-free region. An argon-ion laser, model 171-19, manufactured by Spectra Physics, was used. It has an all-line output of 18 W, and the use of an internal prism allows selection of 10 discrete wavelengths in the range 528.7–454.5 nm which lies within the visible region of the electromagnetic spectrum. With suitable optical components in the laser, 2.5 W of radiation in the ultraviolet is obtainable within a wavelength range 363.8–351 nm. No attempt was made in the work at Swansea to select individual wavelengths in the ultraviolet; the light was taken as effectively being one line of wavelength 358 nm.

In order to introduce the laser radiation into the second field-free region of the spectrometer, the magnet flight tube was modified as shown in Fig. 1. The window W_1 is made of quartz and transmits over 90% of the beam through the working wavelength range of the laser. In order to bring the ion and laser beams into approximate coincidence, a beam locator mounted at the end of the magnet flight tube is used. It consists of a block of stainless steel, with a 3 × 5-mm hole in it, which can be lowered so that a large fraction of the ion beam passes through it. By using the steering mirrors M_1 and M_2, the laser beam can be directed to pass through the hole. The locator is withdrawn, and fine adjustments to the position of the laser beam are made until the photofragment-ion current is maximized.

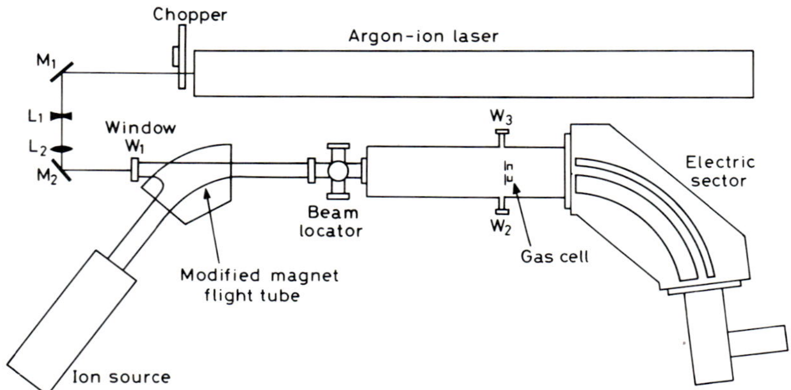

FIG. 1 Schematic drawing of the ZAB-2F mass spectrometer modified at Swansea (Mukhtar *et al.*, 1981a) for studies of photodissociation of ions.

The photofragment ions are transmitted through the electric sector (radius 380 mm) to an electron multiplier by scanning the voltages applied to the sector plates. The masses of the ions making up the peaks in the resulting spectra are identified from the position of the peaks on the electric sector voltage scale.

For some species of parent ions, the same fragment ions can be formed, both from metastable ions and from photodissociating ions. In order to detect and measure the photofragment-ion currents only, the technique of phase-sensitive detection was used in which the laser beam was mechanically chopped and the fragment-ion currents resulting from photodissociation recorded using a lock-in amplifier.

A reversed-geometry, double-focusing mass spectrometer has also been modified at the University of Bonn for studies of photodissociation of ions (Wagner-Redeker and Levsen, 1981). The modifications are very similar to those carried out on the ZAB-2F at Swansea; longitudinal irradiation of ions in the second field-free region is performed, the laser beam is mechanically chopped, and the photofragment ions are detected and measured using a lock-in amplifier. A Spectra Physics model 171-19 argon-ion laser was used.

A conventional-geometry (i.e., the electric sector precedes the magnetic sector), double-focusing mass spectrometer has also been modified at Swansea for photodissociation studies (Mukhtar *et al.*, 1980; Harris *et al.*, 1981). The arrangement is shown in Fig. 2. Radiation from the laser passes into the source housing (via a quartz window W_1) through a hole in the repeller plate, through the source, and into the first field-free region. Photoexcited ions which dissociate in this region give photofragment ions which are

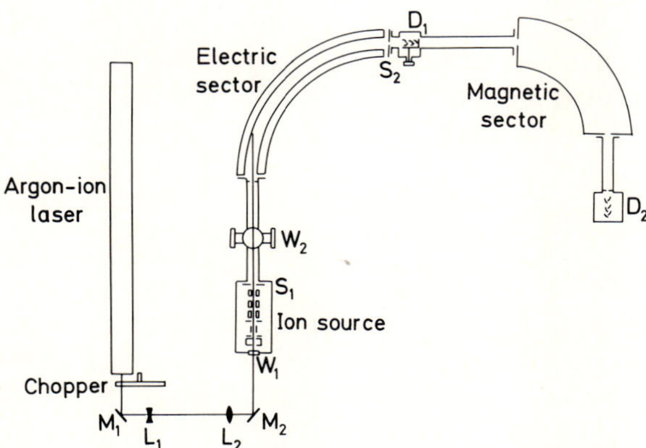

FIG. 2 Schematic drawing showing how laser radiation has been introduced into the source and first field-free region of a conventional-geometry, double-focusing mass spectrometer (Mukhtar *et al.*, 1980; Harris *et al.*, 1981).

transmitted through the electric sector by scanning the voltages applied to its plates. They are detected by the electron multiplier D_1. In addition, the photodissociation of slow-moving ions within the ion chamber can be studied using a procedure which is described in detail in Section VII.

At Texas A & M University, a commercial, conventional-geometry, double-focusing mass spectrometer—the Kratos MS902—has been modified recently for photodissociation work (Krailler *et al.*, 1982a,b). Longitudinal irradiation of the ions in the first field-free region was achieved by passing radiation through the source in a manner similar to that described above. In addition, crossed-beam work has been made possible by mounting windows on the vacuum housing of the first field-free region. A very powerful 30 W all-line, argon-ion laser manufactured by Coherent Radiation was used; the beam was chopped mechanically and the photofragment-ion currents recorded using a lock-in amplifier.

At the University of California, Santa Barbara, infrared radiation from a CO_2 laser has been introduced into a commercial ZAB-2F spectrometer by removing the intermediate slit mechanism and mounting a suitable window (Bowers, 1982). This particular crossed-beam arrangement irradiates the ion beam in the z-drection (parallel to the lengths of the slits), ensuring maximum sensitivity in single-pass work.

III. DISTINGUISHING ISOMERIC IONS

Photodissociation of ions in ICR spectrometers has proved to be very effective in distinguishing isomeric ions (Dunbar, 1979, 1981; Dunbar, this volume, Chapter 20). With such work proving so successful, it was logical for the groups who have coupled lasers to sectored mass spectrometers to see if their systems were sufficiently well-suited and sensitive enough also to distinguish isomeric ions. Various techniques were used, which are described next together with the results obtained.

A. *Differences in the Dependences of Translational Energy Released on Radiation Wavelength*

When an excited polyatomic ion dissociates, part of the energy in excess of the critical energy for the reaction is given to the fragments in the form of translational energy. One of the great advantages of working with ions having high translational energy, such as those in most sectored instruments, is that the translational energy released on dissociation in the center-of-mass frame of reference is considerably amplified in the laboratory frame of reference (Cooks *et al.*, 1973). This effect will be described more fully in Section V. It is sufficient to note here that, as a result of the amplification, fragment-ion peaks are considerably broadened, making it possible to determine with good accuracy values for T_t, the translational energy released. This is the parameter which has been investigated as a function of photoexcitation energy (Mukhtar, 1981b) to see if isomeric ions show different dependences. In this work, carried out on the ZAB-2F at Swansea, results for the three isomeric ions of xylene were obtained. Previous investigations have shown that it is very difficult to distinguish one from the others. From the earliest days of mass spectrometry, for example, it was known that the 70-eV mass spectra for the isomers are identical (Stenhagen *et al.*, 1974). More recently, this was shown to be the case for doubly charged mass spectra (Ast, 1972) and also for ion–kinetic-energy spectra resulting from dissociation of the metastable ions, the peaks in these latter spectra showing no difference in shapes or intensities (Mukhtar *et al.*, 1981b). Thus, xylene isomeric ions generated in an electron-impact (EI) source appear to be indistinguishable. When $[M - H + O]^-$ ions are formed in a chemical-ionization (CI) source, however, it is possible to distinguish the *m*-isomer from the other two (Bruins *et al.*, 1978).

In the work carried out by Mukhtar and his associates, the isomeric ions were generated in an EI source using 70-eV ionizing electrons. The photodissociation reactions studied involved the loss of the methyl radical, and T_t values (determined from peak widths at 10% height) were obtained for each of the isomeric ions at six different laser wavelengths corresponding to photon energies of 2.41, 2.49, 2.54, 2.60, 2.71, and 3.47 eV. The results are shown in Fig. 3, from which it can be seen that the three isomeric ions show

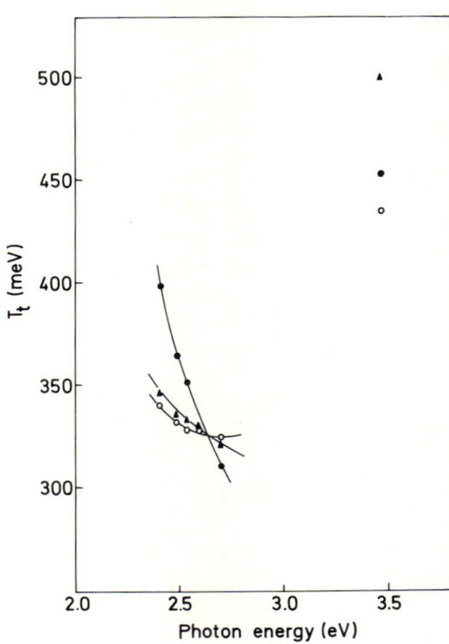

FIG. 3 Translational energy released as a function of photon energy for the three isomeric molecular ions of xylene (Mukhtar *et al.*, 1981b): (○) ion of ortho-isomer; (▲) ion of meta-isomer; and (●) ion of para-isomer.

differences in the dependence of T_t on photon energy, these being particularly large for ultraviolet radiation.

The fact that the photodissociation behavior of a mixture of stable and metastable ions (such as that obtained from a 70-eV EI source) is different for the different isomers proves that at least some of the ions have retained distinguishing structural features. It seems probable, in view of the identical behavior of the metastable ions in the mixture, that the critical energy for isomerization to a common structure lies below the lowest critical energy for dissociation. The differences in the photodissociation characteristics shown in Fig. 3 indicate that some of the ions having less than the minimum energy for isomerization can absorb photons and are raised to an energy sufficiently high that dissociation occurs in a time sufficiently short that appreciable isomerization cannot occur.

B. *Relative Cross Sections for Photodissociation and Relative Abundances of Photofragment-Ion Currents*

These quantities have been measured by Wagner–Redeker and Levsen (1981) at Bonn to see if five isomeric ions $[C_5H_8]^{+\cdot}$ from various precursors could be distinguished. Previous work had shown that fragment-ion spectra from collision-induced dissociation were the same for each. The photodissociation of ionized 1,2-pentadiene (**1**), 1,3-pentadiene (**2**), 2-pentyne, (**3**) and cyclopentene (**4**) were studied, as these precursors have different chromophoric groups (cumulated and conjugated double bonds, a triple bond, or a double bond). The behavior of the above four compounds was compared with that of $[C_5H_8]^{+\cdot}$ ions (**5**) generated by dissociative ionization of 3-octyne.

The relative cross sections for photodissociation σ_d were obtained by determining the total fragment-ion currents—corrected for reduced multiplier response—relative to the precursor ion current and then normalizing to constant light intensity. It is assumed that fragmentation-rate constants are such that all ions that absorb a photon dissociate before leaving the field-free region. The results obtained are shown in Fig. 4. It is apparent that σ_d and its dependence on wavelength differ significantly both in the UV and visible regions for all isomers. In the UV, σ_d increases markedly for **2**, peaking below 340 nm. This peak was attributed to the $\Pi \rightarrow \Pi^*$ transition, which appears to be shifted to even shorter wavelengths in the case of **1** and **3**. In the visible region, σ_d increases rapidly with decreasing wavelength, peaking at > 530 nm. This peak is thought to correspond to the ion specific $\Pi \rightarrow \Pi$ transition to the half-filled highest occupied orbital.

The pronounced differences in the dependences of σ_d on wavelength observed for the $[C_5H_8]^{+\cdot}$ isomers show that they can be unequivocally dis-

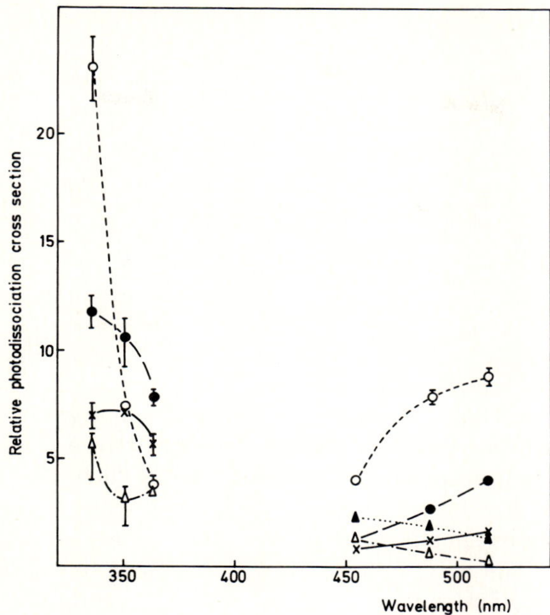

FIG. 4 Relative photodissociation cross sections of $[C_5H_8]^{+\cdot}$ ions as a function of wavelength (Wagner–Redeker and Levsen, 1981). $[C_5H_8]^{+\cdot}$ ions were generated from (×) 1,2-pentadiene; (○) 1,3-pentadiene; (△) 2-pentyne; (▲) cyclopentene; and (●) 3-octyne.

tinguished. Confirmation of the differences was obtained by Wagner-Redeker and Levsen by examining the photofragment-ion spectra of the five ions. These spectra were obtained at two wavelengths: 515 nm and 351 nm. Such spectra should reflect ion structure in much the same way as collision-induced spectra have done for numerous other ion species. Photoexcitation, of course, has the distinct advantage that discrete energies are deposited in the precursor ions. The results for the $[C_5H_8]^{+\cdot}$ ions show an interesting effect; the spectra of the ions **1**, **2**, **3**, and **4** photoexcited through 3.5 eV (wavelength 351 nm) are significantly different, but they are the same when radiation of 515-nm wavelength (2.4-eV photoexcitation energy) is used. Although there is limited knowledge available about the critical energies involved, the authors conclude that the isomerization barrier between **2** and the other isomers is between 0.5 and 1.6 eV.

In a recent investigation (Morgan *et al.*, 1982), techniques similar to those just described have been used at Swansea to investigate the properties of $[M - OH]^+$ ions from isomeric ethylnitrobenzene molecular ions. Previously, Butcher and Thomas (1979), in a study of alkylbenzene molecular ions, showed that $OH^{\cdot}$ loss takes place if there are α-hydrogens present in nitrobenzenes substituted with alkyl groups in the *o* position and also to a lesser extent in the corresponding *m*- and *p*-isomers. They suggested that the loss mechanism for the *m*- and *p*-alkylnitrobenzene molecular ions might involve a slow isomerization to a common structure, viz., that of the *o*-isomer, from which $OH^{\cdot}$ abstraction could occur. If such an isomerization takes place, it would be expected that the resulting $[M - OH]^+$ fragment ions would have a common structure. In two subsequent investigations (Baldwin, 1980; Baldwin *et al.*, 1982), it was found that the unimolecular dissociation spectra of $[M - OH]^+$ ions from the *m*- and *p*-isomers were the same but that that of the $[M - OH]^+$ ions from the *o*-isomer was different. A similar situation pertained to the collision-induced spectra of the $[M - OH]^+$ ions. In their investigation, Morgan *et al.* (1982) used radiation of one wavelength (514.5 nm) to obtain photofragment-ion spectra of the $[M - OH]^+$ ions from the three ethylnitrobenzene isomeric ions. The spectra deriving from the *m*- and *p*-isomer forms showed a single peak at *m*/*z* 104, which was formed as the result of $NO^{\cdot}$ loss from the $[M - OH]^+$ ions. The spectrum obtained from $[M - OH]^+$ ions from the *o*-isomer, on the other hand, had only a small peak corresponding to *m*/*z* 104 and several larger peaks at other masses. From the relative magnitudes of the *m*/*z* 104 peaks, it was concluded that even if some isomerization were taking place not more than 1% of the $[M - OH]^+$ ions from the *o*-isomer have the same structure as the corresponding ions from the *m*- and *p*-isomers.

In a second part of their investigation, Morgan and his associates measured the relative photodissociation cross sections of $[M - OH]^+$ ions as a function of radiation wavelength. The experimental technique is different from that used by Wagner–Redeker and Levsen (1981) in that effectively the loss of $[M - OH]^+$ ions following photoexcitation and dissociation is measured. This is done automatically by the lock-in amplifier, which gives a negative peak when the parent $[M - OH]^+$ peak is scanned. Each time a pulse of radiation is allowed to pass through the interaction region by the chopper, a fraction of the $[M - OH]^+$ ions is photodissociated resulting in a loss of ions from the $[M - OH]^+$ ion beam. The lock-in amplifier is thus receiving a signal which is decreasing in phase with the chopping frequency and it thus gives a negative output signal. The results from each measurement normalized for laser powers and main-beam currents are shown in Fig. 5, from which it can be seen that the absorption characteristics of the $[M - OH]^+$ ions from *m*- and *p*-ethylnitrobenzene are similar but that that from the *o*-isomer is very different from the other two. This study confirms, therefore, that it is unlikely that the losses of OH from the isomeric molecular ions of ethylnitrobenzene occur via a common ion structure.

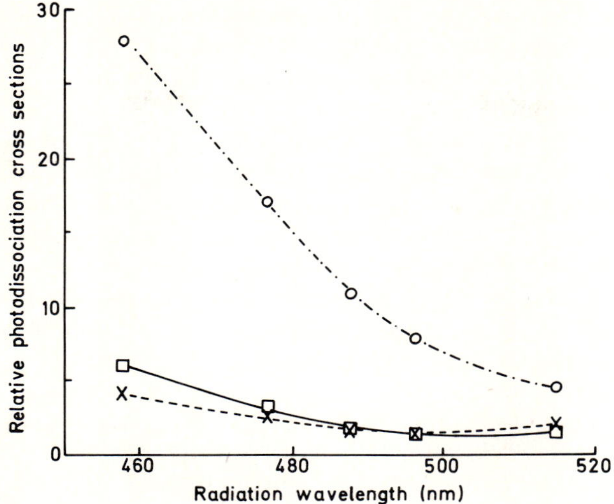

FIG. 5 Relative photodissociation cross sections as a function of wavelength for $[M - OH]^+$ ions generated from the isomers of ethylnitrobenzene (Morgan *et al.*, 1982): (○) $[M - OH]^+$ ions from *o*-ethylnitrobenzene; (□) $[M - OH]^+$ ions from *m*-ethylnitrobenzene; and (×) $[M - OH]^+$ ions from *p*-ethylnitrobenzene.

Krailler *et al* (1982, 1983) studied the photodissociation of the fragment ions of benzene and other $C_6H_6^{+\cdot}$ radical cations. This investigation was not concerned with distinguishing isomeric ions but, since relative photodissociation cross sections were measured as a function of photon energy, it is convenient to describe it in this section. Krailler and his associates found that ions such as $C_4H_4^{+\cdot}$, $C_4H_3^+$, $C_4H_2^{+\cdot}$ absorb radiation strongly at 514 nm and 488 nm. The object of the investigation was to obtain results which could be discussed in relation to the photodissociation mechanism for fragment ions proposed to explain the extensive fragmentation observed in the multiphoton ionization of molecules of benzene and 2,4-hexadiyne.

IV. THE EFFECT ON COMPETING DISSOCIATION REACTIONS OF CHANGING THE PHOTOEXCITATION ENERGY

This type of experiment depends on the fact that, if an excited ion has two or more dissociation pathways open to it, the relative yields of fragment ions will depend on the rates of the individual dissociation reactions. Since for the reactions the rates change in different ways as the ions' internal energy changes, it is to be expected that in photodissociation experiments the relative yields will depend on the values of the photoexcitation energy. Thus, using an argon-ion laser, relative yields can be measured as a function of photoexcitation energy in the range 2.41 eV–3.47 eV.

Sectored mass spectrometers are particularly well suited for investigations of competing dissociation reactions. Parent ions are photoexcited in field-free regions through which they pass only once. These regions are usually highly evacuated so that the subsequent dissociation of the ions is unaffected by collisions with neutral molecules. Further, the number density of ions is such that photoexcited ions or photofragment ions are unlikely to experience significant ion–ion interactions. The photoexcited ions thus dissociate in a collision-free environment, which is important if competitive unimolecular reactions are to be studied.

The first report of such investigations on sectored spectrometers was that of Mukhtar *et al.* (1980), who used the conventional-geometry spectrometer at Swansea (see Fig. 2). Nitrobenzene molecular ions were studied; the metastable constituents dissociate primarily via a reaction involving the loss of NO. Ions photoexcited through 2.41 eV, on the other hand, dissociate primarily via a reaction involving the loss of $NO_2^{\cdot}$ Thus, in the photofragment-ion spectrum, the $[M - NO^{\cdot}]^+$ peak is not present but the $[M - NO_2^{\cdot}]^+$ peak is large.

A more interesting molecular ion for such experiments is that of *n*-butylbenzene, the photodissociation of which was also studied (Mukhtar *et al.*, 1981a), using the ZAB-2F spectrometer at Swansea. This ion, when photoexcited through 2.41 eV, gave roughly equal yields of the photofragment ions 91^+ and 92^+ as a result of the competing reactions:

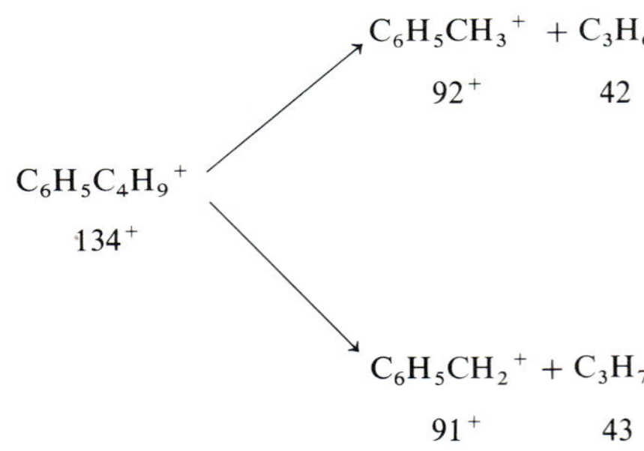

Unimolecular dissociation of the metastable 134^+ ions involves primarily the loss of 42 to give 92^+ fragment ions. As the photoexcitation energy was increased from 2.41 eV to 3.47 eV, the relative yields of the 91^+ and 92^+ fragment ions changed as shown in Table I.

TABLE I

Relative Yields of the Photofragment Ions 91⁺ and 92⁺ from n-butylbenzene Molecular Ions for Various Photoexcitation Energies

Wavelength (nm)	Photoexcitation energy (eV)	Ratio $[91^+/92^+]$
514.5	2.41	1.02 ± 0.05
488.0	2.54	1.30 ± 0.05
457.9	2.71	1.69 ± 0.05
358	3.47	6.90 ± 0.30

In order to obtain expressions for the rates of the competing reactions as a function of the ions' internal energy, these data have been analyzed (Griffiths *et al.*, 1981c) by adapting the procedure developed for the results of photodissociation of the nitrobenzene ion (Griffiths *et al.*, 1981a).

To understand the analytical procedure, consider the case of an ion M^+ which can dissociate via two competing reactions to give fragment ions A^+ and B^+. If the reactions have critical energies E_A and E_B, respectively, it was shown that the number of M^+ ions arriving at any point in the field-free region, t seconds after the ions' formation, is $dN(E)K$, where

$$K = \exp\{-[k_A(E - E_A) + k_B(E - E_B)]t\} \tag{1}$$

The terms $k_A(E - E_A)$ and $k_B(E - E_B)$ are the rate constants, respectively, for the reactions giving A^+ and B^+, and $dN(E)$ is the number of parent ions created at $t = 0$ having internal energies between E and $E + dE$.

If photon absorption occurs for a fraction $f(E)$ of ions passing this point, it was shown that the number of ions that fragment to form A^+ in the second field-free region of a ZAB-2F as the result of photodissociation is

$$\int_0^{E_{\max}} \frac{dN(E)f(E)k_A(\mathscr{E}_p + E - E_A)}{k_A(\mathscr{E}_p + E - E_A) + k_B(\mathscr{E}_p + E - E_B)} K(1 - \mathrm{M}), \tag{2}$$

where $M = \exp\{-[k_A(\mathscr{E}_p + E - E_A) + k_B(\mathscr{E}_p + E - E_B)][t_2 - t_1]\}$, $\mathscr{E}_p$ is the photon energy, and t_2 is the time at which the ions leave the field-free region.

The total number of ions available for the "normal" metastable ion dissociation in the field-free region is reduced, however, when the laser light is on due to the effective removal of a fraction $f(E)$ of such ions from the beam.

The reduction is

$$\int_{E_A}^{E_{max}} \frac{dN(E)f(E)k_A(E - E_A)}{k_A(E - E_A) + k_B(E - E_B)} \quad (K - L), \tag{3}$$

where $L = \exp\{-[k_A(E - E_A) + k_B(E - E_B)]t_2\}$.

The total effect of the laser light on the A^+ fragment-ion current can then be obtained by subtracting expression (3) from (2). A similar procedure allows the calculation of the B^+ fragment-ion yield.

One of the main problems in the analysis was that the initial energy distribution $dN(E)$ is unknown. A reasonable estimate of its approximate form was made from the photoelectron spectrum of *n*-butylbenzene (Nagy-Felsobuki and Peek, 1979). The uncertainty may, however, not be quite as critical as would appear at first sight. The distribution changes markedly during the lifetime of the ions and, by the time they have reached the second field-free region of the ZAB-2F, it has changed so much that minor differences in the initial form chosen do not appear to be too critical.

Another problem is whether $f(E)$ should be treated as a constant, i.e., independent of the internal energy of the ions prior to excitation. Again, the photoelectron spectrum helps. It exhibits a wide peak of overlapping bands extending from ~ 1.6 to 8.3 eV above the ionization energy. Since the majority of ions reaching the second field-free region will have internal energies less than the lowest critical energy of the reactions considered (1.4 eV for the formation of 92^+), it follows that all of them will have a good, if not exactly equal, chance of absorbing radiation in the energy range of 2.41–3.47 eV used in the work of Griffiths and his associates.

The elementary rate equation $k(E) = \nu[(E - E_c)/E]^{s-1}$ derived from simple quasi-equilibrium theory (QET), was used. In it, ν is a frequency factor, E_c is the critical energy for the dissociation reaction under consideration, and s is the effective number of oscillators. The values of E_c for the reactions giving 92^+ and 91^+ from the *n*-butylbenzene molecular ion have been measured (Brown, 1970) as 1.4 eV and 2.4 eV, respectively. Since a bond cleavage is likely to occur in the formation of 91^+, the value of ν for the reaction was assumed to have a value typical of a molecular vibration, i.e., 3×10^{13} sec^{-1}. The value of ν for the reaction giving 92^+ was treated as an unknown in the analysis, as was s. For longitudinal overlap of the photon and ion beams, photon absorption occurs at all values of t between the time of the ion entering the second field-free region and that of it leaving. A numerical integration program was used to evaluate the difference between equations such as (2) and (3) and the difference summed over a large number of values of t between the extremes. In practice, satisfactory convergence was obtained with 40 equal increments of t between the entry and exit times.

For photoexcitation energies in the range 2.41–3.47 eV, the analysis outlined above gave ratios of $91^+/92^+$, in agreement with the experimental values of Table I when the following rate equations were used:

(a) For the formation of 92^+ ions,

$$k(E) = 6 \times 10^9 \left(\frac{E - 1.4}{E}\right)^{17};$$

(b) For the formation of 91^+ ions,

$$k(E) = 3 \times 10^{13} \left(\frac{E - 2.6}{E}\right)^{17}.$$

V. TRANSLATIONAL ENERGY RELEASED AND THE QET

When a polyatomic ion dissociates, the energy in excess of the critical energy for the reaction may be distributed among the fragments as vibrational, rotational, translational, and electronic energy. When the dissociating ion has high translational energy, the translational energy given to the fragments in the center-of-mass frame of reference is amplified in the laboratory frame of reference (Cooks *et al.*, 1973). For ions having several kiloelectron volts of translational energy, the amplification factor can have values of the order of several hundred with the result that fragment-ion kinetic energy peaks are broadened to such an extent that their width at various fractions of the peak heights can be measured with considerable accuracy. Analysis of the shapes of the peaks gives information on the distribution of translational energies released.

The high accuracy possible in such measurements is invaluable in experiments in which the release of translational energy is studied as a function of the total internal energy of the dissociating ion. In the case of photodissociation, the excited ions' internal energy can be systematically changed by changing the photoexcitation energy. These changes are known precisely, even though the internal energies of the ions prior to photoexcitation are not.

In investigations carried out at Swansea (Griffiths *et al.*, 1981b; Mukhtar *et al.*, 1981c), a large number of ions have been photoexcited, and the average translational energy released, $\bar{T}_t$, has been measured as a function of photoexcitation energy $\mathscr{E}_p$. Examples of fragment-ion peak profiles and how they change with photoexcitation energy are shown in Fig. 6.

For most of the data, $\bar{T}$ was determined from peak widths at 22% of peak heights using well-known equations originally derived for metastable

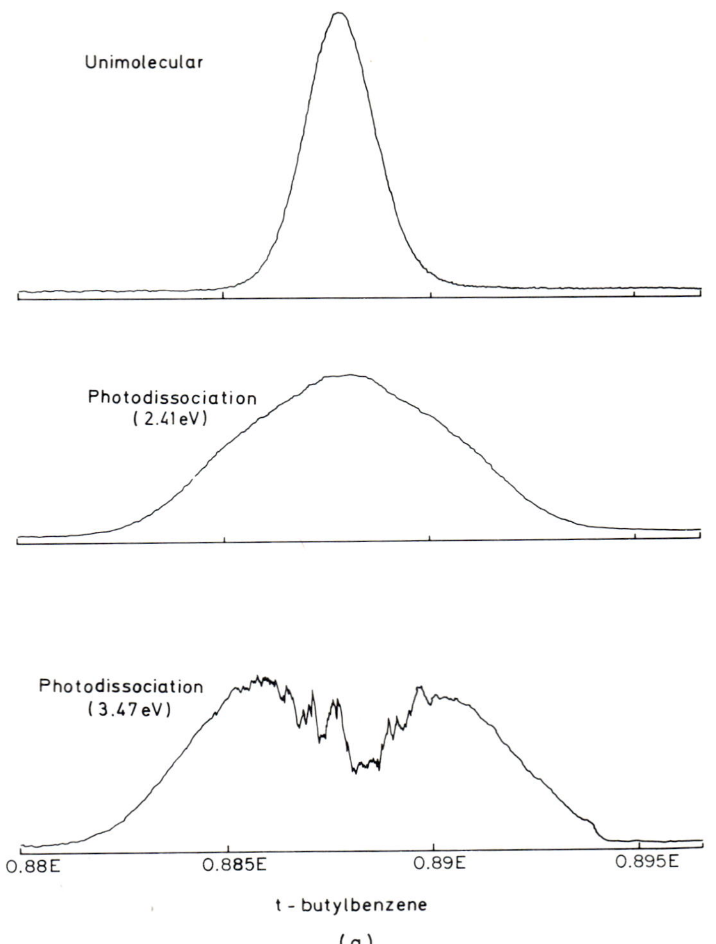

FIG. 6 Unimolecular and photo-induced peak profiles measured by Mukhtar *et al.* (1981c) for dissociation of *t*-butylbenzene ($[C_6H_5C_4H_9]^{+\cdot} \rightarrow [C_6H_5C_3H_6]^+ + CH_3^{\cdot}$) and *sym*-triazine ($[C_3H_3N_3]^{+\cdot} \rightarrow [C_2H_2N_2]^{+\cdot} + HCN$).

reactions (Cooks *et al.*, 1973). This procedure is known to give average values for T_t when the peaks have Gaussian or pseudo-Gaussian profiles. Even when this is not the case, the procedure has been shown (Mukhtar *et al.*, 1981c) not to introduce large errors into the estimates of $\bar{T}_t$.

For many of the ions studied, a linear increase of $\bar{T}_t$ with $\mathscr{E}_p$ was observed over the range of $\mathscr{E}_p$ from 2.41 to 3.47 eV. For others, however, nonlinear effects were observed, some (such as the xylene ions; see Section

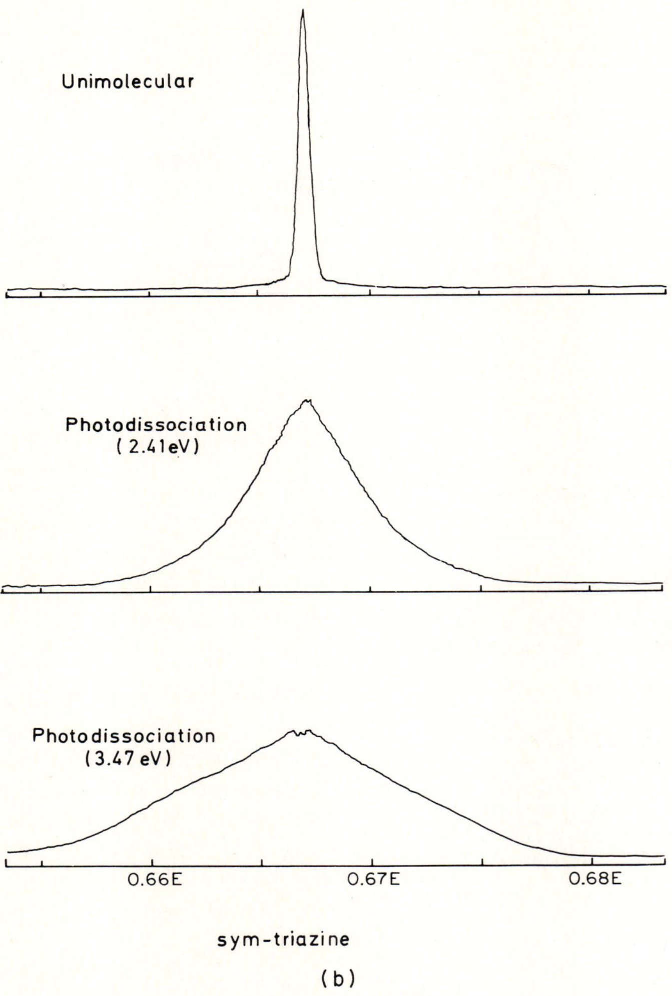

FIG. 6 (*continued*)

III.A) showing a decrease of $\bar{T}_t$ with increase of $\mathscr{E}_p$, or, at the largest $\mathscr{E}_p$, a value of $\bar{T}_t$ which did not fit into the pattern of results seen at lower values of $\mathscr{E}_p$. Clearly, it will be of considerable interest to study these effects in greater detail, but to date only the linearly increasing relationships have been analyzed in terms of the QET (Mukhtar *et al.*, 1981c).

The results for seven alkylbenzene ions were chosen to illustrate the linear dependence of $\bar{T}_t$ on $\mathscr{E}_p$. These are shown in Fig. 7. For each of three of

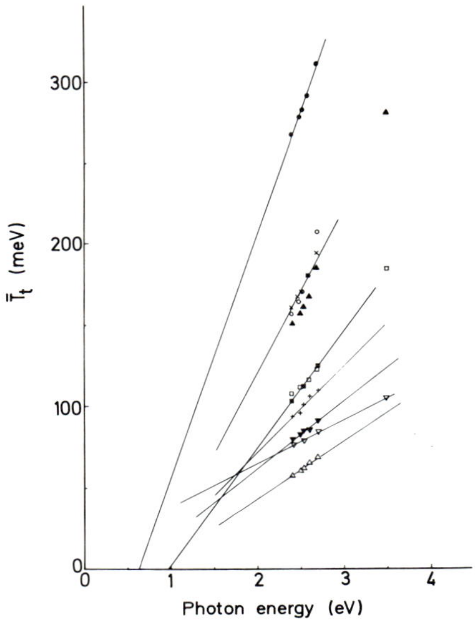

FIG. 7 Values of average translational energy released as a function of photon energy for the reactions listed in Table II (Mukhtar *et al.*, 1981c). Reaction (1) (□), (2) (▲), (3) (▽), (4) ○, (5) ■, (6) ×, (7) ▼, (8) ●, (9) +, and (10) △.

the ions, two reaction pathways were studied involving the loss of different neutral fragments (Table II).

It is not immediately obvious that the linearity observed would be expected when the overall photoexcitation/dissociation process is considered. The ions which enter the interaction region have a range of internal energies

TABLE II

Parent Ions and Reactions Giving the Results Shown in Fig. 7 and the Gradients $\Delta\bar{T}_t/\Delta\mathscr{E}_p$

Reaction	Parent ion	Neutral loss	$(\Delta\bar{T}_t/\Delta\mathscr{E}_p)(\times 100)$
(1)	ethylbenzene	$CH_3^{\cdot}$	7.3
(2)	*n*-propylbenzene	$CH_3^{\cdot}$	10.0
(3)	*n*-propylbenzene	$C_2H_5^{\cdot}$	2.8
(4)	*iso*-propylbenzene	$CH_3^{\cdot}$	11.0
(5)	*n*-butylbenzene	$C_2H_5^{\cdot}$	7.2
(6)	*sec*-butylbenzene	$CH_3^{\cdot}$	11.0
(7)	*sec*-butylbenzene	$C_2H_5^{\cdot}$	4.5
(8)	*t*-butylbenzene	$CH_3^{\cdot}$	14.8
(9)	*sec*-pentylbenzene	$C_2H_5^{\cdot}$	5.3
(10)	*sec*-pentylbenzene	$C_3H_7^{\cdot}$	3.5

because they are generated in a 70-eV EI source. In fact, metastable ion fragmentation takes place, indicating that a range of internal energies exists up to and including those of the metastable "window." It is likely, however, that the ions are in their ground electronic state. When photons having energies in the range 2.41–3.47 eV are absorbed by the ions, excitation to a higher electronic state is involved. Dissociation is assumed to follow the usual QET process of radiationless transition to the ground electronic state, the ion being in a high, vibrationally excited state before it fragments.

Because the ions entering the interaction region have a distribution of internal energies, it is unlikely that all of them will have the same probability of being excited to a given higher electronic state. The situation can be put in better perspective, however, by considering photoelectron spectra for the alkylbenzenes. Nagy–Felsobuki and Peel (1979) have measured such spectra for three of the samples used in the work of Mukhtar *et al.* (1981c) (*n*-butylbenzene, *sec*-butylbenzene, and *t*-butylbenzene). These show a low-energy peak extending from zero to about 1.3 eV (measuring energies relative to ionization energies) and a very wide peak of overlapping bands extending from ~1.6 eV to 8.3 eV. It is known from the analysis of fragment-ion abundances described in Section IV that the majority of *n*-butylbenzene ions reaching the interaction have energies less than 1.4 eV. The same is true of the other ions because of the relatively low critical energies for the dissociation reactions of the metastable ions (Rosenstock *et al.*, 1977). Thus, it is argued that all the ions reaching the second field-free region have a good, if not equal, chance of being excited into the broadband extending upward in energy from 1.6 eV since photons in the energy range 2.41–3.47 eV are used. Also, because of the broad nature of the second photoelectron peak, it is probable that the average internal energy of the photoexcited ions increases linearly with increase in photoexcitation energy. This appears to be the case for the ions studied.

The statistical theory of Klots (1971, 1972) has been applied by Mukhtar *et al.* (1981c) to calculate $\bar{T}_t$ as a function of the excess energy $\mathscr{E}^{\#}$ above the critical energy for the following ions and reactions:

$$[C_6H_5C_2H_5]^{+\cdot} \rightarrow [C_6H_5CH_2]^+ + CH_3^{\cdot}$$
$$[C_6H_5C_3H_7]^{+\cdot} \rightarrow [C_6H_5C_2H_4]^+ + CH_3^{\cdot}$$
$$[C_6H_5C_4H_9]^{+\cdot} \rightarrow [C_6H_5C_3H_6]^+ + CH_3^{\cdot}$$
$$[C_4H_{10}]^{+\cdot} \rightarrow [C_3H_7]^+ + CH_3^{\cdot}.$$

The results obtained are plotted in Fig. 8 together with some of the experimental data. In order to render the theoretical and experimental results comparable, the $\mathscr{E}_p$ values would have to be converted to $\mathscr{E}^{\#}$ values. It was not possible to do this since the internal energy distribution of the ions prior to photoexcitation was not known. If such a conversion could be carried

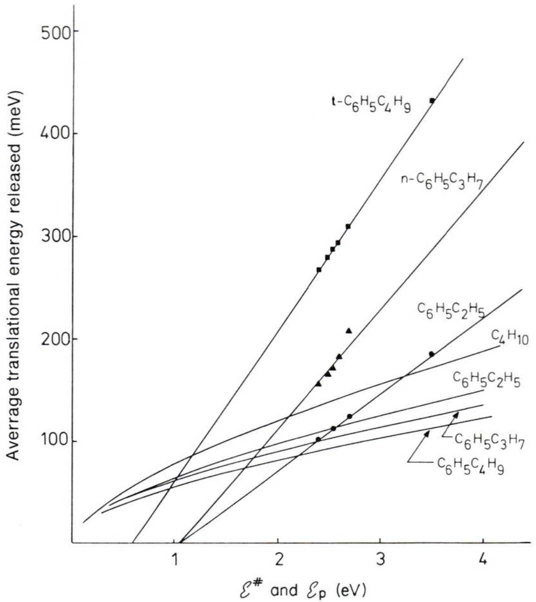

FIG. 8 Comparison of some of the measured average translational energies released (Mukhtar *et al.*, 1981c) with those obtained from calculations based on the statistical theory of Klots (1971, 1972).

out, however, the effect would be to move all the experimental values in Fig. 8 to lower energy values. With this in mind, it can be seen that the experimental values for $\bar{T}_t$ are in general larger than the calculated ones. Further, the increases in experimental values of $\bar{T}_t$ with increases in $\mathscr{E}^{\#}$ are considerably larger than those predicted from theory.

VI. EFFECT OF SOURCE TEMPERATURE ON THE PHOTODISSOCIATION OF IONS

From the work described in Sections IV and V, it is clear that effects can be observed which are sensitively dependent on the energy through which the ions are photoexcited. Both the yields of photofragment ions from competing dissociation reactions of photoexcited ions as well as values of $\bar{T}_t$ measured for a given reaction can show significant changes with small changes in $\mathscr{E}_p$; this occurs in many cases, even if the change is as small as 0.1 eV. Because of this sensitivity, Griffiths *et al.* (1982) decided to find out if the experimental techniques developed on double-focusing mass spectrometers at Swansea would allow changes in the internal energy of ions that had occurred prior to photoexcitation to be detected and measured. In their experiments, the photoexcitation energy was held constant by working at one laser-light wavelength; changes in the internal energy of the parent ions were effected by changing the temperature of the EI source in which the ions were formed over the range 70–240°C.

In their first series of experiments, the effect on the competing dissociation reactions of photoexcited *n*-butylbenzene ions was investigated. The reactions giving 91^+ and 92^+ photofragment ions have been discussed in Section IV; typical spectra measured at one wavelength (488.0 nm) and with the source at each of four temperatures are shown in Fig. 9. Since one photoexcitation energy is used, and it is known that the ratio $91^+/92^+$ increases as the total average internal energy of the excited ions increases, the results of Fig. 9 show that the average internal energy of the ions prior to excitation is increased as the source temperature is increased. Similar data obtained at two other separate wavelengths and the four source temperatures are included in Fig. 10. Several estimates were made by Griffiths and his associates of the change in average internal energy of the ions for a 100°C change in temperature. (This temperature change was chosen so that comparisons could be made with calculated data from known thermodynamic data, as will be discussed later in this section.) For example, when the data obtained for a photoexcitation energy of 2.71 eV are considered, it is seen that a change in temperature from 150 to 240°C changes the ratio $91^+/92^+$ from 1.53 to 2.33. By joining up sequential points at 240°C, it is estimated that the same change in the ratio is produced by changing the photoexcitation energy from 2.48 to 2.71 eV. Thus, a change in temperature of 90°C

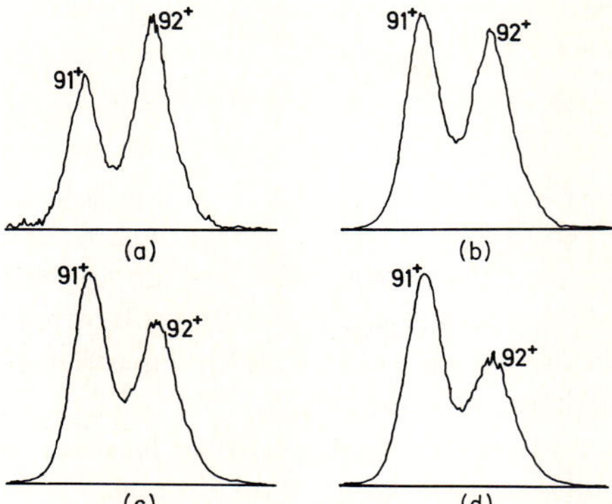

FIG. 9 Peaks due to the photofragment ions 91^+ and 92^+ resulting from dissociation of *n*-butylbenzene molecular ions photoexcited by radiation of 488.0-nm wavelength (Griffiths *et al.*, 1982). Source temperatures: (a) 80°C, (b) 150°C, (c) 190°C, and (d) 240°C.

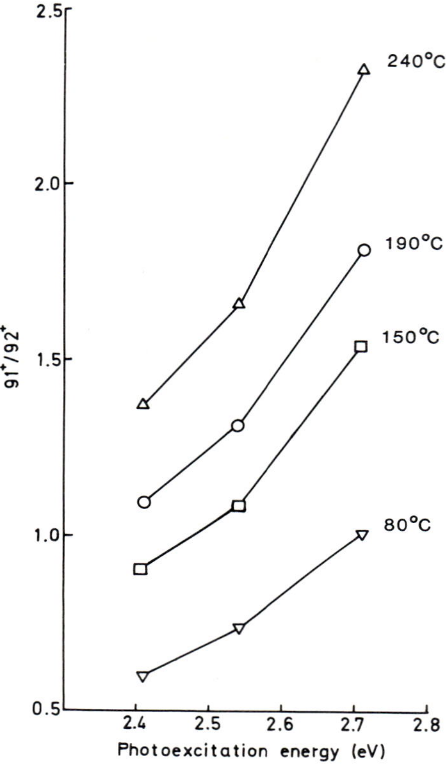

FIG. 10 Ratios $91^+/92^+$ determined from the photofragment-ion spectra of *n*-butylbenzene molecular ions measured at three photoexcitation energies and for four source temperatures (Griffiths *et al.*, 1982).

with a constant photoexcitation energy produces the same effect as a change in photoexcitation of about 0.23 eV at a constant temperature. Assuming a linear relationship with temperature (over a small range), the corresponding change for a 100°C change is 0.26 eV. Several estimates were made in this way, which indicate that the average internal energy of the ions increases by about 0.27 eV for a 100°C increase in source temperature. As is shown later on in this section, this is in reasonable agreement with changes predicted from known thermodynamic data.

In their second series of experiments, Griffiths *et al.* (1982) measured $\bar{T}_t$ as a function of source temperature, again keeping the photoexcitation energy constant by working at a single wavelength. In order to obtain good signal-to-noise ratios so that accurate peak-width measurements could be made, it was necessary to work with relatively large time constants on the lock-in amplifier; thus, peaks had to be scanned slowly. Since the source was not temperature-controlled, these measurements were taken at two temperatures only: (a) with the source heaters off, the source being heated by the filament only, and (b) with the source heaters on. In this way, data were obtained for a temperature difference of about 120°C. Dissociation reactions for the molecular ions of *t*-butylbenzene, nitrobenzene, and *n*-propylbenzene were investigated; the results are shown in Figs. 11 and 12. They all show that, at one photoexcitation energy, $\bar{T}_t$ increases as the source temperature increases. This work thus confirmed that heating the source resulted in an increase in the average internal energy of ions reaching the interaction region. From the data of Figs. 11 and 12, it was estimated that for a 100°C change in source temperature, the average internal energy of the ions changed by about 0.21 eV. This value is similar to that calculated from the results of the first series of experiments. The experiments are, however, very different—one type involving the competing dissociation reactions

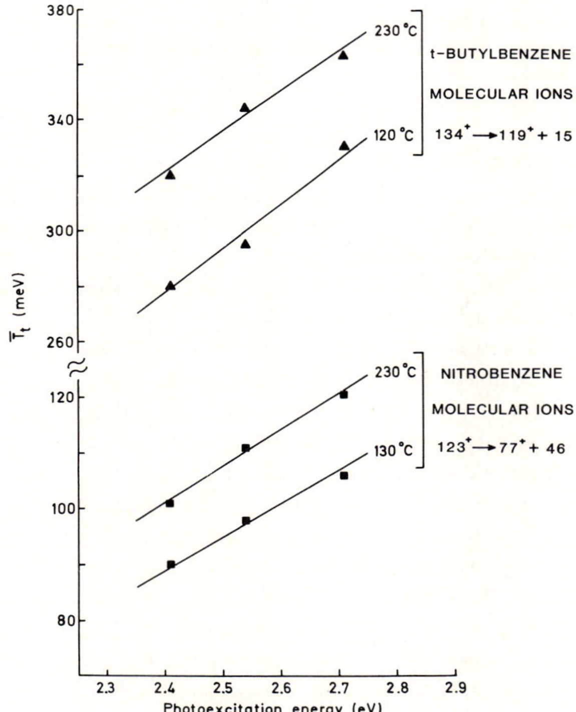

FIG. 11 Translational energy released as a function of photoexcitation energy with source temperatures held constant at the values shown (Griffiths *et al.*, 1982).

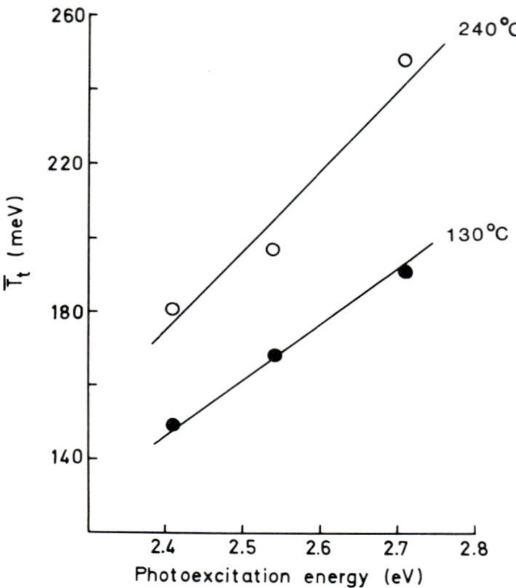

FIG. 12 $\bar{T}_t$ values obtained for two source temperatures at various photoexcitation energies when molecular ions of *n*-propylbenzene dissociate with the loss of the methyl radical (Griffiths *et al.*, 1982).

of photoexcited ions and the other the relative translational energy of the fragments released during a specific dissociation reaction.

In discussing these effects, Griffiths *et al.* (1982) argue that once the ions are formed they are unlikely to experience any subsequent thermal excitation. Since the ions generated in the ion chamber move under the action of a weak extracting field to the exit aperture, they do not make contact with the walls and are thus unaffected by wall temperature. Further, the sample pressures are sufficiently low ($<10^{-3}$ torr) in an EI source to make it unlikely that ion–molecule interactions occur to any significant extent. It follows that changes in internal energies with source temperature of ions extracted from the source must be due to changes in internal energies of the sample molecules. Calculated values obtained by Griffiths and his associates were thus obtained on the neutral molecules. Thermodynamic data for three of the samples studied are available in the literature (Rossini *et al.*, 1953) over the temperature range used. Heat content functions $(H - H_0)/T$ have been used, where $(H - H_0)$ is the heat content in the standard state at T K less the heat content in the standard state at 0 K. The data for 400 and 500 K (i.e., 127 and 227°C, which are within the temperature range of the experiments) are shown in Table III together with

(a) the change in the *total* energy of the molecules for the 100-K change from 400 to 500 K, and

TABLE III

Thermodynamic Data[a] *for Three Samples*[b,c]

Sample	$(H - H_0)/T$ (cal K^{-1} $mole^{-1}$) 400 K	$(H - H_0)/T$ (cal K^{-1} $mole^{-1}$) 500 K	Change in total energy for temperature change of 100 K (eV)	Change in average internal energy for temperature change of 100 K (eV)
n-butylbenzene	30.93	36.85	0.25	0.23
n-pentylbenzene	35.00	41.63	0.29	0.27
n-propylbenzene	27.00	32.2	0.22	0.20

[a] From Rossini *et al.* (1953).

[b] Samples studied by Griffiths *et al.* (1982).

[c] Calculated values of the changes in total and average internal energies of the molecules for a temperature change of 100 K are shown.

(b) the change in the average *internal* energy of the molecules over the same temperature range.

From Table III it can be seen that changes in the average internal energy of the three molecules are between 0.2 and 0.27 eV and are thus similar to the experimentally measured changes in the internal energies of the ions studied over the same temperature range. The calculations thus substantiate that the changes in the relative fragment-ion yields and variations in the translational energy released during photodissociation, as a function of source temperature, are due to changes in the internal energies of the molecules prior to ionization. Such small changes of a few tenths of an eV are only revealed because of the high sensitivity of the photodissociation results to the average internal energy of ions photoexcited through accurately known and constant photoexcitation energies.

VII. PHOTODISSOCIATION OF IONS IN LOW-ENERGY BEAMS WITHIN A SECTORED MASS SPECTROMETER

As has been discussed earlier in this chapter, the translational energy released when ions of high translational energy dissociate is amplified in the laboratory frame of reference giving rise to broad fragment-ion peaks. These result in poor resolution in fragment-ion spectra, and when peaks overlap it is often difficult to obtain detailed information on the individual reactions giving rise to them. For some ions, the loss of H or H_2

is often copious but the reactions can be difficult to investigate because of the proximity of the fragment-ion peak to the large precursor-ion peak. Clearly, what is desirable in many experiments is higher resolution in recording fragment-ion spectra. With this in mind, Kingston *et al.* (1982) studied the photodissociation of slow-moving ions within the ion chamber of a source. The mass spectrometer used is that illustrated in Fig. 2. The ions, which were generated in the usual way by ionizing sample molecules using electrons, were irradiated, as they drifted slowly toward the exit aperture of the ion chamber, by the laser light introduced via the window in the source housing. Photofragment ions formed by dissociation of photoexcited ions *within the chamber* were accelerated through the full potential difference V applied between the chamber and the earthed source slit. These were transmitted to the multiplier D_2 by operating the mass spectrometer in the way used to obtain a mass spectrum, i.e., the appropriate voltages were applied to the plates of the electric sector, so that ions of energy eV were transmitted, and the field of the magnetic sector was scanned. Ions which had not interacted with photons—both molecular and fragment ions—were also transmitted. These, however, were not recorded by the lock-in amplifier used in the usual phase-sensitive detection system. Examples of photofragment-ion mass spectra obtained using this experimental procedure are shown in Fig. 13.

Obvious striking features of these spectra are the very large negative signals obtained from the lock-in amplifier at various magnetic-field values. Such peaks, which are referred to as "negative" peaks, result from a depletion of ions which occur in phase with the chopping frequency. Consider, for example, the case of nitrobenzene and the situation with the magnetic field set to transmit the molecular ion 123^+. With the laser off, a certain dc current corresponding to the number of 123^+ ions arriving at the detector per second is obtained. With the laser on, a fraction of the 123^+ ions in the ion chamber will be photoexcited and dissociate there. Thus, the 123^+ ion current at the detector is reduced when the laser is on. Since the laser beam is effectively switched on and off by the chopper, the number of 123^+ ions reaching the detector will be reduced in phase with the chopping frequency. The lock-in amplifier records this as a negative signal and gives a negative output voltage. Since the photon energies were too low to photoionize the sample molecules and the photon fluxes were too low to consider the possibility of multiphoton ionization, the laser beam cannot generate molecular ions but only remove them by the process of photoexcitation followed by dissociation. The magnitude of the negative peak at the molecular ion mass position reflects,, therefore, the efficiency with which photodissociation takes place within the ion chamber. This is, in fact, very high with an estimated 4% of the 123^+ ions photodissociating, resulting in the excellent signal-to-noise characteristics evident in the spectrum.

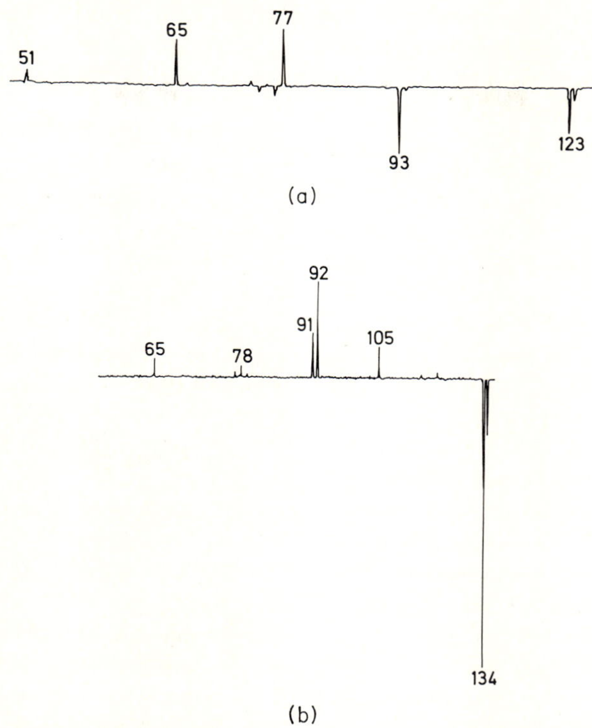

FIG. 13 Photofragment-ion mass spectra resulting from dissociation of ions within the ion chamber of the mass spectrometer illustrated in Fig. 2 (Kingston *et al.*, 1982). (a) Nitrobenzene; (b) *n*-butylbenzene.

The magnitude of the negative molecular ion peak is a measure of the cross section for photoabsorption leading to dissociation. Relative values of this cross section can be measured accurately because of the large decreases in currents which give rise to the negative peaks. Results obtained for nitrobenzene and toluene ions are shown in Fig. 14. The procedure used in this work with low-energy ions is superior to that involving the summation of fragment-ion peaks, which introduces cumulative errors. Furthermore, no broadening of peaks occurs because dissociation of ions having low translational energies is involved. The efficiency of the experimental procedure is such that photodissociation of ions within the ion chamber has also been demonstrated using an Hg-Xe lamp as a source of light (Kingston, 1982). There is every possibility that high optical-resolution experiments, similar to those carried out on a triple quadrupole spectrometer using a tunable dye laser (Goss *et al.*, 1981a,b; Morrison *et al.*, 1979; McGilvery and Morrison,

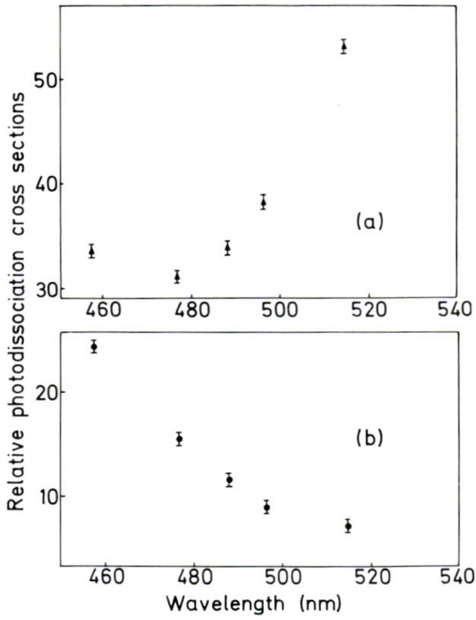

FIG. 14 Relative photodissociation cross sections as a function of wavelength obtained (Kingston *et al.*, 1982) by irradiating slow-moving ions within the ion chamber of a mass spectrometer source (Fig. 2). (a) Nitrobenzene molecular ions; (b) toluene molecular ions.

1979) could be performed when slow-moving ions within the source of a double-focusing mass spectrometer are irradiated.

It can be seen from spectrum (b) of Fig. 13 that the 91^+ and 92^+ photofragment-ion peaks from *n*-butylbenzene are completely separated. This was achieved with the mass resolution of the spectrometer set at a modest 500 and is in marked contrast to the situation with spectra obtained from ions having high translational energy in which the 91^+ and 92^+ peaks are not separable (see, for example, Fig. 9). The effect of photoexcitation energy on competing dissociation reactions of the slow-moving *n*-butylbenzene ions was investigated and, as for high-velocity ions, the ratio $91^+/92^+$ was found to increase as the radiation wavelength was reduced. However, the situation is somewhat confused since, in these experiments, there was no initial parent-ion selection and thus there was the possibility that the changes observed were not entirely due to the photodissociation of the molecular ions. 91^+ and 92^+ ions formed by unimolecular dissociation of excited ions following ionization may themselves absorb radiation, which could affect the relative heights of the photofragment-ion peaks. Evidence for this came from results obtained at the highest photon energy where the 92^+ photofragment-ion peak was found to be negative, indicating that more of the 92^+ ions were being lost by photodissociation than were being generated by photodissociation of the molecular ion.

Source temperature effects were also observed in the work with slow ions. Using one wavelength in the visible region of the spectrum, it was found that the ratio $91^+/92^+$ increased with increases in source temperature, confirming previous conclusions (Section VI) that the average internal energy of ions prior to photoexcitation is dependent on source temperature.

One particularly interesting experiment, which became possible because of the high efficiency with which the low-velocity ions in the ion chamber can be photodissociated, was to obtain evidence that Ar^+ ions can be photoionized using photons of 2.41-eV energy. This experiment had previously been tried with high-velocity ions without success, probably because of lower photoexcitation rates. Charge-stripping experiments (Ast *et al.*, 1972) had demonstrated that when argon atoms are ionized using 70-eV electrons, the ions Ar^+ are formed not only in the ground state but also in long-lived excited states. For ions in one of these states, 10 eV is required to form doubly charged ions but others exist in Rydberg states so highly excited that <1 eV is required to ionize them. The object of the experiment with the slow-moving Ar^+ ions was to see if those of them which are in highly excited, long-lived states could be ionized with 2.41-eV laser radiation to give doubly charged ions. The absorption of radiation was weak but sufficient to show a negative peak at mass 40 in the photoinduced spectrum, which indicated that Ar^+ ions were being lost from the beam as a result of photoabsorption. The only way in which such ions can be lost is by further ionization, which must be taking place as the result of absorption of 2.41-eV photons. This was confirmed by the presence of a small positive peak at mass 20 due to Ar^{2+}. The experiment, therefore, provides confirmatory evidence that Ar^+ ions can be formed in long-lived, highly excited states.

ACKNOWLEDGMENTS

Work done in the authors' laboratory described in this chapter has been supported by the Science and Engineering Research Council, The Royal Society, and the University College of Swansea.

REFERENCES

Ast, T. (1972). Ph.D. Thesis, Purdue Univ.

Ast, T., Beynon, J. H., and Cooks, R. G. (1972). *J. Am. Chem. Soc.* **94**, 6611.

Baldwin, M. A. (1980). *Org. Mass Spectrom.* **15**, 109.

Baldwin, M. A., Carter, D. M., and Gilmore, J. (1982). *Org. Mass Spectrom.* **17**, 45.

Beynon, J. H., and Gilbert, J. R. (1979). *Gas Phase Ion Chem.* **2**, 153.

Bowers, M. T. (1982). Private communication.

Brown, P. (1970). *Org. Mass Spectrom.* **3**, 1175.

Bruins, A. P., Ferrer-Correia, A. J., Harrison, A. G., Jennings, K. R., and Mitchum, R. K. (1978). *Adv. Mass Spectrom.* **7**, 355.

Butcher, A. R., and Thomas, C. B. (1979). *Org. Mass Spectrom.* **14**, 448.

Cooks, R. G. (1978). *In* "Collision Spectroscopy" (R. G. Cooks, ed.), Chapter 7. Plenum Press, New York.

Cooks, R. G., Beynon, J. H., Caprioli, R. M., and Lester, G. R. (1973). "Metastable Ions." Elsevier, Amsterdam.

Dunbar, R. C. (1979). *Gas Phase Ion Chem.* **2**, 181.

Dunbar, R. C. (1981). *In* "Specialist Periodical Reports—Mass Spectrometry" (R.A.W. Johnstone, ed.), Volume 6, Chapter 4. Royal Society of Chemistry, London.

Goss, S. P., McGilvery, D. C., Morrison, J. D., and Smith, D. L. (1981a). *J. Chem. Phys.* **75**, 1820.

Goss, S. P., Morrison, J. D., and Smith, D. L. (1981b). *J. Chem. Phys.* **75**, 757.

Griffiths, I. W., Harris, F. M., Mukhtar, E. S., and Beynon, J. H. (1981a). *Int. J. Mass Spectrom. Ion Phys.* **38**, 127.

Griffiths, I. W., Mukhtar, E. S., Harris, F. M., and Beynon, J. H. (1981b). *Int. J. Mass Spectrom. Ion Phys.* **38**, 333.

Griffiths, I. W., Harris, F. M., Mukhtar, E. S., and Beynon, J. H. (1981c). *Int. J. Mass Spectrom. Ion Phys.* **41**, 83.

Griffiths, I. W., Mukhtar, E. S., Harris, F. M., and Beynon, J. H. (1982). *Int. J. Mass Spectrom. Ion Phys.* **43**, 283.

Harris, F. M., Mukhtar, E. S., Griffiths, I. W. and Beynon, J. H. (1981). *Proc. R. Soc. London Ser. A* **374**, 461.

Huber, B. A., Miller, T. M., Cosby, P. C., Zeman, H. D., Leon, R. L., Moseley, J. T., and Peterson, J. R. (1977). *Rev. Sci. Instrum.* **48**, 1306.

Kingston, E. E. (1982). Private communication.

Kingston, E. E., Morgan, T. G., Harris, F. M., and Beynon, J. H. (1982). *Int. J. Mass Spectrom. Ion Phys.* **43**, 262.

Klots, C. E. (1971). *J. Phys. Chem.* **75**, 1826.

Klots, C. E. (1972). *Z. Naturforsch.* **27a**, 553.

Krailler, R. E., McLaurin, A. E., and Russell, D. H. (1982). Paper presented at *Ann. Conf. Mass Spectrom. Allied Top., 30th, Honolulu, Hawaii.*

Krailler, R. E., Wade, D. M., and Russell, D. H. (1983). *J. Chem. Phys.* (in press).

McGilvery, D. C., and Morrison, J. D. (1978). *Int. J. Mass Spectrom. Ion Phys.* **28**, 81.

Miller, T. M. (1981). *Adv. Electron. Electron Phys.* **55**, 119.

Morgan, T. G., Kingston, E. E., Harris F. M., and Beynon, J. H. (1982). *Org. Mass Spectrom.* **17**, 594.

Morrison, J. D., McGilvery, D. C., and Smith D. L. (1979). *J. Chem. Phys.* **70**, 4761.

Mukhtar, E. S., Griffiths, I. W., Harris, F. M., and Beynon, J. H. (1980). *Org. Mass Spectrom.* **15**, 51.

Mukhtar, E. S., Griffiths, I. W., Harris, F. M., and Beynon, J. H. (1981a). *Int. J. Mass Spectrom. Ion Phys.* **37**, 159.

Mukhtar, E. S., Griffiths, I. W., Harris, F. M., and Beynon, J. H. (1981b). *Org. Mass Spectrom.* **16**, 51.

Mukhtar, E. S., Griffiths, I. W., March, R. E., Harris, F. M. and Beynon, J. H. (1981c). *Int. J. Mass Spectrom. Ion Phys.* **41**, 61.

Nagy-Felsobuki, E. and Peel, J. B. (1979). *J. Electron Spectrosc. Relat. Phenom.* **16**, 397.

Rosenstock, H. M., Draxl, K., Steiner, B. W., and Herron, J. T. (1977). *J. Phys. Chem. Ref. Data, Suppl. 1* **6**.

Rossini, F. D., Pitzer, K. S., Arnett, F. L., Braun, R. M., and Pimentel, G. C. (1953). "Selected Values of Physical and Thermodynamic Properties of Hydrocarbons and Related Compounds." Carnegie Press, Pittsburgh, Pennsylvania.

Stenhagen, E., Abrahamsson, S., and McLafferty, F. W. (1974). "Registry of Mass Spectra." Wiley (Interscience), New York.

van Asselt, N.P.F.B., Maas, J. G., and Los, J. (1974). *Chem. Phys.* **5**, 429.

Wagner-Redeker, W., and Levsen, K. (1981). *Org. Mass Spectrom.* **16**, 538.

Chapter 20
Photodissociation in the ICR ion trap

ROBERT C. DUNBAR
DEPARTMENT OF CHEMISTRY
CASE WESTERN RESERVE UNIVERSITY
CLEVELAND, OHIO

ISBN 0-12-120803-6

I. INTRODUCTION

The past few years, even the brief period since the publication of Volume 2 in this treatise, have seen a blossoming and maturing of photodissociation study of gas phase ions. It is no longer necessary to point out that photodissociation offers a versatile, convenient, and fruitful source of information about the optical spectroscopy and the electronic and vibrational energy levels of ions. The number of instruments and techniques being applied has multiplied, with large bodies of results now available from ion beams, drift tubes, tandem quadrupoles, and rf quadrupole traps, in addition to the ion cyclotron resonance (ICR) work which was practically the whole story not long ago. It is now widely understood and appreciated that the spectral peaks observed in photodissociation give information about ion excited states complementary to that available from UV photoelectron spectra of neutrals, from optical emission spectra of excited ions, from laser-induced fluorescence of ions, and from various photoionization coincidence experiments.

From being practically the exclusive source of photodissociation information, the ICR ion trap has evolved to a position where its own particular strengths are being exploited. Such areas as the kinetic energy distribution of photofragments, photodissociation of cluster ions, and photodissociation spectroscopy at extreme photon-energy resolution are being most vigorously pursued by beam or drift tube techniques which are generally better suited to these purposes.

The focus here will be on illustrating the kinds of experimental information most actively sought by current practitioners of ICR photodissociation methods. Both anion photodetachment and infrared photodissociation in the ICR spectrometer are omitted from this discussion, although they are closely related, since they are fully covered elsewhere in this volume. Among recent reviews, several describing aspects of drift tube photodissociation (Moseley, 1982), ICR photodissociation (Dunbar, 1979b,d, 1980, 1981, 1982a), ion beam dissociation (Moseley and Durup, 1980, 1981), ion spectroscopy (Maier, 1980, 1982; Saykally and Woods, 1981), and infrared dissociation (Woodin *et al.*, 1979) may be interesting to the reader of this chapter.

II. METHODS

The ICR photodissociation experiment is not elaborate. The ICR ion trap contains the ions (which are usually electron-impact generated) for a period limited only by collisions: At magnetic fields around 1 T, the containment time is of the order of a few hundred ion–molecule collisions, with substantial gains being possible at higher fields. Ion trapping times of minutes are easily achieved at low pressures (around 10^{-8} torr), but most photodissociation is done with a trapping time of a few seconds.

In the usual pulsed mode, ions are generated by an electron beam pulse, trapped and irradiated, and then detected by activating a marginal oscillator or bridge detector, or by initiating a Fourier transform excite–detect sequence. The output is a record of the abundance of ions of given mass at the moment of ion detection. By varying the time of the detect pulse, the history of a chosen ion mass may be traced from the time of ion formation out to long times. Figure 1 displays a pulse sequence; chosen for illustration is the pulse sequence for one of the experiments discussed later, in which the light was delivered in the form of a train of short pulses.

The term *photodissociation spectroscopy* denotes the sort of photodissociation experiment in which a chosen parent ion is generated and irradiated for a constant length of time; the extent of dissociation is traced as a function of wavelength of the irradiating light to produce a spectrum for the photodissociation process. Since dissociation must be preceded by light absorption, we are justified in drawing appropriate conclusions about the optical absorption properties of the ion from the spectrum. Figure 2 shows an illustrative photodissociation spectrum: The ion in this case is the parent ion obtained from methyl-cyclopentadiene. This spectrum is actually of some interest, as discussed below because it is identical with that obtained from 1,3-cyclohexadiene ion, suggesting rearrangement of the methyl-cyclopentadiene ion structure to the more stable cyclohexadiene structure (Hays, 1980).

In a simple, uncomplicated ion photodissociation experiment, the photodissociation cross section of the ion is obtained from the experimental quan-

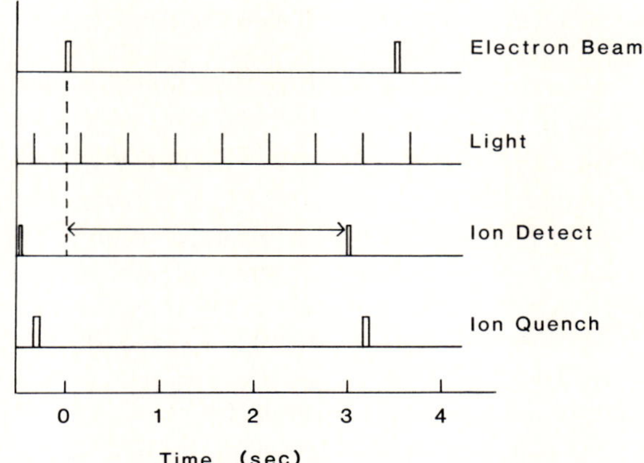

FIG. 1 A typical ICR photodissociation pulse sequence. In this case, the laser is repetitively pulsed at a rate of about 2 sec^{-1}, and the ions are trapped and irradiated for 3 sec.

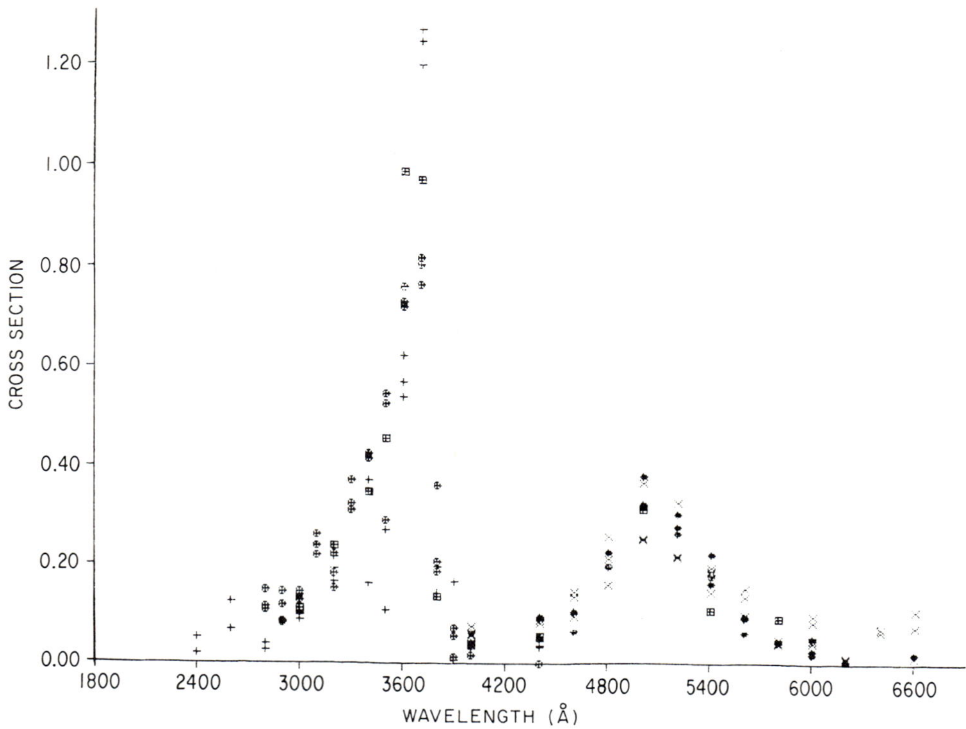

FIG. 2 ICR photodissociation spectrum showing the disappearance rate of the parent ion of methyl-cyclopentadiene as a function of wavelength.

tity D (sometimes called simply the *dissociation*), defined by

$$D = -\ln \frac{[\text{light-on signal}]}{[\text{light-off signal}]}, \tag{1}$$

where the light-on and light-off signals are the ICR signals for the chosen ion after irradiation for time t. The photodissociation cross section S is given by $S = D/(It)$, where I is the light intensity. Even when complications like multiphoton processes or interfering ion–molecule chemistry intervene, the quantity D is often reported, but must be related to more fundamental quantities by more elaborate analysis. We will discuss one illustration of this in the case of two-photon kinetics.

The term *time-resolved photodissociation* (Fu *et al.*, 1976) has been used to denote experiments in which the abundance of the irradiated ion is monitored as a function of time after turning on the light source. A homogeneous population of ions dissociating without complications in a steady light should decay by a simple exponential law, and deviations from this are useful indicators of ion inhomogeneity. Many applications of this capability are implied in the ion structure studies described in the next section. (*Time-resolved* is a term which might equally be applied to the experiment in which the dissociation of an ion following a short light pulse is followed on a time scale of microseconds or faster; but this experiment has not yet been done in the ICR spectrometer, and the term has not been used in this sense.)

Exhaustive photodissociation (Fu *et al.*, 1976) is a similar idea: A population of ions is strongly irradiated for a sufficient time that all of the susceptible ions are destroyed. If there remains a residue of undissociated ions, they are usually assigned as differing structurally from the ions which did dissociate.

Complications have often arisen when the photodissociation product ion reacts with the parent neutrals to regenerate parent ion: This will always interfere with quantitative measurements and may even make the observation of dissociation difficult. This can be obviated by working at sufficiently low pressure, but a more elegant solution often applied is to eject the product ions from the cell, using ICR double-resonance techniques, as rapidly as they are formed.

III. SPECTROSCOPY AND ION STRUCTURES

A. *General Discussion*

The aspect of ion photodissociation which has been most widely recognized and exploited is the linking of spectra and ion structures. Often the point of view is that the photodissociation spectra are a reliable fingerprint of the structures of different isomeric ions; more demanding have been efforts to assign specific structures of ions based on systematic understanding of their spectral features.

Before surveying the possibilities of these ideas, we might digress to consider whether we can make a firm and unique association between a given ion structure and an observed photodissociation spectrum.

In the first place, as has been noted innumerable times, the photodissociation spectrum does not necessarily equal the optical absorption spectrum. For one thing, if the photon absorbed has an energy below the one-photon thermochemical threshold for fragmentation, dissociation can only be seen through two-photon processes and will be much reduced or suppressed. Freiser showed an illustration of this kind of spectral distortion for $C_6H_5NOH^+$ (Reents and Freiser, 1980) (where, in fact, the one-photon threshold seemed to be blue-shifted by a small dissociation barrier). Similar truncation of the red portion of spectra was also suggested for some of the alkane ion spectra (Benz and Dunbar, 1979).

It is also possible for dissociation to be reduced if the photoexcited ion dissipates its energy by fluorescence. Several groups have enlarged our experience of such processes to the point that we know that fluorescence seldom competes with dissociation of polyatomic ions. Maier (1980) has enumerated a series of ions, such as hexatriene, octatetraene, many acetylene and diacetylene derivatives, and some smaller ions, for which both fragmentation and fluorescence are observed. Dujardin *et al.* (1980) have added a number of the fluorobenzene ions to this list. A larger number of ions are known (Maier, 1980), mostly derivatives of fluorobenzene and chlorobenzene, which fluoresce, but for which dissociation has not yet been observed: Most of the latter ions probably either do dissociate or else decay by internal conversion, but fail to dissociate for thermochemical reasons. The important point is that there are probably no known cases of polyatomic ions which fluoresce so rapidly as to suppress completely the process of internal conversion with subsequent fragmentation (if fragmentation is possible). So fluorescence effects may at worst reduce peak heights, but are not expected to suppress peaks in the photodissociation spectrum.

The above considerations all affect the interpretation of photodissociation spectra in terms of spectroscopic transitions in the ions but make no difference when the spectra are used in a fingerprint mode to characterize isomeric structures. A troublesome question which does bear on these fingerprinting applications is whether the state of internal excitation of the ion might distort its spectrum. It is certainly true that if an ion contains excess vibrational energy, hot bands will shade the peaks to the red (see, for instance, Morgenthaler and Eyler, 1981a); and van Velzen and van der Hart (1981a) have suggested that the peak intensity of $C_2H_4O^+$ dissociation is reduced by a factor of about 4 in going from cool ions to highly vibrationally excited ions. In general, the possibility of effects like these has been ignored in qualitative interpretations of photodissociation spectra—first, because they will probably not distort spectra beyond recognition and, second, because on the usual time scale of ICR photodissociation experiments, ions will relax most or all of their excess vibrational energy by collisional and infrared-radiative cooling processes.

Photodissociation spectra which show well-resolved vibrational structure will, of course, be most satisfactory in ensuring the identification of a particular ion structure from its spectrum. However, outside of very small ions, few photodissociation spectra have been obtained which show resolved vibrations. These exceptional cases include laser spectra for CH_3I^+ (McGilvery and Morrison, 1977), hexatriene ion (Dunbar and Teng, 1978), CO_3^- (Moseley *et al.*, 1976), $C_3H_5Cl^+$ (Orth and Dunbar, 1982), and $Fe(CO)_3^-$ (Rynard and Brauman, 1980). Little advantage has been taken of these possibilities, in part because of the difficulty of the experiments, and in part because many radical cations have their vibrational spectroscopic structure obscured by various combinations of spectral congestion and lifetime broadening by rapid internal conversion.

The task of relating ion structure and spectroscopy through a systematic connection between an ion's optical absorptions (reflected by its photodissociation spectrum) and its electronic structure, orbitals, and chromophores is challenging and interesting. These questions will only be touched on tangentially here; more of our own reflections on them can be found in the references (for instance, Dunbar, 1976, 1978, 1979c; Dunbar and Teng, 1979; Teng and Dunbar, 1978; Dunbar *et al.*, 1979a; Benz and Dunbar, 1979).

B. *Specific Illustrations*

The blossoming of the field of ion structure determination by photodissociation has generated a large literature, of which we will survey a recent sampling, arranged arbitrarily in order of increasing number of double bonds in the ions.

1. *Cycloalkanes*

The spectral characteristics of saturated hydrocarbon ions on one hand and alkene ions on the other are highly characteristic and strongly differentiated, so the question of whether cycloalkanes open their rings upon ionization is an especially attractive textbook-type case for ion structure distinction. It has been recognized for some time (Benz and Dunbar, 1979; van Dishoek *et al.*, 1979) that the parent ions of the alkanes are strongly colored, with their absorptions in the visible region. Cyclohexane ion and the larger cycloalkane ions show this same characteristic pattern (see Fig. 3), indicating unequivocally that they retain their ring structure upon ionization. The various alkenes, in contrast, characteristically have weak visible absorption and very strong UV peaks (Kramer and Dunbar, 1973; Dunbar, 1976; van Velzen and van der Hart, 1981b).

Recently, van Velzen and van der Hart (1981b) have looked at the smaller cycloalkane ions, whose behavior is more complicated. As suggested in reaction (2),

$$\text{cyclo-}(CH_2)_3^{+\cdot} \xleftarrow[60\%\ (13\ \text{eV})]{90\%\ (9\ \text{eV})} \text{cyclo-}(CH_2)_3 \xrightarrow[40\%\ (13\ \text{eV})]{10\%\ (9\ \text{eV})} CH_2{=}CH{-}CH_3^{+\cdot} \qquad (2)$$

electron-impact ionization of cyclopropane near threshold 9 eV yields about 90% of ring-closed ions, which are characterized by light absorption in the

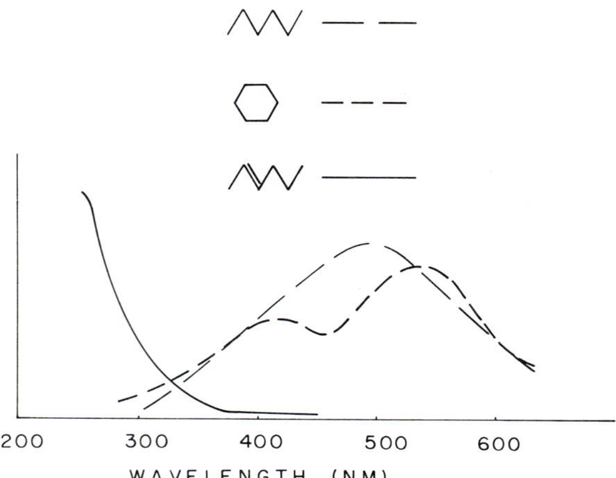

FIG. 3 Photodissociation spectra of cyclohexane, cyclohexene, and *n*-hexane parent ions.

300–500nm region. At 13-eV ionizing energy, the fraction of ring-closed ions falls to about 60%; the remaining (ring-opened) ions are recognized by their intense UV absorption below 300 nm. A fortunate chemical fractionation of the two $C_3H_6^+$ structures simplified the difficult job of distinguishing the spectra: The ring-opened (propene) isomer was found to react with neutral cyclopropane on essentially every collision while the ring-closed structure was much less reactive; the spectra of the two structures could be sorted out by comparing spectra at low pressure (3×10^{-10} torr) with the simpler spectrum arising from residual ring-closed ions at higher pressure. Since the time scale of these experiments was of the order of 20 sec, it is clear that the cyclopropane radical ion is stable against rearrangement for at least tens of seconds.

This is in contrast to the cyclobutane situation: van Velzen and van der Hart (1981b) found that, at most, a very small fraction of the ions formed had a stable ring-closed $C_4H_8^+$ structure at any electron energy or pressure. The photodissociation spectrum of the cyclobutane ionization products looked like a mixture of acyclic butene ion structures. No obvious reason comes to mind for the unexpected observation that the cyclopropane ion is easily formed and stable while the cyclobutane ion is not.

2. $C_2H_4O^+$

For such a small ion, the $C_2H_4O^+$ ion presents a rather obscure ion-structure picture. The photodissociation study of van Velzen and van der Hart (1981a) has at least partly clarified matters. It seems clear that acetaldehyde retains its structure upon ionization: While van Velzen and van der Hart did not explain their observation of no measurable photodissociation for this ion, it seems reasonable to account for it on the basis that the lowest electronic transition of this and other carbonyl-localized radical ions is forbidden by the orbital symmetries around the carbonyl. For carbonyls of low symmetry such as acetaldehyde, the $n \rightarrow \pi$ transition is forbidden by local symmetry and might be at most weakly allowed by overall broken symmetry; for ions of high symmetry such as formaldehyde, the transition should be forbidden.

The $C_2H_4O^+$ from ethyl vinyl ether likewise gave no photodissociation, from which of course one cannot deduce its structure, except to say that the ion from ethylene oxide, which gave strong photodissociation, must be different. This latter ion was proposed to have the symmetric $CH_2OCH_2^+$ structure, and the ion from 1,3-dioxolane, which gave an identical spectrum, was assigned the same structure. Reasonable (but perhaps not definitive) arguments were made against the retention of a cyclic structure in the ethylene oxide ionization: The behavior of this ion is complicated, with its apparent initial formation in a highly excited state which relaxes during the course of the photodissociation experiment. A general summary of this situation might be that photodissociation has added a good deal of hard evidence, but we may expect more chapters in this story.

3. $C_3H_4^+$

In contrast to the preceding complexities, the picture drawn from photodissociation of ions from allene and propyne is a simple and sensible one (van Velzen and van der Hart, 1981c), as diagrammed in reaction (3):

$$\begin{array}{ccccc} & \overset{100\%}{\nearrow} & \underset{(550\ \text{nm})}{CH_2{=}C{=}CH_2^{+\cdot}} & \overset{86\%}{\nwarrow} & \\ CH_2{=}C{=}CH_2 & & & & CH_3{-}C{\equiv}CH. \\ & & \underset{(300\ \text{nm})}{CH_3{-}C{\equiv}CH^{+\cdot}} & \underset{14\%}{\swarrow} & \end{array} \tag{3}$$

The ions from allene show a satisfactory peak near 550 nm and are presumed to retain the parent structure. The ions from propyne consist of about 86% unrearranged parent ions, and about 14% allene ions, based on the propyne ion isomer absorption at 300 nm. In support of this, van Velzen and van der Hart obtained a time-resolved photodissociation curve showing a nice double-exponential ion decay, the allene-ion component dissociating rapidly, and the propyne-ion component much more slowly.

4. *Alkenes*

Alkene ions were among the first whose photodissociation spectra were systematically characterized, and the alkene radical ion chromophore is a well-recognized one (Dunbar, 1976; von Velzen and van der Hart, 1981b): These ions characteristically have a very strong dissociation feature in the UV rising steeply as the wavelength decreases below 300 nm or so. This strong far-UV feature is universal among ions with an isolated double bond; but, in addition, some of the ions have been found to have a much weaker dissociation in the visible region, whose position and strength have not been very well correlated with any particular structural feature of the alkenes. Van Velzen and van der Hart (1981b) have cast some new light on this question with their very low-pressure work on the alkene systems. In hexene ions, as a test case, they found that two distinct ions were formed at electron energies between 9.5 and 25 eV. The ions denoted A were very reactive and could be observed only at very low pressure: They showed a modest photodissociation at 480 nm, in addition to the far-UV feature. (It was not explicitly shown that the A ions dissociate in the far UV, but they evidently do so since the UV intensity remains strong even when the A ions are predominant.) The other parent-ion component, the B ions, make up about 35% of the total. They have strong UV absorption but little dissociation in the visible, and their reactivity is so low that they become predominant in spectra taken at higher pressures.

This situation is found to be common among the ions of various C_5 and C_6 alkenes: Varying fractions of reactive A ions and unreactive B ions are formed by electron impact, these two classes of parent ions differing in their visible-region photodissociation properties. It is far from clear what feature of the parent ions distinguishes the A and B populations: Van Velzen and van der Hart argue convincingly that the absence of observable relaxation on the time scale of seconds rules out internal excitation as the differentiating characteristic. The most plausible remaining possibility, although one which is not very satisfactory, is that some combination of rotational isomerism about bonds and positional isomerism of the double bond accounts for the difference.

Cyclohexene shows the same pattern: In this case, van Velzen and van der Hart argue that one of the two populations (the nonreactive one) has a ring-opened structure, but this seems uncertain on the grounds that the postulated ring-opened ions have a photodissociation spectrum entirely unlike that of any known hexadiene ion; it seems more likely that for cyclohexene, as for the acyclic alkene ions, the spectroscopically distinct ion populations are distinguished by some structural distinction less dramatic than ring open versus ring closed.

5. *Chloropropenes*

The four chloropropene ions 1–4 are readily distinguishable from their spectra in the red (Orth and Dunbar, 1978, 1982), as shown in Fig. 4. Three

of the ions, **1**–**3**, show clearly resolved vibrational progressions, probably arising from a skeletal vibration involving the chlorine, appearing on a weak electronic band assigned as a $\pi^{-1} \rightarrow n^{-1}$ transition. For allyl chloride, **4**, the

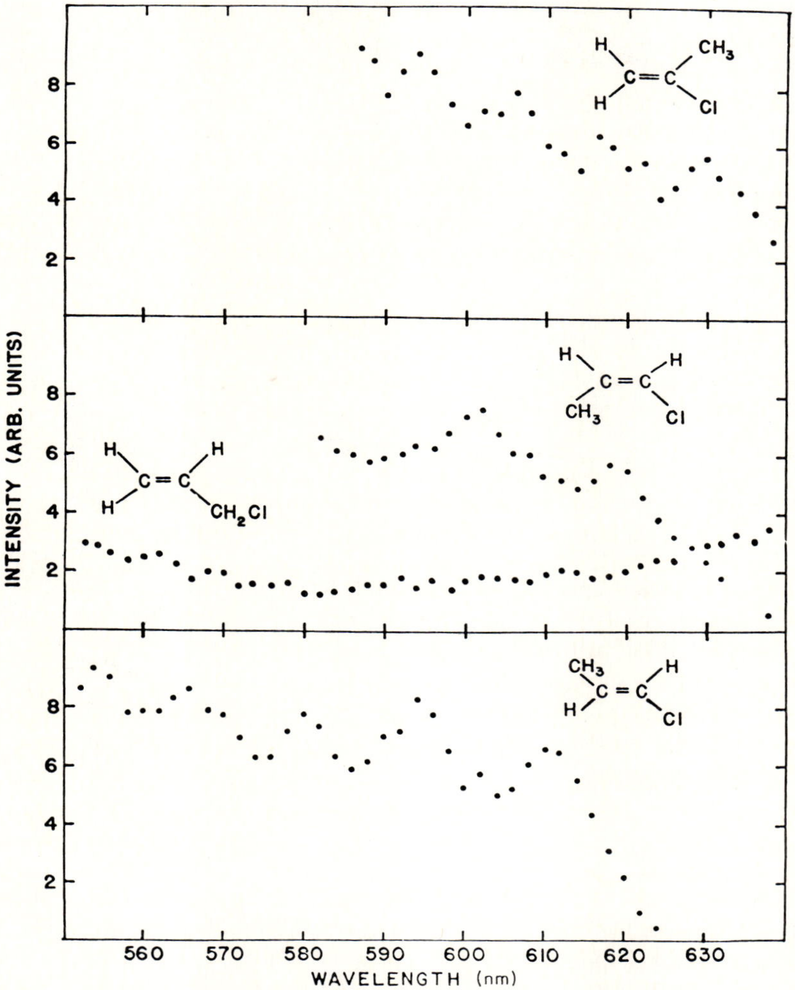

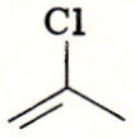

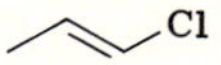

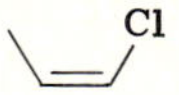

FIG. 4 Laser photodissociation spectra of four chloropropene ions [Reprinted with permission from Orth and Dunbar (1982). Copyright 1982 American Chemical Society.]

weaker overlap of the π system with the chlorine n orbitals changes the electronic structure in such a way that the $\pi^{-1} \rightarrow n^{-1}$ transition is shifted from the red to the infrared; for this ion, only a very weak, structureless spectrum is seen in the visible region.

One of the interesting features here is that the cis and trans isomers of 1-chloropropene are clearly distinct and noninterconverting. This seems to be the closest information yet available on the question of whether an isolated double bond in a radical cation is sufficient to inhibit cis-trans isomerization. [The early report (Kramer and Dunbar, 1973) of slightly different spectra for *cis*- and *trans*-2-butene ions has not stood up (van Velzen and van der Hart, 1981b), and the unsubstituted alkenes have proven too intractable for such a fine spectroscopic distinction to be made.] While it might be argued that interaction with the chlorine strengthens the π system in the chloropropene case, still this is fairly good evidence that removing one of the two π electrons does not permit free rotation about the double bond.

The excellent spectra of the chloropropene ions also made possible some detective work on the products of the ion–molecule reaction (Orth and Dunbar, 1982):

Cl + $C_3H_6^{+\cdot}$ ⟶ $Cl^{+\cdot}$ + C_3H_8 (4a)

$h\nu$

$C_3H_5^{+}$ + Cl.

Cl $h\nu$ $Cl^{+\cdot}$

+ $C_3H_6^{+\cdot}$ ⟶ + C_3H_8 (4b)

Photodissociation spectra of the product $C_3H_5Cl^+$ ions showed that the carbon skeleton of the initial chloropropane ends up in the ionic product. For reaction (4a), with 1-chloropropane as reactant, the product ion's spectrum looked like a mixture of *cis*- and *trans*-1-chloropropene ions while, for reaction (4b), the products gave the spectrum of the 2-chloropropene structure. Thus, the reaction behaves (in a formal sense) as an H_2^- abstraction from chloropropane.

6. *Dienes*

The diene radical ion chromophore is among the most intense and easily recognized (Dunbar, 1976). These ions universally show a pair of peaks: One in the visible around 500 nm with cross section of the order of 5×10^{-18} cm^2, and one in the UV around 350 nm with cross section typically an order of magnitude larger. Some recent work on some of the simpler dienes has expanded the kinds of information about these ions that can be extracted from the photodissociation spectra.

The spectra of the various isomeric hexadiene ions (Benz *et al.*, 1981) provide one of the most satisfactory examples of the ability of photodissociation spectra to give a straightforward mapping of ion structure/rearrangement chemistry. The rearrangement pattern shown in reaction (5),

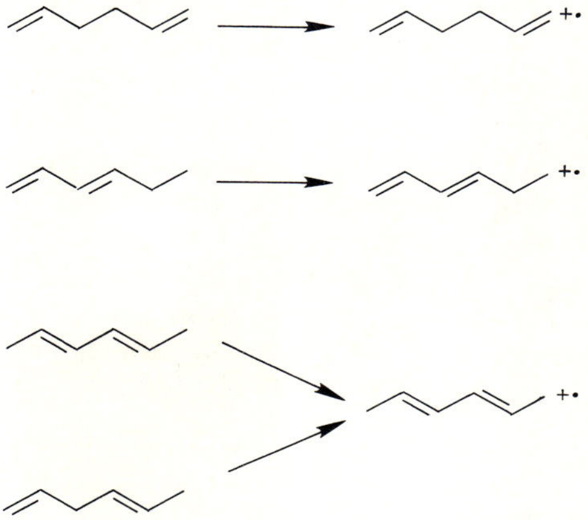

+•

+•

(5)

+•

is clear from the observation that 1,5-hexadiene ion has a characteristic alkene-type spectrum whereas the spectrum of the ions obtained from 1,4-hexadiene is identical with that from 2,4-hexadiene. Both 2,4-hexadiene ions and 1,3-hexadiene ions give spectra which match the characteristic pattern for conjugated diene ions, but there is enough difference between them that complete rearrangement to the more stable 2,4-hexadiene structure can be ruled out for the 1,3-hexadiene ions.

More subtle is the question of cis–trans isomerization in these ions. The three cis–trans isomers of 2,4-hexadiene ion are just barely distinguishable from their low-resolution spectra, but the laser photodissociation at about 1-nm resolution resolves the three structures comfortably (Fig. 5). Observing clearly distinct peaks for these three isomers rules out their interconversion by rotation about the double bonds (which is far easier than in the neutral diene). Molecular orbital calculations prove to be useful in predicting the peak shifts due to cis–trans isomerism in these ions: The ordering of the photodissociation peaks for the three cis–trans isomers and the spacing between them of about 7 nm is reproduced exactly by a Koopman's theorem calculation using MINDO/3.

Separate cis–trans isomers of 1,3-hexadiene have not been studied, but the visible-light spectrum of the ions from a mixed-isomer sample shows two peaks, which may well correspond to the two noninterconverting isomers.

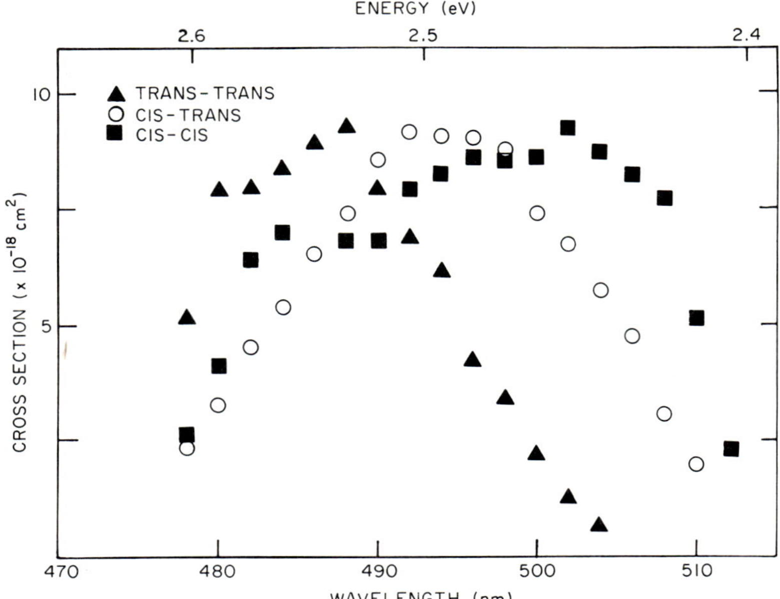

FIG. 5 Laser photodissociation spectra of the three cis-trans isomers of 2,4-hexadiene parent ion. [Reprinted with permission from Benz *et al.* (1981). Copyright 1981 American Chemical Society.]

The visible-region spectrum of the 1,4-hexadiene sample is also interesting in showing three peaks, which suggest the formation of the trans–trans, cis–trans, and cis–cis isomers of 2,4-hexadiene ions in a ratio of 3:2:1 when the initially formed 1,4-hexadiene ions rearrange by bond migration to the conjugated structure.

The ions from the octadienes show an entirely similar pattern (Dunbar and Fitzgerald, 1984). The 1,7- and 2,6-octadienes yield ions with characteristic alkene spectra, indicating no rearrangement, while the 1,4- and 2,4-ions yield characteristic diene spectra, indicating rearrangement of the 1,4-structure into conjugation. The 1,3-octadiene ion spectrum is similar to but distinguishable from these latter spectra.

7. *Cyclic Dienes*

The photodissociation peaks of the typical open-chain diene ion are at 345 nm and 490 nm. Cyclizing seems to give a significant red shift to the UV peak while the effect on the visible peak is unpredictable (Hays, 1980).

Thus, 1,3-cyclohexadiene has its peaks at 360 nm and 489 nm, compared with about 325 nm and 495 nm for an average of hexadiene ion isomers. 1,3-cyclooctadiene shows peaks at 360 nm and 560 nm, compared with 330 nm and 500 nm for the straight-chain octadienes. Curiously, the UV peak of 1,3-cyclohexadiene is in exactly the same place (although only half as intense) as the UV peak of 1,3,5-hexatriene ion; this must, however, be coincidental, since the red shift on going from diene to triene is well understood on the basis of the lengthened π system. There is no possibility that the cyclohexadiene ring opens to the triene structure, since the spectra of these two ions are completely orthogonal in the visible region.

These highly characteristic spectra make isomer distinctions very easy in the cyclic diene ions. Both 1,4-cyclohexadiene and methylcyclopentadiene give parent ions having spectra identical in peak positions and intensities with the 1,3-cyclohexadiene spectrum, and there is no doubt that in both cases the ions are rearranging to the most stable 1,3-cyclohexadiene structure: This rearrangement is downhill by about 40 kcal/mole in the methylcyclopentadiene case. 1,4-cyclooctadiene ion similarly rearranges completely to a 1,3-cyclooctadiene structure [reaction (6)]:

10 eV
+•
collision
(6)
10 eV
+•

1,5-cyclooctadiene, on the other hand, yields a spectrum that looks like a mixture of 1,5- and 1,3- isomers; recent work with a gated light source has shown that the absorption due to the 1,3-isomeric component of the mixture decreases with lower pressure and shorter reaction time. Apparently, this is a case similar to that of benzyl chloride, in which the parent ion is formed by electron impact with unrearranged parent structure and then undergoes a collision-induced rearrangement to the more stable structure.

8. Hexatriene

The parent ion of 1,3,5-hexatriene occupies a special place in the evolution of ICR photodissociation spectroscopy because it provided the first spectrum at laser resolution showing extensive resolved vibrational fine structure: This is still one of the richest of recorded large-ion spectra (Dunbar and Teng, 1978). Resolution sufficient to resolve vibrations makes very fine structure comparisons possible, and, in the hexatriene ion case, the resolution on the 0–0 transition at 631 nm is good enough to distinguish the cis and trans isomers clearly, using the 1.1-nm difference in peak position known from fluorescence work (Allan and Maier, 1976). To our surprise, the spectra of ions obtained from cis and trans isomers were identical and definitely gave the 0–0 peak expected for the cis isomer; this is a striking contrast to what is observed in fluorescence, where the emission spectra of the isomers are clearly different. Morgenthaler and Eyler (1981a) have used their powerful pulsed-laser pulsed-ICR techniques to show that even 10 msec after ionization, the ions obtained from the cis and trans isomers already give identical spectra. The most sensible picture to explain this is that the trans ions rearrange to the cis structure in a time long compared to fluorescence (10^{-8} sec) but short compared with the ICR measurement at 10^{-2} sec. As the coincidence results of Allan *et al.* (1980) make clear, the decay processes of these ions are very complex, and such a relatively slow rearrangement might not be surprising.

9. *Benzyl Chloride*

The structural study of the ion generated from benzyl chloride has turned out to be a complex and puzzling question focusing the development of new techniques and ideas. The first observations (Fu *et al.*, 1976) used the original time-resolved photodissociation approach, in which a population of ions was built up in the ion trap and was then followed in its dissociation as a function of time until either all the ions were dissociated or those remaining were photoinert. Benzyl chloride was one of the first cases found in which part of the ion population did not photodissociate, and it was concluded that electron-impact ionization forms a mixture of structures, 30% of which (probably unrearranged parent ions) dissociated rapidly at 600 nm and 70% of which (perhaps having a chlorotoluene structure) were photoinert at this wavelength.

This attractive picture was overturned by Morgenthaler and Eyler's (1981b) observation that, in their pulsed-laser instrument, the ions behaved as a homogeneous population which dissociated cleanly at 600 nm. The violent contradiction between these two apparently valid experiments was resolved with a new set of experiments (Honovich and Dunbar, 1982a) using pulsed-ion production followed by dissociation with gated lasers at varying times after ion formation. It was found that the two ion structures are indeed present but are not both formed by the initial electron impact. The chemistry is as shown in reaction (7):

$$\text{C}_6\text{H}_5\text{CH}_2\text{Cl} \xrightarrow[\text{impact}]{\text{electron}} \text{C}_6\text{H}_5\text{CH}_2\text{Cl}^{+\cdot} \xrightarrow{\text{collisions}} \text{Rearranged parent ion}; \quad \text{C}_6\text{H}_5\text{CH}_2\text{Cl}^{+\cdot} \xrightarrow{h\nu\ (\text{red})} \text{Fragments}; \quad \text{Rearranged parent ion} \xrightarrow{h\nu\ (\text{blue})} \text{Fragments} \tag{7}$$

Electron impact forms the red-absorbing ions observed in both previous studies. At high pressures and long times, a collisional process converts these into ions of different structure which dissociate in the blue but not in the red, requiring about 50 collisions for the conversion to happen. Morgenthaler and Eyler, working at short times and modest pressures, naturally saw no collisionally converted ions.

10. $C_6H_6O^+$

A major study (and a notable collaboration) (van Velzen *et al.*, 1981) has untangled much structural chemistry in the $C_6H_6O^+$ system. Two ion structures, the phenol structure **5** [see reaction (8)]

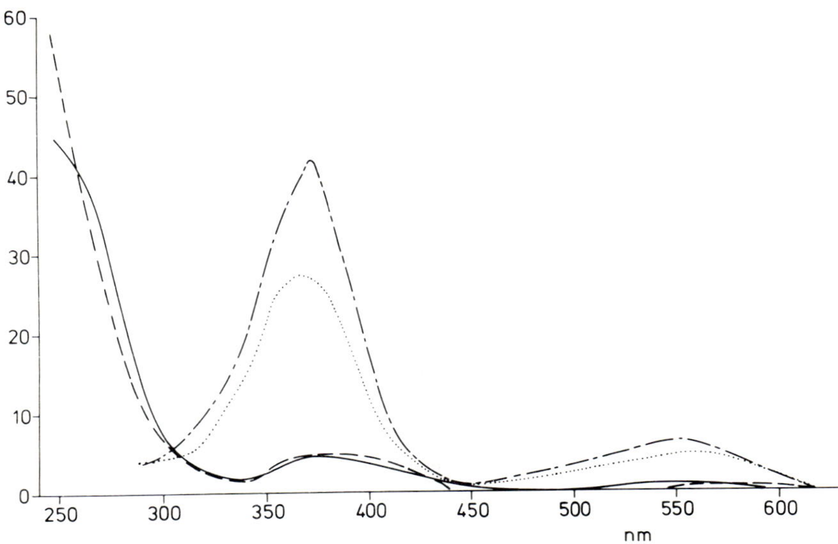

FIG. 6 Photodissociation spectra of the *m/e* 94 ions from phenol (—), phenetole (---), bicyclo(2.2.2)oct-2-ene-5,7-dione (····), and 2-phenoxyethyl chloride (– · –). (The spectra of the ions from phenol and phenetole have been multiplied by a factor of 5.) [Reprinted with permission from van Velzen *et al.* (1982). Copyright 1982 American Chemical Society.]

and the dienone structure **6**, were clearly identified by their photodissociation spectra—structure **5** giving characteristically strong far-UV dissociation and structure **6** giving strong near-UV and weaker visible peaks, as shown in Fig. 6. The *m/e* 94 ion was generated from a number of neutral precursors, and the extents of production of the two ion isomers were found to be as shown schematically in reaction (8). The evidence was against slow tautomerism of **5** and **6**, indicating that upon ionization a given molecule decides rapidly between possible ion isomers.

11. Nitrosated Ions

The group at Purdue has made an extensive and ground-breaking study of ions containing the NO^+ group. An interesting result was the observation of quite distinct spectra (as well as distinct ion–molecule chemistry) for the $C_6H_6NO^+$ ions obtained alternatively by protonation of nitrosobenzene (**7**) or by nitrosation of benzene by NO^+-donating ions (**8**) (Reents and Freiser, 1980). The protonated nitrosobenzene ion was shown by deuterium-labeling to be protonated on the NO group, as indicated by structure **7**, rather than on the ring, as suggested in structure **9**. The nitrosated benzene ion, on the other hand, is probably a π complex, representing the arrested first step in the solution nitrosation of aromatics.

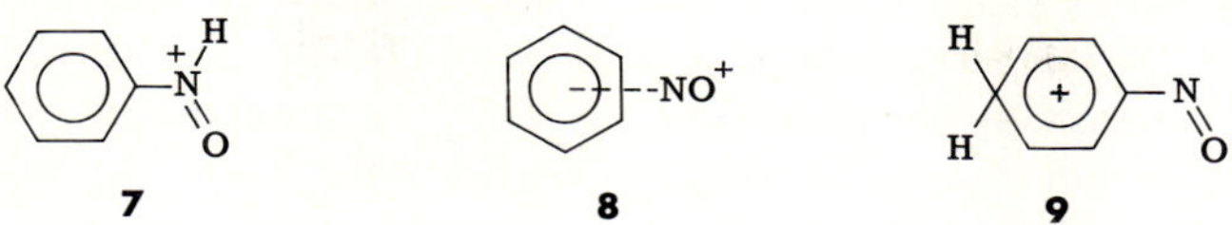

This latter ion was found to be representative of a class of ions obtained by nitrosation of various organic bases (Reents and Freiser, 1981). The spectra of many of these NO^+ complexes were obtained and discussed systematically in terms of a charge–transfer transition in the complex. The charge–transfer spectra were very similar for the complexes formed within a given class of bases but differed between classes. An interesting and useful distinction was discovered between the spectra of complexes involving similar molecules, but differing in the site of the charge. Thus, the spectrum of the 3-pentanone complex $(C_2H_5)_2CO{\cdot}NO^+$, where the charge lies on NO, showed strongest absorption in the far UV while the di-isopropyl ketone complex $(i\text{-}C_3H_7)_2CO^+$ NO, where the charge is on the ketone, gave a very different spectrum, with a maximum around 360 nm. This suggests a capability which may be unique to photodissociation spectroscopy for probing the site of charge in a complex ion.

C. *Spectroscopy of Acrolein Derivatives*

The spectroscopy of gas phase organic ions does not yet approach the depth of systematic understanding that has been built up with neutrals. For radical ions the multitude of low-lying states makes the problems hard, but with closed-shell ions we can hope to carry over much accumulated understanding from organic spectroscopy. A suggestion that this may be fruitful comes from results on some protonated derivatives of acrolein. The spectroscopy of the neutral α,β-unsaturated ketone chromophore has been extensively worked out, and the effects of substituents on the UV peak positions are well understood. It seemed likely that protonation would not have a disruptive effect on the electronic properties of the π system, since oxygen protonation involves interaction with the nonbonding O electrons in a nodal region of the π system.

Freiser and Beauchamp (1977) pointed out that the photodissociation spectra of some protonated molecules closely resemble the UV spectra of the neutrals, and they argued that the electronic character of the transitions was the same except for an energy shift due to the proton. The effect of the proton was understood as arising from different degrees of perturbation of the spectroscopic ground and excited states by the proton; using a Forster cycle argument, they interpreted the peak shift on protonation in terms of a quantity ΔPA, the difference in proton affinity between ground and excited states. Given this basis for expecting the spectroscopy of neutrals and their protonated counterparts to be similar, we investigated the spectra of a series of closely related acrolein derivatives to see if the spectroscopic substituent effects could be understood in the same way as in the neutral spectra (Honovich and Dunbar, 1981).

Table I shows the observed positions of the UV peaks in the series of neutrals and their protonated counterpart ions. Also tabulated is the peak

TABLE I

Observed Positions of UV Peaks in the Series of Neutrals and Their Protonated Counterpart Ions

	λ_{max}		
	Ion	Neutral	ΔPA(eV)
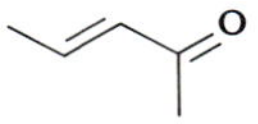	262	210	1.17
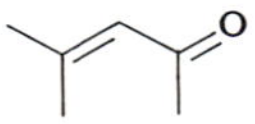	280	224	1.11
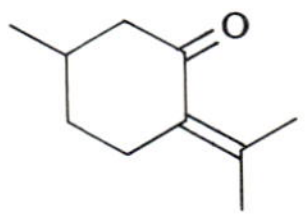	308	237	1.20
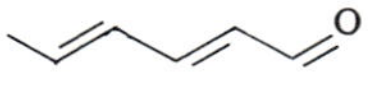	340	246	1.39
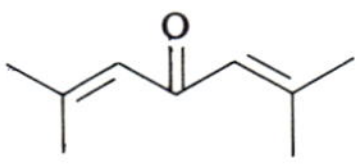	339	248	1.34
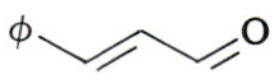	378 410	262	1.45 1.68
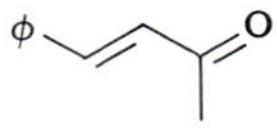	392 410	265	1.52

shift expressed as ΔPA. The substituent effects in the ions are seen to follow the same general pattern as in the neutrals. If we group the ions into three classes—(1) the simple acrolein derivatives, (2) those with extended conjugation, and (3) those with phenyl conjugation—it is seen that within each class the ΔPA values are very close. It appears that the substituent effects are indeed quite independent of protonation of the chromophore and that the effect of the proton "label" is to give an overall red shift which is constant within molecules of a given type. If the promise of these results is borne out, it may become possible to use photodissociation spectra of protonated molecules to make the same sort of structural distinctions as have been made using the systematic spectra of the neutral chromophoric molecules. In a slightly lighthearted effort to emphasize these possibilities, we presented a set of "Woodward–Fieser" parameters for the ion substituent effects analogous to the well-known parameters for the neutrals (Honovich and Dunbar, 1981).

IV. REARRANGEMENTS

Since we have in hand a powerful tool for examining ion structures, it is interesting to give some attention to the various rearrangement reactions that ions undergo. Much of what we have learned is spelled out in the ion-by-ion accounts above, but here we can summarize our understanding.

It is almost always found that, upon electron impact, a double bond will migrate by one carbon position along a carbon chain to come into conjugation [as, for instance, in reaction (5)]. In addition to examples discussed above for pentadienes, hexadienes, octadienes, cyclohexadienes, and cyclooctadienes, we can note the further case of **10** to **11** (Fu and Dunbar, 1978). The only counterexample so far reported is the substituted styrene **12**, which rearranges into conjugation only partially if at all.

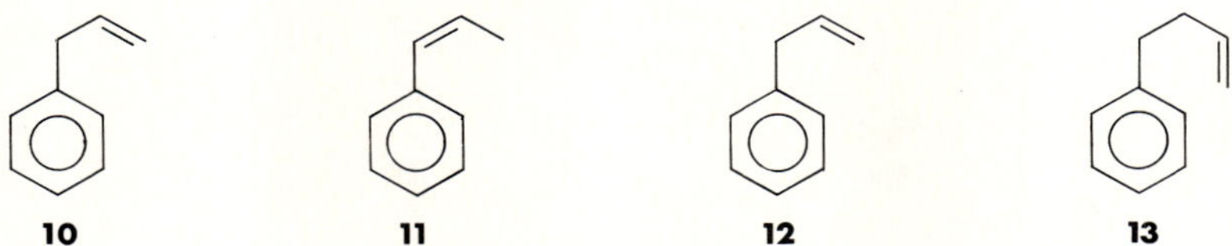

On the other hand, electron-impact ionization does not guarantee migration of a double bond by two or more positions into conjugation. The straight-chain 1,5- or 2,6- dienes show no sign of bond migration, nor does the substituted styrene **13**. 1,5-cyclooctadiene appears also to follow this generalization, although its picture is somewhat clouded by the collision-induced rearrangement mentioned above.

By and large, other skeletal rearrangements seem not to occur as often as one might expect upon electron-impact ionization to form parent ion. This is an important contrast with the mass spectroscopy of hydrocarbon molecules, where the observation of rearrangement prior to fragmentation is very common. The fundamental difference, of course, is that in ICR photodissociation studies, only those ions are sampled which were formed with *insufficient* internal energy to fragment, while in conventional mass spectroscopy only those ions are characterized which were formed with *sufficient* excitation for fragmentation. Among the ions which might be expected to rearrange to more stable structures, but which are indicated by photodissociation to retain their parent structure, we can mention cyclopropyl benzene (Fu and Dunbar, 1978); cyclooctatetraene and barrelene (Dunbar *et al.*, 1979b); the cycloheptatriene, norbornadiene, and methylenecyclohexadiene structures of $C_7H_8^+$ (Dunbar and Fu, 1973; Dunbar and Klein, 1977); benzyl chloride, bromide, and fluoride (Fu *et al.*, 1976); and cyclohexadiene (Hays and Dunbar, 1979), as well as the cis-trans isomers of hexadienes and chloropropenes.

Some of what is known about cis–trans isomerism in radical cations was discussed above. A classic test system for problems of this sort is the cis and trans stilbenes, **14** and **15**, and recently Gooden and Brauman (1982) have succeeded in obtaining excellent photodissociation spectra of the parent ions from both isomers. The spectra of the two were identical, both showing a strong, sharp peak at 460 nm and a broad, weak absorption in the red. For these ions, high-quality matrix spectra are also available, and comparison shows that the gas phase ions have the trans structure: The peak in the matrix spectrum of **15** is similar to the photodissociation spectra (with a matrix shift of 10 nm to the red) while the matrix spectrum of **14** is quite different. Clearly, the cis ion rearranges to trans in the gas phase.

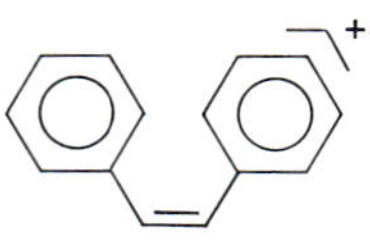

+•

14

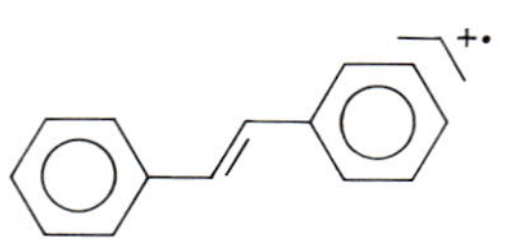

+•

15

V. FRAGMENTATION PRODUCTS

The study of ions by mass spectroscopy is based, to an overwhelming extent, on the varying patterns of fragmentation products observed for different excited ions. So, it may be surprising that fragmentation patterns in photodissociation have played a minor role in the uses found for this technique. Partly, this is a question of opportunities not yet grasped: Fragmentation patterns of ions with well-controlled internal energy, such as photon excitation provides so conveniently, will eventually be a fruitful source of new information about the details of ion-fragmentation mechanisms. Partly, also, this reflects a characteristic of photon excitation—i.e., in a typical experiment, the photon energy gives the ion only 1 or 2 eV of internal energy above the fragmentation threshold and for many (but not all!) ions, the lowest-energy fragmentation channel is predominant at these low excitation levels. Thus, we make the generalization that photodissociation usually gives only the most stable products.

There are some interesting exceptions. Butene ion loses $H^{\cdot}$ and $CH_3^{\cdot}$ with nearly equal activation energy (Riggin *et al.*, 1976). A wavelength-dependent study of this competition showed the applicability of statistical (quasi-equilibrium theory) concepts to the photodissociation chemistry, and gave an early impetus to exploiting the precise energy control of photon excitation for controlled-ion fragmentation studies. Among other ions that have shown comparable yields of two or more different product ions at ordinary photon energies, trifluorotoluene and para-fluoro-trifluorotoluene both give an array of different product ions with near-UV photons (Cody and Freiser, 1979).

Cassady *et al.* (1981) reported a striking example of the dependence of fragmentation on parent-ion structure: They prepared $C_7H_7O^+$ ions from several precursors with benzaldehyde, cresol, and tropone structures. The ion generated from benzaldehyde gave a spectrum with distinctive peaks at 300 and 370 nm, and gave m/e 79 ($C_6H_7^+$) as the only product. The ions showing this behavior were assigned structure **16**. The parent ions from the other precursors gave only a peak at 300 nm and gave product ions at both m/e 79 and m/e 91 ($C_7H_7^+$). This behavior was assigned to structure **17**. In addition, nondissociating ions were observed in most cases, which were assigned yet a third structure **18**.

H–C⁺–OH (attached to benzene ring) **16**   OH (on tropylium ring, +) **17**   HO–C₆H₄–$\overset{+}{C}H_2$ **18**

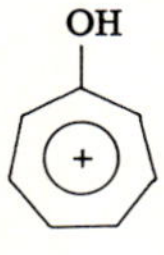

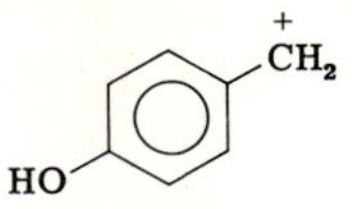

One of the characteristics of ion trap photodissociation experiments, distinguishing them from ion beam experiments, is the high degree of thermalization of the ion population. A comparison of photodissociation of a representative ion, *n*-butylbenzene parent ion, in the ICR ion trap and in a fast ion beam brings out the way this affects the sorts of fragmentation patterns observed. Beynon's group (Griffiths *et al.*, 1981a–c; Mukhtar *et al.*, 1981a,b) has photodissociated ions formed by 70-eV electron-impact ionization, and they observe comparable yields of *m/e* 91 and 92 fragments. Their RRKM (Rice–Ramsperger–Kassel–Marcus) calculations, assuming ions with internal energy up to 1.4 eV, agree with the observations. Photodissociation in the Fourier transform ICR instrument gives very different results (Chen and Dunbar, 1984). At low light intensity, *m/e* 92 is the only product at all wavelengths; with more light, *m/e* 91 is formed by the further dissociation of the *m/e* 92 ion, according to the scheme

$$m/e\ 134 \xrightarrow{h\nu} m/e\ 92 \xrightarrow{h\nu} C_7H_7^+\ (m/e\ 91) \tag{9}$$

The contrast between the beam results and the ICR results is entirely understandable on the basis of the higher internal energy of the ions in the beam. In the ICR experiment, the ions are near thermal—first, because they are formed at an electron energy near threshold and, second, because they have ample time to relax by collisional and radiative relaxation processes on the time scale of the dissociation experiment (1 sec).

The most intriguing prospects of multiple-fragment ion dissociations are raised by those cases where the products are wavelength-dependent because of dissociation from excited electronic states. Morgenthaler and Eyler (1979) found that dissociation of $C_2H_5Cl^{+\cdot}$ yields the two products corresponding to loss of $Cl^{\cdot}$ and of HCl from the parent. The branching ratio was strongly wavelength-dependent and was strikingly at odds with the predictions of RRKM theory, giving strong evidence that energy equilibration within the ion does not precede dissociation. One or more excited electronic states of the parent ion are certainly involved directly in the dissociation.

Brauman's group (Gooden and Brauman, 1977; Rynard *et al.*, 1982) have reported an exciting set of results on photodissociation of butyrophenone ion and related ions: Two products are observed in different wavelength regions, and the two fragmentation paths can be rationalized in a very convincing way in terms of ground and excited electronic state dissociation paths. Since there is a degree of uncertainty about the experimental interpretation (Kim *et al.*, 1978), we will just note here the very attractive prospect that photochemical understanding at this level may begin to evolve for ions as it has already for many neutral organic molecules.

VI. RELAXATION STUDIES USING TWO-PHOTON DISSOCIATION

A. *Two-Photon Dissociation Quenching*

Freiser and Beauchamp (1975) opened a wealth of new possibilities for ion photochemistry with the discovery that continuous-wave (CW) light of modest power could drive two-photon ion dissociation according to the sequential kinetic scheme

$$A^{+} \xrightarrow[IS_1]{h\nu} A^{+\dagger} \xrightarrow{\text{fast}} A^{+*} \xrightarrow[IS_2]{h\nu} \text{Fragments} \qquad (10)$$

(return paths from A^{+*} to A^{+}: k_cP and k_r)

where $A^{+\dagger}$ is an electronically excited ion; A^{+*} is a vibrationally excited ground-state ion; I is the light intensity; S_1 and S_2 are optical absorption cross sections; P is the neutral pressure; k_c is the bimolecular collisional quenching rate constant; and k_r is the infrared radiative relaxation rate constant. The photon absorption cross sections S_1 and S_2 need not be equal, but it has often been assumed that since they both refer to the same electronic excitation they will not differ by a large factor; this greatly simplifies interpreting experiments, but one must be cautious not to draw conclusions which would be sensitive to a very possible difference of a factor of 2 or 3. On the other hand, van Velzen and van der Hart (van Velzen, 1981) have shown that, given a sufficient amount of high-quality kinetic data, an analysis separating S_1 and S_2 is possible. For bromobenzene ions, these were calculated to be within a factor of 2 of equality; for benzene ion, they were calculated to be within perhaps a factor of 4 different (with large uncertainty); for cyanobenzene ion, however, they appeared to differ by a factor of 25.

Our interest in this two-photon chemistry has centered on the access it gives us to the two relaxation processes: the collisional relaxation rate constant, represented by k_c, and the radiative relaxation rate constant, k_r.

If we neglect radiative relaxation for the moment, the extent of dissociation after time t is easily written from the solution of the kinetic scheme (10) (see Dunbar and Fu, 1977):

$$\frac{1}{D} = \frac{IS_1 + IS_2 + k_cP}{I^2S_1S_2t}. \qquad (11)$$

Since $1/D$ appears as a linear function of pressure, we can find the slope and intercept of a plot of $1/D$ versus P, giving two equations for the unknowns k_c and IS (assuming $S_1 = S_2 = S$). The collisional relaxation rate for photoexcited bromobenzene ions was measured in this way using a number of different gases as the collisional quenching partner (Kim and Dunbar, 1979). More recent experiments on iodobenzene ion (Lev and Dunbar, 1982a) incorporate an improvement: By using ICR line-broadening techniques to measure the ion-neutral collision rate, and by using two-photon relaxation techniques to measure the collisional relaxation rate in the same instrument, the relaxation rate can be compared with the collision rate free of uncertainties in pressure measurements and in collision-rate calculations from theory. Table II shows the values obtained.

The most efficient quenching is provided by the parent neutral, which takes two or three collisions for iodobenzene ion; this is less efficient than the quenching measured for bromobenzene ion in its parent neutral, where quenching was found to occur as fast as the orbiting collision rate. (The bromobenzene quenching may be enhanced by rapid charge transfer; however, the measurements need to be repeated in this system incorporating concurrent line-broadening measurements.)

Assuming that energy transfer is efficient only in orbiting ion–molecule collisions, the extreme limit of efficient quenching will occur when the internal energies of the ion and neutral are fully equilibrated within each orbiting collision complex. Making the simple calculation based on assuming that the iodobenzene ion and the neutral reach a common vibrational temperature at each collision, we may find the number of collisions needed to bring the iodobenzene ion from its initial photoexcited energy of 2 eV down to below 0.4 eV, which is the energy at which one-photon dissociation is just possible. Calculating this equilibration limit, iodobenzene bath gas requires 2–3 collisions, cyclohexane about 2 collisions, and methane about 15 collisions. From Table II, it is seen that parent neutral as bath gas matches this full equilibration limit. This is not surprising, since the match of vibrational levels, along with the possibility of efficient symmetric charge exchange, make efficient energy transfer very likely within the collision complex. On the other hand, quenching by either cyclohexane or methane bath gases is about a quarter as fast as the full equilibration prediction: These collision complexes are far from reaching statistical sharing of internal energy between ion and neutral.

These quenching results are in line with the findings of Brauman's group, looking at collisional quenching in the multiphoton infrared dissociation of anions. Collisional relaxation of CF_3O^- requires between 7 collisions (CF_2O neutral) and 30 collisions (N_2 neutral) (Jasinski and Brauman, 1980); collisional quenching of CH_3OHF^- in $HCOOCH_3$ requires tens of collisions (Rosenfeld *et al.*, 1979). Radiative quenching of CF_3O^-, on the other hand, was quite fast, at 20 sec^{-1}: We will see in the next section that the larger ions we have studied radiate their energy considerably more slowly.

TABLE II

Values Obtained by Using ICR Line-broadening Methods and Two-Photon Relaxation Techniques

Bath gas	Measured momentum transfer collision rate (10^{-10} cm^3 molecule^{-1} sec^{-1})	Derived orbiting collision rate (10^{-10} cm^3 molecule^{-1} sec^{-1})	Measured laser quenching rate (10^{-10} cm^3 molecule^{-1} sec^{-1})	Collisions needed for quenching
C_6H_5I	10.0	18.0	7.0	2.5
CH_4	1.6	19.0	0.4	50.0
cyclo-C_6H_{12}	2.6	7.7	1.1	7.0

B. *Pulsed-Laser Two-Photon Experiments*

When the light source is not continuous but delivers light in spaced pulses, an interesting interplay develops between the occasional photoexcitation of the ions and their steady relaxation by collisions and by infrared fluorescence. Our group (Lev and Dunbar, 1981) explored the effect of changing from continuous to repetitively pulsed radiation in the two-photon dissociation of iodobenzene ion. The effect is simply understood in a qualitative way: If the laser pulse repetition rate is fast compared with all the ion relaxation rates, then the light source excites ions just as if it were actually continuous and the time-average light intensity is the important variable. But if the pulse rate becomes slow compared with the fastest relaxation, then the ions can relax between each pulse and the experiment is equivalent to a series of relaxation-free single pulse experiments for which the pulse energy and number of pulses, but not the pulse shape or repetition rate, are the important variables.

Recently, results of higher precision have been obtained for bromobenzene ion, using a cw laser chopped at rates varying from 1–12 sec^{-1}, and the radiative and collisional parts of the relaxation have been clearly separated (Dunbar, 1983). Using a chopper, the average light intensity is conveniently kept constant, while the repetition rate and pulse length are varied, so that the dissociation should be pulse-rate-independent at high pulse rates and low relaxation rates. Figure 7 shows the strong dependence of dissociation on pulse rate at rather high pressure, compared with the theoretical curve calculated using the collisional relaxation rate already determined from the quenching studies. At high pulse rates (4 sec^{-1} and above) the dissociation is strongly pressure-dependent in the $2 - 30 \times 10^{-8}$ torr region. However, at 1 sec^{-1}, there is little pressure dependence, and it is clear that, at this low repetition rate, relaxation by infrared radiation is largely completed between each pulse. At 2 and 3 sec^{-1}, the pressure dependence is intermediate, indicating that at these pulse rates radiative relaxation is partial, allowing some collisional relaxation in addition to the radiative relaxation. Using a radiative relaxation rate of around 2 sec^{-1}, the basic kinetic scheme [reaction (10)] gives a good account of the pressure and pulse rate dependences in this experiment, and we believe this firmly establishes the radiative rate k_r in the range 1–3 sec^{-1}.

C. *Threshold Photodissociation*

Van Velzen and van der Hart (van Velzen, 1981) realized that the visible photodissociation band of bromobenzene ion spans the one- to two-photon threshold, and showed that as the threshold region is traversed, between 400 and 600 nm, the pressure dependence of the kinetics changes from the

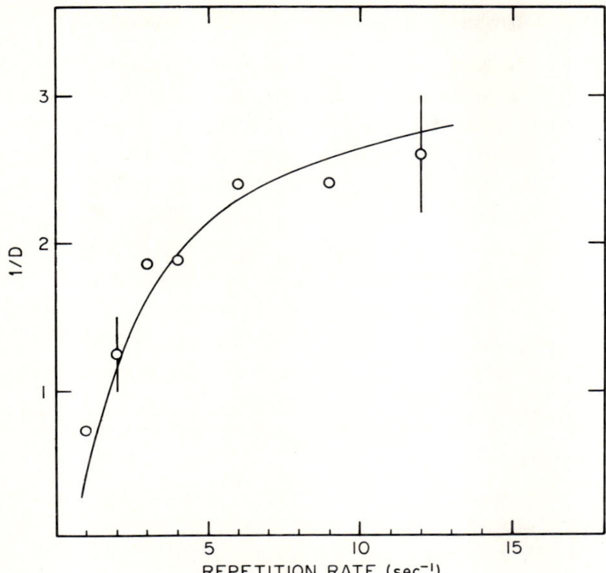

FIG. 7 Photodissociation of bromobenzene ion as a function of pulse rate at 2×10^{-7} torr. The solid line is the theoretical curve assuming a radiative relaxation rate of 1.5 sec^{-1}. [Reprinted with permission from Dunbar (1983). Copyright 1983 American Chemical Society.]

strong dependence expected for a two-photon process to the pressure-independence characteristic of one-photon dissociation. Although they proposed a picture of this chemistry involving two different kinds of excited intermediate $C_6H_5Br^+$ ions, we believe that the kinetics in the transition region can be accounted for more simply in terms of a mixed population of parent ions: At wavelengths in the threshold region, the thermal spread of ion internal energies will give some ions which are above, and others below, the internal excitation needed for one-photon dissociation.

Figure 8 shows some illustrative data from a recent examination of iodobenzene ion (Honovich and Dunbar, 1982b), for which the photodissociation band also spans the one-photon threshold. The curves are drawn from numerical solution of the mixed kinetic equations assuming a fraction f_1 of ions above the one-photon threshold (and taking into account a small kinetic shift arising from the slow dissociation of ions very near threshold). The values of f_1 derived from this analysis are in agreement with a thermal distribution of ions at 300 K if the one-photon threshold is at 2.48 eV. Since the threshold kinetic behavior as a function of wavelength is very sensitive to the value of the one-photon threshold energy, this is a promising route to obtaining thermochemical information for ion fragmentations.

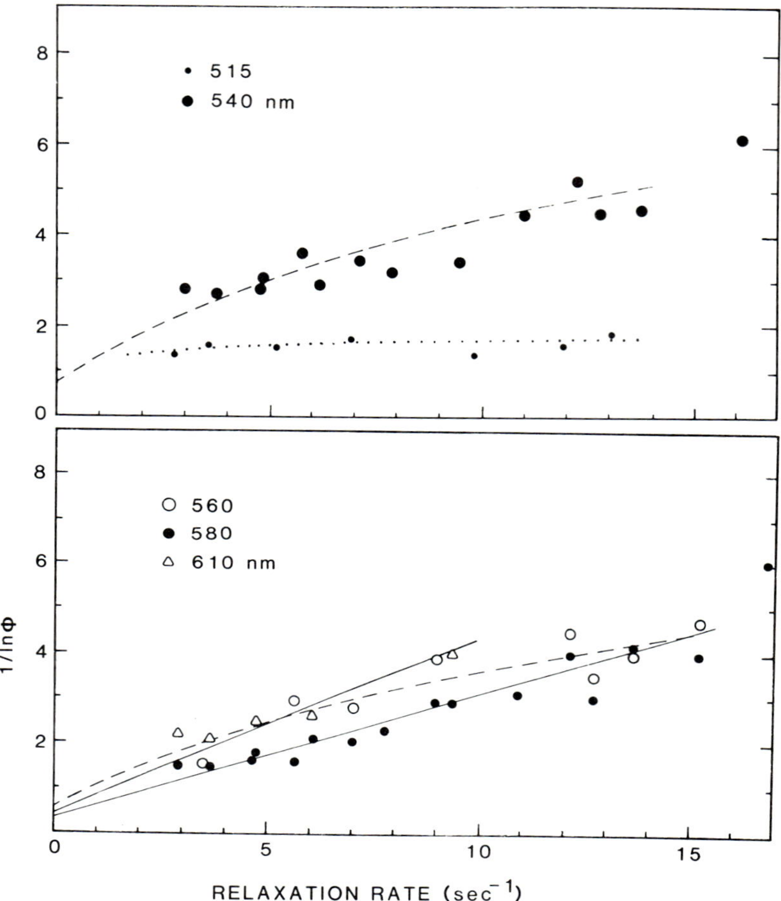

FIG. 8 Dissociation versus pressure for iodobenzene ion at several wavelengths. The theoretical curves show the best choice for the fraction f_1 of one-photon ions in the population to fit the experiment: (—), $f_1 = 0$; (--), $f_1 = 0.04$; and (· · · · ·), $f_1 = 0.25$. [From Honovich and Dunbar (1982c).]

VII. TWO-PHOTON KINETICS AND REARRANGEMENT PROCESSES

Having laid the foundations for an understanding of the complexities of two-photon dissociation kinetics, we can return to a question of ion rearrangement chemistry, exploiting some new ideas from the two-photon kinetic studies. Consider the dissociation of iodotoluene ions [reaction (12)]:

$\lambda < 1200$ nm → 19 or 20 (12)

$\lambda \gtrsim 530$ nm → 21

Dissociation to yield the low-energy tropylium ion product **19** has its one-photon threshold at about 1200 nm, so dissociation is energetically feasible at all visible wavelengths; the benzyl ion product **20** is only slightly more energetic and is similarly accessible. However, either of these products involves a major rearrangement from the initial structure. On the other hand, the direct-cleavage tropyl ion product **21** has its calculated one-photon threshold near 530 nm and is not accessible as a one-photon product at long wavelengths. (Since the tolyl ion energetics have not been well characterized, this threshold is uncertain but should be very similar to the threshold for iodobenzene ion dissociation at about 530 nm.) Therefore, when iodotoluene parent ion is irradiated in its strong visible absorption band, it is possible to see whether dissociation can or cannot be accompanied by rearrangement simply by characterizing the dissociation kinetics as one photon or two photon.

Figure 9 shows the pressure dependence of the iodotoluene photodissociation at 580 nm and at 458 nm (Dunbar *et al.*, 1982). At 458 nm (and also at 514 nm), the dissociation is pressure-independent, as expected for a normal one-photon process; but at 580 nm, the strong pressure dependence clearly signals a two-photon process. It is obvious that following photon absorption at 580 nm (2.14 eV) the ion finds no route to the energetically feasible dissociation products **19** or **20**, but must await the arrival of another photon to drive the dissociation to **21**. Light at 514 or 458 nm is above the one-photon threshold for dissociation to **21**, so the lack of pressure dependence in these cases is expected.

Bromotoluene ion provides an interesting contrast. The energetic scheme is very similar to reaction (9), except that the one-photon threshold for the dissociation to tropyl ion **21** is about 2.8 eV (440 nm). However, no pressure dependence is observed for photodissociation, even at wavelengths longer

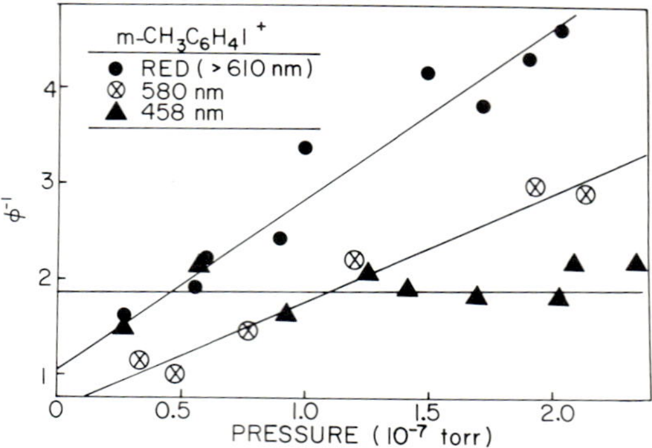

FIG. 9 Pressure dependence of the iodotoluene dissociation at wavelengths of 458, 580, and 610$^+$ nm. [From Dunbar *et al.* (1982).]

than 610 nm. This labels the dissociation as a one-photon process at these wavelengths and must mean that depositing 2 eV of internal energy in the bromotoluene ion is sufficient to allow rearrangement to either a tropylium or a benzyl fragment ion structure. It is interesting to ask whether the intact parent ion rearranges or whether there is a concerted rearrangement/dissociation process, but it is not clear how to go about deciding this question. These first results point the way to a potentially useful new route of attack on questions of rearrangement processes in ions having well-defined excess internal energy.

VIII. IR-VISIBLE TWO-LASER PHOTODISSOCIATION

A. *Kinetics and Mechanism*

An interesting variation on the two-photon dissociation idea gives us access to some new aspects of the infrared spectroscopy and photochemistry of ionic molecules. An infrared laser can be used to deliver the energy represented by the S_1 step in reaction (10), following the new scheme

$$A^+ \xrightarrow[\text{(vis)}]{IS} A^{+\dagger} \longrightarrow A^{+*} \xrightarrow[\text{(vis)}]{IS_2} \text{Fragments} \qquad \left(A^{+*} \xrightarrow[\text{(IR)}]{I_{IR}S_{IR}} A^+;\quad A^{+*} \xrightarrow{k_cP + k_r} A^+\right) \tag{13}$$

This effect is clearly observed as a strong increase in dissociation (typically up to a factor of 3 increase) when a CO_2 laser of several watts power is added to the visible irradiation (Dunbar *et al.*, 1980).

We know that the IR radiation serves to give initial excitation to the ions, rather than serving to push excited ions over the brink of dissociation, as a result of recent sequential-excitation experiments (Honovich and Dunbar, 1982d): if the ions (iodobenzene or bromobenzene ions were studied) are first irradiated with the IR laser and then immediately with the visible laser, the IR enhancement is practically undiminished; while if the visible-laser irradiation precedes the IR irradiation, no enhancement is seen, nor does IR irradiation alone have any effect. [Wight and Beauchamp (1981) have reported a very small IR enhancement effect in cyanobenzene ion, which seems to require *simultaneous* irradiation by both lasers and very likely is a different sort of effect than the large effect discussed here.]

For iodobezene ion, which has been studied the most extensively, photodissociation data were taken as a function of pressure and of the intensities of both lasers. Given the five independently adjustable rate constants in the kinetic scheme (13) (with the added complication that these constants may be dependent on the internal energy level of the ion), it is not surprising that agreement can be forced between theory and experiment, but it was surprising that we found only a narrow choice of these parameters which were both reasonable and which gave a satisfactory fit of the kinetic equations to all of the data (Lev and Dunbar, 1982b). Based on extensive exploration of the parameter values, several features of the iodobenzene kinetics were concluded to be essential features of a successful model:

(a) Most or all of the ions are susceptible to both infrared and visible up-pumping processes at the wavelengths used.

(b) The infrared photoexcitation has a bottleneck somewhere below 4000 cm^{-1} and is also too slow to compete with relaxation processes near the dissociation threshold at 20,000 cm^{-1}.

(c) Infrared radiative relaxation is significant at pressures of a few times 10^{-8} torr, but collisional relaxation dominates above about 10^{-7} torr. The radiative relaxation time constant is around 500 msec.

(d) The visible photoexcitation cross section at 610 nm decreases by a factor of 2 to 5 with increasing internal energy from 0 to 20,000 cm^{-1}.

B. *Infrared Spectroscopy*

Since the IR enhancement effect depends on the absorption of IR photons by the ion, the wavelength dependence of the effect gives an indirect but effective route to finding the IR absorption peaks of gas phase ions, which is extremely difficult information to come by in any more direct way.

Figure 10 shows an example of the spectra we have been able to obtain (Honovich and Dunbar, 1982c): As can be seen, the spectrum does indeed show wavelengths of high and low IR enhancement, and the main obstacle

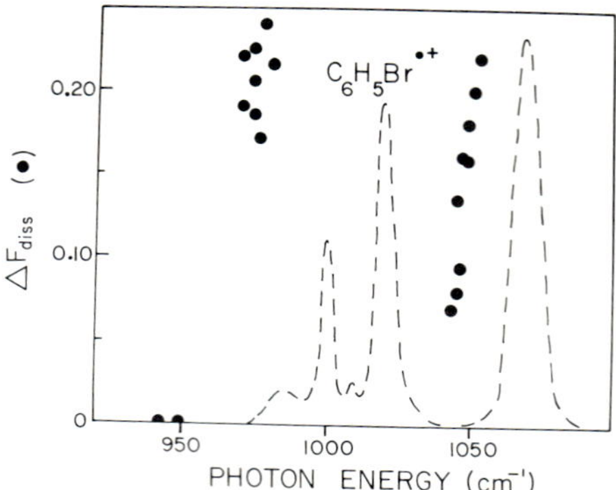

FIG. 10 IR enhancement of bromobenzene ion photodissociation (610 nm) as a function of IR photon energy. [Reprinted with permission from Honovich and Dunbar (1982c). Copyright 1982 American Chemical Society.]

to measuring a complete IR spectrum is the sparse wavelength coverage of the CO_2 laser. We have measured spectra of this sort for several halobenzene and halotoluene ions in the IR region characteristic of ring vibrational modes. These ions all show sharply wavelength-dependent effects. In most cases there is strong IR absorption in the ion at wavelengths longer than any of the ring vibrations in the neutral molecules. This seems to reflect some red shift of the ring vibrational frequencies upon ionization, which is not surprising since it is a bonding ring π electron which is removed. Beauchamp's group has reported (as discussed elsewhere in this volume) the observation of infrared absorption in many ions having low-energy dissociation pathways: The IR-enhanced two-photon approach described here gives a complementary alternative which is favorable for ions having a high dissociation threshold.

C. *Relaxation of IR-Excited Ions*

The experiment of Fig. 11 gives an excellent way to observe the relaxation of excited ions (Honovich and Dunbar, 1982d): After the end of the IR excitation period, the ions begin to dissipate their excess internal energy by collisions and by radiation, and the amount of energy remaining after time T is probed by the visible-laser dissociation pulse. Data as a function of delay time T are shown for iodobenzene ion in Fig. 12, along with the theoretical expectation for a relaxation time constant of 500 msec. At a pressure of 2×10^{-8} torr, the relaxation does not vary much with pressure and is undoubtedly due mostly to infrared fluorescence. Relaxation becomes faster at higher pressures, and the collisional relaxation rate k_c can be ex-

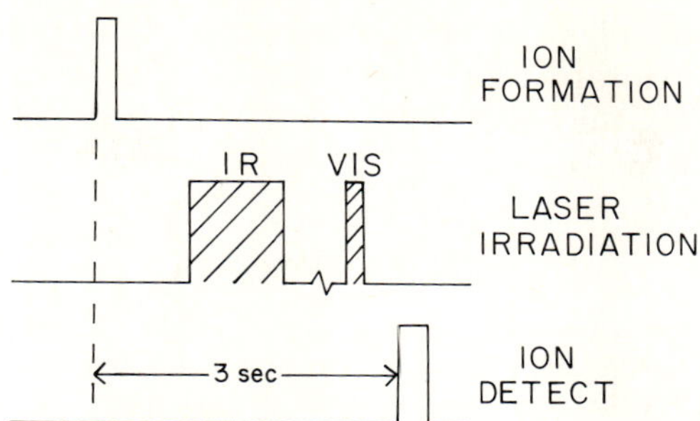

FIG. 11 Pulse sequence for timed two-laser photodissociation experiment, [Reprinted with permission from Honovich and Dunbar (1982d). Copyright 1982 American Chemical Society.]

tracted separately from the radiative component k_r. The rates for iodobenzene measured in this way are $k_r = 2\ \text{sec}^{-1}$ and $k_c = 6 \times 10^{-10}\ \text{cm}^3$ molecule^{-1} sec^{-1}.

This is a collisional rate slightly faster than half the orbiting collision rate and is thus slightly faster than that found in the quenching experiments described above. The relaxation rates for IR-excited ions might be expected to be somewhat faster than the corresponding rates in visible two-photon processes, since it is likely that the average energy of the IR-excited ions is less than that of the visible-photon-excited ions. This seems to be true in the

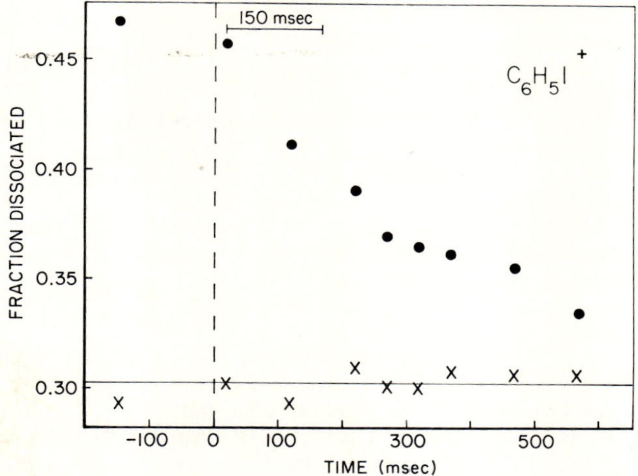

FIG. 12 Two-laser photodissociation of iodobenzene ion as a function of the delay between the two lasers (as in the sequence of Fig. 11). The × values are visible dissociation in the absence of IR irradiation. [Reprinted with permission from Honovich and Dunbar (1982d). Copyright 1982 American Chemical Society.]

bromobenzene ion, where the IR-enhancement experiment described here gives a radiative relaxation rate of 4.5 sec^{-1}, while the pulsed-laser visible work discussed above definitely indicates a k_r rate less than 3 sec^{-1} for the visible two-photon process.

ACKNOWLEDGMENTS

The author acknowledges with gratitude the support of both the National Science Foundation and the donors of the Petroleum Research Fund, administered by the American Chemical Society, during the preparation of this chapter and during the course of much of the research described.

REFERENCES

Allan, M., and Maier, J. P. (1976). *Chem. Phys. Lett.* **43**, 94.
Allan, M., Dannacher, J., and Maier, J. P. (1980). *J. Chem. Phys.* **73**, 3114.
Benz, R. C., and Dunbar, R. C. (1979). *J. Am. Chem. Soc.* **101**, 6363.
Benz, R. C., Claspy, P. C., and Dunbar, R. C. (1981) *J. Am. Chem. Soc.* **103**, 1799.
Cassady, C. J., Freiser, B. S., and Russell, D. (1981). Presented at the *Ann. Conf. Mass Spectrom. Allied Top., 29th Minneapolis, Minnesota.*
Chen, J. H., and Dunbar, R. C. (1984). To be published.
Cody, R. B., and Freiser, B. S. (1979). *Anal. Chem.* **51**, 547.
Dujardin, G., Leach, S., and Taieb, G. (1980). *Chem. Phys.* **46**, 407.
Dunbar, R. C. (1976). *Anal. Chem.* **48**, 723.
Dunbar, R. C. (1978). *J. Chem. Phys.* **68**, 3125.
Dunbar, R. C. (1979a). *Int. J. Mass Spectrom. Ion Phys.* **29**, 77.
Dunbar, R. C. (1979b). In "Gas Phase Ion Chemistry" (M. T. Bowers, ed.), Vol. 2, Ch. 14. Academic Press, New York.
Dunbar, R. C. (1979c). *J. Phys. Chem.* **83**, 2376.
Dunbar, R. C. (1979d). *In* "Kinetics of Ion–Molecule Reactions" (P. Ausloos, ed.) p. 463. Plenum Press, New York.
Dunbar, R. C. (1980). *In* "Physical Methods of Modern Chemical Analysis" (T. Kuwana, ed.), Vol. 2, p. 277. Academic Press, New York.
Dunbar, R. C. (1981). *In* "Mass Spectrometry" (R. A. W. Johnstone, ed.), Vol. 6, p. 100. The Royal Society, London.
Dunbar, R. C. (1982). *In* "Lecture Notes in Chemistry: Ion Cyclotron Resonance Spectrometry II" (H. Hartmann and K.-P. Wanczele, eds.), p. 1. Springer-Verlag, Berlin and New York.
Dunbar, R. C. (1983). *J. Phys. Chem.* **87**, 3105.
Dunbar, R. C., and Fitzgerald, G. B. (1984). To be published.
Dunbar, R. C., and Fu, E. W. (1973) *J. Am. Chem. Soc.* **95**, 2716.
Dunbar, R. C., and Fu, E. W. (1977). *J. Phys. Chem.* **81**, 1531.
Dunbar, R. C., and Klein, R. (1977). *J. Am. Chem. Soc.* **99**, 3744.
Dunbar, R. C., and Teng, H. H. (1978). *J. Am. Chem. Soc.* **100**, 2279.
Dunbar, R. C., Fu, E. W., and Olah, G. A. (1977). *J. Am. Chem. Soc.* **99**, 7502.
Dunbar, R. C., Teng, H. H., and Fu, E. W. (1979a). *J. Am. Chem. Soc.* **101**, 6506.
Dunbar, R. C., Kim, M. S., and Olah, G. A., (1979b). *J. Am. Chem. Soc.* **101**, 1368.
Dunbar, R. C., Hays, J. D., Honovich, J. P., and Lev, N. B. (1980). *J. Am. Chem. Soc.* **102**, 3950.
Dunbar, R. C., Honovich, J. P., and Segall, J. (1982). To be published.

Freiser, B. S., and Beauchamp, J. L. (1975). *Chem. Phys. Lett.* **35**, 35.
Freiser, B. S., and Beauchamp, J. L. (1977). *J. Am. Chem. Soc.* **99**, 3214.
Fu, E. W., Dymerski, P. P., and Dunbar, R. C. (1976). *J. Am. Chem. Soc.* **98**, 337.
Fu, E. W., and Dunbar, R. C. (1978). *J. Am. Chem. Soc.* **100**, 2283.
Gooden, R., and Brauman, J. I. (1977). *J. Am. Chem. Soc.* **99**, 1977.
Gooden, R., and Brauman, J. I. (1982). *J. Am. Chem. Soc.* **104**, 1483.
Griffiths, I. W., Mukhtar, E. S., Harris, F. M., and Beynon, J. H. (1981a). *Int. J. Mass. Spectrom. Ion Phys.* **38**, 333.
Griffiths, I. W., Mukhtar, E. S., March, R. E., Harris, F. M., and Beynon, J. H. (1981b). *Int. J. Mass. Spectrom. Ion Phys.* **39**, 125.
Griffiths, I. W., Harris, F. M., Mukhtar, E. S., Beynon, J. H. (1981c). *Int. J. Mass. Spectrom. Ion Phys.* **41**, 83.
Hays, J. D. (1980). Ph.D. Thesis, Case Western Reserve Univ.
Hays, J. D., and Dunbar, R. C. (1979). *J. Phys. Chem.* **83**, 3183.
Honovich, J. P., and Dunbar, R. C. (1981) *J. Phys. Chem.* **85**, 1558.
Honovich, J. P., and Dunbar, R. C. (1982a). *Int. J. Mass Spectrom. Ion Phys.* **42**, 333.
Honovich, J. P., and Dunbar, R. C. (1982b) *Int. J. Mass Spectrum. Ion Phys.* (in press).
Honovich, J. P., and Dunbar, R. C. (1982c). *J. Phys. Chem.* **87**, 3755.
Honovich, J. P., and Dunbar, R. C. (1982d). *J. Am. Chem. Soc.* **104**, 6220.
Jasinski, J. M., and Brauman, J. I. (1980). *J. Chem. Phys.* **73**, 6191.
Kim, M. S., and Dunbar, R. C. (1979). *Chem. Phys. Lett.* **60**, 247.
Kim, M. S., Dunbar, R. C., and McLafferty, F. W. (1978). *J. Am. Chem. Soc.* **100**, 4600.
Kramer, J. M., and Dunbar, R. C. (1973). *J. Chem. Phys.* **59**, 3092.
Lev, N. B., and Dunbar, R. C. (1981). *Chem. Phys. Lett.* **84**, 483.
Lev, N. B., and Dunbar, R. C. (1982a). *J. Phys. Chem.* **87**, 1924.
Lev, N. B., and Dunbar, R. C. (1982b). *Chem. Phys.* (in press).
Maier, J. P. (1980). *Chimia* **34**, 219.
Maier, J. P. (1982). *Accounts Chem. Res.* **15**, 18.
McGilvery, D. C., and Morrison, J. D. (1977). *J. Chem. Phys.* **67**, 368.
Morgenthaler, L. N., and Eyler, J. R. (1979). *J. Chem. Phys.* **71**, 1486.
Morgenthaler, L. N., and Eyler, J. R. (1981a). *J. Chem. Phys.* **74**, 4356.
Morgenthaler, L. N., and Eyler, J. R. (1981b). *Int. J. Mass. Spectrom. Ion Phys.* **37**, 153.
Moseley, J. T., Cosby, P. C., and Peterson, J. R. (1976). *J. Chem. Phys.* **65**, 2512.
Moseley, J. T., and Durup, J. (1980). *J. Chem. Phys.* **77**, 673.
Moseley, J. T. (1982). *In* "Applications of Atomic Collision Physics" (H.S.W. Massey, E. W. McDaniel, and B. Bederson, Eds.), Vol. 5, Ch. 10. Academic Press, New York.
Moseley, J. T., and Durup, J. (1981). *Ann. Rev. Phys. Chem.* **32**, 53.
Mukhtar, E. S., Griffiths, I. W., March, R. E., Harris, F. M., and Beynon, J. H. (1981a). *Int. J. Mass. Spectrom. Ion Phys.* **41**, 61.
Mukhtar, E. S., Griffiths, I. W., Harris, F. M., and Beynon, J. H. (1981b). *Int. J. Mass Spectrom. Ion Phys.* **37**, 159.
Orth, R. G., and Dunbar, R. C. (1978). *J. Am. Chem. Soc.* **100**, 5949.
Orth, R. G., and Dunbar, R. C. (1982). *J. Am. Chem. Soc.* **104**, 5617.
Reents, W. D., Jr., and Freiser, B. S. (1980). *J. Am. Chem. Soc.* **102**, 271.
Reents, W. D., Jr., and Freiser, B. S. (1981). *J. Am. Chem. Soc.* **103**, 2791.
Riggin, M. T., Orth, R. G., and Dunbar, R. C. (1976). *J. Chem. Phys.* **65**, 3365.
Rosenfeld, J. M., Jasinski, J. M., and Brauman, J. I. (1979). *J. Am. Chem. Soc.* **101**, 3999.
Rynard, C. M., and Brauman, J. I. (1980). *Inorg. Chem.* **19**, 3544.
Rynard, C. M., Gooden, R., and Brauman, J. I. (1982). To be published.

Saykally, R. J., and Woods, R. C. (1981). *Ann. Rev. Phys. Chem.* **32**, 403.
Teng, H. H., and Dunbar, R. C. (1978). *J. Chem. Phys.* **68**, 3133.
van Dishoek, E. F., van Velzen, P. N. T., and van der Hart, W. J. (1979). *Chem. Phys.* **62**, 135.
van Velzen, P. N. T. (1981). Ph.D. Thesis, Univ. of Leiden.
van Velzen, P. N. T., and van der Hart, W. J. (1981a). *Chem. Phys. Lett.* **83**, 55.
van Velzen, P. N. T., and van der Hart, W. J. (1981b). *Chem. Phys.* **61**, 335.
van Velzen, P. N. T., and van der Hart, W. J. (1981c). *Org. Mass Spectrom.* **16**, 237.
van Velzen, P. N. T., van der Hart, W. J., van der Greef, J., Nibbering, N. M. M., and Gross, M. L. (1982). *J. Am. Chem. Soc.* **104**, 1298.
Woodin, R. L., Bomse, D. S., and Beauchamp, J. L. (1979). *In* "Chemical and Biochemical Applications of Lasers" (C. B. Moore, ed.), Vol. 4, p. 1. Academic Press, New York.
Wight, C. A., and Beauchamp, J. L. (1981). *Chem. Phys. Lett.* **77**, 30.

Chapter 21

Electron photodetachment from gas phase molecular anions

*PAUL S. DRZAIC**

JEFFREY MARKS

JOHN I. BRAUMAN

DEPARTMENT OF CHEMISTRY
STANFORD UNIVERSITY
STANFORD, CALIFORNIA

*Present address: Raychem Corporation, Menlo Park, California.

ISBN 0-12-120803-6

I. INTRODUCTION AND SCOPE

Electron photodetachment from gas phase molecular anions has proved to be a remarkable tool in the study of both anions and their neutral photoproducts. A variety of thermochemical and spectroscopic constants can be measured for systems that would prove difficult to study by other means. Among the techniques that have been developed for the study of electron photodetachment are those employing ion beams, trapped ion cells, and drift tubes. The advent of tunable laser systems has increased the precision of measurements made in photodetachment experiments, allowing for the detection of vibrational and even rotational transitions in some systems. In addition, the discovery that electron detachment can be induced by exciting vibrational modes of anions with infrared lasers promises to open up new areas of study involving electron attachment and detachment processes.

Upon irradiation with light, a molecular anion may either detach an electron [reaction (1)] or dissociate [reaction (2)]:

$$A^- + h\nu \rightarrow A + e^- \quad (1)$$

$$A^- + h\nu \rightarrow B^- + C. \quad (2)$$

In photodetachment spectroscopy, process (1) is monitored as a function of the wavelength of the incident light. This can be accomplished by detecting the decrease in the concentration of A^- or an increase in either A or the number of free electrons. The related technique of negative ion photoelectron spectroscopy analyzes the energy of detached electrons produced by the intersection of a fixed frequency light source and an ion beam.

Photodetachment spectroscopy has been used to determine a number of thermochemical and spectroscopic parameters for a variety of atoms and molecules. These include the electron affinity of the neutral photoproduct, the energies and transition moments of electronically excited states of anions, spin–orbit splittings in radicals, and vibrational and rotational spacings in both neutrals and anions. While most negative anions of organic molecules exhibit only photodetachment, a number of inorganic anions also photodissociate upon irradiation with visible light. Photodissociation experiments are useful in probing excited states of anions, as well as providing information important in modeling the effects of radiation on the atmosphere.

The emphasis of this review is on the photodetachment of molecular negative ions with a view toward illustrating the chemistry and spectroscopy that may be learned. Individual sections are concerned with modeling the behavior of photodetachment cross sections, vibronic transitions in photodetachment, and the observation of excited electronic states via photodetachment spectroscopy. The use of electron affinities derived from photodetachment experiments to determine homolytic bond dissociation energies is discussed. A summary of the published results concerning infrared multiphoton-induced electron detachment is also included. Finally, we conclude with a listing of molecular electron affinities measured since the tabulation by Janousek and Brauman (1979).

A number of reviews on negative ion photochemistry have appeared recently. Among these are those by Mead *et al.* (this volume), Miller (1981), Janousek and Brauman (1979), Corderman and Lineberger (1979), and Dunbar (1979). Photodetachment put in perspective with other electron attachment and detachment processes has been discussed by Berry and Leach (1981). These articles provide rich sources of information regarding the state of the art in negative ion spectroscopy. The reader interested in the general areas of atomic photodetachment, electron affinities, negative ion photodissociation, and complementary spectroscopic techniques should investigate these articles.

II. OPTICAL PHOTODETACHMENT

A. *Experimental Considerations*

A number of experimental techniques have been applied to the study of negative ion spectroscopy. Among the devices used to generate and study the anions used in these experiments are molecular beams, drift tubes, and trapped-ion cells. A number of these techniques have been illustrated in the reviews by Miller (1981) and Janousek and Brauman (1979). Each technique offers its own advantages for different types of negative ions. For the study of photodetachment from polyatomic anions, ion cyclotron resonance (ICR) spectrometry has been the most extensively applied method. The advantages offered by ICR spectrometry include the ability to generate a large variety of molecular ions under relatively controlled conditions, the long trapping times of ions, and high sensitivity. In keeping with the theme of this chapter, we briefly outline the application of ICR spectrometry to molecular anion photodetachment studies.

The technique of ion cyclotron resonance spectrometry allows one to generate gas phase ions and trap them for periods of 100 msec to several seconds (Lehman and Bursey, 1976). An ICR spectrometer consists essentially of a cell encased in a high-vacuum can and placed between the poles of a magnet (Fig. 1). Ions are formed either directly by electron impact or by secondary ion–molecule collisions. An ion in a magnetic field will exhibit a cyclotron motion with a frequency dependent on the mass of the ion and the strength of the magnetic field. By adjusting the magnetic-field strength, the cyclotron frequency of an ion of a particular m/e ratio can be brought

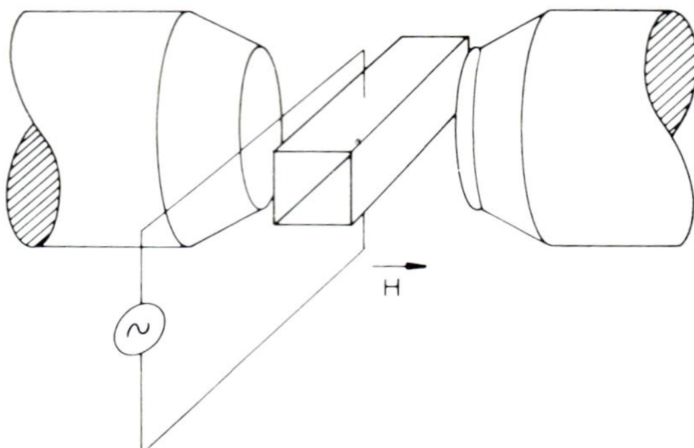

FIG. 1 Schematic of ICR spectrometer ($\omega = v/r = qH/mc$). Marginal oscillator or capacitance bridge circuit may be used as detector.

into resonance with an oscillating radio frequency signal placed on the upper or lower cell plates. This resonance condition allows ions of a given m/e to be detected using either a marginal oscillator or a capacitance bridge circuit, without the necessity of physically collecting the ions (Warnick *et al.*, 1974; McIver *et al.*, 1980). Photodetachment experiments are performed in an ICR by monitoring the decrease in an ion signal upon irradiation with light. Both broadband arc lamp/monochromator systems and tunable dye lasers have been used as photon sources in photodetachment experiments.

Most optical photodetachment experiments have been performed using an ICR operating in a steady-state, or "drift," mode. Ions are formed continuously by electron bombardment, and ideally the application of an electric field perpendicular to the magnetic field induces the ions to drift down the length of the cell. In practice, the motion of the ions is much more complicated than this. This is best exemplified by the observation that changing the direction of the magnetic field often has only a minimal effect on the size of the ion signal!

In order to translate an observed decrease of an ion signal upon irradiation into a photodetachment cross section, it is necessary to have a model for the behavior of ions in the ICR. It appears that the most appropriate description involves a steady-state model. In this model the ions are assumed to enter and leave the ICR cell at the same average rate, but the probability of a given ion being lost is independent of the amount of time spent there (Dunbar, 1973). In determining the cross section, the pseudo-first-order rate constant for ion loss is taken to be the sum of an intrinsic rate constant for nonphotochemical ion loss plus one which involves the product of the light intensity, ion lifetime, cross section, and an ion–light overlap factor.

This leads to expression (3) for the relative photodetachment cross section $\sigma(\lambda)$:

$$\sigma(\lambda) \propto \frac{F(\lambda)}{g\rho(\lambda)[1 - F(\lambda)]}. \tag{3}$$

Here, $F(\lambda)$ represents the fractional signal decrease in the ion signal upon irradiation, g is the geometrical overlap factor for the light beam and the ICR cell, and $\rho(\lambda)$ is the photon flux. If a broadband light source is used, the resulting cross sections $\sigma(\lambda)$ may be further deconvoluted to compensate for instrumental broadening (Reed and Brauman, 1974).

B. *Interpretation of Photodetachment Cross Sections*

The photodetachment process is a transition from a bound electron in the anion to a free continuum electron plus neutral. In general, the cross section near threshold is an increasing function of photon energy. This results in an integral spectrum, with the cross section at a given wavelength being the sum of all allowed transitions. Any electronic features or new onsets are superimposed on a rising background. While certain systems provide photodetachment spectra with sharp onsets and are straightforward to interpret, other systems show more gradual onsets, making it difficult to assign the threshold for a new onset. A theoretical description of photodetachment cross sections is essential in these cases in order to glean useful information from the experimentally observed cross sections. Theoretical models for both atomic and molecular photodetachment cross sections have been developed and have proved useful in the interpretation of experimental cross sections.

Atomic photodetachment was first described as a special case of a more general formalism developed by Wigner (1948). Wigner showed that at threshold the cross section for optically detaching an electron depends only on the long-range interaction between the two product particles. In contrast to photoionization, where the long-range potential falls off as $1/r$ due to the Coulombic force between electron and cation, the strongest long-range electronic interaction in atomic photodetachment is a centrifugal force contributed by the $l(l + 1)/r^2$ term contained in the Schrödinger radial wave equation. Thus, the behavior at threshold is seen to depend on the angular momentum of the electron in the final state, and is given by Eq. (4)

$$\sigma(E) \propto k^{2l+1}. \tag{4}$$

Here, k is the reduced linear momentum of the photoelectron and l is the relative angular momentum between the two final particles. For atoms, the usual dipole selection rule $\Delta l = \pm 1$ applies, and l in Eq. (4) can be related to the angular momentum of the atomic orbital from which the electron was detached. At energies above threshold, the cross section can be described as a power series in k. This is given by Eq. (5)

$$\sigma(E) \propto k^{2l+1}(a_0 + a_1k^2 + a_2k^4 + \cdots). \tag{5}$$

The cross-section behavior may be recast in terms of ΔE, the energy above the photodetachment threshold ($k = (2m\,\Delta E)^2/\hbar$). Equation (6) can be used to find the dependence of the slope of the cross-section curve at threshold as a function of photon energy. Thus, it is seen that with $l = 0$, the slope of the cross section at threshold is infinite while for higher values of l the slope is finite:

$$\sigma(E) \propto \Delta E^{l+1/2}. \tag{6}$$

The problem of describing the photodetachment cross section of diatomic anions has been treated by Geltman (1958). The photodetachment process was treated as a one-electron problem in which an electron moves in a cylindrically symmetric field. While no use of the Wigner formalism was used, Geltman derived an equation functionally equivalent to (4). In this case, l is related to the axial angular momentum of an electron in the highest occupied molecular orbital (HOMO) of the anion.

The extension of the Wigner result to include polyatomic anions was accomplished by Reed *et al.* (1976). In general, the behavior of the cross section near threshold for a polyatomic anion is given by Eq. (7)

$$\sigma(E) = \mathrm{A}_{l^*}k^{2l^*+1}[1 + O(k^2)], \tag{7}$$

where l^* is the smallest integer value l for which A_{l^*} is not identically zero for symmetry reasons. This result leads to a simple algorithm based on group theory for determining the slope of the cross section at threshold. To perform the algorithm, one finds the irreducible representation of the anion's HOMO for the point group of the anion. One then assigns a value l' based on that irreducible representation. For example, $l' = 0$ if the irreducible representation is totally symmetric; $l' = 1$ if it transforms as x, y, or z; and $l' = 2$ if it is labeled by xy, xz, yz, $x^2 - y^2$, or z^2. The allowed l^* values for photodetachment are then these $l' \pm 1$.

This algorithm reproduces Geltman's results for diatomics (Reed *et al.*, 1976). Application to larger systems is straightforward. For example, photodetachment from NH_2^- (C_{2v} symmetry) is calculated to occur from a HOMO with B_1 symmetry. B_1 transforms as x; thus, that orbital has $l' = 1$. This implies that the outgoing wave can have $l^* = 0$, with a very sharp onset. This is borne out by experiment.

In a number of systems, more than one l^* is allowed at threshold. This may occur if the HOMO of the anion contains contributions of different irreducible representations. In general, the lowest l^* value will dominate at threshold, as it rises the most quickly with increasing photon energy. However, if the contributions in the HOMO are unequal, higher l^* may dominate. This appears to be the case in the SiH_3^- and GeH_3^- systems (Reed and Brauman, 1974). Although both $l' = 0$ and $l' = 1$ are allowed, the $l' = 1$ (p wave) component dominates the cross-section behavior near threshold. This indicates that the anion's HOMO is primarily an s-type orbital on the central atom, with only a small p-orbital contribution.

Although this algorithm accurately predicts which l' are allowed at threshold, it does suffer from certain limitations. It cannot describe the cross-section behavior above threshold. As mentioned, it cannot by itself predict the relative intensities of competing allowed transitions. Finally, for systems of low symmetry, most transitions become allowed, giving the algorithm limited usefulness in these cases.

Reed *et al.* (1976) have addressed these problems by producing a one-electron formalism to calculate photodetachment cross sections as a function of energy. The heart of this calculation is the evaluation of the matrix element shown in Eq. (8)

$$\sigma(E) \propto |\langle\psi_i|\mathbf{r}|\psi_f\rangle|^2, \tag{8}$$

where ψ_i, the initial state of the anion, is assumed to be the HOMO of the anion. This HOMO is taken to be a linear combination of atomic orbitals produced by a molecular orbital calculation. The orbital exponents of the atomic orbitals in the HOMO are corrected to account for the diffuseness of the orbitals in the anion. These corrections are obtained by empirically fitting calculated cross sections for atomic photodetachment to experimental determinations. Again, $\mathbf{r}$ is the dipole operator and ψ_f, the final state wave function, is taken to be the detached electron described as a plane wave. This plane wave is orthogonalized to the HOMO of the anion.

Agreement of this calculation with experiment is striking in a number of cases. Figures 2, 3, and 4 compare calculated and experimental results for the anions OH^-, cyclopentadienyl, and cyanomethyl. The agreement of the experimental and calculated results is usually good up to 1 eV above threshold, although the formalism might not be expected to work as well at energies far above threshold.

An illustration of the advantage of this one electron formalism over the simple threshold law is given in the case of HC_2^- (Janousek *et al.*, 1979). It is instructive to compare the predicted cross-section behavior of HC_2^- with the related species C_2^- and CN^-. Application of the group theoretical

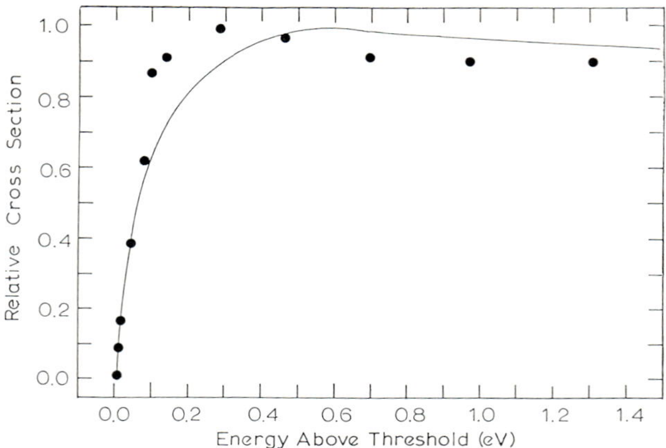

FIG. 2 Relative experimental photodetachment cross sections for OH^- [(●), Branscomb (1966)] plotted as a function of energy above threshold. The normalized theoretical cross section is labeled by (—). [Reprinted with permission from Reed *et al.* (1976). Copyright 1976 American Institute of Physics.]

law results in $l^* = 0$ for HC_2^- and CN^-, indicating a sharp onset at threshold. These symmetry rules also predict $l^* = 1$ for C_2^-, resulting in a more gradual onset.

This threshold behavior may be contrasted with the cross-section behavior predicted by the formalism by Reed *et al.* (1976) at energies above threshold. The results of these calculations are shown in Fig. 5. It is seen that although

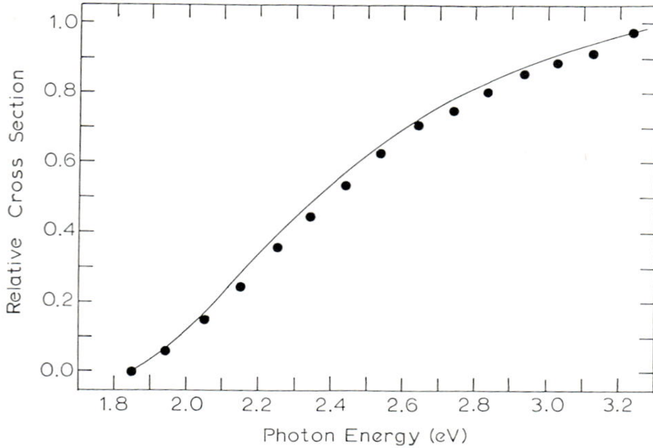

FIG. 3 Relative experimental [(●), Richardson *et al.* (1973)] photodetachment cross sections for $C_5H_5^-$ plotted as a function of incident photon energy. [Reprinted with permission from Reed *et al.* (1976). Copyright 1976 American Institute of Physics.]

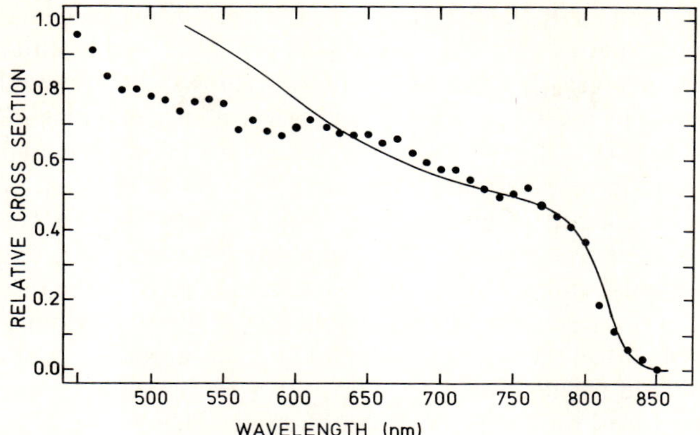

FIG. 4 Relative experimental (●) and theoretical (—) photodetachment cross sections for cyanomethyl anion. [Reprinted with permission from Zimmerman and Brauman (1977a). Copyright 1977 American Chemical Society.]

HC_2^- is allowed to exhibit s-wave behavior at threshold, the calculation predicts that p-wave behavior will dominate above threshold. While in a strict symmetry sense the wave function of the highest occupied molecular orbital in HC_2^- is unsymmetrical (as in CN^-), the system is actually only slightly perturbed from the symmetric C_2^- HOMO, and the predicted cross section for HC_2^- is similar to that predicted for C_2^-. The experimental cross section for HC_2^- agrees well with these predictions.

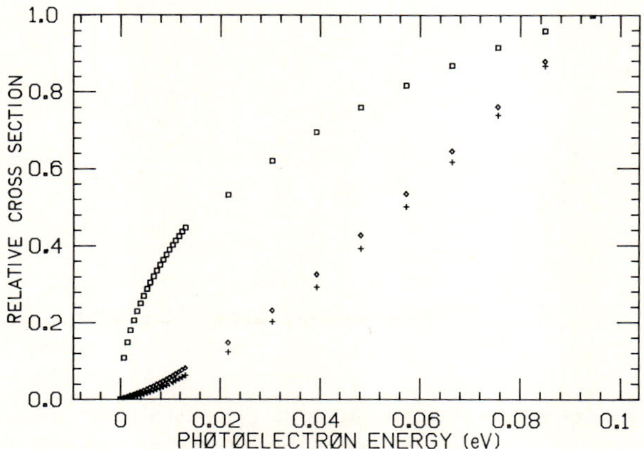

FIG. 5 Relative theoretical photodetachment cross sections for CN^- (□), HC_2^- (◇), and C_2^-(+). [Reprinted with permission from Janousek and Brauman (1979). Copyright 1979 American Institute of Physics.]

The applicability of these procedures is nicely illustrated in the photodetachment of species with very high symmetry. The perinaphthenyl anion has a lowest allowed l^* value of 3. Figure 6 shows experimental and calculated cross sections for perinaphthenyl anion photodetachment (Gygax *et al.*, 1979). The use of the calculated cross section allows the photodetachment threshold to be determined to within ± 0.1 eV in this case, a precision that would be difficult to obtain using only the very shallow cross-section value measured experimentally. The same analysis was performed for the cyclooctatetraenyl anion, $C_8H_8^-$, but the determination of the photodetachment onset in this case is much less certain due to the poor Franck–Condon factors expected for the transitions involving the ground vibrational state of the anion and the neutral.

In certain cases, the behavior of the photodetachment cross section near threshold deviates measurably from that predicted by the threshold laws just discussed. Such deviations are indicative of a long-range neutral-electron interaction comparable in strength to the potential due to the relative angular momentum of the two particles. For atoms, it is well known that an interaction due to the polarizability of the neutral atom can affect thresh-

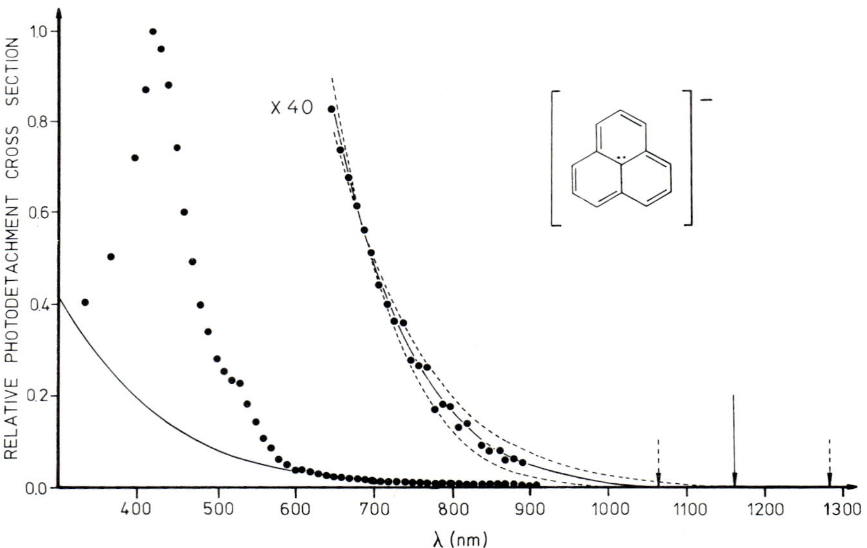

FIG. 6 Relative experimental (●) and theoretical (—) photodetachment cross sections for the perinaphthenyl anion plotted as a function of incident photon wavelength. The large arrow points to the threshold at 1.07 eV. Dashed lines and arrows indicate the theoretical curves with thresholds offset by ± 0.1 eV from the best fit. [Reprinted with permission from Gygax *et al.* (1979). Copyright 1979 American Chemical Society.]

old behavior. This effect has been treated theoretically by O'Malley (1965). For polyatomic species, molecular dipoles can provide an additional means of interaction. Such effects have been modeled in the photodetachment of OH^- and OD^- (Schulz *et al.*, 1982) and implicated in the photodetachment of CH_3S^- and CD_3S^- (Janousek and Brauman, 1980). These effects are more fully discussed in the chapter by Mead *et al.* in this volume.

C. *Vibronic Transitions in Photodetachment*

The one-electron picture for photodetachment by Reed *et al.* (1976) describes a cross section for only the electronic portion of the photodetachment transition for the anion. If the geometry of the anion and the neutral differ significantly, poor Franck–Condon overlap will smear out the observed cross section by allowing nondiagonal transitions between the ground state anion and vibrationally excited neutrals. These will occur at energies greater than that of the adiabatic transition $A(v' = 0, 0, 0 \ldots) \leftarrow A(v'' = 0, 0, 0 \ldots)$. In cases of extremely poor overlap, the adiabatic transition may be dominated by higher energy onsets or may not even be discernible.

An example of poor Franck–Condon overlap factors suppressing the adiabatic photodetachment threshold is seen in the case of the CF_3^- anion (Richardson *et al.*, 1975a). The electron affinity of the CF_3 radical can be calculated by a thermochemical cycle to be 2.01 eV. Yet, the photodetachment threshold for CF_3^- corresponds to a photon energy of 2.82 eV. Complete neglect of differential orbitals (CNDO) calculations indicate a substantial geometry change in going from anion to neutral; the F—C—F bond angle corresponds to 116° in the anion but only 105° in the neutral. The high-energy photodetachment threshold can be explained by assuming that the Franck–Condon overlap becomes substantial only for a transition to a highly vibrationally excited neutral.

Poor Franck–Condon overlap was also invoked to explain the large discrepancy between experimental and theoretical cross sections for SF_6^- photodetachment (Drzaic and Brauman, 1982). No photodetachment was observed from SF_6^- for photon energies up to 4.0 eV. Other experimental methods have placed the electron affinity of SF_6 as being roughly 0.5 eV. A calculation of the photodetachment cross section for an octahedral SF_6^- shows a large cross section at energies within 1 eV of threshold. This lack of agreement between experiment and theory is explained by postulating that the structure of SF_6^- is distorted significantly from that of the neutral. This result is used with other evidence to suggest that SF_6^- may exist as an ion–molecule complex rather than a discrete, octahedral molecule.

Alkoxide anions produce experimental cross sections which do not show the sharp onset predicted by theory, but instead rise more slowly (Janousek *et al.*, 1978). This has been interpreted as the result of several closely spaced vibronic transitions near threshold. In fact, a number of vibrational onsets have been picked out in *t*-butoxide in a high-resolution dye laser experiment (vide infra). Geometry optimization calculations using CNDO suggest a significant change in bond lengths and angles in methoxyl and *t*-butoxyl anions and radicals. A similar effect of vibronic transitions smearing out the adiabatic onset is observed in the photodetachment of a number of enolate anions (Zimmerman *et al.*, 1977).

Many times it is difficult to assign the onsets for vibronic transitions in a photodetachment experiment. Since the spectrum is integral in nature, onsets appear as a change in slope of the cross section rather than as distinct peaks. Often these slope changes are quite gradual and difficult to spot by inspection. To aid in the evaluation of vibronic transitions, a procedure has been developed which transforms the integral photodetachment spectrum into a derivative spectrum. This is accomplished by fitting the points of the integral spectrum to a smoothed cubic spline and differentiating this function. This procedure has been used to locate vibronic and rotational fine structure in alkoxides (Janousek *et al.*, 1978), thioalkoxides (Janousek *et al.*, 1980), and HS^- (Janousek and Brauman, 1981).

It is often important to include the effects of vibronic transitions in fitting calculated photodetachment cross sections to experimentally derived cross sections. One example of this is provided by Lineberger and co-workers, who obtained the laser photodetachment spectrum of NO_2^- in an ion beam (Herbst *et al.*, 1974). Using calculated Franck–Condon factors and known vibrational frequencies, they were able to model the NO_2^- photodetachment spectrum as a function of temperature. This analysis led them to conclude that the temperature of the NO_2^- in their ion beam was between 1500 and 2000 K. This analysis was also used to determine the 0–0 photodetachment transition and a NO_2 electron affinity of 2.36 ± 0.10 eV. A similar analysis was performed by Woo *et al.* (1981) on the photodetachment spectrum of NO_2^- generated in a drift tube. The ions in this experiment were assumed to be thermalized, and the structural constants of NO_2^- were determined from the best fit of their calculated curve.

Vibronic structure has been observed in the negative ion photoelectron spectra of a number of species. The frequencies and intensities of these vibrational transitions often make it possible to determine structural constants in neutral photoproducts. In the case of CH_3^-, the measurement of the vibrational frequencies using photoelectron spectroscopy (PES) allowed the determination of the relative geometry of the radical and the anion (Ellison *et al.*, 1978). This was achieved by calculating the expected photoelectron spectrum for the transition from pyramidal CH_3^- to a planar CH_3. This calculated spectrum was in excellent agreement with the experimental results. A partial list of negative ions that have shown vibrational structure in their photoelectron spectra are $Ni(CO)_n^-$ ($n = 1$–3) (Stevens *et al.*, 1982); $Fe(CO)_n^-$ ($n = 1$–4) (Engelking and Lineberger, 1979); CH_3O^-, CD_3O^-, and CH_3S^- (Engelking *et al.*, 1978); PO^-, PH^-, and PH_2^- (Zittel and Lineberger, 1976); SiH_2^- and SiH^- (Kasdan *et al.*, 1975a); S_2^- (Celotta *et al.*, 1974); O_2^- (Celotta, 1972); NO^- (Siegel *et al.*, 1972); and SO^- (Lineberger *et al.*, 1974).

The vibronic transitions discussed so far have resulted from photodetachment from a vibrationally cold anion to a vibrationally excited neutral. A related phenomenon appears if the anions in a photodetachment experiment are not all in their ground vibrational states. It is possible in these cases to induce transitions from these excited anions to ground-state neutrals. These "hot bands" have the potential of obscuring the adiabatic photodetachment onset by providing transitions at energies lower than the adiabatic process. In ICR experiments, hot bands are always a potential problem, as ions are produced by processes that are exothermic to varying degrees. Although an ion typically undergoes 10–100 collisions during its lifetime in the ICR, recent experiments have indicated that ion–molecule collisions are often inefficient in relaxing vibrationally excited ions (Jasinski and Brauman, 1980). Even drift tubes, with their relatively high background pressures, can produce ions which exhibit hot band behavior in photodetachment experiments (vide infra).

So far, attempts to affect photodetachment cross sections by changing the internal energy of the ions have met with only limited success. One means of affecting the internal energy of an ion is by altering the process by which it is made. In the case of *t*-butoxide anion, the photodetachment cross section for the ion formed via proton abstraction by F^- from *t*-butyl alcohol (up to 9-kcal/mole excess energy available) is the same as that formed by electron impact on *t*-butyl peroxide (up to 40-kcal/mole internal energy available) (Janousek *et al.*, 1978). Enolate anions formed from photon abstraction by F^- show a small wavelength shift in the photodetachment onset when different F^- sources are used. F^- formed from electron impact on SO_2F_2 may have significantly less translational energy than F^- formed from NF_3 (Franklin, 1976; Zimmerman *et al.*, 1977, ref. 26). Photodetachment cross sections using SO_2F_2 as the fluoride source appear slightly shifted to the blue relative to the F^-/NF_3 spectra (Zimmerman *et al.*, 1977). In another experiment, fine structure observed in the photodetachment of CH_3S^- generated by electron impact on CH_3SSCH_3 was broadened considerably when the mercaptide anion was generated via proton abstraction from CH_3SH by F^- (Janousek and Brauman, 1980).

Although in many cases ions are thermalized rapidly, there are examples of ions maintaining significant excitation even after hundreds of collisions.

This appears to be the case for NO_2^-. Photodetachment in both ICR (Richardson *et al.*, 1974) and in ion beam (Herbst *et al.*, 1974) experiments showed two onsets for photodetachment. The relative intensities of the two onsets in the ICR remained roughly the same over a relatively wide pressure range (8×10^{-8}–5×10^{-6} torr), arguing against a vibrationally activated NO_2^- being responsible for the long wavelength onset. In addition, a theoretical analysis of the ion beam data indicated that an unreasonably high ion temperature (5000 K) would be necessary to produce the observed ion beam photodetachment spectrum. These results led to the postulate that a higher energy peroxy (N–O–O) isomer of NO_2^- was responsible for the long wavelength onset.

However, Mosely and co-workers performed a series of drift tube experiments (pressure = 0.1 torr) that indicated the first onset was due to hot band transitions (Huber *et al.*, 1977). By increasing the number of ion–molecule collisions that NO_2^- undergoes prior to photodetachment, they were able to induce a decrease in the relative intensity of the long wavelength onset. These results indicate that NO_2^- can maintain significant internal excitation even after a large number of collisions. A more recent drift tube photodetachment study by Woo *et al.* (1981) failed even to detect a long wavelength onset, presumably because the ions were more successfully thermalized. The observation that NO_2^- can maintain significant vibrational excitation after hundreds of collisions is unusual, but it points out the potential problems that a distribution of internal energies of ions can cause in the interpretation of photodetachment cross sections.

It should be noted that even thermalized molecules can exhibit hot bands if they possess low-frequency vibrational modes. At room temperature, vibrational modes with frequencies $\leqslant 400\ \text{cm}^{-1}$ are populated to a significant extent ($>10\%$). In the case of *t*-butoxide photodetachment using a high-resolution dye laser, the region near threshold shows a number of vibrational onsets (Fig. 7). As mentioned, attempts to increase the internal energy of the *t*-butoxide anion by altering its means of formation were unsuccessful. Lacking data on the dependence of these transitions on internal energy, it is difficult to assign any one onset as the definitive adiabatic transition.

Hot bands play an important role in a controversial series of experiments dealing with the species CH_2^-. Negative ion photoelectron experiments by Engelking *et al.* (1981) measure the photon energy needed for the transition from the anion to the $^3B_1$ ground state and the $^1A_1$ first excited state of the neutral. Due to a substantial geometry change, a vibrational progression is expected and observed in the transition to the triplet ground state. One dominant peak is also obtained for the anion to singlet transition, where only a slight geometry change is expected. Based on assignments of the vibrational progressions, a methylene singlet–triplet splitting of 19.5 kcal/

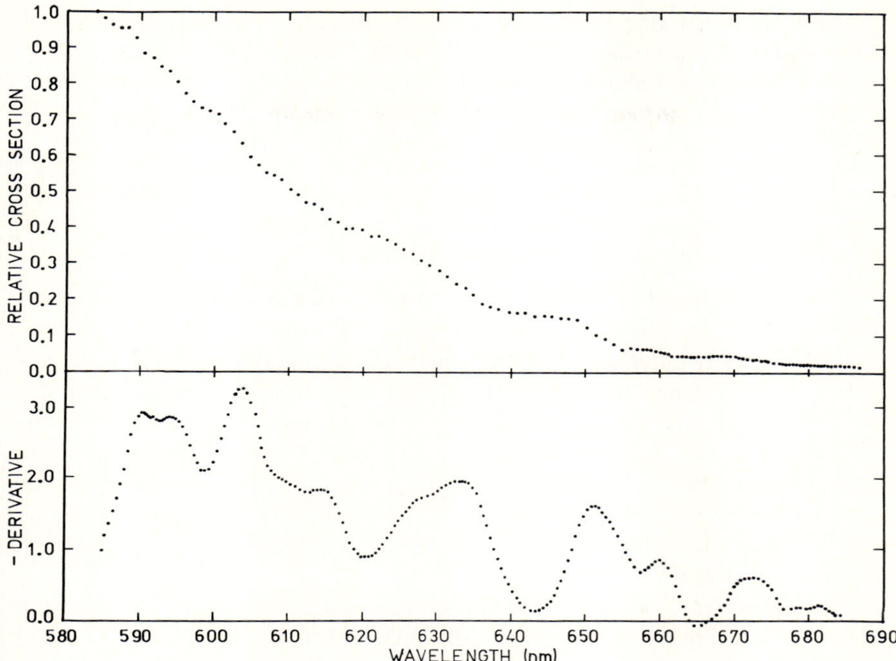

FIG. 7 Experimental laser photodetachment cross section *tert*-butoxyl anion and smoothed derivative of the observed cross section. [Reprinted with permission from Janousek *et al.* (1978). Copyright 1978 American Chemical Society.]

mole was determined. This is at odds with a large number of experimental and theoretical studies which place the splitting at roughly 8 kcal/mole (Hayden *et al.*, 1982, and references therein). Interestingly enough, it is possible to get an 8 kcal/mole splitting from the photoelectron data if a number of the observed transitions are assumed to be hot bands, and the fourth ("D") onset assigned to the 0–0 transition. However, numerous attempts to change the internal energy of the methylene anion prior to photodetachment failed to change the relative intensities of the observed transitions, making it difficult to rationalize the presence of hot bands in this series of experiments.

Finally, in small systems, rotational transitions may be spaced far enough apart to be resolved with tunable laser light sources or by high-resolution photoelectron spectrometers. Individual rotational transitions have been observed by Lineberger and co-workers in the systems C_2^- (Jones *et al.*, 1980) and OH^- and OD^- (Schulz *et al.*, 1982). These allow for the determination of rotational spectroscopic constants for these systems. In comparison, lower resolution experiments allow for the observation of averages of rotational branches in hydroxide (Hotop *et al.*, 1974); the HS^- anion (Janousek and Brauman, 1981); and the OH^-, SH^-, and SD^- anions (Breyer *et al.*, 1981). In these latter experiments, theoretical photodetachment cross sections were calculated and compared to the experimental results in order to make an estimate of the actual rotational temperature of the ions.

D. *Electronic Transitions in Photodetachment*

Photodetachment spectroscopy can provide useful information concerning the electronic structure and energetics of negative ions. Because of solvation and counterion effects on electronic transitions, gas phase experiments are necessary to assign the excited-state transiton energy and the electron affinity of the excited state unambiguously. Electronic states in negative ions will generally be higher in energy than the thermodynamic threshold for electron detachment, although certain ions have been shown to possess bound excited states. In the following section, examples of anions whose electronic structure have been probed by photodetachment spectroscopy are illustrated.

The presence of an electronically excited state above the threshold for electron detachment provides an additional pathway for electron loss, i.e., excitation to an excited state followed by autodetachment. This alternate channel may show up as a resonance in the photodetachment cross section. The width of this resonance is related to the lifetime of the excited state of the anion. The intensity of the resonance will depend on the transition probability for accessing the excited state relative to direct photodetachment. In addition, multiple resonances in photodetachment may appear if different vibrational or rotational levels of the excited state are accessible or if more than one electronic excited state exists.

Broad resonances far above the photodetachment threshold have been observed in a number of large delocalized anions (Richardson *et al.*, 1975b). The photodetachment spectrum of phenoxide anion is shown in Fig. 8. The photodetachment spectrum shows a sharp increase in cross section below 400 nm, followed by a subsequent decrease. The sharp increase in cross section cannot be explained by formation of an excited state of the neutral, as the lowest observed excitation of the phenoxyl radical is about 2 eV above the ground state (Ward, 1968). Furthermore, it is unlikely that this onset is due to a channel leading to an excited state of the neutral photoproduct, as this would not lead to the decrease in cross section observed at shorter wavelengths. Therefore, it is reasonable to assign this resonance to an $n \rightarrow \pi^*$ or $\pi \rightarrow \pi^*$ transition in the phenoxide ion.

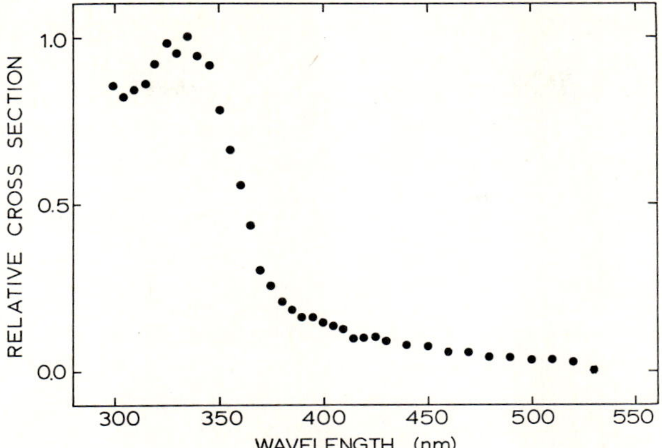

FIG. 8 Relative cross section for photodetachment from phenoxide anion. Photodetachment threshold appears at 525 nm. [Reprinted with permission from Richardson *et al.* (1975d). Copyright 1975 American Institute of Physics.]

This resonance corresponds well to the lowest energy transition observed in the solution absorption spectrum of phenoxide (Tichy and Zahradnik, 1969). The maximum in the gas phase is red-shifted about 50 nm from that in solution. This red shift is consistent with the expectation that solvation effects can increase the energy of electronic transitions by destabilizing an electronic excited state relative to the ground state (Amos and Burrows, 1973).

An example of the correspondence between the resonances in photodetachment spectra and solution phase absorption spectra is the perinaphthenyl anion. The photodetachment spectrum of the perinaphthenyl anion is shown in Fig. 6 (Gygax *et al.*, 1979). The photodetachment cross section for this ion peaks at about 420 nm and then rapidly decreases. The solution absorption spectrum (Fig. 9) shows a sharp resonance at 430 nm that is very similar to the photodetachment resonance. Unlike the phenoxide spectrum, the maximum in the perinaphthenyl anion is essentially unchanged in the gas phase spectrum relative to the solution spectrum. In the perinapthenyl anion there is no hydrogen bonding with the solvent and apparently no differential stabilization of the excited and ground states.

Broad resonances have also been observed in heptatrienyl and pentadienyl anions (Zimmerman *et al.*, 1978a). In these systems the resonances are assigned as $\pi \rightarrow \pi^*$ transitions. The maximum for the heptatrienyl anion is shifted to the red of that for the pentadienyl anion. A similar red shift is observed in spectra of neutral polyenes of different sizes, as a result of the

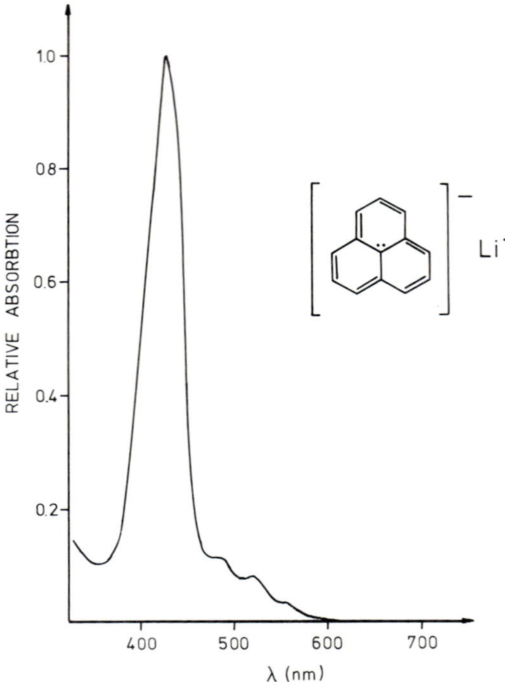

FIG. 9 Absorption spectrum of the lithium salt of perinaphthene in ether obtained by reaction with butyllithium. [Reprinted with permission from Gygax *et al.* (1979). Copyright 1979 American Chemical Society.]

increase in conjugation in the larger systems. The transitions in the ions are considerably lower in energy than those in neutral polyenes of similar conjugation length. The red shift in the anions was attributed in part to the diffuseness of the electron distribution in the anions compared to the neutrals.

However, a red shift has also been observed in the solution absorption spectra of protonated polyenes (Sorensen, 1965). The maxima in the cation spectra correspond closely to the maxima in the anion spectra for ions with the same number of *p* orbitals in the polyene system. The red shift in both the anions and the cations appears to be a consequence of these systems possessing an odd number of orbitals in their respective π systems (Navangul and Blatz, 1973). Calculations at the self-consistent field (SCF) level have estimated the transition energies for a number of odd and even orbital polyene systems, and the values derived support the experimentally observed results (Navangul and Blatz, 1978).

The red shift associated with the odd versus even polyenes can be explained by considering the symmetry and the orbital pairing properties of alternate π systems. The lowest energy $\pi \rightarrow \pi^*$ transition in an odd, charged polyene system transfers an electron to an orbital with a considerably different electron distribution. In contrast, the comparable transition in an even, uncharged polyene involves two orbitals of identical electron distribution (in the Hückel molecular orbital approximation). Thus, electron repulsion interactions will be lowered more for the transitions in the odd systems compared to the even ones, and the transition energy will be correspondingly lower.

Transitions to an autodetaching electronic state have been observed by Lineberger and co-workers for the C_2^- system (Jones *et al.*, 1980). As shown in Fig. 10, the C_2^- photodetachment spectrum does not look like the normal integral spectrum usually observed in photodetachment, but instead appears as a series of clusters of sharp peaks (see Fig. 5 for the calculated C_2^- photodetachment spectrum). These clusters have been interpreted as different vibronic transitions to an autodetaching excited electronic state of C_2^-. Individual peaks correspond to rotational fine structure in a vibronic transition. An extensive analysis of these vibrational and rotational transitions indicates that the lower vibrational levels of this electronically excited state lie below the thermodynamic threshold for electron detachment, providing evidence for one of the few known bound electronic excited states in anions. This analysis also provides rotational constants for both the ground and

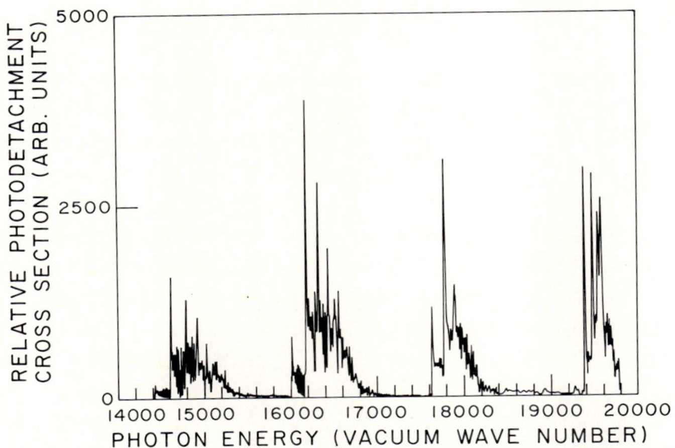

FIG. 10 C_2^- low-resolution ($\bar{\nu} \sim 10\ cm^{-1}$) photodetachment cross section, plotted against photon energy. [Reprinted with permission from Jones *et al.* (1980). Copyright 1980 American Institute of Physics.]

excited states of the anion. Observation of the direct photodetachment process in this system is suppressed due to a very strong transition moment to the excited state, as well as by extremely poor Franck–Condon overlap for diagonal transitions from the anion to the neutral.

Narrow resonances have also been seen in the threshold region of several enolate anions. The photodetachment spectrum for acetophenone enolate is shown in Fig. 11 (Zimmerman and Brauman, 1977b). The threshold region resolution is about 3 cm^{-1} and was measured using a tunable dye laser. The resonances appearing in the threshold region of these systems appear to correspond to a vibrational progression of an excited electronic state superimposed upon the rising background for direct photodetachment. The width of the threshold resonances is much narrower than the resonances previously observed in large delocalized anions.

The energy of the threshold resonances does not correspond to what would be expected for either a $\pi \rightarrow \pi^*$ or a $n \rightarrow \pi^*$ transition. However, there is a large broad resonance in the blue end of the spectrum which may be assigned as one of these transitions. This assignment was substantiated by studying a series of substituted acetophenone enolates, with electron-donating and electron-accepting substituents placed on the phenyl ring (Jackson *et al.*, 1979). These experiments show a blue shift in the high-energy maximum for para-substituted F, CH_3, and *t*-butyl acetophenone enolates, with the magnitude of the shift following the order F > CH_3 > *t*-butyl. In contrast, a para-CN substituent induces a red shift in the maximum

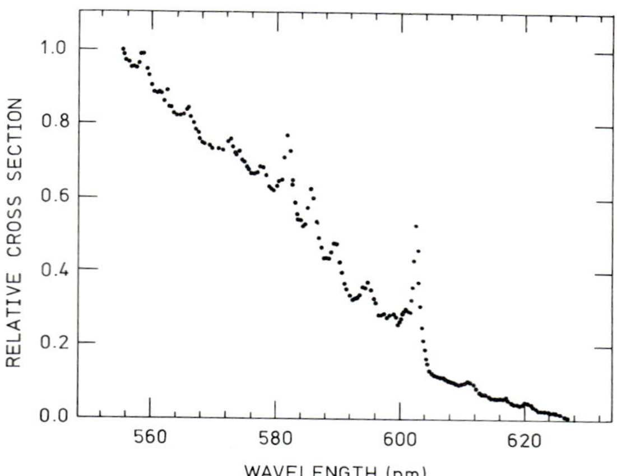

FIG. 11 Relative laser photodetachment cross section for acetophenone enolate anion. Resolution better than 0.1 nm. [Reprinted with permission from Zimmerman and Brauman (1977b). Copyright 1977 American Institute of Physics.]

of the high-energy resonance. The direction of these shifts agrees with what would be expected for $\pi \rightarrow \pi^*$ or $n \rightarrow \pi^*$ transitions due to inductive effects on the π-electron system caused by substitution of the phenyl ring.

The narrow low-energy resonances follow the photodetachment threshold upon substitution in the phenyl ring. These shifts in the threshold upon substitution are opposite to those observed in the maximum of the broad resonances and make it unlikely that the threshold resonances correspond to an $n \rightarrow \pi^*$ or $\pi \rightarrow \pi^*$ transition. A Fano line-shape analysis (Fano, 1961) of the peak widths places the lifetime of the excited state between 10^{-12} and 10^{-13} sec, the period of a few molecular vibrations. This lifetime is considerably longer than that expected for an unbound electronic excited state.

It was not possible to make a vibrational assignment of the resonances in acetophenone because of the large number of possible modes. A simpler system that shows resonances is the acetaldehyde enolate anion (Jackson *et al.*, 1981a). Comparison of the spectra of the deuterated and nondeuterated anions indicates that the first resonance corresponds to the transition from the lowest vibrational level of the ground state to the lowest level of the excited state (0–0 transition). Because of the small size of the molecule, it was possible to calculate the vibrational frequencies of the acetaldehyde enolate radical. Reasonable agreement was obtained between the calculated and observed vibrational frequencies.

Since these resonances in enolate anions do not correspond to any known state of the neutral species, it is necessary to consider that a new low-lying electronic state exists in the anion. One possible explanation would be a diffuse electronic state in which the electron is bound by the dipole moment of the radical. The possibility of such a dipole-supported state has been discussed in the theoretical literature for many years (Turner, 1977). Such a state would be analogous to a Rydberg state of a neutral system but of much lower energy. The dipole moment necessary to support a bound electron is 1.625 Debye for the stationary system (Turner and Fox, 1966) and up to 30% larger for a rotating system (Garrett, 1970). Garrett (1979) has more recently suggested that for systems with large dipole moments, bound states must necessarily exist.

To explore the possibility of the existence of such a state, an ab initio calculation on acetaldehyde enolate anion has been performed (Wetmore *et al.*, 1980). The results of this calculation predict an excited state with energy very close to the experimental value of the 0–0 transition observed by Jackson *et al.* The wave function resulting from this calculation corresponds to a diffuse nonbonding orbital that has a large amount of oxygen 3s character but a node near the carbonyl carbon (Fig. 12). The potential surface of this state appears similar to that of the neutral radical. These results indicate that this state is well described as a radical binding a loose electron

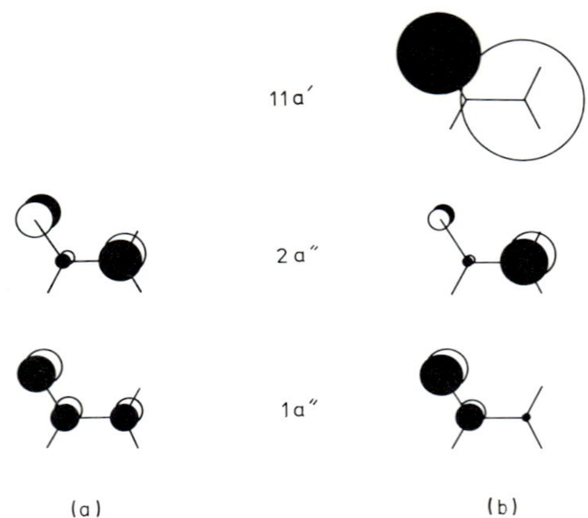

FIG. 12 Highest occupied molecular orbitals of acetaldehyde enolate: (a) ground anion state; (b) lowest excited state. [Reprinted with permission from Wetmore *et al.* (1980). Copyright 1980 American Chemical Society.]

and support the hypothesis that these resonances may correspond to an electronically excited, dipole supported state.

Negative ion spectroscopy has also been applied to probe other aspects of electronic structure in anions and neutrals. For example, photodetachment and photoelectron spectroscopy have been used to measure the spin–orbit splitting in a number of species. Spin–orbit levels of radicals provide different final states for photodetachment transitions, providing multiple onsets in the photodetachment spectra or doubling of the peaks in the photoelectron spectra.

Photodetachment spectroscopy was first used to measure the spin–orbit splitting in the SeH system (Smyth and Brauman, 1972a). For this molecule, the ground-state spin–orbit splitting is sufficiently large (1815 cm^{-1}) that the relatively low resolution (14.1 nm) of the experiment was sufficient to resolve the splitting. No off-diagonal vibrational transitions were observed, indicating that the geometry of the neutral is very close to that of the ion.

A high-resolution laser photodetachment study has been carried out on the SH^- ion (Janousek and Brauman, 1981). The spectrum in this system is too complex to extract the spin–orbit splitting directly, but it was possible to use a previously measured value (Ramsay, 1952) to construct a theoretical cross section that shows good agreement with experiment. A similar analysis of the photoelectron spectra of the OH^-, SH^-, and SD^- anions was performed by Breyer *et al.* (1981).

Another set of molecules, in which the spin–orbit splitting has been measured using photodetachment spectroscopy, are the CD_3S and CH_3S radicals

(Janousek and Brauman, 1980). In the CD_3S^- and CH_3S^- anions, several vibrational onsets are observed in the photodetachment spectrum. In both systems there is what appears to be a doubling of the channel openings. This doubling of the onsets corresponds to the spin–orbit splitting of the $^2E_{3/2}$ and $^2E_{1/2}$ states of the radical and corresponds to a spin–orbit splitting of 280 cm^{-1}.

E. Electron Affinities and Thermochemistry

One of the more important thermochemical parameters that may be determined in a photodetachment experiment is the electron affinity of the neutral photoproduct. Molecular electron affinities are important in determining the thermochemistry of a large number of gas phase reactions, as well as in understanding trends in the relative stabilities of anions and neutrals in series of related chemical species. A discussion and tabulation of molecular electron affinities have appeared in Volume 2 of this treatise (Janousek and Brauman, 1979). In this section, we discuss some interesting aspects of molecular electron affinities determined by negative ion photodetachment. Applications of electron affinities to the determination of neutral bond strengths will be illustrated. In addition, a table of updated molecular electron affinities appears at the end of this chapter.

The growing number of measurements of molecular electron affinities and gas phase acidities of related species have allowed homolytic bond dissociation energies to be determined by use of a thermochemical cycle. This cycle is illustrated in reaction (9)

$$\begin{array}{rl} A^- \rightarrow A + e^- & \text{(Electron affinity) of A} \\ AH \rightarrow A^- + H^+ & \text{Proton affinity of } A^- \\ H^+ + e^- \rightarrow H & -\text{Ionization potential H} \\ \hline AH \rightarrow A + H & \Delta H^0 = \text{Bond dissociation energy of A—H.} \end{array} \tag{9}$$

This method of determining bond strengths is attractive in that the thermochemical properties of unstable radicals may be studied if their corresponding ions are relatively stable. Alternative methods that directly involve reactions of unstable radicals are often difficult to carry out. A comparison of a number of bond strengths determined via this thermochemical cycle with bond strengths measured by thermokinetic means show very good agreement in most cases (Brauman, 1980).

Recent photodetachment and gas phase basicity studies on the azide anion N_3^- have resulted in some thermochemical values that differ significantly from previous literature values determined by indirect means. Photodetachment of the azide anion N_3^- produces a photodetachment spectrum with two apparent onsets, occurring at 900 and 460 nm (Jackson *et al.*, 1981b). Double resonance and pulsed ICR experiments indicate that the long wavelength onset is due to a precursor of N_3^- with an *m*/*e* of 87. This corresponds to the species $(CH_3)_3SiN^-$; it appears that nearly all of the azide anion formed is due to the process shown in reaction (10)

$$\begin{aligned} (CH_3)_3SiN_3 + e^- &\rightarrow (CH_3)_3SiN^- \\ (CH_3)_3SiN^- + (CH_3)_3SiN_3 &\rightarrow N_3^-. \end{aligned} \tag{10}$$

The short wavelength onset can be ascribed to N_3^- photodetachment. This gives an N_3 electron affinity of 62.1 ± 2.8 kcal/mole.

This electron affinity, when coupled to a recent proton affinity measurement for the azide anion, leads to some serious discrepancies with indirect determinations of N_3 and HN_3 thermochemistry (Pellerite *et al.*, 1981). The proton affinity of N_3 was bracketed as 344 ± 2 kcal/mole. The use of N_3^- generated from three separate sources gave consistent results, arguing against internal energy effects playing a significant role in these experiments. Previous indirect determinations of HN_3 and N_3 thermochemistry, made by analyzing Born–Haber cycles for alkali metal azides and calculated lattice energies (Jenkins and Pratt, 1977), place the N_3^- proton affinity as 330.5 kcal/mole. Since the gas phase relative proton affinity scale is internally consistent and well established relative to absolutely determined acidities (Bartmess and McIver, 1979), it appears that these indirect determinations are in error.

Combination of the N_3 electron affinity with the N_3^- basicity measurement allows the N—H bond strength in HN_3 to be placed at 92 ± 5 kcal/mole, much higher than the 79.4-kcal/mole value determined with the older thermochemical data. This higher bond strength value is more in line with the N—H bond dissociation energies determined for species such as ammonia (H_2N—H, 107 kcal/mole; see Brauman, 1980) and NH_2 (HN—H, 95 kcal/mole; see Okabe, 1978).

Both a thermochemical cycle and a novel use of an excited state in photodetachment were used to determine the electron affinity of the hydroperoxyl radical (HOO) (Bierbaum *et al.*, 1981). The direct observation of the adiabatic photodetachment process for HOO^- lies outside the wavelength range available in the crossed ion beam/dye laser apparatus used. However, the 2A′ excited state of HOO radical is known to lie 0.872 eV above the ground state. The photodetachment onset of the channel resulting in the excited neutral product was found near 2.06 eV. This gives an HOO electron affinity of 1.19 ± 0.01 eV, in agreement with the value of 1.16 ± 0.15 eV determined using gas phase acidity measurements and a thermochemical cycle.

The effects of ring strain on the relative stabilities of anions and their radical counterparts have been probed by photodetachment experiments

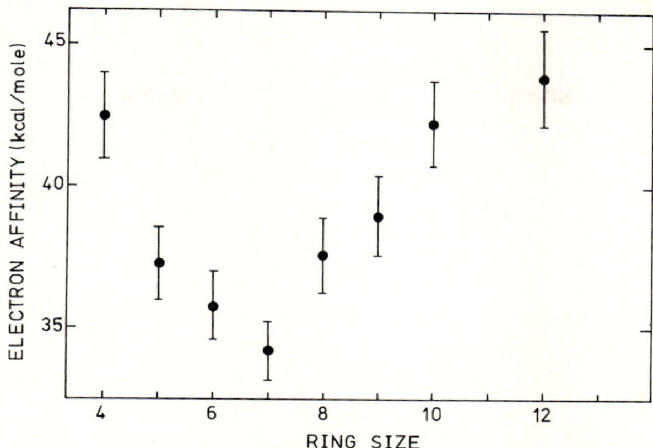

FIG. 13 Electron affinities of cyclic enolate radicals, as a function of the number of carbon atoms in the cyclic system. [Reprinted with permission from Zimmerman *et al.* (1978b). Copyright 1978 American Chemical Society.]

(Zimmerman *et al.*, 1978b). Photodetachment onsets were measured for a number of cyclic enolates with ring sizes varying from 4 to 12. The relative electron affinities of these species are shown in Fig. 13. It is seen that the electron affinity decreases markedly in going from a four-membered system to a seven-membered ring. This effect is interpreted in terms of the dependence of the relative stabilities of the anion and radical being dependent on the O—C—C bond angle. This bond angle in these enolates in dictated by the size of the ring. Geometry optimization by CNDO suggest an optimum O—C—C bond angle of 132° in the anion, but only 123° for the radical. Thus, for the four-membered ring system, the anion is stabilized by the large O—C—C bond angle enforced by the ring, and the electron affinity is large. As the size of the ring is increased however, this bond angle becomes smaller, with a concomitant decrease in the stability of the anion and of the electron affinity. The increase in the electron affinities for the eight-membered and larger rings is interpreted as being the result of the relaxation of angle constraints in a more flexible ring system coupled with an increase in the polarizability of the larger hydrocarbon system.

The determination of electron affinities and gas phase acidities has permitted an analysis of the effects of alkyl substitution in a series of alcohols and mercaptans. These electron affinities, gas phase acidities, and resultant bond strengths are listed in Table I. It is seen that, in each series, the bond strength is independent of the alkyl substituent, indicating that alkyl groups offer no special stabilization for the radicals in these systems. This is consistent with thermokinetic determinations (Shaw, 1974).

TABLE I

Electron Affinities, Gas Phase Acidities, and Calculated Bond Dissociation Energies for Selected Alcohols and Mercaptans[a]

	Electron affinity of radical RO or RS (kcal/mole)	Gas phase acidity (kcal/mole)	RO–H and RS–H bond strength (kcal/mole)
CH_3OH	36.7 ± 0.09	379.2 ± 2	101.3 ± 1.0
$(CH_3)_3COH$	43.1 ± 0.2	373.3 ± 2	102.8 ± 1.0
	43.8 ± 0.2		102.1 ± 1.0
$(CH_3)_3CCH_2COH$	44.5 ± 1.4	371.8 ± 2	102.2 ± 1.4
CH_3SH	42.91 ± 0.09	359.0 ± 2	88.1 ± 2
CD_3SH	42.80 ± 0.14	359.3 ± 2	88.3 ± 2
C_2H_5SH	45.04 ± 0.14	357.4 ± 2	88.6 ± 2
n-C_3H_7SH	46.1 ± 0.5	356.4 ± 2	88.7 ± 2
i-C_3H_7SH	46.6 ± 0.5	355.6 ± 2	88.4 ± 2
t-C_4H_9SH	47.7 ± 0.5	354.7 ± 2	88.6 ± 2

[a] Electron affinities from Janousek *et al.* (1978, 1980). Gas phase acidities from Bartmess and McIver (1979).

Trends in the electron affinities for these alcohols and mercaptans may be described by a perturbation molecular orbital treatment, which has been thoroughly discussed elsewhere (Janousek *et al.*, 1978, 1980). Briefly, the stability of the anion depends on the extent of delocalization from the filled 2p orbitals on O (or 3p on S) into the unfilled π^* orbitals in the alkyl system. This correlates well with the semiclassical notion of alkyl polarizability, with charge density being spread more readily onto larger alkyl systems, resulting in a more stable ion.

III. INFRARED MULTIPHOTON-INDUCED ELECTRON DETACHMENT

A. *Introduction*

The development of high-power infrared lasers has spawned a great deal of work concerning the effects of intense infrared radiation on molecules. In the past few years, these studies have been extended to include ionic systems, and the infrared multiphoton dissociation of a number of anions and cations has been probed. Among the reactions that have been observed are simple fragmentation processes, multistep reactions, and rearrangements indicating that cyclic transitions states may be involved (see Thorne and Beauchamp, Chapter 18).

Recently, benzyl anion trapped in an ion cyclotron resonance (ICR) spectrometer was shown to undergo electron detachment following irradiation by a high-power pulsed carbon dioxide laser (Rosenfeld *et al.*, 1979). Since then, a number of anions have shown similar behavior upon infrared irradiation. Electron detachment is thermodynamically favored over dissociation in a number of cases, as the energy required to remove an electron from an anion (typically < 3 eV) is often less than that required to break a bond. The reverse process, nondissociative electron attachment, is a well-known phenomenon and has been the focus of a great deal of study. This makes it likely that infrared multiphoton-induced electron detachment is also a generalized phenomenon and may be observable in a large number of systems.

This section contains a review of the reported observations of infrared multiphoton-induced electron detachment, along with the information that may be drawn from these experiments. Possible mechanisms for these processes are discussed and related to some theoretical work that has been done in this area, as well as to experimental work in the related field of nondissociative electron attachment.

B. *Experimental Considerations*

Observations of infrared multiphoton-induced electron detachment from anions have been observed using ion cyclotron resonance spectrometers in conjunction with tunable carbon dioxide lasers. The experimental arrangements are the same as those used in other infrared laser studies of ions using ICR spectrometers (Bomse *et al.*, 1979; Jasinski *et al.*, 1982). Briefly, the ICR spectrometers are used in the pulsed mode as trapped ion cells. Ions are formed by a short (~ 10 msec) grid pulse of electrons and are trapped for up to several seconds in the ICR cell. Irradiation by a CO_2 laser takes place at an arbitrary time after ion formation, with the ions detected after another arbitrary amount of time. Ions are then removed from the cell, and the cycle is repeated.

The infrared sources in these experiments have consisted of either a pulsed CO_2 TEA laser or a continuous output CO_2 laser. Typically, the CO_2 TEA laser used in these experiments puts out a pulse of several joules over several microseconds, corresponding to a photon flux of megawatts per centimeter squared. The temporal profile of the pulse consists of a 200-nsec spike followed by a 3-μsec tail, each containing roughly 50% of the total energy of the pulse. In comparison, experiments using continuous output CO_2 lasers have used fluxes of about 30 W/cm^2. The effective infrared intensity seen by the ions in the ICR is doubled by using a mirror at the end of the ICR cell to reflect the infrared radiation back through the cell. CO_2 lasers are tunable over a series of lines spaced roughly 2 cm^{-1} apart over the range of 923–987 and 1027–1088 cm^{-1}.

Infrared multiphoton-induced electron detachment in an ICR spectrometer may be observed in a number of ways. Indirect evidence is provided by a decrease in a particular ion signal without observation of any photoproducts following irradiation. This may be misleading, however, as photoproducts are sometimes not trapped well or may undergo further decomposition. More direct evidence is obtained by adding a small amount (10^{-8} torr) of CCl_4 to the ICR cell. Carbon tetrachloride readily captures a free electron to form chloride anion, so electron detachment may be detected as an increase in the chloride ion signal following infrared irradiation.

C. *Experimental Results*

Infrared multiphoton-induced electron detachment has been observed in a number of hydrocarbons using a high-intensity pulsed carbon dioxide TEA laser (Meyer *et al.*, 1982). These species include the benzyl, allyl, 2,4-hexadienyl, 1,3-cycloheptadienyl, and anilide anions. Electron detachment from the indenyl anion was not observed. These compounds were all formed via proton abstraction from the parent neutrals and are similar in that the negative charge of the ion is delocalized over the π system of the molecule. Of the systems where electron detachment was observed, the anilide system possessed the highest electron affinity: 1.70 eV. If the anilide ion is formed in its ground vibrational state, infrared-induced electron detachment indicates that at least 14 CO_2 laser photons were absorbed.

The cross section for electron detachment in these species displayed some dependence on the wavelength of the incident infrared radiation. However, these infrared multiphoton-induced electron detachment spectra cannot be correlated with the infrared absorption spectra of the parent neutrals. One example is the anilide ion, which undergoes extensive electron detachment when irradiated by CO_2 laser radiation in the range of 930–990 cm^{-1}, although the parent aniline shows no absorption features in the 900–1100 cm^{-1} range of its gas phase infrared spectrum.

Wight and Beauchamp (1981) have used infrared multiphoton-induced electron detachment spectra to distinguish between three $C_7H_7^-$ isomers in the gas phase. Using a "low-power" (20 W/cm^2) cw CO_2 laser, different wavelength dependencies were seen in the infrared electron detachment spectra of benzyl and cycloheptatrienyl anions. Irradiation of the $C_7H_7^-$ species formed by proton abstraction from norbornadiene by NH_2^- produced no detectable electron detachment. These results indicate that the three isomers are structurally distinct in the gas phase.

The reaction of CH_3O^- with norbornadiene results in a small amount of a $C_7H_7^-$ product, although direct proton abstraction by methoxide is estimated to be 23 kcal/mole endothermic. However, rearrangement and deprotonation to the benzyl and cycloheptatrienyl systems are both exothermic. Irradiation of the $C_7H_7^-$ species formed in this manner results in a spectrum which appears to be a composite of the benzyl and the cycloheptatrienyl anion electron detachment spectra previously measured. This suggests that the norbornadiene system can rearrange to both of the other $C_7H_7^-$ isomers upon reaction with methoxide.

Infrared multiphoton-induced electron detachment has been observed for the SF_6^- system using a high-power CO_2 TEA laser (Drzaic and Brauman, 1981). This is the first instance of infrared-induced electron detachment from a system whose electron attachment properties have been studied (vide infra). SF_6^- shows an interesting dependence of the extent of its electron detachment on both the time separating the initial electron grid pulse forming SF_6^- and laser irradiation. There is also a dependence on the neutral background pressure in the ICR. These data are shown in Fig. 14. At 10^{-7} torr, it is seen that extensive detachment can be induced by firing the laser 200 msec after the electron grid pulse but that longer delays reduce the amount of electron detachment seen. At 600 msec, no electron detachment is observed. The same qualitative behavior is seen at 10^{-6} torr, but with the time scale compressed in which electron detachment may be observed.

This time and pressure dependence of the electron detachment properties of SF_6^- is most readily explained by postulating that the extent of electron detachment depends on the degree of vibrational excitation in SF_6^- at the

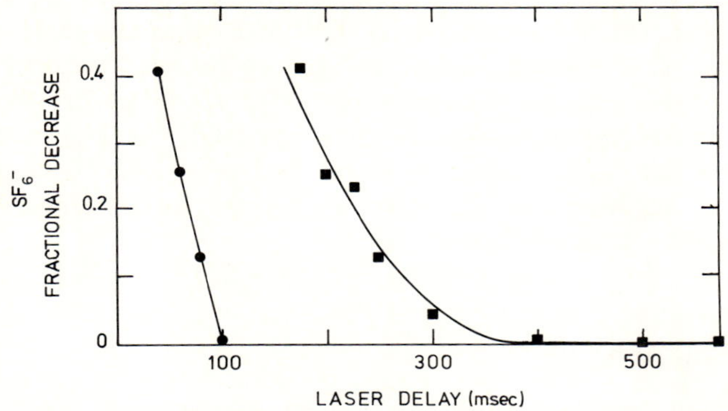

FIG. 14 Relative SF_6^- signal decrease versus CO_2 laser time delay after electron "grid" pulse. Pressure refers to background pressure of $SF_6 \cdot 10^{-6}$ torr data multiplied two times for clarity (●, 10^{-6} torr; ■, 10^{-7} torr). [Reprinted with permission from Drzaic and Brauman (1981). Copyright 1981 North-Holland Publishing Company.]

time of the laser pulse. At early times, the SF_6^- in the ICR may contain a significant amount of vibrational energy, with a correspondingly large density of states. At lager times, radiative and collisional deactivation produces an SF_6^- with much less vibrational energy and a correspondingly lower state density. At low energies, SF_6^- may not contain an infrared absorption mode resonant with the CO_2 laser and does not efficiently absorb the infrared radiation. At higher energies, the high density of states makes it more likely that there will exist two vibrational levels with the proper energy difference to absorb a CO_2 laser photon. This picture is supported by the pressure dependence on electron detachment, as a higher ion–molecule collision rate would deactivate excited SF_6^- more rapidly.

These and other experiments indicate that the infrared multiphoton spectrum and chemistry of an ion may depend strongly on the internal energy of the ion. For example, Rosenfeld *et al.* (1982) observed that both the wavelength and the fluence dependence of the infrared multiphoton-induced dissociation of the methanol solvate of the fluoride anion CH_3OHF^- depends on the internal energy of the ion. Other work has suggested that vibrationally excited molecules may undergo infrared multiphoton decomposition under conditions where the ground-state system will not (Ambartzumian *et al.*, 1977). Thus, caution must be used in applying infrared multiphoton chemistry as a means of identification of a molecule. Such a determination could be misleading if the internal energy of the molecule under study is not well defined.

An unusual feature of the SF_6^- study is in the comparison of these results with the infrared multiphoton decomposition observed for SF_6 neutral. SF_6 has been one of the most extensively studied systems in infrared multiphoton decomposition due to its large absorption cross section for CO_2 laser photons (Schulz *et al.*, 1980, and references therein). Bond cleavage 90 kcal/mole endothermic readily proceeds upon irradiation of SF_6 with a carbon dioxide laser. In contrast, vibrationally deactivated SF_6^- cannot absorb the 13 kcal/mole needed for electron detachment to be observed. This indicates that the infrared absorption characteristics of SF_6 and SF_6^- differ markedly, supporting the contention that the structure of SF_6^- differs significantly from that of SF_6 (Drzaic and Brauman, 1982).

D. *Discussion of Results*

Infrared multiphoton-induced electron detachment represents a set of reactions quite distinct from photodetachment using visible light. The optical photodetachment process is well described as a direct, one-electron transition between a bound electron in the anion and a free, continuum electron plus neutral. This transition is found to follow dipole selection rules, as in other optical processes. Being distinctly an electronic process, models invoking the Born–Oppenheimer principle work well in describing photodetachment cross sections (Section II.B).

In contrast, the mechanism for infrared multiphoton induced electron detachment has not been determined. In some systems, it appears that the detachment process involves a transition from the vibrationally excited potential surface of the anion to a low vibrational level on the neutral potential surface. This is in direct analogy to electron attachment via a nuclear-excited Feshbach resonance, in which electron capture by a neutral results in a vibrationally excited anion. Such processes would not be expected to be well described by a simple one-electron picture and, by necessity, violate the Born–Oppenheimer principle invoking separability between electronic and nuclear coordinates.

This process is almost certainly the case for the SF_6^- infrared-induced electron detachment. The reverse process, electron attachment, has been extensively studied by a number of workers (Christophorou *et al.*, 1971; Fehsenfeld, 1970). For electron attachment, the evidence points to formation of a vibrationally excited SF_6^- following electron capture. This anion will then either autodetach after a characteristic time (tens of microseconds) or can be vibrationally relaxed, allowing for an anion with essentially an infinite lifetime (Foster and Beauchamp, 1975). By this picture, excitation of SF_6^- by the CO_2 laser places the energy of the anion above the autodetachment limit, and the electron is quickly detached.

It is difficult to determine the relative efficiencies of electron detachment for different species using the techniques described so far. From nondissociative electron attachment studies, it is seen that the negative ion lifetime prior to autodetachment is typically on the order of tens of microseconds for "long-lived" anions (Christophorou, 1978). Although the rate of autodetachment at the thermodynamic threshold may vary by many orders of magnitude for different species, even slow autodetachment is fast on the time scale of the ICR spectrometers used in these experiments. In addition, vibrational relaxation is not a factor in pulsed-laser experiments, since radiative (milliseconds) and collisional (tens of milliseconds) relaxation are both slow compared to the length of the laser pulse and the autodetachment rate. Thus, ions that absorb sufficient energy to detach will show similar efficiencies in these experiments. Differences in electron detachment cross sections for ions and any wavelength dependence in these experiments will be due only to differences in the absorption rate of infrared photons.

There has not been enough work done to distinguish effects due to the infrared pumping rate on electron detachment. Assuming a cross section for absorption of a single photon of 10^{-19} cm^2, the pumping rate in a CO_2

TEA laser experiment is on the order of 10^7 sec^{-1}. In contrast, the continuous output laser pumping rate is most likely 10^2–10^3 sec^{-1}, on the same time scale as vibrational relaxation rates in the ICR. Vibrational relaxation rates are thought to have an effect on the differences in cross sections measured in infrared multiphoton decomposition studies comparing both high- and low-power CO_2 lasers (Jasinski *et al.*, 1982).

A set of propensity rules for infrared multiphoton-induced electron detachment involving a nuclear-excited Feshbach resonance type picture has been developed by Simons (1981). In his analysis, he treats the case where electron detachment involves only the ground electronic states of both the anion and the neutral. In his analysis, he predicts that the process of moving from the potential surface of the anion to the potential surface of the neutral should be favored in cases in which (1) the potential surfaces come close together and (2) the potential surfaces have a similar slope as a function of some internal nuclear motion coordinate. This analysis holds well in cases like SF_6^-, in which the geometry of the anion and the neutral are expected to vary considerably (Hay, 1982; Drzaic and Brauman, 1982) and the potential surfaces of the two species may intersect.

However, it is difficult to evaluate the applicability of these rules based on current experimental data. For example, Simons' rules predict inefficient electron detachment from the anilide anion ($C_6H_5NH^-$), since the potential surfaces of the anion and the neutral are not expected to approach closely to one another. Yet, extensive detachment from the anilide system has been induced by infrared excitation (Meyer *et al.*, 1982). Due to the temporal resolution ($\sim$10 msec) of the ICR spectrometer used in these experiments, it is not possible to compare the electron detachment efficiencies of anilide and a system like SF_6^-, other than to say that both are fast on the ICR time scale. Experiments measuring the absolute rates for infrared induced electron detachment are required to evaluate the applicability of these propensity rules.

The mechanism discussed so far treats the autodetachment process as a nonradiative transition from a vibrationally excited anionic potential surface to a surface where the electron is no longer bound and the molecule contains less vibrational energy. Such a mechanism is quite common, as witnessed by the large number of species that are observed to attach electrons nondissociatively via a nuclear-excited Feshbach resonance. However, it is possible to imagine a mechanism in which the infrared-driven electron detachment process involves a photon-mediated electronic transition. In this mechanism, the transition from the anion's potential surface to that of the neutral would no longer be nonradiative in nature, but instead would be an electromagnetically induced electronic transition, analogous to optical photodetachment. Whether or not this mechanism is operative will depend on the relative rates of the nonradiative versus photon-mediated transition from the potential surface of the anion to that of the neutral.

There are a number of other routes leading to infrared multiphoton-induced electron attachment that are possible. One interesting possibility would involve an intramolecular rearrangement resulting in a product anion with enough internal energy to detach an electron by one of the above mechanisms. This pathway is not easily ruled out, and the thermochemistry of such side reactions needs to be explored in order to evaluate the likelihood of such an event. This and other mechanisms may surface as the phenomenon of infrared multiphoton-induced electron detachment is explored.

In conclusion, a large amount of work remains to be performed before an understanding of these processes can be reached. Although it appears that infrared multiphoton-induced electron detachment should be a very general process, the actual number of systems that have been observed to detach is small. Whether one or more of the above mechanisms may be responsible for electron detachment remains to be seen. Other experimental methods brought to bear on this problem may prove very useful. For example, Simons (1981) has pointed out that mechanisms for electron detachment may be probed by comparing vibrational transition propensities to the detached electron kinetic energy distribution. Also, work on smaller systems with well-mapped-out potential surfaces will aid in testing the propensity rules that Simons has generated. Determination of the actual rates of vibrational excitation and electron detachment would certainly be of interest. These rates may be compared with the large number of autodetachment rate constants measured for species exhibiting nondissociative electron attachment (Christophorou, 1978). It is not known whether these electron detachment processes will show dependencies on laser fluence, laser intensity, and background pressure similar to that already observed for infrared multiphoton decomposition of ions. Finally, future applications of this process in probing the structure of gas phase ions (as in the $C_7H_7^-$ isomers) and the effects of vibrational excitation on ions (as in SF_6^-) will certainly appear.

IV. SUMMARY OF EXPERIMENTALLY MEASURED MOLECULAR ELECTRON AFFINITIES

In Table II we present a compilation of experimentally determined electron affinities. This list is updated from the compilation by Janousek and Brauman (1979). We have restricted our list to include techniques that yield only the most direct, unambiguous determinations. These include photoelectron spectroscopy, photodetachment spectroscopy, charge transfer, and collisional ionization. Other techniques that have been used to determine

TABLE II

Experimentally Determined Molecular Electron Affinities (EA)

Molecule	Electron affinity (eV)	Method[a] reference
	Diatomics	
As_2	0	PD (Feldmann *et al.*, 1977)
AsH	1.0 ± 0.1	PD (Rackwitz *et al.*, 1977)
BeH	0.70 ± 0.1	PD (Rackwitz *et al.*, 1977)
Br_2	2.51 ± 0.10	ECT (Chupka *et al.*, 1971)
Br_2	2.55 ± 0.10	ECT (Baede, 1972)
Br_2	2.62 ± 0.2	ECT (Hughes *et al.*, 1973)
Br_2	2.6 ± 0.2	ECT (Dispert and Lacmann, 1977b)
C_2	$\leqslant 3.54 \pm 0.05$	PD (Feldmann, 1970)
C_2	$\geqslant 3.374$; $\leqslant 3.408$	LPD (Jones *et al.*, 1980)
CH	0.74 ± 0.05	PD (Feldmann, 1970)
CH	1.238 ± 0.008	PES (Kasdan *et al.*, 1975b)
CS	0.205 ± 0.021	PES (Burnett *et al.*, 1982)
CaH	0.93 ± 0.05	PD (Rackwitz *et al.*, 1977)
CN	3.82 ± 0.02	PD (Berkowitz *et al.*, 1969)
Cl_2	2.38 ± 0.10	ECT (Chupka *et al.*, 1971)
Cl_2	2.45 ± 0.15	ECT (Baede, 1972)
Cl_2	$\geqslant 2.32$	CT (Dunkin *et al.*, 1972)
Cl_2	2.32 ± 0.1	ECT (Hughes *et al.*, 1973)
Cl_2	2.4 ± 0.2	ECT (Dispert and Lacmann, 1977b)
F_2	3.08 ± 0.10	ECT (Chupka *et al.*, 1971)
FeO	1.492 ± 0.020	PES (Engelking and Lineberger, 1977a)
IBr	2.7 ± 0.2	ECT (Chupka *et al.*, 1971)
IBr	2.55 ± 0.10	ECT (Baede, 1972)
I_2	2.58 ± 0.10	ECT (Chupka *et al.*, 1971)
I_2	2.55 ± 0.05	ECT (Baede, 1972)
I_2	2.42 ± 0.2	ECT (Hughes *et al.*, 1973)
LiCl	0.61 ± 0.02	PES (Carlsten *et al.*, 1976)
MgH	1.05 ± 0.06	PD (Rackwitz *et al.*, 1977)
NH	0.381 ± 0.014	PES (Engelking and Lineberger, 1976)
NH	0.38 ± 0.03	PES (Celotta *et al.*, 1974)
NO	$0.024 + 0.010 - 0.005$	PES (Siegel *et al.*, 1972)
NO	$\geqslant 0.09$	ECT (Berkowitz *et al.*, 1971)
NO	$\leqslant 0.46$	CT (Fehsenfeld *et al.*, 1966)
NO	0.8 ± 0.2	ECT (Refaey, 1976)
NO	0.015 ± 0.1	ECT (Hughes *et al.*, 1973)
NO	$\geqslant 0.1 \pm 0.1$	CI (Nalley *et al.*, 1973)
NS	1.194 ± 0.011	PES (Burnett *et al.*, 1982)
OD	1.825486 ± 0.000037	LPD (Schulz *et al.*, 1982)
OD	1.823 ± 0.002	LPD (Hotop *et al.*, 1974)

TABLE II (*continued*)

Molecule	Electron affinity (eV)	Method[a] reference
	Diatomics	
OH	1.827608 ± 0.000021	LPD (Schulz *et al.*, 1982)
OH	1.83 ± 0.04	PD (Branscomb, 1966)
OH	1.825 ± 0.002	LPD (Hotop *et al.*, 1974)
OH	1.829 + 0.010 − 0.014	PES (Celotta *et al.*, 1974)
O_2	0.440 ± 0.008	PES (Celotta *et al.*, 1972)
O_2	0.45 ± 0.10	ECT (Tiernan *et al*, 1971)
O_2	⩾0.48	ECT (Berkowitz *et al.*, 1971)
O_2	0.5 ± 0.1	ECT (Baede, 1972)
O_2	0.15 ± 0.05	PD (Burch *et al.*, 1958)
O_2	⩾0.46 ± 0.05	CI (Nalley and Compton, 1971)
PH	1.028 ± 0.010	PES (Zittel and Lineberger, 1976)
PH	1.00 ± 0.06	PD (Rackwitz *et al.*, 1977)
P_2	⩽0.65	PD (Feldmann *et al.*, 1977)
PO	1.092 ± 0.010	PES (Zittel and Lineberger, 1976)
SD	2.315 ± 0.005	PES (Breyer *et al.*, 1981)
SH	2.319 ± 0.010	PD (Steiner, 1968)
SH	2.28 ± 0.15	PD (Brauman and Smyth, 1969)
SH	2.314 ± 0.003	LPD (Janousek and Brauman, 1981)
SH	2.317 ± 0.005	PES (Breyer *et al.*, 1981)
SO	1.126 ± 0.013	PES (Lineberger, 1974)
SO	⩽1.09 ± 0.05	PD (Feldmann, 1970)
SO	⩾1.0; ⩽1.27	CT (Kraus *et al.*, 1961)
S_2	1.663 ± 0.040	PES (Celotta *et al.*, 1974)
SeH	2.21 ± 0.03	PD (Smyth and Brauman, 1972b)
SiH	1.277 ± 0.009	PES (Kasdan *et al.*, 1975a)
ZnH	⩽0.95	PD (Rackwitz *et al.*, 1977)
	Triatomics	
AsH_2	1.27 ± 0.03	PD (Smyth and Brauman, 1972c)
C_2H	⩽3.73 ± 0.05	PD (Feldmann, 1970)
C_2H	2.94 ± 0.10	PD (Janousek *et al.*, 1979)
CD_2 (triplet)	0.205 ± 0.015	PES (Engleking *et al.*, 1981;
CH_2 (triplet)	0.210 ± 0.03	Engelking *et al.*, 1981; Zittel *et al.*, 1976)
CH_2 (triplet)	⩽0.6	PD (Feldmann *et al.*, 1977)
COS	0.46 ± 0.2	CI (Compton *et al.*, 1975)
CS_2	1.0 ± 0.2	CI (Compton *et al.*, 1975)

(*continued*)

TABLE II (*continued*)

Molecule	Electron affinity (eV)	Method[a] reference
	Triatomics	
CS_2	0.50 ± 0.2	ECT (Hughes *et al.*, 1973)
CS_2	0.62 ± 0.2	CI (Dispert and Lacmann, 1975)
FeCO	1.26 ± 0.02	PES (Engelking and Lineberger, 1979)
N_3	2.70 ± 0.12	PD (Jackson *et al.*, 1981b)
NH_2	0.744 ± 0.022	PD (Smyth and Brauman, 1972c)
NH_2	0.74	LPD (Smyth *et al.*, 1971)
NH_2	0.76 ± 0.04	PD (Feldmann, 1971)
NH_2	0.779 ± 0.037	PES (Celotta *et al.*, 1974)
N_2O	$\leqslant 1.465$	CT (Paulson, 1966)
NO_2	3.10 ± 0.05	PD (Warneck, 1969)
NO_2	$\geqslant 2.32$; $\leqslant 2.5$	CT (Dunkin *et al.*, 1972)
NO_2	2.5 ± 0.1	ECT (Baede, 1972)
NO_2	$\geqslant 2.04$	ECT (Berkowitz *et al.*, 1971)
NO_2	$\geqslant 3.6$	CT (Curran, 1962)
NO_2	$\geqslant 1.8$	CT (Ferguson *et al.*, 1969)
NO_2	2.1	ECT (Refaey, 1976)
NO_2	2.28 ± 0.1	ECT (Hughes *et al.*, 1973)
NO_2	$\geqslant 2.5 \pm 0.1$	CI (Nalley *et al.*, 1973)
NO_2	2.36 ± 0.10	LPD (Herbst *et al.*, 1974)
NO_2	2.50 ± 0.05	CI (Leffert *et al.*, 1973)
NO_2	2.275 ± 0.025	LPD (Woo *et al.*, 1981)
O_2H	1.19 ± 0.01	LPD (Bierbaum *et al.*, 1981)
O_3	1.92 ± 0.13	PD (Wong *et al.*, 1971)
O_3	$\geqslant 1.96$	ECT (Berkowitz *et al* 1971)
O_3	2.14 ± 0.15	CI (Rothe *et al.*, 1975)
O_3	$\leqslant 2.09$	LPD (Sinnott and Beatty, 1971)
O_3	$\geqslant 1.8234$	CT (Dotan *et al.*, 1977)
O_3	$\geqslant 2.26 + 0.04 - 0.06$	ECT (Lifshitz *et al*, 1978)
O_3	2.1028 ± 0.0025	LPD (Novick *et al.*, 1979)
PH_2	1.25 ± 0.03	LPD (Smyth and Brauman, 1972a)
PH_2	1.26 ± 0.03	LPD (Smyth *et al.*, 1971)
PH_2	1.271 ± 0.010	PES (Zittel and Lineberger, 1976)
SO_2	$\leqslant 1.0 \pm 0.1$	PD (Feldmann, 1970)
SO_2	$\geqslant 1.0$; $\leqslant 1.27$	CT (Kraus *et al.*, 1961)
SO_2	1.05	ECT (Refaey and Franklin, 1976)
SO_2	1.14 ± 0.15	CI (Rothe *et al.*, 1975)
SO_2	0.99 ± 0.1	ECT (Hughes *et al.*, 1973)

TABLE II *(continued)*

Molecule	Electron affinity (eV)	Method[a] reference
	Triatomics	
SO_2	1.097 ± 0.036	PES (Celotta *et al.*,1974)
S_3	2.0 ± 0.15	PD (Feldmann *et al.*, 1977)
SiH_2	1.124 ± 0.020	PES (Kasdan *et al.*, 1975a)
	Tetra-atomics	
CCH_2	$\leqslant 0.43$	CT (Lindinger *et al.*, 1975)
CH_3	$\leqslant 0.5$	PD (Feldmann *et al.*, 1977)
CH_3	0.08 ± 0.03	PES (Ellison *et al.*, 1978)
CO_3	$2.69 + 0.07 - 0.14$	PD (Hong *et al.*, 1977)
GeH_3	$\leqslant 1.74 \pm 0.04$	PD (Reed and Brauman, 1974)
PBr_3	1.6	CI (Mathur *et al.*, 1976a)
PBr_2Cl	1.6	CI (Mathur *et al*, 1976a)
PCl_2Br	1.5	CI (Mathur *et al.*, 1976a)
PCl_3	0.8	CI (Mathur *et al.*, 1976a)
$POCl_2$	3.8	CI (Mathur *et al.*, 1976a)
SO_3	$\geqslant 1.70 \pm 0.15$	CI (Rothe *et al.*, 1975)
SiH_3	$\leqslant 1.44 \pm 0.03$	PD (Reed and Brauman, 1974)
	Larger Polyatomics	
CF_3Br	0.91 ± 0.2	CI (Compton *et al.*, 1978b)
CF_3I	1.4 ± 0.2	CI (Tang *et al.*, 1976)
CF_3I	1.57 ± 0.2	CI (Compton *et al.*, 1978b)
CH_3I	0.2 ± 0.1	ECT (Moutinho *et al.*, 1974)
CH_3NO_2	$0.44 \pm 0.1–0.2$	CI (Compton *et al.*, 1978b)
CD_3O	1.552 ± 0.022	PES (Engelking *et al.*, 1978)
CH_3O	$\leqslant 1.59 \pm 0.04$	PD (Reed and Brauman, 1975)
CH_3O	1.570 ± 0.022	PES (Engelking *et al.*, 1978)
CH_3O	1.59 ± 0.04	PD (Janousek *et al.*, 1978)
CD_3S	1.856 ± 0.006	LPD (Janousek et al., 1981)
CH_3S	1.882 ± 0.024	PES (Engelking *et al.*, 1978)
CH_3S	1.861 ± 0.004	LPD (Janousek *et al.*, 1981)
$CO_3[H_2O]$	2.1 ± 0.2	PD (Burt, 1972)
C_2H_2FO (acetyl fluoride enolate)	2.22 ± 0.09	PD (Zimmerman *et al.*, 1977)
C_2H_2N (cyanomethyl)	1.507 ± 0.018	PD (Zimmerman and Brauman, 1977b)
C_2H_5N (ethyl nitrene)[b]	0.56 ± 0.01	PD (Rynard, 1981)
C_2D_3O (acetaldehyde-d_3 enolate)	1.819 ± 0.005	LPD (Jackson *et al.*, 1981a)
C_2H_3O (acetaldehyde enolate)	1.83 ± 0.06	PD (Zimmerman *et al.*, 1977)
C_2H_3O (acetaldehyde enolate)	1.824 ± 0.005	LPD (Jackson *et al.*, 1981a)
C_2H_5S (ethyl sulfide)	1.953 ± 0.006	LPD (Janousek *et al.*, 1980)
C_3H_5 (allyl)	0.550 ± 0.054	PD (Zimmerman and Brauman, 1977b)

(continued)

TABLE II (*continued*)

Molecule	Electron affinity (eV)	Method[a] reference
	Larger Polyatomics	
$C_3H_2F_3O$ (1,1,1-trifluoroacetone enolate)	2.58 ± 0.12	PD (Zimmerman *et al.*, 1977)
C_3H_5O (acetone enolate)	1.79 ± 0.06	PD (Zimmerman *et al.*, 1977)
C_3H_5O (propionaldehyde enolate)	1.69 ± 0.05	PD (Zimmerman *et al.*, 1977)
$C_3H_5O_2$ (methyl acetate enolate)	1.80 ± 0.06	PD (Zimmerman *et al.*, 1977)
C_3H_7S (*n*-propyl sulfide)	2.00 ± 0.02	PD (Janousek *et al.*, 1980)
C_3H_7S (*i*-propyl sulfide)	2.02 ± 0.02	PD (Janousek *et al.*, 1980)
C_4H_6DO (2-butanone-3-*dl* enolate)	1.67 ± 0.05	PD (Zimmerman *et al.*, 1977)
$C_4H_5D_2O$ (2-butanone-3,3-*d*2 enolate)	1.75 ± 0.06	PD (Zimmerman *et al.*, 1977)
$C_4F_4O_3$ (tetrafluorosuccinic anhydride)	0.5 ± 0.2	CI (Cooper and Compton, 1974)
C_4H_4N (pyrrolate)	2.39 ± 0.13	PD (Richardson *et al.*, 1975a)
$C_4H_2O_3$ (maleic anhydride)	$.4 \pm 0.2$	CI (Compton *et al.*, 1974)
C_4H_5O (cyclobutanone enolate)	1.84 ± 0.06	PD (Zimmerman *et al.*, 1978a)
C_4H_7O (butyraldehyde enolate)	1.67 ± 0.05	PD (Zimmerman *et al.*, 1977)
C_4H_9O (*t*-butoxyl)	$< 1.87 + 0.04$	PD (Reed and Brauman, 1975)
C_4H_9O (*t*-butoxyl)	1.87 or 1.90 ± 0.01	LPD (Janousek *et al.*, 1978)
C_4H_9S (*n*-butyl sulfide)	2.03 ± 0.02	PD (Janousek *et al.*, 1980)
C_4H_9S (*t*-butyl sulfide)	2.07 ± 0.02	PD (Janousek *et al.*, 1980)
$C_5F_6O_3$ (hexafluoroglutaric anhydride)	1.5 ± 0.2	CI (Cooper and Compton, (1974)
C_5H_5 (cyclopentadienyl)	$< 1.84 \pm 0.03$	PD (Richardson *et al.*, 1973)
C_5H_5 (cyclopentadienyl)	1.786 ± 0.020	PES (Engelking and Lineberger, 1977b)
C_5H_7 (pentadienyl)	0.91 ± 0.03	PD (Zimmerman *et al.*, 1978b)
C_5H_7O (cyclopentanone enolate)	1.62 ± 0.06	PD (Zimmerman *et al.*, 1978a)
C_5H_9O (3-pentanone enolate)	1.69 ± 0.05	PD (Zimmerman *et al.*, 1977)
$C_5H_{11}O$ (neopentoxyl)	$< 1.93 \pm 0.06$	PD (Reed and Brauman, 1975)
$C_5H_{11}O$ (neopentoxyl)	1.93 ± 0.05	LPD (Janousek *et al.*, 1978)
$C_5H_{11}S$ (*n*-pentyl sulfide)	2.09 ± 0.02	PD (Janousek *et al.*, 1980)
$C_6Br_4O_2$ (*p*-bromanil)	2.44 ± 0.2	ECT (Cooper *et al.*, 1978)
$C_6Cl_4O_2$ (*p*-chloranil)	2.76 ± 0.2	ECT (Cooper *et al.*, 1978)
$C_6F_4O_2$ (*p*-fluoranil)	2.92 ± 0.2	ECT (Cooper *et al.*, 1978)
C_6F_6 (hexafluorobenzene)	$> 1.8 \pm 0.3$	ECT (Lifshitz *et al.*, 1973)
C_6F_{10} (perfluorocyclohexene)	$> 1.4 \pm 0.3$	ECT (Lifshitz *et al.*, 1973)
C_6H_7 (methylcyclopentadienyl)	$< 1.67 \pm 0.04$	PD (Richardson *et al.*, 1973)
C_6N_4 (TCNE)	2.3 ± 0.3	PD (Lyons and Palmer, 1976)
C_6H_4ClO (*o*-chloro pehoxide)	$\leqslant 2.58 \pm 0.08$	PD (Richardson *et al.*, 1975e)
$C_6H_5NO_2$ (nitrobenzene)	$> 0.7 \pm 0.2$	ECT (Lifshitz *et al.*, 1973)

TABLE II (*continued*)

Molecule	Electron affinity (eV)	Method[a] reference
	Larger Polyatomics	
$C_6H_4O_2$ (*p*-benzoquinone)	1.89 + 0.2 − 0.3	CI (Cooper *et al.*, 1975)
C_6H_5O (phenoxide)	⩽2.36 ± 0.06	PD (Richardson *et al.*, 1975c,e)
C_6H_5S (thiophenoxide)	⩽2.47 ± 0.06	PD (Richardson *et al.*, 1975e)
C_6H_9O (cyclohexanone enolate)	1.55 ± 0.05	PD (Zimmerman *et al.*, 1978a)
$C_6H_{11}O$ (pinacolone enolate)	1.88 ± 0.06	PD (Zimmerman *et al.*, 1977)
$C_6H_{11}O$ (3,3-dimethylbutanal enolate)	1.82 ± 0.06	PD (Zimmerman *et al.*, 1977)
$C_6H_5CH_2$ (benzyl)	0.88 ± 0.01	PD (Drzaic and Brauman, 1983)
$C_6H_5CH_2$ (benzyl)	<0.88 ± 0.06	PD (Richardson *et al.*, 1975d)
C_6H_5N (phenyl nitrene)	1.46 ± 0.02	PD (Drzaic and Brauman, 1983)
C_6H_5NH (anilide)	1.70 ± 0.03	PD (Drzaic and Brauman, 1983)
C_7F_8 (octafluorotoluene)	>1.7 ± 0.3	ECT (Lifshitz *et al.*, 1973)
C_7H_9 (heptatrienyl)	1.27 ± 0.03	PD (Zimmerman *et al.*, 1978b)
C_7H_7O (*o*-methyl phenoxide)	⩽2.36 ± 0.06	PD (Richardson *et al.*, 1975e)
C_7H_9O (2-norbornanone enolate)	1.61 ± 0.05	PD (Zimmerman *et al.*, 1978a)
$C_7H_{11}O$ (cycloheptanone enolate)	1.48 ± 0.04	PD (Zimmerman *et al.*, 1978a)
$C_7H_{11}O$ 2,5-dimethylcyclopentanone enolate)	1.49 ± 0.04	PD (Zimmerman *et al.*, 1978a)
$C_7H_{13}O$ (4-heptanone enolate)	1.72 ± 0.06	PD (Zimmerman *et al.*, 1977)
$C_7H_{13}O$ (di-isopropyl ketone enolate)	1.46 ± 0.04	PD (Zimmerman *et al.*, 1977)
C_7FH_6O (*m*-fluoroacetophenone enolate)	2.218 + 0.005 − 0.010	LPD (Jackson *et al.*, 1979)
C_7FH_6O (*p*-fluoroacetophenone enolate)	2.176 + 0.005 − 0.010	LPD (Jackson *et al.*, 1979)
C_8H_7O (acetophenone enolate)	2.06 ± 0.08	PD (Zimmerman *et al.*, 1977)
C_8H_7O (acetophenone enolate)	2.057 + 0.005 − 0.010	LPD (Jackson *et al.*, 1979)
C_8H_7O (phenylacetaldehyde enolate)	2.10 ± 0.08	PD (Zimmerman *et al.*, 1977)
$C_8H_{13}I$ (cyclooctanone enolate)	1.63 ± 0.06	PD (Zimmerman *et al.*, 1978a)
$C_8H_{13}O$ (cyclooctanone enolate)	1.63 ± 0.06	PD (Zimmerman *et al.*, 1978a)
$C_8H_{13}O$ (cyclooctanone enolate)	1.63 ± 0.06	PD (Zimmerman *et al.*, 1978a)
$C_9H_{15}O$ (cyclononanone enolate)	1.69 ± 0.06	PD (Zimmerman *et al.*, 1978a)
C_9H_9O (*m*-methylacetophenone enolate)	2.030 + 0.005 − 0.010	LPD (Jackson *et al.*, 1979)
C_9H_9SiN (trimethylsilylnitrene)	1.43 ± 0.10	PD (Jackson *et al.*, 1981b)
$C_{10}H_{17}O$ (cyclodecanone enolate)	1.83 ± 0.06	PD (Zimmerman *et al.*, 1978a)
$C_{12}H_4N_4$ (TCNQ)	2.8 ± 0.1	CI (Klots *et al.*, 1974)

(*continued*)

TABLE II (*continued*)

Molecule	Electron affinity (eV)	Method[a] reference
	Larger Polyatomics	
$C_{12}H_4N_4$ (TCNQ)	$2.8 + 0.05 - 0.3$	CI (Compton and Cooper, 1977)
$C_{12}H_9$ (perinaphthenyl)	1.07	PD (Gygax *et al.*, 1979)
$C_{12}H_{15}O$ (*t*-butylacetophenone enolate)	$2.032 + 0.005 - 0.010$	LPD (Jackson *et al.*, 1979)
$C_{12}H_{21}O$ (cyclododecanone enolate)	1.90 ± 0.07	PD (Zimmerman *et al.*, 1978a)
$Fe(CO)_2$	1.22 ± 0.02	PES (Engelking and Lineberger, 1979)
$Fe(CO)_3$	1.8 ± 0.2	PES (Engelking and Lineberger, 1979)
$Fe(CO)_4$	2.4 ± 0.3	PES (Engelking and Lineberger, 1979)
HNO_3	0.57 ± 0.15	CI (Mathur *et al.*, 1976b)
H_3O_2 (hydrated hydroxyl radical)	$<2.95 \pm 0.15$	PD (Golub and Steiner, 1968)
Ni(CO)	0.804 ± 0.012	PES (Stevens *et al.*, 1982)
$Ni(CO)_2$	0.643 ± 0.014	PES (Stevens *et al.*, 1982)
$Ni(CO)_3$	1.077 ± 0.013	PES (Stevens *et al.*, 1982)
$POCl_3$	1.4	CI (Mathur *et al.*, 1976a)
SF_4	2.35 ± 0.1	CT (Babcock and Streit, 1981)
SF_6	0.75 ± 0.1	CI (Leffert *et al.*, 1974)
SF_6	$>0.6 \pm 0.1$	ECT (Lifshitz *et al.*, 1973)
SF_6	$0.54 + 0.1 - 0.17$	CI (Compton and Cooper, 1973)
SF_6	>0.43; <1.465	CT (Fehsenfeld, 1970)
SF_6	0.46 ± 0.2	CI (Compton *et al.*, 1978a)
SF_6	1.0 ± 0.2	CT (Streit, 1982)
SeF_6	2.9 ± 0.2	CI (Compton *et al.*, 1978a)
TeF_6	$3.34 + 0.1 - 0.17$	CI (Compton and Cooper, 1973)
TeF_6	3.3 ± 0.2	CI (Compton *et al.*, 1978a)
UF_6	$\geqslant 5.1$	CI (Compton, 1977)
WF_6	3.7 ± 0.3	CI (Dispert and Lacmann, 1977a)

[a] Collisional ionization (CI), charge transfer (CT), endothermic charge transfer (ECT), photoelectron spectroscopy (PES), photodetachment spectroscopy (PD), laser photodetachment spectroscopy (LPD).

[b] An earlier determination of the ethyl nitrene electron affinity placed it at $\leqslant 1.87 \pm 0.16$ eV (Richardson *et al*,. 1974). However, subsequent work has shown that the actual species studied in those experiments was acetaldehyde enolate, a species with the same *m*/*e* as ethyl nitrene. Experiments using a different precursor support the new, lower value of 0.56 ± 0.01 eV (Rynard, 1981).

electron affinities but are not included in the table are electron swarm, surface ionization, dissociative electron attachment, and thermochemical cycles. Electron affinities measured by all types of methods are listed in the compilation by Rosenstock *et al.* (1977).

V. SUMMARY

Electron photodetachment spectroscopy is very useful in studying the structure and energetics of both anions and their neutral photoproducts. The shape of photodetachment cross sections is dependent on the symmetry of the highest occupied molecular orbital of the anion and may be successfully modeled. Electronic, vibrational, and even rotational transitions are often observed in photodetachment studies and may be used to gather spectroscopic information on the systems studied. Photodetachment experiments frequently yield the electron affinity of the neutral photoproduct, and this quantity may be related by a thermochemical cycle to gas phase acidities and bond dissociation energies. Finally, infrared multiphoton excitation has been used to induce electron detachment from anions. While the exact mechanism of this process is not known for certain at this point, electron detachment from vibrationally excited molecules appears to be a general process. It appears that infrared multiphoton-induced electron detachment will be of much interest in the future.

ACKNOWLEDGMENTS

We gratefully acknowledge the National Science Foundation for its research support, and its fellowship support for PSD. We thank the Exxon Foundation for its fellowship support for JM. We also thank Professor W. C. Lineberger for many helpful discussions.

REFERENCES

Ambartzumian, R. V., Furzikov, N. P., Gorokhov, Y. A., Letokhov, V. S., Makarov, G. N., and Puretzky, A. A. (1977). *Opt. Lett.* **1**, 22.

Amos, A. T., and Burrows, B. L. (1973). *Adv. Quant. Chem.* **7**, 289.

Babcock, L. M., and Streit, G. E. (1981), *J. Chem. Phys.* **75**, 3864.

Baede, A. P. M. (1972). *Physica* (*Utrecht*) **59**, 541.

Bartmess, J. E., and McIver, R. T., Jr. (1979). *In* "Gas Phase Ion Chemistry" (M. T. Bowers, ed.), Vol. 2, pp. 88–123. Academic Press, New York.

Berkowitz, J., Chupka, W. A., and Walter, T. A. (1969). *J. Chem. Phys.* **50**, 1497.

Berkowitz, J., Chupka, W. A., and Gutman. D. (1971). *J. Chem. Phys.* **55**, 2733.

Berry, R. S., and Leach, S. (1982). *Adv. Electron. Electron Phys.* **57**, 1.

Bierbaum, V. M., Schmitt, R. J., and DePuy, C. H. (1981). *J. Am. Chem. Soc.* **103**, 6262.

Bomse, D. S., Woodin, R. L., and Beauchamp, J. L. (1979). *J. Am. Chem. Soc.* **101**, 5503.

Branscomb, L. M. (1966). *Phys. Rev.* **148**, 11.

Brauman, J. I. (1980). *In* "Frontiers of Free Radical Chemistry" (W. A. Pryor, ed.), pp. 23–30. Academic Press, New York.
Brauman, J. I., and Smyth, K. C. (1969). *J. Am. Chem. Soc.* **91**, 7778.
Breyer, F., Frey, P., and Hotop, H. (1981). *Z. Phys.* **A300**, 7.
Burch, D. S., Smith, S. J., and Branscomb, L. M. (1958). *Phys. Rev.* **112**, 171.
Burnett, S. M., Feigerle, C. S., Stevens, A. E., and Lineberger, W. C. (1982). *J. Phys. Chem.* **86**, 4486.
Burt, J. A. (1972). *J. Chem. Phys.* **57**, 4649.
Carlsten, J. L., Peterson, J. R., and Lineberger, W. C. (1976). *Chem. Phys. Lett.* **37**, 5.
Celotta, R. J., Bennett, R. A., Hall, J. L., Siegel, M. W., and Levine, J. (1972). *Phys. Rev. A* **6**, 631.
Celotta, R. J., Bennett, R. A., and Hall, J. L. (1974). *J. Chem. Phys.* **60**, 1740.
Christophorou, L. G. (1978). *Adv. Electron. Electron Phys.* **46**, 55.
Christophorou, L. G., McCorkle, D. L., and Carter, J. G. (1971). *J. Chem. Phys.* **54**, 253.
Chupka, W. A., Berkowitz, J., and Gutman, D. (1971). *J. Chem. Phys.* **55**, 2724.
Compton, R. N. (1977). *J. Chem. Phys.* **66**, 4478.
Compton, R. N., and Cooper, C. D. (1973). *J. Chem. Phys.* **59**, 4140.
Compton, R. N., and Cooper, C. D. (1977). *J. Chem. Phys.* **66**, 4325.
Compton, R. N., Reinhardt, P. W., and Cooper, C. D. (1974). *J. Chem. Phys.* **60**, 2953.
Compton, R. N., Reinhardt, P. W., and Cooper, C. D. (1975). *J. Chem. Phys.* **63**, 3821.
Compton, R. N., Reinhardt, P. W., and Cooper, C. D. (1978a). *J. Chem. Phys.* **68**, 2023.
Compton, R. N., Reinhardt, P. W., and Cooper, C. D. (1978b). *J. Chem. Phys.* **68**, 4360.
Cooper, C. D., and Compton, R. N. (1974). *J. Chem. Phys.* **60**, 2424.
Cooper, C. D., Naff, W. T., and Compton, R. N. (1975). *J. Chem. Phys.* **63**, 2752.
Cooper, C. D., Frey, W. F., and Compton, R. N. (1978). *J. Chem. Phys.* **69**, 2367.
Corderman, R. R., and Lineberger, W. C. (1979). *Ann. Rev. Phys. Chem.* **30**, 347.
Curran, R. K. (1962). *Phys. Rev.* **125**, 910.
Dispert, H., and Lacmann, K. (1975). *Hahn-Meitner-Inst. Kernforsch. Berlin* **HMI-B 198**, 34.
Dispert, H., and Lacmann, K. (1977a). *Chem. Phys. Lett.* **45**, 311.
Dispert, H., and Lacmann, K. (1977b). *Chem. Phys. Lett.* **47**, 533.
Dotan, I., Davidson, J. A., Streit, G. E., Albritton, D. L., and Fehsenfeld, F. C. (1977). *J. Chem. Phys.* **68**, 247.
Drzaic, P. S., and Brauman, J. I. (1981). *Chem. Phys. Lett.*, **83**, 508.
Drzaic, P. S., and Brauman, J. I. (1982). *J. Am. Chem. Soc.* **104**, 13.
Drzaic, P. S., and Brauman, J. I. (1983). Manuscript in preparation.
Dunbar, R. C. (1973). *J. Am. Chem. Soc.* **95**, 6191.
Dunbar, R. C. (1979). *In* "Gas Phase Ion Chemistry" (M. T. Bowers, ed.), Vol. 2, pp. 182–220. Academic Press, New York.
Dunkin, D. B., Fehsenfeld, F. C., and Ferguson, E. E. (1972). *Chem. Phys. Lett.* **15**, 257.
Ellison, G. B., Engelking, P. C., and Lineberger, W. C. (1978). *J. Am. Chem. Soc.* **100**, 2556.
Engelking, P. C., and Lineberger, W. C. (1976). *J. Chem. Phys.* **65**, 4323.
Engelking, P. C., and Lineberger, W. C. (1977a). *J. Chem. Phys.* **66**, 5054.
Engelking, P. C., and Lineberger, W. C. (1977b). *J. Chem. Phys.* **67**, 1412.
Engelking, P. C., and Lineberger, W. C. (1979). *J. Am. Chem. Soc.* **101**, 5569.
Engelking, P. C., Ellison, G. B., and Lineberger, W. C. (1978). *J. Chem. Phys.* **69**, 1826.
Engelking, P. C., Corderman, R. R., Wendoloski, J. J., Ellison, G. B., ONeil, S. V., and Lineberger, W. C. (1981). *J. Chem. Phys.* **74**, 5460.
Fano, U. (1961). *Phys. Rev.* **124**, 1866.
Fehsenfeld, F. C. (1970). *J. Chem. Phys.* **53**, 2000.
Fehsenfeld, F. C., Ferguson, E. E., and Schmeltekopf, A. L. (1966). *J. Chem. Phys.* **45**, 1844.
Feldmann, D. (1970). *Z. Naturforsch. Teil A* **25**, 621.

Feldmann, D. (1971). *Z. Naturforsch. Teil A* **26**, 1100.
Feldmann, D., Rackwitz, R., Kaiser, H. J., and Heinicke, E. (1977). *Z. Naturforsch. Teil A* **2**, 600.
Ferguson, E. E., Fehsenfeld, F. C., and Schmeltkopf, A. L. (1969). *Adv. Chem. Ser.* **80**, 83.
Foster, M. S., and Beauchamp, J. L. (1975). *Chem. Phys. Lett.* **31**, 482.
Franklin, J. L. (1976). *Science* **193**, 725.
Garrett, W. R. (1970). *Chem. Phys. Lett.* **5**, 393.
Garrett, W. R. (1979). *J. Chem. Phys.* **71**, 651.
Geltman, S. (1958). *Phys. Rev.* **112**, 176.
Golub, S. and Steiner, B. (1968). *J. Chem. Phys.* **49**, 5191.
Gygax, R., McPeters, H. L., and Brauman, J. I. (1979). *J. Am. Chem. Soc.* **101**, 2567.
Hay, P. J. (1982). *J. Chem. Phys.* **76**, 502.
Hayden, C. C., Neumark, D. M., Shobatake, K., Sparks, R. K., and Lee, Y. T. (1982). *J. Chem. Phys.* **76**, 3607.
Herbst, E., Patterson, T. A., and Lineberger, W. C. (1974). *J. Chem. Phys.* **61**, 1300.
Hong, S. P., Woo, S. B., and Helmy, E. M. (1977). *Phys. Rev. A* **15**, 1563.
Hotop, H., Patterson, T. A., and Lineberger, W. C. (1974). *J. Chem. Phys.* **60**, 1806.
Huber, B. A., Cosby, P. C., Peterson, J. R., and Moseley, J. T. (1977). *J. Chem. Phys.* **66**, 4520.
Hughes, B. M., Lifshitz, C., and Tiernan, T. O. (1973). *J. Chem. Phys.* **59**, 3162.
Jackson, R. L., Zimmerman, A. H., and Brauman, J. I. (1979) *J. Chem. Phys.* **71**, 2088.
Jackson, R. L., Hiberty, P. C., and Brauman, J. I. (1981a). *J. Chem. Phys.* **74**, 3705.
Jackson, R. L., Pellerite, M. J., and Brauman, J. I. (1981b). *J. Am. Chem. Soc.* **103**, 1802.
Janousek, B. K., and Brauman, J. I. (1979). *In* "Gas Phase Ion Chemistry" (M. T. Bowers, ed.), Vol. 2, pp. 53–86. Academic Press, New York.
Janousek, B. K., and Brauman, J. I. (1980). *J. Chem. Phys.* **72**, 694.
Janousek, B. K., and Brauman, J. I. (1981). *Phys. Rev. A* **23**, 1673.
Janousek, B. K., Zimmerman, A. H., Reed, K. J., and Brauman, J. I. (1978). *J. Am. Chem. Soc.* **100**, 6142.
Janousek, B. K., Brauman, J. I., and Simons, J. (1979). *J. Chem. Phys.* **71**, 2057.
Janousek, B. K., Reed, K. J., and Brauman J. I. (1980). *J. Am. Chem. Soc.* **102**, 3125.
Jasinski, J. M., and Brauman, J. I. (1980). *J. Chem. Phys.* **73**, 6191.
Jasinski, J. M., Rosenfeld, R. N., Meyer, F. K, and Brauman, J. I. (1982). *J. Am. Chem. Soc.* **104**, 658.
Jenkins, H. D., and Pratt, K. F. (1977). *J. Phys. Chem. Solids.* **38**, 573.
Jones, P. L., Mead, R. D., Kohler, B. E., Rosner, S. D., and Lineberger, W. C. (1980). *J. Chem. Phys.* **73**, 4419.
Kasdan, A., Herbst, E., and Lineberger, W. C. (1975a). *J. Chem. Phys.* **62**, 541.
Kasdan, A., Herbst, E., and Lineberger, W. C. (1975b). *Chem. Phys. Lett.* **31**, 78.
Klots, C. E., Compton, R. N., and Raaen, V. F. (1974). *J. Chem. Phys.* **60**, 1177.
Kraus, K., Muller-Duysing, W., and Neuert, H. (1961). *Z. Naturforsch. Teil A* **16**, 1385.
Leffert, C. B., Jackson, W. M., and Rothe, E. W. (1973). *J. Chem. Phys.* **58**, 5801.
Leffert, C. B., Tang, S. Y., Rothe, E. W., and Cheng, T. C. (1974). *J. Chem. Phys.* **61**, 4929.
Lehman, T. A., and Bursey, M. M. (1976). "Ion Cyclotron Resonance Spectrometry." Wiley, New York.
Lifshitz, C., Tiernan, T. O., and Hughes, B. M. (1973). *J. Chem. Phys.* **59**, 3182.
Lifshitz, C., Wu, R. L. C., Tiernan, T. O., and Terwilliger, D. T. (1978). *J. Chem. Phys.* **68**, 247.
Lindinger, W., Albritton, D. L., Fehsenfeld, F. C., and Ferguson, E. E. (1975). *J. Chem. Phys.* **63**, 3238.
Lineberger, W. C. (1974). *In* "Laser Spectroscopy" (R. Brewer and A. Mooradian, eds.), pp. 581–595. Plenum Press, New York.
Lyons, L. E., and Palmer, L. D. (1976). *Aust. J. Chem.* **29**, 1919.
McIver, R. T., Jr., Ledford, E. B., Jr., and Hunter, R. L. (1980). *J. Chem. Phys.* **72**, 2535.

Mathur, B. P., Rothe, E. W., Tang, S. Y., and Reck, G. P. (1976a). *J. Chem. Phys.* **65**, 565.
Mathur, B. P., Rothe, E. W., Tang, S. Y., Mahajan, K., and Reck, G. P. (1976b). *J. Chem. Phys.* **64**, 1247.
Meyer, F. K., Jasinski, J. M., Rosenfeld, R. N., and Brauman, J. I. (1982). *J. Am. Chem. Soc.* **104**, 663.
Miller, T. M. (1981). *Adv. Electron. Electron Phys.* **55**, 119.
Moutinho, A. M. C., Aten, J. A., and Los, J. (1974). *Chem. Phys.* **5**, 84.
Nalley, S. J., and Compton, R. N. (1971). *Chem. Phys. Lett.* **9**, 529.
Nalley, S. J., Compton, R. N., Schweinler, H. C., and Anderson, V. E. (1973). *J. Chem. Phys.* **59**, 4125.
Navangul, H. V., and Blatz, P. E. (1973). *J. Am. Chem. Soc.* **95**, 1508.
Navangul, H. V., and Blatz, P. E. (1978). *J. Am. Chem. Soc.* **100**, 4340.
Novick, S. E., Engelking, P. C., Jones, P. L., Futrell, J. H., and Lineberger, W. C. (1979). *J. Chem. Phys.* **70**, 2652.
Okabe, H. (1978). "Photochemistry of Small Molecules," p. 261. Wiley (Interscience), New York.
O'Malley, T. F. (1965). *Phys. Rev.* **137**, A1668.
Paulson, J. F. (1966). *Adv. Chem. Ser.* **58**, 28.
Pellerite, M. J., Jackson, R. L., and Brauman, J. I. (1981). *J. Phys. Chem.* **85**, 1624.
Rackwitz, R., Feldmann, D., Kaiser, H. J., and Heinicke, E. (1977). *Z. Naturforsch. Teil A* **32**, 594.
Ramsay, D. A. (1952). *J. Chem. Phys.* **20**, 1920
Reed, K. J., and Brauman, J. I. (1974). *J. Chem. Phys.* **61**, 4830.
Reed, K. J., and Brauman, J. I. (1975). *J. Am. Chem. Soc.* **97**, 1625.
Reed, K. J., Zimmerman, A. H., Andersen, H. C., and Brauman, J. I. (1976). *J. Chem. Phys.* **64**, 1368.
Refaey, K. M. A. (1976). *Int. J. Mass Spectrom. Ion Phys.* **21**, 21.
Refaey, K. M. A., and Franklin, J. L. (1976). *J. Chem. Phys.* **65**, 1994.
Richardson, J. H., Stephenson, L. M., and Brauman, J. I. (1973). *J. Chem. Phys.* **59**, 5068.
Richardson, J. H., Stephenson, L. M., and Brauman, J. I. (1974). *Chem. Phys. Lett.* **25**, 321.
Richardson, J. H., Stephenson, L. M., and Brauman, J. I. (1975a). *Chem. Phys. Lett.* **30**, 17.
Richardson, J. H., Stephenson, L. M., and Brauman, J. I. (1975b). *J. Am. Chem. Soc.* **97**, 2967.
Richardson, J. H., Stephenson, L. M., and Brauman, J. I. (1975c). *J. Am. Chem. Soc.* **97**, 1160.
Richardson, J. H., Stephenson, L. M., and Brauman, J. I. (1975d). *J. Chem. Phys.* **62**, 1580.
Richardson, J. H., Stephenson, L. M., and Brauman, J. I. (1975e). *J. Chem. Phys.* **63**, 74.
Rosenfeld, R. N., Jasinski, J. M., and Brauman, J. I. (1979). *J. Chem. Phys.* **71**, 1030.
Rosenfeld, R. N., Jasinski, J. M., and Brauman, J. I. (1982). *J. Am. Chem. Soc.* **104**, 658.
Rosenstock, H. M., Droxl, K., Steiner, B. W., and Herron, J. T. (1977). *J. Phy. Chem. Ref. Data Suppl. 1* **6**.
Rothe, E. W., Tang, S. Y., and Reck, G. P. (1975). *J. Chem. Phys.* **62**, 3829.
Rynard, C. M. (1981). Ph.D. Thesis, Stanford Univ.
Schulz, P. A., Sudbo, A. S., Grant, E. R., Shen, Y. R., and Lee, Y. T. (1980). *J. Chem. Phys.* **72**, 4985.
Schulz, F. A., Mead, R. D., Jones, P. L., and Lineberger, W. C. (1982). *J. Chem. Phys.* **77**, 1153.
Shaw, R. (1974). "Chemistry of the Thiol Group" (S. Patai, ed.), pp. 151–161. Wiley, New York.
Siegel, M. W., Celotta, R. J., Hall, J. L., Levine, J., and Bennett, R. A. (1972). *Phys. Rev. A* **6**, 607.
Simons, J. (1981). *J. Am. Chem. Soc.* **103**, 3971.
Sinnott, G., and Beatty, E. C. (1971). *Proc. Int. Conf. Phys. Elec. At. Collis. 7th, Amsterdam* Abstr., p. 176.
Smyth, K. C., and Brauman, J. I. (1972a). *J. Chem. Phys.* **56**, 5993.
Smyth, K. C., and Brauman, J. I. (1972b). *J. Chem. Phys.* **56**, 1132.
Smyth, K. C., and Brauman, J. I. (1972c). *J. Chem. Phys.* **56**, 4620.
Smyth, K. C., McIver, R. T., Brauman, J. I., and Wallace, R. W. (1971). *J. Chem. Phys.* **54**, 2758.
Sorensen, T. S. (1965). *J. Am. Chem. Soc.* **87**, 5075.

Steiner, B. (1968). *J. Chem. Phys.* **49**, 5097.
Stevens, A. E., Feigerle, C. S., and Lineberger, W. C. (1982). *J. Am. Chem. Soc.* **104**, 5026.
Streit, G. E. (1982). *J. Chem. Phys.* **77**, 826.
Tang, S. Y., Mathur, B. P., Rothe, E. W., and Reck, G. P. (1976). *J.* Chem. Phys. **64**, 1270.
Tichy, M., and Zahradnik, R. (1969). *J. Phys. Chem.* **73**, 534.
Tiernan, T. O., Hughes, B. M., and Lifshitz, C. (1971). *J. Chem. Phys.* **55**, 5692.
Turner, J. E. (1977). *Am. J. Phys.* **45**, 758.
Turner, J. E., and Fox, K. (1966). *Phys. Lett.* **23**, 547.
Ward, B. (1968). *Spectrochim. Acta* **24A**, 813.
Warneck, P. (1969). *Chem. Phys. Lett.* **3**, 532.
Warnick, A., Anders, L. R., and Sharp, T. E. (1974). *Rev. Sci. Instrum.* **45**, 929.
Wetmore, R. W., Schaefer, H. F., III, Hiberty, P. C., and Brauman, J. I. (1980). *J. Am. Chem. Soc.* **102**, 5470.
Wight, C. A., and Beauchamp, J. L. (1981). *J. Am. Chem. Soc.* **103**, 6499.
Wigner, E. P. (1948). *Phys. Rev.* **73**, 1002.
Wong, S. F., Vorburger, T. V., and Woo, S. B. (1971). *Bull. Am. Phys. Soc.* **16**, 213.
Woo, S. B., Helmy, E. M., Mauk, P. H., and Paszek, A. P. (1981). *Phys. Rev. A* **24**, 1380.
Zimmerman, A. H., and Brauman, J. I. (1977a). *J. Am. Chem. Soc.* **99**, 3565.
Zimmerman, A. H., and Brauman, J. I. (1977b). *J. Chem. Phys.* **66**, 5823.
Zimmerman, A. H., Reed, K. J., and Brauman, J. I. (1977). *J. Am. Chem. Soc.* **99**, 7203.
Zimmerman, A. H., Gygax, R., and Brauman, J. I. (1978a). *J. Am. Chem. Soc.* **100**, 5595.
Zimmerman, A. H., Jackson, R. L., Janousek, B. K., and Brauman, J. I. (1978b). *J. Am. Chem. Soc.* **100**, 4674.
Zittel, P. F., and Lineberger, W. C. (1976). *J. Chem. Phys.* **65**, 1236.
Zittel, P. F., Ellison, G. B., ONeil, S. V., Herbst, E., Lineberger, W. C., and Reinhardt, W. P. (1976). *J. Am. Chem. Soc.* **98**, 3731.

Chapter 22
Photodetachment in negative ion beams

ROY D. MEAD
AMY E. STEVENS
*W. C. LINEBERGER**
DEPARTMENT OF CHEMISTRY
UNIVERSITY OF COLORADO
AND
JOINT INSTITUTE FOR LABORATORY ASTROPHYSICS
UNIVERSITY OF COLORADO AND NATIONAL BUREAU OF STANDARDS
BOULDER, COLORADO

*Present address: Department of Chemistry, University of Colorado, Boulder, Colorado 80309.

ISBN 0-12-120803-6

I. INTRODUCTION

This chapter surveys photodetachment in negative ion beams and the chemical information that can be obtained from such studies. The study of negative ions in the gas phase had its origin when Wildt (1939) recognized that H^- photoabsorption could be a major source of opacity in the solar infrared spectrum. Quantitative experimental studies of photodetachment, however, began in the 1950s, spurred by developments in ultrahigh vacuum technology and in crossed-beam experiments. This early work, pioneered by Lewis Branscomb and his colleagues, is summarized in an excellent review (Branscomb, 1962). The past decade has provided major advances both in photon and ion sources for negative ion spectroscopic experiments and in the development of ion trapping devices. These developments have enabled qualitatively new studies of negative ions which previously could not have been attempted.

While neutral molecules possess an infinite number of bound states, negative ions have typically only one or a few bound states; and spectroscopic investigations are generally limited to the study of bound–free transitions, namely, photodetachment. Two major experimental approaches have been developed: photodetachment threshold spectroscopy and photoelectron spectroscopy of negative ions. While data from either type of experiment may be analyzed to obtain electron affinities and molecular structural information, each has its particular strengths and weaknesses. Photodetachment threshold spectroscopy, for example, can be utilized to determine electron affinities with accuracy limited essentially by the wavelength resolution of the light source, while similar determinations using photoelectron spectroscopy are limited by the much poorer resolution of electron energy analyzers. A photodetachment threshold measurement obviously requires the availability of an intense light source tunable in the wavelength region corresponding to threshold, while the negative ion photoelectron spectroscopy studies require only that the wavelength of the intense light source be shorter than the wavelength corresponding to the photodetachment threshold.

In this review, we emphasize threshold photodetachment behavior for atomic and diatomic negative ions and photoelectron spectroscopic investigations of molecular anions. A detailed discussion of threshold behavior for polyatomic anions and molecular electron affinities appears in another chapter in this volume (Drzaic *et al.*, Chapter 21); this contribution is complementary to the work reported here.

There are a number of earlier reviews which stress various aspects of this field. Miller (1981) provides an excellent overview of photodetachment and photodissociation of negative ions. Janousek and Brauman (1979) and

Corderman and Lineberger (1979) reviewed electron affinities and negative ion spectroscopic methods; both of these reviews are updated and extended in this volume. A comprehensive treatise on negative ion properties, and the role of gas phase negative ions in many fields, is provided in the most recent revision of Massey's text (Massey, 1976), while Berry and Leach (1982) provide an in-depth discussion of electron attachment and detachment processes.

There are essentially three types of information that can be obtained from photodetachment spectroscopy of negative ions: electron affinities and energy levels for negative ions; dynamical information, such as electron–molecule scattering cross sections; and information on the structure of neutral species produced in the photodetachment process. In the remainder of this chapter, each of these categories will be examined via case studies. We briefly examine the limits of resolution attainable in tunable laser photodetachment experiments. Finally, a summary of the current preferred atomic electron affinities is presented, and some speculations are offered concerning future developments in photodetachment spectroscopy.

The tunable light photodetachment experiments employ a variety of methods to produce negative ions and to allow their interaction with the photon field. The presently employed methods include photodetachment in high-pressure drift tubes (Moseley, 1982; Woo *et al.*, 1981), photodetachment in ion cyclotron resonance traps (Jopson and Larsen, 1981; Blumberg *et al.*, 1979; Drzaic *et al.*, this volume, Chapter 21), photodetachment in crossed-beam devices (Slater *et al.*, 1978; Jones *et al.*, 1980), and photodetachment in coaxial ion-laser beam devices (Moseley and Durup, 1981; Hefter *et al.*, 1983; Carrington and Kennedy, this volume, Chapter 26). Experiments utilizing the first two techniques are discussed by Drzaic *et al.* (this volume, Chapter 21), while experiments using the latter appear in this chapter.

Negative ion photoelectron spectrometers are currently in use by three groups (Breyer *et al.*, 1978; Ellis and Ellison, 1983; Engelking *et al.*, 1981). Each of these devices utilizes hemispherical electron energy analyzers (resolution 6–50 meV) and employs the intracavity radiation of an ArII laser as the fixed frequency light source. An extensive discussion of each of these devices is given by Miller (1981).

In the remainder of this chapter, we consider applications of each of these experimental techniques. The crossed-beam tunable laser device provides qualitative information on the effect of long-range forces on threshold behavior while coaxial beam devices provide extremely high resolution ($\leqslant 15$ MHz) and detailed information on the autodetachment mechanism. The photoelectron spectroscopy machines provide important information on "dark" states of neutral radicals and permit the systematic study of periodic properties and trends in chemical properties.

II. THRESHOLD PHOTODETACHMENT

Tunable laser photodetachment experiments using ion beams (Slater *et al.*, 1978; Jones *et al.*, 1980) measure the relative cross section for photodetachment as a function of photon energy. Typically, a fast negative ion beam crosses a tunable, pulsed or continuous wave (cw) laser beam, and the neutrals and/or electrons produced are counted as a function of the photon energy. In some cases, low-resolution electron energy discrimination is employed (Slater *et al.*, 1978) or internal state distributions of the neutrals are determined (Rouze and Geballe, 1983). The resolution of such crossed tunable laser-ion beam experiments can be as high as one part in 10^5, and coaxial beam techniques (discussed in Section III) can attain a resolution better than one part in 10^7.

In a well-known paper published in 1948, Eugene Wigner derived the form of the photodetachment cross section for energies near threshold. Valid for final states with two particles whose longest range interaction is the centrifugal barrier, his result gives the variation of the cross section with energy (its shape) but not its absolute magnitude. The law is

$$\sigma \propto (\Delta E)^{l+1/2} \qquad \text{or} \qquad \sigma \propto k^{2l+1}, \tag{1}$$

where σ is the cross section, ΔE is the energy of the departing electron (the energy above the minimum required for reaction to occur) and k and l are the linear and orbital angular momenta of the outgoing electron. The reader should refer to the contribution by Drzaic *et al.* (Chapter 21) in this volume for a detailed discussion of molecular applications of the law. This section will discuss limitations on the law when long-range forces are active.

The physical origin for this predictable threshold behavior is the variation of the electron's potential energy with distance from the neutral. The essence of the Wigner derivation is that, at threshold, the dynamics are solely governed by the longest range potential interaction between the electron and the neutral atom. In the case described by Eq. (1), this interaction is the centrifugal potential, proportional to $l(l + 1)/r^2$. Departures from the law can occur above threshold, where shorter-range terms in the potential become important or when forces with the same range as the centrifugal force are active. Examples of such forces are electron–quadrupole and electron–dipole interactions, respectively.

A. *Effects of Long-Range Forces*

The Wigner law [Eq. (1)] is a *threshold* law, giving the shape of the cross section as the energy of the departing electron approaches zero. A threshold law may be viewed as only the first term of a power series expansion in momentum. The power series may be written

$$\sigma \propto k^{2l+1}(1 + ak^2 + bk^3 + \cdots), \tag{2}$$

where the higher terms may be disregarded for sufficiently small k. The range of validity of the threshold law can be determined only by calculating the next term in the power series or by directly measuring the cross section. The law is then considered "valid" if the deviation from the law in some energy range is less than some arbitrarily chosen fraction.

A useful intuitive picture relating force range and validity of threshold laws at finite energy comes from the correspondence of momentum and inverse distance:

$$k \leftrightarrow 1/r. \tag{3}$$

Near threshold, $k \to 0$, so the potential as $r \to \infty$ determines the threshold law. Potentials acting at smaller r will influence the cross section at larger k (or E).

1. Polarizability

Neutral polarizability gives a $1/r^4$ long-range potential, causing departures from Wigner's law, whose severity depends upon the magnitude of α, the neutral dipole polarizability. The cross-section shape in the presence of neutral polarizability has been derived* (O'Malley, 1965). Valid for spherically symmetric potentials (atoms with no permanent quadrupole moment), it is given by

$$\sigma \propto k^{2l+1}\left[1 - \frac{4\alpha k^2 \ln k}{a_0(2l+3)(2l+1)(2l-1)} + \mathrm{O}(k^2)\right]. \tag{4}$$

Since $E \propto k^2$, we see the leading term is just the Wigner law. If the polarizability of the neutral state is known, the second term can be trivially calculated. Unfortunately, the third term depends upon the details of the potential at very short range, which cannot be computed easily.

This form was tested experimentally by Hotop and Lineberger (1973), who found it could fit their Au^- photodetachment data. They fixed the polarizability at a theoretically determined value and then allowed the coefficient of the k^2 term to vary. Their results qualitatively confirmed the cross section given by Eq. (4). Atomic polarizability decreases cross sections for s-wave ($l = 0$) channels, but increases those for p wave and higher channels. More detailed treatments of polarizability are given by Watanabe and Green (1980) and by Norcross and Taylor (1983).

* O'Malley's 1965 paper contains a sign error in his Eq. (6). The correct sign is used here.

O'Malley (1965) also deals with the effects of quadrupole moments on cross sections. The treatment is more involved than for the polarizability problem. The complications arise from the angular dependence of the forces. The result also has k^2 and $k^2 \ln k$ terms, but with different coefficients. Although quadrupole moments occur for all states with $J \geqslant 1$, their effects can often be neglected in comparison with polarizability effects.

The Wigner law is dramatically modified for photodetachment in the presence of strong magnetic fields (Blumberg *et al.*, 1979; Jopson and Larson, 1981). Magnetic fields reduce the dimensionality of the phase space in the final state and introduce a quantization of the kinetic energy transverse to the field.

2. *Negative Ion Resonances*

When a negative ion has levels embedded in the continuum, cross sections for photodetachment can show resonances at energies corresponding to excitations of the ion (Miller, 1981). Both total and partial cross sections can be affected, and threshold laws may hold only for very small energy ranges. The alkali negative ions Rb^- (Frey *et al.*, 1978) and Cs^- (Mead *et al.*, 1983a) exhibit dramatic structure in their cross sections just below the channel openings for forming ^2P excited neutral states. Due to the existence of quasi-bound negative-ion configurations close in energy to the ^2P channel openings, the total cross sections drop to zero and then rise abruptly to display a Wigner cusp at the new channel opening. Theoretical works by Moores and Norcross (1974) and Lee (1975) use multichannel quantum defect theory to model successfully the behavior of these systems.

3. *Interpretation of Spectra*

The aforementioned effects cause wide variation in the range of validity for threshold laws. For example, the Wigner law has been found to hold in Au^- (Hotop and Lineberger, 1973) over a range of 100 meV, with deviations of only a few percent. Extremely rapid breakdown of the law is seen in detachment to the $^2P_{1/2}$ state of Cs (Mead *et al.*, 1983a), where large deviations from Wigner's law are seen only 25 μeV above threshold! This is not to say that Wigner's law is incorrect, but rather the condition that E be taken sufficiently close to zero has not been satisfied. These variations in range of validity complicate the extraction of constants (such as electron affinities) from data, since the precise functional form of the cross section is not known. One justifiable procedure involves fitting some theoretical form (i.e., the Wigner law) to data sets of successively smaller energy range. For data sets whose range of energies is sufficiently close to threshold, the constants obtained should all be consistent.

4. *Intensities and Selection Rules*

While absolute cross sections cannot be calculated directly, relative intensities of different atomic fine structure transitions can be computed according to the formalism of Rau and Fano (1971). Their method involves a frame transformation from an L–S coupled negative ion system to a *j–j* coupled neutral plus free electron. Experimental confirmations of their result have been provided by Lineberger and Woodward (1970) and by Hotop *et al.* (1973a).

The photodetachment selection rules for molecules are simply derived from the conservation of angular momentum and parity. The conservation of angular momentum requires that the change in angular momentum quantum number for an s-electron detachment is restricted to $\Delta J = \pm\frac{3}{2}$ or $\pm\frac{1}{2}$. Conservation of parity in photodetachment restricts the final state to a parity opposite the initial negative ion parity. Intensities of the thresholds in OH^- have been measured (Schulz *et al.*, 1982) and show behavior in accord with the view that the photodetachment process proceeds through an intermediate (e^- + neutral) complex. The agreement between the theory as developed by Walker (1973) and Schulz *et al.* (1983) and the observed intensity ratios confirms that threshold channel strengths in molecular photodetachment are governed by the same mechanism as in atomic photodetachment.

As a result of studies in a number of laboratories, one can say that threshold behavior is now understood for many atomic and molecular negative ions. A major class of systems not yet understood, however, is that which forms neutrals with dipole moments, since the Wigner law breaks down completely in the presence of dipole forces. The next section discusses OH^- photodetachment experiments elucidating the threshold behavior for such cases.

B. *Photodetachment from Dipolar Molecules*

When potentials with the same range as the $l(l+1)/r^2$ centrifugal potential interaction are present, the Wigner law does not apply, even as $E \to 0$. This result may be of significance for many molecular systems, since the potential between a fixed polar molecule and an electron has a $1/r^2$ radial dependence. Solution of the radial potential problem provides guidelines for the real system, but is not completely applicable to molecules because the dipole potential has a $\cos\theta$ angular dependence and the molecular rotation effects may be important. Gailitis and Damburg (1963) and O'Malley (1965) extended Wigner's work to include a spherically symmetric long-range potential, $V(r) = -ed/r^2$, where e is the electron charge and d is the coefficient of the r^{-2} potential. The cross section near threshold then has the form

$$\sigma \sim \begin{cases} (E-E_{\text{thr}})^{[(l+1/2)^2-2d]^{1/2}} & \text{for } \left(l+\dfrac{1}{2}\right)^2-2d>0 \\ \left\{\sinh^2\dfrac{\pi v}{2}+\cos^2\left[\dfrac{v}{2}\ln(E-E_{\text{thr}})+\delta\right]\right\}^{-1} & \text{for } \left(l+\dfrac{1}{2}\right)^2-2d\equiv -v^2<0, \end{cases} \tag{5}$$

where δ is a phase factor sensitive to the short-range behavior of the potential. The major question is whether the interaction between the departing electron and the dipolar molecule can be represented by a central r^{-2} attractive potential. Engelking (1982) has extended this approach and applied it to OH^- photodetachment. All of these studies suggest that the cross section near threshold might rise more sharply than $\sigma \propto (\Delta E)^{1/2}$ and the exact shape should depend on the rotational quantum number of the resulting polar molecule. The most comprehensive experimental work on dipolar systems has been done over the past ten years on OH^-.

1. *Experimental Work on* OH^-

The unresolved manifold of rotational transitions in OH^- causes the photodetachment cross section to have the shape measured by Hotop *et al.* (1974) shown in Fig. 1. Their interpretation of the spectrum is that the sharp increase in photodetachment at 14,700 cm^{-1} is caused by the pileup of many

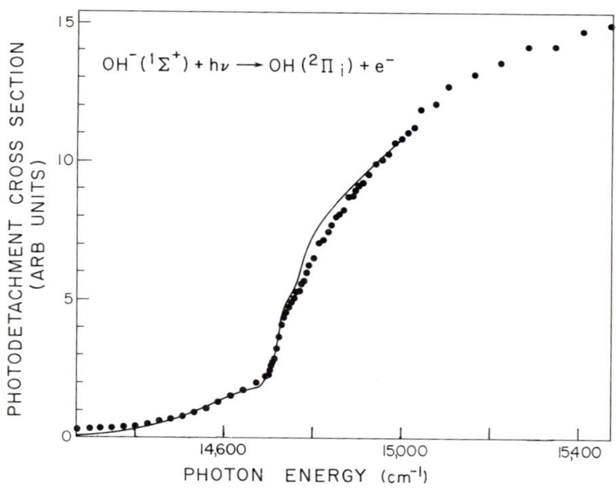

FIG. 1 OH^- photodetachment cross section in the energy range 14,300–15,400 cm^{-1}. The sharp onset near 14,700 cm^{-1} corresponds to the opening of the Q3 branch. From Hotop *et al.* (1974).

different rotational thresholds in the Q branch. The monotonically increasing photodetachment cross section at lower energies is caused by P-branch transitions and, at higher energies, by R-branch transitions.

The most detailed experimental study of photodetachment from dipoles was done recently by Schulz *et al.* (1982), who studied the shapes and intensities of individual rotational thresholds in OH^-. Utilizing a crossed tunable dye laser–ion beam apparatus, the major improvements over previous OH^- photodetachment experiments were the cooler, more stable ion source and the cw intracavity laser which resulted in a signal-to-noise ratio of 200 and resolution of 0.2 cm^{-1}.

Schulz *et al.* (1982) reinvestigated a portion of the region shown in Fig. 1 at much higher resolution. Figure 2 shows a part of the OH^- photodetachment spectrum measured by them. Each abrupt rise is in fact associated with the threshold for detachment from a single rotational state of the negative ion, producing a single rotational state of the neutral molecule. The photon energy at such a threshold is the energy difference between the corresponding states of the negative ion and neutral. The number of thresholds observed is large enough that unique assignments of the individual thresholds are readily obtained.

From the data, energy levels of the OH^- and OD^- ions are obtained. From the analysis, the electron affinities of OH and OD were found to be 14,741.03 ± 0.17 cm^{-1} and 14,723.92 ± 0.30 cm^{-1}, respectively. The band molecular constants for the negative ions were also obtained. In addition, the equilibrium constants from a composite fit of OH^- and OD^- data were found to be $B_e(OH^-) = 19.127(12)$ cm^{-1}, $D_e(OH^-) = 2.067(38) \times 10^{-3}$ cm^{-1}, $\alpha_e = 0.773(24)$ cm^{-1}, and $\omega_e = 3680(37)$ cm^{-1}. Thus, $OH^-(OD^-)$ has a

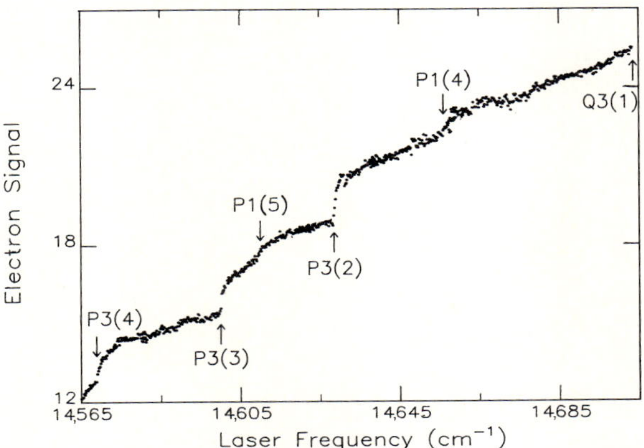

FIG. 2 OH^- photodetachment spectrum in the energy range 14,565–14,705 cm^{-1}. Point spacing is 0.2 cm^{-1} at a resolution of 0.4 cm^{-1} [From Schulz *et al.* (1982).]

slightly shorter bond than OH(OD). The negative ion has a slightly smaller vibrational frequency than the neutral.

With the resolution available in this experiment, the shape of the photodetachment cross section for an *individual* rotational threshold could be observed. To do this, the cross section arising from all the already open, lower-energy channels must be subtracted from the observed cross section. This can be done if it is assumed that the cross section for the open channels does not change abruptly near the new threshold. There is no indication of any resonance or cusp structure in OH^- photodetachment, nor is any expected. Thus, the simple procedure of subtracting a linearly rising cross section from the data should be adequate to obtain a state-selected partial cross section.

Figure 3 (Schulz *et al.*, 1982) shows the P3(2) threshold so derived from Fig. 2, with scales expanded to show the threshold more clearly. The measured cross section is compared with a function corresponding to the Wigner threshold law convolved with the resolution of the apparatus. By comparison, the actual cross section looks almost like a step function and is showing the influence of a long-range electron–dipole interaction. Examination of other rotational thresholds shows that the cross-section shape, though independent of the OH^- rotational state, depends on the rotational state of the OH. The energy dependence is sharper for lower rotational states and more like a square root dependence for higher rotational states. This behavior is qualitatively consistent with the notion of the dipole averaging out for high rotational levels. The applicability of this classical picture must be considered more closely.

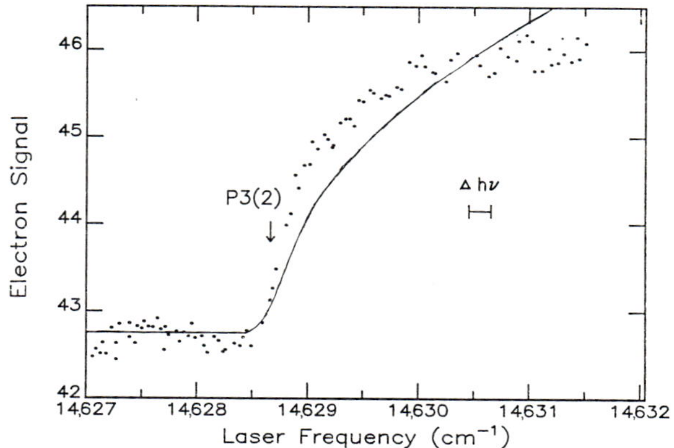

FIG. 3 Detailed view of the P3(2) threshold with point spacing of 0.05 cm^{-1} at a resolution of 0.2 cm^{-1}. The solid line shows a $\Sigma \propto (\Delta E)^{1/2}$ threshold law convolved with the instrumental resolution. It is seen that for an energy range as small as 3 cm^{-1}, the cross section rises much more sharply than the Wigner (square root) threshold law. [From Schulz *et al.* (1982).]

2. Theory

In OH^-, the highest occupied orbital is a π orbital that looks like an atomic p orbital. An electric dipole transition can occur to the s-wave continuum, which, in the absence of a molecular dipole, gives a cross section of the form $\sigma \propto (\Delta E)^{1/2}$. Shapes different from this may be expected in OH^- photodetachment because there is an anisotropic r^{-2} potential arising from the electron–dipole interaction in the final state.

The experiment provides information about the potential as far as 300 Å away from the dipole, since farther from the dipole the potential energy is less than the apparatus resolution. Simple calculations show that the departure time for the electron (the time required for the electron to travel this 300 Å) is much longer than a molecular rotation time. Thus, by examining the shape of the threshold cross section for the first few wave numbers above the threshold, the long-range interaction between an electron and a dipole rotating rapidly on the time scale in which the electron leaves is probed.

Another physically meaningful method of viewing the charge–dipole interactions is by considering the energy levels in the OH molecule and how they are affected by the detached electron. Dipole moments are only possible for states of mixed parity because states of definite parity have equal probability density upon inversion through the origin. A first-order dipole moment requires degeneracy of states of opposite parity or a field strong enough to mix the nondegenerate levels. For electron energies below 1 meV, there is no possibility for first-order mixing of OH rotational states of opposite parity because the separation of rotational states is much larger than 1 meV. However, for a molecule like OH that has orbital angular momentum, the near degeneracy of Λ-doubled levels with opposite parity produces a dipole moment in an external field. For detached electrons with kinetic energies less than the splitting between levels of opposite parity, the parity states are not mixed in first order and there is no long-range $1/r^2$ interaction. The threshold cross-section shape then is expected to follow the Wigner law. On the other hand, for electron kinetic energies larger than the splitting of parity states, the first-order mixing of parity states gives rise to a dipole. If the electron kinetic energies are still small enough that mixing of rotational states is unimportant, the threshold cross-section shape may be solved using a time-independent potential in a close-coupling calculation. This approach has been taken by Engelking (1982) to derive the threshold law. Since his theory makes a specific prediction on the behavior of the threshold law, a comparison with the experiment can be made.

The cross section of Fig. 3 shows clearly the effects of a strong electron–dipole interaction in that it is definitely different from Wigner behavior and is consistent with the $\sigma \propto (\Delta E)^{0.28}$ predicted by Engelking (1982). Though the shape of the cross section is consistent with his theory, the signal-to-noise ratio does not allow a definite conclusion to be made concerning the range of validity of the threshold law. Still, the agreement should be considered support for the close-coupling approach used by Engelking.

III. ULTRAHIGH RESOLUTION PHOTODETACHMENT

Fundamental limitations of the crossed-beam photodetachment technique preclude resolution much better than one part in 10^5 (Jones *et al.*, 1980). The recent advent of coaxial-beam techniques, in which the ion and photon beams are merged, has produced an improvement to one part in 10^7. The velocity spread of the ions is greatly reduced along the beam direction, greatly reducing the Doppler broadening of the ion signal. In fact, this is one of the few methods in which subdoppler resolution is obtainable using single-photon spectroscopic techniques (Moseley and Durup, 1981; Carrington and Kennedy, this volume, Chapter 26; Hefter *et al.*, 1983). The first study of negative ions using the coaxial-beams technique has been of perturbations and autodetachment in C_2^-.

A. *Spectroscopy of* C_2^-

Most negative ions show a photodetachment cross section devoid of sharp structure (Fig. 1 presents a typical example), as is typical of bound–free processes. A notable exception to this rule is demonstrated in the photodetachment cross section of C_2^-, shown at low resolution in Fig. 4. The structure shown is due to photoabsorbtion to produce a C_2^{-*} quasi-bound state

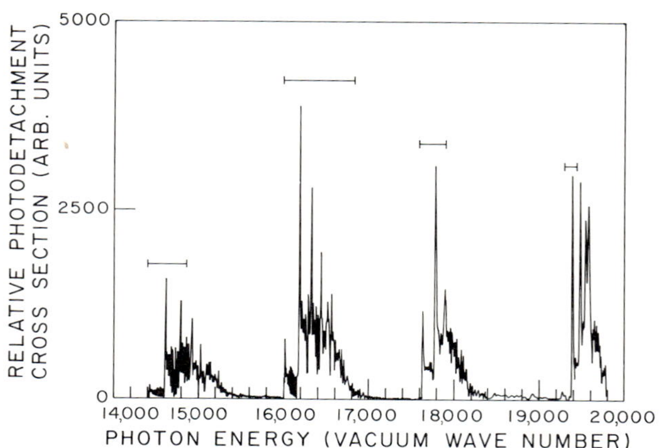

FIG. 4 C_2^- low-resolution ($\Delta\nu \cong 10\ \text{cm}^{-1}$) data showing from left to right the $\Delta v = -3$, -2, -1, and 0 progressions of the C_2^- $B\,^2\Sigma_u^+ - X\,^2\Sigma_g^+$ transition in autodetachment. Each progression begins on a transition with final state $v' = 5$. [From Jones *et al.* (1980).]

above the onset of the (C_2 + e) continuum, which then decays nonradiatively to form a neutral plus an electron.

The relevant potential curves for C_2^- and C_2 are shown in Fig. 5. The transitions seen involve high vibrational levels of the ion ground state and high vibrational levels of excited ion states. Isotopic substitution experiments show that all of the structure seen in Fig. 4 is from Σ–Σ transitions terminating on levels with $v' \geqslant 5$. The four groups of lines correspond to transitions with $\Delta v = -3, -2, -1$, and 0, respectively. The small features to the

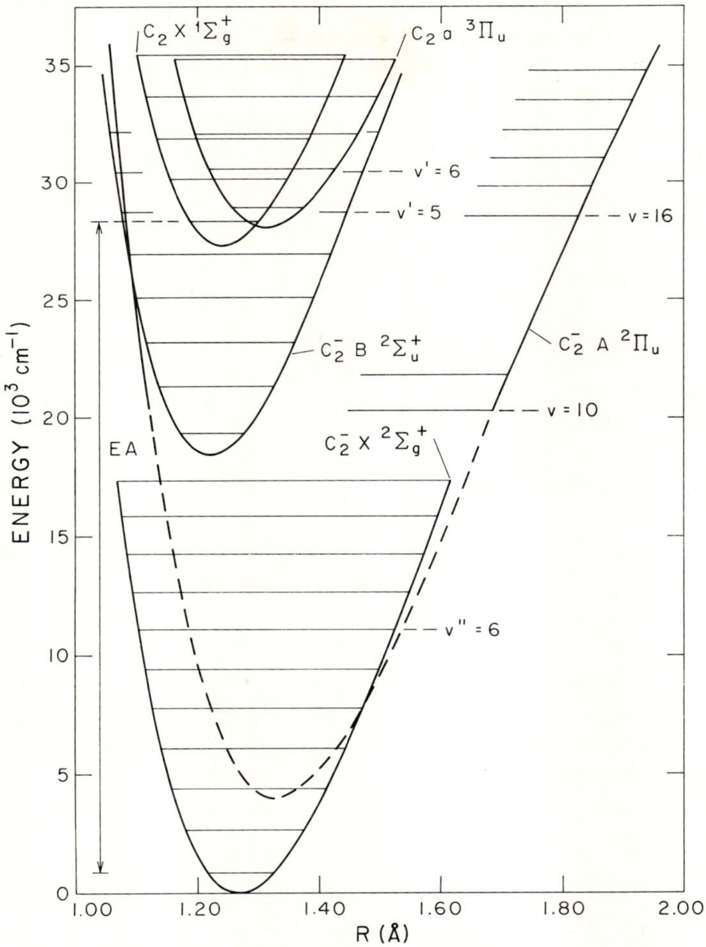

FIG. 5 Potential curves for the X, A, and B states of C_2^- and the X and a states of C_2. The potential for the A state of C_2^- was determined from its perturbations of the B state. [From Mead *et al.* (1984).]

low-energy side of each group of lines have upper state $v' = 5$, while the larger features have $v' = 6, 7, 8$ in sequence. The observation of such structure motivated examination with the crossed-beam instrument at higher resolution, producing data exemplified by Fig. 6. With the 200-fold energy scale expansion from Fig. 4 to Fig. 6, individual rotational lines are now visible, and assignment of the rotational structure allowed rotational constants to be obtained. These data, taken together with those of Herzberg and Lagerqvist (1968), produced the best set of potential curves and spectroscopic constants available for any negative ion. Still, there was unresolved substructure in the spectra and many lines were unassignable. Lifetimes for the upper states could not be determined directly, since the instrumental width far exceeded the uncertainty broadening of the lines.

Use of the coaxial instrument resolves further structure (Mead *et al.*, 1984), as seen in Fig. 7, whose energy scale is a further 600-fold expansion from Fig. 6! Because the C_2^- levels belong to doublet states, each rotational state is split into two levels by spin–rotation interaction, producing two strong lines and one satellite line. Though the instrumental width is less than 30 MHz (0.001 cm^{-1}), these lines have widths of 60 MHz due to the finite lifetimes of the upper states. All quantum levels for these states (with the exception of the degenerate magnetic sublevels) are now resolved. Examination of $^{13}C_2^-$ spectra (Fig. 8) reveals further structure due to hyperfine interactions. This structure is also readily resolved by the coaxial-beam apparatus.

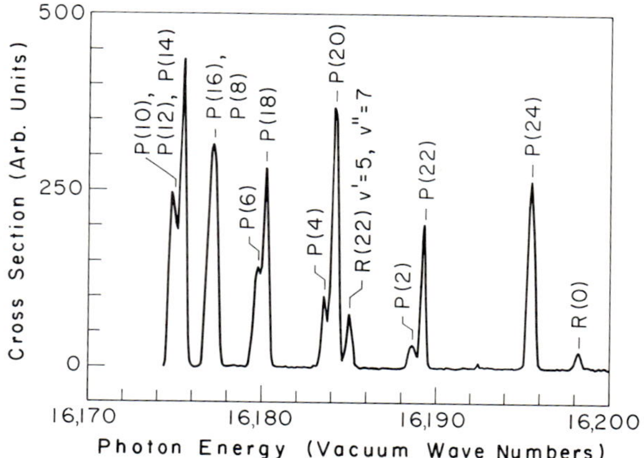

FIG. 6 High-resolution ($\Delta\nu \cong 0.4$ cm^{-1}) data showing the (6, 8) C_2^- B–X band head region with rotational assignments. Although rotational lines were resolved, the spin–rotation splitting of the lines was not. [From Jones *et al.* (1980).]

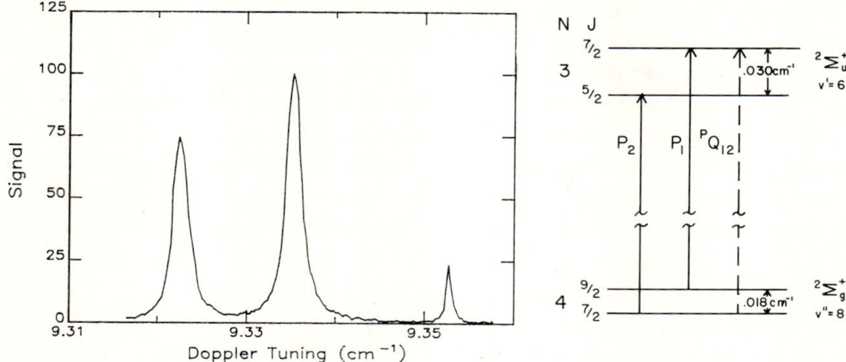

FIG. 7 Ultrahigh resolution ($\Delta\nu \cong 0.001\ \text{cm}^{-1}$) Doppler-tuned spectrum (Mead *et al.*, 1984) of the P(4) line in Fig. 6 with a diagram of the relevant energy levels. The laser frequency used in this scan was 16,174.88 cm^{-1}, so the lines are near 16,184.2 cm^{-1}. Due to the finite lifetimes of the upper states, the lines are about 0.002 cm^{-1} wide.

B. *Perturbations and Analysis of Spectra*

In analogy to N_2^+, CN, and CO^+, a low-lying $^2\Pi$ state of C_2^- is expected to exist, but has not been observed. The spectra shown in the figures are due to B $^2\Sigma_u^+ - \text{X}\ ^2\Sigma_g^+$ transitions, and no direct observation of the A $^2\Pi_u$ state has yet been made. Nevertheless, the A and B states interact, and the presence of highly perturbed lines in the new data allow the character of the A state

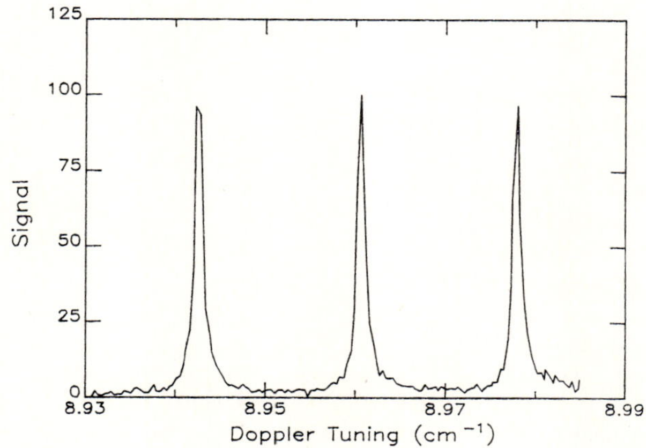

FIG. 8 Ultrahigh resolution ($\Delta\nu \cong 0.001\ \text{cm}^{-1}$) Doppler-tuned spectrum (Mead *et al.*, 1984) of the $v' = 6$, $v'' = 8$ $P_1(17)$ line in $^{13}C_2^-$. What would have been a single rotational line is split into three hyperfine components by the nuclear spin in ^{13}C.

to be determined. One effect of the perturbation can be seen in the splittings between spin–rotation lines like those shown in Fig. 7. Since the splitting between the paired levels of equal rotational quantum number N is due to interaction of electron spin and molecular rotation, the splitting should increase linearly with rotational quantum number. Plotting the splittings versus rotational quantum number, as is done in Fig. 9, reveals a linear trend, but with divergences due to $^2\Pi$ perturbations near $N' = 37$ and again near $N' = 67$. The analysis of these perturbed spectra is based on modern total-fitting techniques, as discussed by Albritton *et al.* (1976) and Gottscho *et al.* (1979).

The designations $^2\Pi_u$ and $^2\Sigma_u^+$ used for the excited states of C_2^- refer to zero-order states of the ion, as some of the quantum numbers given are not perfectly good and there are interactions between rotation and electronic motion. Levels of the two zero-order states with the same J mix by spin–orbit and rotation–electronic coupling, producing the real energy levels of the molecule. In this sense, the potential curves of Fig. 5 reflect an idealized case, where the interaction between the states has been removed so as to produce distinct potential curves.

The analysis begins with the expressions for the unperturbed energy levels of the $^2\Sigma_u^+$ and $^2\Pi_u$ states and for the matrix elements coupling these states (Coxon, 1975; Zare *et al.*, 1973; Hougen, 1970). These quantities are assembled into a Hamiltonian matrix, with the zero-order energies on the diagonal and the coupling matrix elements in the off-diagonal positions. Diagonaliza-

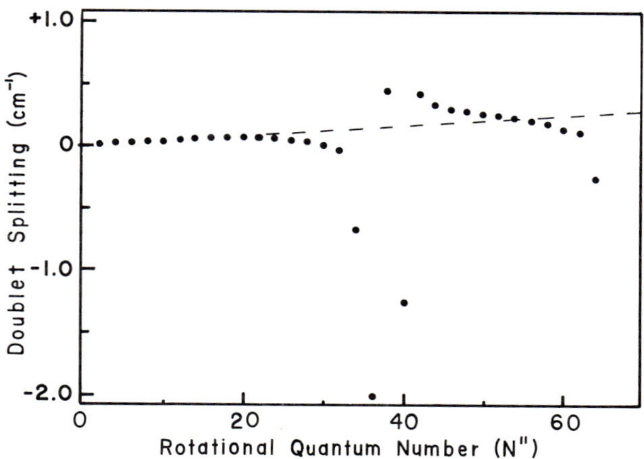

FIG. 9 Splittings between doublets (like the pair of strong lines in Fig. 7) versus rotational quantum number in the (6, 8) band of the C_2^- B–X transition. The usual linear dependence of the doublet splitting is interrupted by perturbations by the A $^2\Pi_u$ state. [From Mead *et al.* (1984).]

tion of the matrix (or, equivalently, extraction of the eigenvalues) then yields the true energy levels of the system.

In actual practice, a set of lines which is only weakly perturbed is first fit to a zero-order $^2\Sigma$ energy-level model, using conventional least squares techniques. Then, more strongly perturbed lines are added to the set, and a fit involving diagonalization of the full $^2\Sigma$–$^2\Pi$ Hamiltonian is done. The new constants are used to assign additional lines in the spectra, so that the lines may be included in subsequent fits. The intermediate results are essential for further assignments of spectral lines, as some lines are displaced from their zero-order positions by more than ten wave numbers! For most spectral regions, all lines are eventually assigned after many iterations through the prediction–assignment–fitting sequence.

The products of the fitting procedure are spectroscopic constants for the Σ states and the Π state, as well as the strengths of the couplings between them. Potential curves can then be constructed from the constants. Data from isotopically substituted C_2^- allow unequivocal assignments of vibrational quantum numbers and provide a consistency check on the analysis. The sophisticated calculations of Zeitz *et al.* (1979) are shown to predict accurately the characteristics of the C_2^- states, including the previously unobserved A $^2\Pi_u$ state.

It should be emphasized that the $^2\Pi$ state of C_2^- has been discovered and characterized *without having to observe direct transitions to the state itself*. This accomplishment depended wholly on the availability of high-resolution, high-accuracy data from the coaxial photodetachment apparatus and upon the advanced analysis techniques developed over the past decade.

C. *Lifetimes for Autodetaching C_2^- Levels*

The work of Jones *et al.* (1980) on a crossed-beam photodetachment apparatus bounded the lifetimes for the autodetaching levels of C_2^-. If the lifetimes had been very short ($\leqslant 10^{-11}$ sec), the lines would have appeared wider than the 0.4 cm^{-1} instrumental width of that study. Arguments based upon the product collection times of the apparatus and known fluorescence lifetimes for C_2^- provided additional limits, bounding the lifetime of the $v' = 5$ level between 10^{-6} and 10^{-7} sec, while the $v' \geqslant 6$ levels had upper and lower bounds of 10^{-7} and 10^{-11} sec, respectively.

Direct observation of uncertainty broadening of C_2^- levels has recently been reported by Hefter *et al.* (1983), who used the ultrahigh resolution coaxial-beam apparatus to study linewidths of C_2^- transitions. Lifetimes were obtained for many rotational and vibrational states, allowing a complete picture of the influence of nuclear motion upon autodetachment to be formed. Lifetimes were shown to decrease rapidly with vibrational excitation and to decrease at higher rotational levels. In addition, autodetachment lifetimes for states of predominantly $^2\Pi$ character were found to be much greater than for states of $^2\Sigma$ character. These data will allow a better understanding of autoionization processes, particularly since rotationally resolved lifetime data have been so difficult to obtain for autoionization. Study of autodetachment, the equivalent process in negative ions, is possible without resorting to the vacuum ultraviolet light sources usually required to study autoionizing states of neutrals, and is uncomplicated by the presence of Rydberg states.

IV. PHOTOELECTRON SPECTROSCOPY OF MOLECULAR IONS

In laser photoelectron spectroscopy, the laser wavelength is fixed and spectra are obtained by analysis of the kinetic energies of the photodetached electrons. Typically, the 488.0-nm (2.540-eV) and occasionally the 363.8-nm (3.408-eV) lines of an argon-ion laser operating intracavity are employed as photon sources, and electrons from near-zero energy to the photon energy are energy analyzed and collected. Since the electron energy analysis, not the laser frequency, determines the resolution, the resolution is much lower than in the tunable laser photodetachment studies described earlier. Breyer *et al.* (1981) report a resolution of about 5–6 meV; at this resolution, the P, Q, and R rotational branches (but not individual rotational transitions) in the photoelectron spectra of OH^-, SH^-, and SD^- are resolved. More typical resolutions are the 20–25 meV obtained by Ellis and Ellison (1983) and the 45–50 meV of the instrument of Lineberger and co-workers. Photoelectron spectroscopy thus provides a means of rapid acquisition of a relatively low-resolution spectrum over a wide energy range and, as such, provides a complementary technique to that of photodetachment spectroscopy.

Photoelectron spectroscopy is typically employed to determine the electron affinities of neutral species; these are useful in determining thermochemistry of both ions and neutrals. The spectra also identify the electronic state and (for molecular species) vibrational structure of negative ions and often provide information on electronic states (and molecular parameters) for the neutral, as well. This last aspect is particularly valuable in identifying previously unobserved optically "dark" states of neutral molecules. Finally, the relative ease of data acquisition allows for comparison of several related species and identification of periodic trends.

Many of the electron affinities of atoms, and the state assignment of the corresponding negative ions, have been determined by this method; these are summarized in Table II, in Section V. Photoelectron spectra of a wide variety of molecular ions have also been obtained. Drzaic *et al.* (this volume,

Chapter 21) discuss several examples and give a tabulation of molecular electron affinities, which includes many references to photoelectron spectroscopy. Current reviews of laser photoelectron spectroscopy are available (Corderman and Lineberger, 1979), so a comprehensive review of photoelectron spectroscopic techniques and spectral analysis will not be given here. Rather, we present several recent examples of spectra of molecular negative ions. The examples chosen are the spectra of HNO^- and DNO^-; the spectra of a series of nickel carbonyl ions, $Ni(CO)_n^-$, $n = 1$–3; and the spectra of a series of transition-metal hydrides MH^-, where M = Mn, Fe, Co, and Ni.

A. *Photoelectron Spectroscopy of* HNO^- *and* DNO^-

The spectra of HNO^- and DNO^- reported by Ellis and Ellison (1983) provide a very nice example of the use of photoelectron spectra in characterization of a previously unobserved dark state of a triatomic molecule. Electron photodetachment to three electronic states is observed in the spectrum of HNO^-, shown in Fig. 10. Electronic transition origins (from the vibrationless negative ion to the vibrationless energy level of the neutral) appear at 2.195-, 1.410-, and 0.573-eV electron kinetic energy. The electronic transition with origin at 2.195 eV is assigned as a transition from ground-state HNO^- to the ground-state $X\,^1A'$ of HNO. A vibrational progression extends from this peak to about 1.6 eV, with a 0–1 frequency of 1565 cm^{-1}. The good correlation between this value and the previously measured N—O stretching frequency (Ogilvie, 1974), as well as the small shift of this progression for DNO^-, leads to its assignment as a N—O stretching vibration.

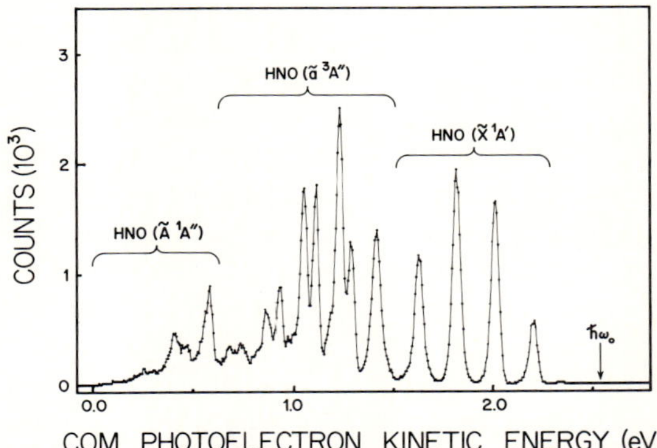

FIG. 10 Photoelectron spectrum of HNO^- obtained with 488-nm excitation. ($M/Z = 31$ and laser 2.54 eV.) The data show clearly transitions to the three lowest electronic states of HNO. [From Ellis and Ellison (1983).]

A small peak to the high electron energy side of the $X\,^1A'$ origin is attributed to a transition from an HNO^- with an excited N—O vibration to the vibrationless level of the HNO neutral. This assignment is confirmed by the deuteride shift, thus determining the N—O stretching frequency of 1153 cm^{-1} for HNO^-. A Franck–Condon factor analysis of this electronic transition determines the N—O bond length in the negative ion to be 1.33 ± 0.02 Å, about 0.12 Å longer than in the neutral. Since vibrational progressions are not observed in either the N—H stretching or H—N—O bending modes, the N—H bond length and H—N—O angle of the negative ion were assumed equal to those of the $X\,^1A'$ neutral, 1.063 Å and 108.60°, respectively.

The vibrational energy levels for $X\,^1A'$ HNO are indicated in Fig. 11. The notation (v_1, v_2, v_3) refers to the vibrational quantum number of the v_1 (N—H stretch), v_2 (N—O stretch), and v_3 (H—N—O bend) vibrational modes. Also shown are the energy levels for several levels with excited N—H or H—N—O bending vibrations; no evidence of excitation of the modes

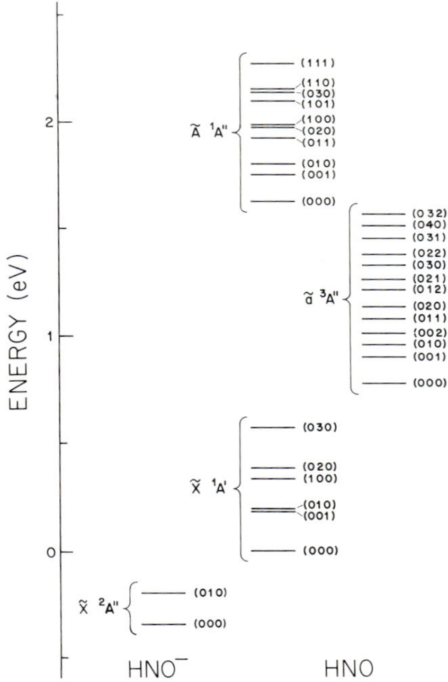

FIG. 11 Electronic and vibrational energy levels for HNO^- and HNO. [From Ellis and Ellison (1983).] The energy levels of $X\,^1A'$ (HNO) and $A\,^1A''$ (HNO) are taken from Ogilvie (1974) and Bancroft *et al.* (1962); the energy levels for a $^3A''$ (HNO) and $X\,^2A''$ (HNO^-) are determined from the photoelectron spectrum of HNO^- shown in Fig. 10.

was seen in the spectra of HNO^- or DNO^-. The electron affinity of HNO obtained from the spectrum determines the location of the (0, 0, 0) energy level of HNO^-, shown in Fig. 11; the observed (0, 1, 0) level is also given.

The observed excitation energy of 13,080 cm^{-1} for the third electronic state of HNO (with origin at 0.573-eV electron kinetic energy) identifies it as the A $^1A''$ state of HNO. The electron analyzer cutoff is at approximately 0.3-eV electron kinetic energy, and thus the total Franck–Condon envelope for this state is not observed. However, the observed vibrational frequencies for the N—O stretch and H—N—O bend are in good agreement with the vibrational frequencies obtained from optical spectra (Bancroft *et al.*, 1962) for the A $^1A''$ state. Appropriate shifts in the spectrum of DNO^- also support this state assignment.

Theoretical calculations (Bruna and Marian, 1979) and indirect experimental evidence (Ishiwata *et al.*, 1978) indicated the a $^3A''$ state of HNO to lie at about 0.8 to 0.9 eV above the X $^1A'$, but this state has not been directly observed. Transition to the a $^3A''$ state is observed in the photoelectron spectrum; the transition origin at 1.410 eV directly establishes the excitation energy of the a $^3A''$ state of HNO to be 0.778 ± 0.020 eV. A vibrational progression at a frequency of 1468 ± 140 cm^{-1} is observed and assigned to the excitation of the N—O stretching mode. Particularly interesting is the superposition of a progression in the H—N—O bending vibration, identified by its frequency of 992 ± 150 cm^{-1}, and deuteride shift. Since the a $^3A''$ state is the triplet coupled state with the same electronic configuration as the A $^1A'$ state, the geometry of the triplet state is expected to be near that of the singlet state, i.e., an N—O bond length of 1.239 Å and an H—N—O bond angle of 114.4° (Bancroft *et al.*, 1962). Although Franck–Condon analyses were not made, observations of vibrational progressions in the N—O stretching mode and H—N—O bending vibrations are as expected for an N—O bond length decrease of 0.09 Å from HNO^-, and an H—N—O bond angle increase of 6° from the negative ion. The observed energy levels for the a $^3A''$ state are also presented in Fig. 11.

In summary, the spectra of Ellis and Ellison (1983) provide the electron affinities of HNO and DNO, and the N—O stretching frequency and geometry of the negative ions. More importantly, the data provide the first direct identification of the optically dark a $^3A''$ state and give quantitative determination of the excitation energy for this state. The observation of N—O stretching and H—N—O bending frequencies for this state also provide a good example of well-resolved vibrational structure, made possible by the 20–25-meV resolution of the analyzer employed by these authors.

As an example of a series of spectra which aid in the determination of electronic and geometric structural information and the thermochemistry of neutral molecules, we present the spectra of the $Ni(CO)_n^-$ in the following section. These spectra also provide a contrast to the spectra of the triatomic HNO^- and show the types of information obtainable for large polyatomic species.

B. *Electronic States, Geometry, and Thermochemistry of* $Ni(CO)_n^-$ *and* $Ni(CO)_n$

Several examples of the photoelectron spectra of larger polyatomic molecular negative ions are afforded by the spectra of the nickel carbonyl series, $Ni(CO)_n^-$, $n = 1–3$, reported by Stevens *et al.* (1982). The spectra of $Ni(CO)^-$ and $Ni(CO)_3^-$ are presented in Figs. 12 and 13; the spectrum of $Ni(CO)_2^-$ is qualitatively similar to that of the tricarbonyl. The electron affinity of each of the corresponding neutrals is determined by the peak labeled X in each spectrum and is summarized in Table I. Evident in each spectrum is a vibrational progression, extending from peak X, in the C—O symmetric stretching vibration; the 0–1 frequency for this vibration is also given in Table I. For the $Ni(CO)_2$ and $Ni(CO)_3$, only a single electronic state is observed in the photoelectron spectrum. In the spectrum of $Ni(CO)^-$, however, a second electronic state A of Ni(CO) is observed at 0.36-eV electron kinetic energy, which corresponds to a 1.37-eV excitation energy from the ground state.

One of the important uses of electron affinities is in the determination of thermochemical bond energies; this has been discussed for the case of determination of homolytic hydrogen bond energies by other authors (Drzaic *et al.*, this volume, Chapter 21; Bartmess and McIver, 1979). The data for

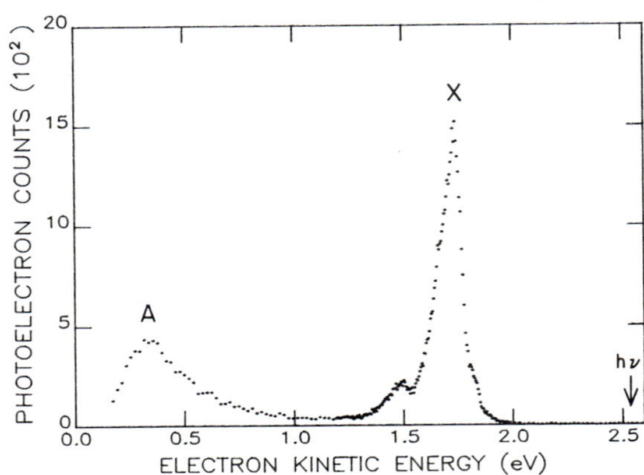

FIG. 12 Photoelectron spectrum of $Ni(CO)^-$ obtained with 488.0-nm (2.540-eV) photons; the peak labeled X is the electron-affinity defining transition. [From Stevens *et al.* (1982). Copyright 1982 American Chemical Society.]

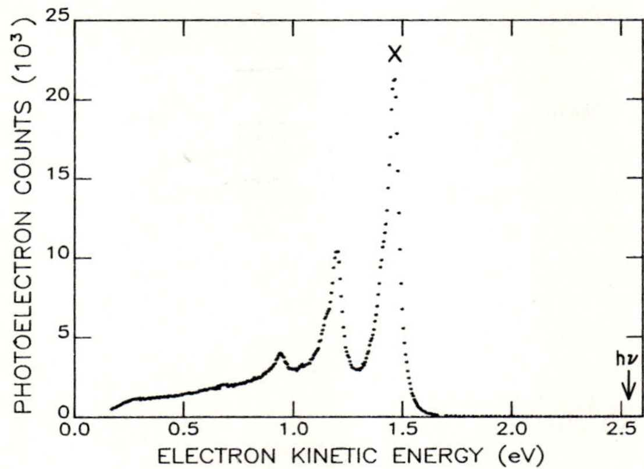

FIG. 13 Photoelectron spectrum of $Ni(CO)_3^-$ obtained with 488.0-nm (2.540-eV) photons. [From Stevens *et al.* (1982). Copyright 1982 American Chemical Society.]

the nickel carbonyls can be used to determine the metal-carbonyl bond energies, as given by Eq. (6), using Ni(CO) as an example:

$$D[Ni\text{–}CO] = D[Ni^-\text{–}CO] + EA[Ni] - EA[Ni(CO)]. \tag{6}$$

The bond energies of the negative ions are determined by the difference in appearance potentials of Ni^- and $Ni(CO)^-$, which are reported by Compton

TABLE I

Electron Affinities (EA), Nickel–Carbonyl Bond Energies, and C—O Vibrational Frequencies for $Ni(CO)_n$[a]

			Symmetric C—O stretching frequency, cm^{-1}	
	$EA[Ni(CO)_n]$ (eV)	$D[Ni(CO)_{n-1}\text{–}CO]$ (kcal/mol)	Stevens *et al.* (1982)	Argon matrix[b]
$Ni(CO)_4$	[c]	25 ± 2	—	2134
$Ni(CO)_3$	1.077 ± 0.013	13 ± 10	2100 ± 80	2120
$Ni(CO)_2$	0.643 ± 0.014	54 ± 15	2100 ± 80	2090
$Ni(CO)$	0.804 ± 0.012	29 ± 15	1940 ± 80	1996
Ni	1.157 ± 0.010[d]	—	—	—

[a] From Stevens *et al.* (1982). Copyright 1982 American Chemical Society.

[b] Calculated from the Cotton–Kraihanzel force constants given by DeKock (1971) and the antisymmetric stretching frequencies reported by Burdett (1973), Breeze *et. al.* (1981), and DeKock (1971).

[c] $Ni(CO)_4^-$ is not bound with respect to CO dissociation.

[d] See Table II in Section V of this chapter.

and Stockdale (1976). The bond energies for the neutral nickel carbonyls are tabulated in Table I.

Examination of Table I reveals that the C—O vibrational frequency increases fairly regularly with the number of carbonyl ligands, although neither the electron affinities nor the bond strengths are observed to have any regular trend. These properties have been interpreted in terms of the electronic and geometric structure of the $Ni(CO)_n$. As an example, molecular orbital calculations (Rives and Fenske, 1981; Dunlap *et al.*, 1982) indicate that Ni(CO) has a $^1\Sigma^+$ ground electronic state, resulting from the bonding of CO to a nickel atom with a d^{10} configuration. The 1.7-eV excitation of the nickel atom to the d^{10} state is evident in the *thermodynamically* weak bond (29 kcal/mol) of Ni(CO). On the other hand, the low C—O vibrational frequency (1996 cm^{-1}) reflects the 68-kcal/mol binding energy of the CO to the *excited* nickel atom. The decrease in electron affinity between Ni and Ni(CO) is also interpreted as a result of the change in electronic structure; the electron affinity of Ni is defined from the $d^8s^2(Ni) \leftarrow d^9s^2(Ni^-)$ transition (Corderman *et al.*, 1979), whereas the electron affinity of Ni(CO) is defined by a $d^{10}(Ni(CO)) \leftarrow d^{10}s^1(Ni(CO)^-)$ s-electron loss and is not directly comparable. The decrease in electron affinity on going to $Ni(CO)_2$ is seen as the result of the sigma-donor property of the CO ligands, which is a repulsive interaction with the additional s electron on the metal. Similar considerations hold for the interpretations of these molecular properties for the other nickel carbonyl species.

Finally, these spectra are quite remarkable for polyatomics, and metal carbonyls in particular, in exhibiting very sharp transitions and well-resolved vibrational structure. Progressions in the C—M—C bending vibrations of metal carbonyls, typically about 100 cm^{-1} in frequency (Jones *et al.*, 1972), are usually the worst culprits for broadening peaks in the spectra. The apparent absence of these progressions gives strong support that, in every case, the symmetry of the molecules does not change on electron loss; i.e., $Ni(CO)^-$ and Ni(CO) are both linear, $Ni(CO)_2{}^-$ and $Ni(CO)_2$ are both linear, and $Ni(CO)_3{}^-$ and $Ni(CO)_3$ are both trigonal planar. A second cause of broad peaks in unsaturated metal carbonyls is the presence of multiple electronic states quite close in energy to the ground state; these are the result of partially filled d shells. Since all the $Ni(CO)_n$ have fully filled d shells, only a single electronic transition occurs in the region of the ground state. The peak A in the spectrum of $Ni(CO)^-$ agrees quite well with the predictions (Rives and Fenske, 1981; Dunlap *et al.*, 1982) of the several excited states of Ni(CO) which have a nearly d^9s^1 configuration on Ni. In fact, peak A is much broader than peak X, as expected for multiple close-lying transitions.

In summary the spectra provide data for understanding electronic and geometric structure, and thermochemistry of a series of large polyatomic molecules and their negative ions. An example of understanding periodic trends in electronic structure of neutral species—in this case, of the diatomic transition-metal hydrides—is presented in the next section.

C. *Electronic Structure of Transition-Metal Hydrides*

A systematic examination of the spectra of the diatomic transition-metal hydrides MH^-, for the cases where M = Mn, Fe, Co, and Ni, is reported by Stevens *et al.* (1983) and Feigerle *et al.* (1983). These studies were designed to establish the photoelectron spectra of the MH^- for which the ground state of the corresponding neutral is known. These data were then used to interpret the spectrum of FeH^-, to aid in the determination of the ground-state electronic structure of the neutral FeH.

The bonding in diatomic transition-metal hydrides MH, for the cases where M = Mn, Fe, Co, and Ni, can be understood by the simplified correlation picture given in Fig. 14. Each of the metal atoms has a ground-state configuration $3d^k4s^2$. Bonding a hydrogen atom to the ground state produces a hydride of configuration $3d^k\sigma^2\sigma^{*1}$, in which one of the 4s electrons forms a bond to hydrogen. The other 4s electron becomes anti-bonding with respect to the M—H bond and high-spin paired with the $3d^k$ electrons. Alternatively, with an excitation to the $3d^{k+1}4s^1$ configuration of the metal atom, the 4s electron forms a strong bond to the hydrogen, resulting in a "low-spin" hydride of $3d^{k+1}\sigma^2$ configuration. For the manganese atom, the excitation from the $3d^54s^2(^6S)$ ground state to the $3d^64s^1(^6D)$ excited state is about 2.1 eV; thus, MnH has a high-spin ground-state $X\,^7\Sigma^+$ (Van Zee

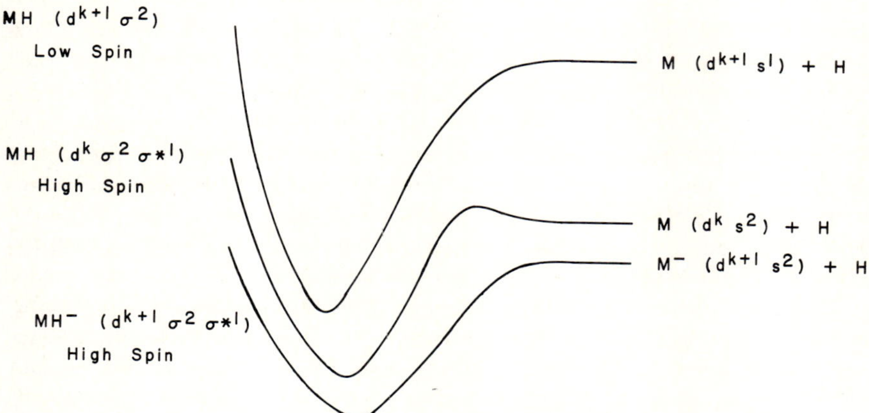

FIG. 14 Qualitative illustration of the bonding potential curves for the high- and low-spin states of the transition-metal hydrides MH, for M = Mn, Fe, Co, and Ni; and for the potential curves for the corresponding negative ions MH^-.

et al., 1978; Hayes *et al.*, 1957; Nevin and Stephens, 1953). At the opposite extreme, the $3d^94s^1(^3D_3)$ state of nickel is only 205 cm^{-1} above the ground-state $3d^84s^2(^3F_4)$. Therefore, NiH has a low-spin ground-state $X\,^2\Delta_{5/2}$ (Åslund *et al.*, 1964) of $3d^9\sigma^2$ configuration. Cobalt hydride is now accepted to have a $^3\Phi_4$ low-spin ground state (Klynning and Kronekvist, 1972). Although the ground states of these molecules have been characterized, neither the low-spin states of MnH nor the high-spin states of NiH or CoD have been observed. The ground state of FeH has not been established, although optical spectra of the gas phase (Smith, 1973; Carroll *et al.*, 1976; Davis and Phillips, 1982) and matrix-isolated molecule (Dendramis *et al.*, 1979) have been reported. The extreme mixing of states for this molecule make calculations of the ground state inconclusive. Various calculations on FeH suggest either a high-spin $^6\Delta$ (Das, 1981; Scott and Richards, 1975) or a low-spin $^4\Delta$ (Jaffe, 1982; Walch and Bauschlicher, 1983) ground state.

The bonding potential curves for the MH^- ions are also given in Fig. 14. For Fe^-, Co^-, and Ni^-, the ground states are all of $3d^{k+1}4s^2$ configuration (Table II in Section V of this chapter); Mn^- is not bound with respect to loss of an electron. The $3d^{k+1}s^1$ states are expected to be too high in energy to be important in bonding; thus, the lowest state for each MH^- is expected to have the high-spin configuration, $3d^{k+1}\sigma^2\sigma^{*1}$. One-electron transitions in the electron photodetachment process would give both the high-spin and low-spin states of the neutral MH; thus, photoelectron spectroscopy provides an ideal way to probe these molecules.

The spectra of MnH^- and MnD^- are given in Fig. 15. Peaks labeled X, A, and B are not shifted in position between MnH^- and MnD^- and thus are assigned as transitions from the $v'' = 0$ vibrational level of the negative ion to the $v' = 0$ vibrational level of the neutral, i.e., (0, 0) transitions. The transition to the ground-state $^7\Sigma^+$ results in peak X, from which an EA(MnH) of 0.869 ± 0.010 eV is determined. Lack of a vibrational progression in this electronic state transition is completely consistent with a predicted $3d^5\sigma^2\sigma^{*1}(X\,^7\Sigma^+\ \mathrm{MnH}) \leftarrow 3d^6\sigma^2\sigma^{*1}(X\,^6\Delta\ \mathrm{MnH}^-)$ transition; i.e., loss of the nonbonding core d electron gives rise to a diagonal transition. A Franck–Condon analysis of this transition, using the molecular parameters of MnH derived from optical spectroscopy (Hayes *et al.*, 1957), and the vibrational frequency for MnH^- from the photoelectron spectra, yields a bond length in MnH^- of 1.82 Å, only 0.09 Å longer than in $X\,^7\Sigma^+$.

The transition to the next excited state of MnH, with an origin at peak A, shows a long vibrational progression which is interpreted as resulting from the loss of the σ^* antibonding electron from MnH^-, to give the low-spin $3d^6\sigma^2$ quintet state of MnH. The higher vibrational frequency (1580 cm^{-1}) and the shorter bond length (1.61 Å, determined from a Franck–Condon analysis) than are seen for $X\,^7\Sigma^+$ MnH are consistent with the stronger bond

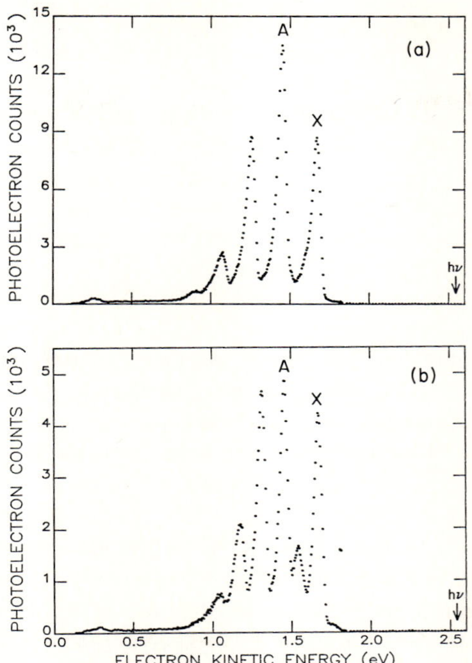

Fig. 15 Photoelectron spectra of (a) MnH^- and (b) MnD^- obtained with 488.0-nm (2.540-eV) photons; peaks labeled X, A, and B correspond to (0, 0) transitions. The data show the importance of isotope shift in determining transition origins. [From Stevens *et al.* (1983).]

expected for a low-spin state. This state is surprisingly close (1725 ± 50 cm^{-1}) in energy to the ground state.

Last, a third state at 11,320 ± 215-cm^{-1} excitation gives rise to peak B in the spectrum. No vibrational data could be obtained for this state nor has it been identified by other experiments.

Preliminary spectra for NiH^- and NiD^- (Feigerle *et al.*, 1983) exhibit a vibrational progression in the transition to the low-spin ground state. The transition to the upper state is a nearly diagonal transition, as expected for an excited high-spin state. The data provide the EA(NiH) of 0.484 ± 0.020 eV and the upper state excitation energy, 11,630 ± 250 cm^{-1}. Preliminary data for CoD^- (Feigerle *et al.*, 1983) also show the vibrational progression in the transition to the low-spin ground state and a diagonal transition to the upper state. The EA(CoD) is found to be 0.674 ± 0.015 eV, and the excitation energy to the upper state is determined as 6710 ± 150 cm^{-1}.

The photoelectron spectrum of FeD^- reported by Stevens *et al.* (1983) is given in Fig. 16. Clearly evident in this spectrum is a vibrational progression associated with the transition to ground-state FeD; the (0, 0) transition

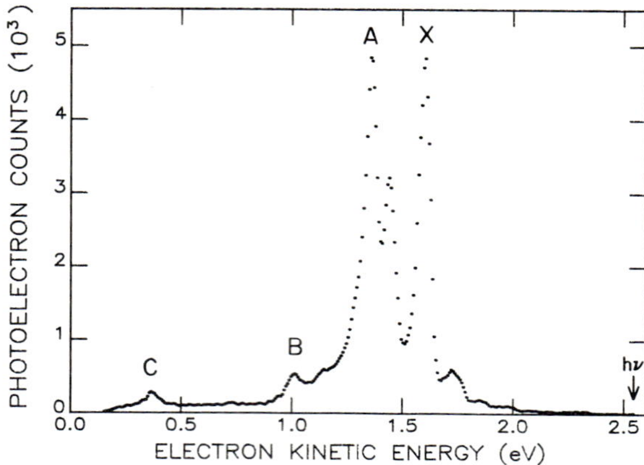

FIG. 16 Photoelectron spectrum of FeD^- obtained with 488.0-nm photons; peaks X, A, B, and C correspond to (0, 0) transitions. [From Stevens *et al.* (1983).]

is labeled as peak X in the spectrum. By comparison to the spectra of the manganese, cobalt, and nickel hydrides, this is strong evidence that FeH has a low-spin $^4\Delta$ ground state with configuration $3d^7\sigma^2$. Peak A is the (0, 0) peak of a diagonal transition and is assigned to the transition to the first high-spin state of FeH. This state, predicted to be a $^6\Delta$ of $3d^6\sigma^2\sigma^{*1}$ configuration, is determined from the spectrum of the hydride to lie at $1945 \pm 55\text{-cm}^{-1}$ excitation energy. Two additional states of FeH, with (0, 0) transitions at peaks B and C, are found to have excitation energies in the hydride of $4925 \pm 90\ \text{cm}^{-1}$ and $9965 \pm 160\ \text{cm}^{-1}$, respectively.

With the series of molecules for which photoelectron spectra are reported, several periodic trends, such as variation of the electron affinity, bond length, or vibrational frequency of the MH with variation in M, can be examined. One example is provided by a comparison of the high-spin to low-spin energy difference in the metal hydride, $E(d^k\sigma^2\sigma^{*1}) - E(d^{k+1}\sigma^2)$, with the corresponding excitation in the metal atom, $E(d^k s^2) - E(d^{k+1}s^1)$. These data are illustrated in Fig. 17 and clearly show that as the d^{k+1} state of the atom becomes more energetically accessible (proceeding from Mn to Ni), the low-spin state of MH drops further in energy below the high-spin state of the MH molecule. Although the data points shown in Fig. 17 lie on a smooth curve, the energetic relationship between molecular and atomic excitation energies is decidedly *not* a linear one. This nonlinearity is an indication of the mixed nature of the MH states.

The spectra presented in this section provide case examples of the type of data obtained from photoelectron spectra, including electronic and vibrational energy levels, geometric information, and thermochemical data for

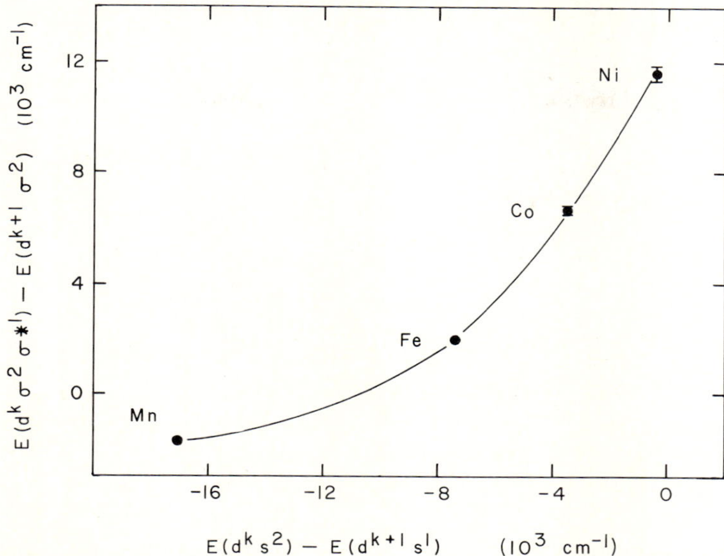

FIG. 17 Comparison of the excitation energy of the lowest-lying high-spin state (from the low-spin state) of MH, $E(d^k\sigma^2\sigma^{*1})-E(d^{k+1}\sigma^2)$, with the excitation energy of the atomic state, $E(d^k s^2)-E(d^{k+1}s^1)$. Data for the MH excitation energies are taken from Stevens *et al.* (1983) and Feigerle *et al.* (1983).

molecular ions and the corresponding neutrals. The utility of photoelectron spectroscopy is also well illustrated by the number of atomic electron affinities and electronic states of atomic negative ions which have been determined by this technique; these are given in the following section.

V. ATOMIC ELECTRON AFFINITIES

A few years ago, Hotop and Lineberger (1975) prepared a review and critical evaluation of atomic electron affinities. This task was made easier because the majority of the most reliable atomic electron affinities had been determined using a few techniques (primarily photodetachment) by workers in only a few laboratories. This review also provided detailed evaluations of the experimental methods and collected data on spin–orbit splittings and electronically excited states of atomic negative ions.

In the seven years since that review was prepared, a large number of atomic negative ions have been studied using either laser photodetachment threshold (LPT) devices or laser photoelectron spectrometers (LPES). As a result of this activity, the accuracy of the recommended electron affinity for nearly half of the elements has been improved.

Table II presents an updated summary of the recommended electron affinities of the elements. It also indicates the electronic configuration of the

TABLE II

Preferred Atomic Electron Affinities

Z	symbol	Configuration Neutral	Configuration Ion	Electron affinity[a] (eV)	Error (eV)	Method[b]	Reference
1	H	1s	$1s^2$	0.754209	0.000003	Var Cal	Pekeris (1958, 1962)
2	He	$1s^2$		Not stable		SEMI EMP	Hotop and Lineberger (1975)[c]
3	Li	2s	$2s^2$	0.6182	0.0005	LPT	Feldmann (1976)
4	Be	$2s^2$		Not stable		SEMI EMP	Hotop and Lineberger (1975)[c]
5	B	$2s^22p$	$2s^22p^2$	0.278	0.010	LPES(O^-)	Feigerle *et al.* (1981a)
6	C	$2p^2$	$2p^3$	1.2629	0.0003	LPT	Feldmann (1977)
7	N	$2p^3$		Not stable		DIS ATT	Mazeau *et al.* (1978)
8	O	$2p^4$	$2p^5$	1.462	0.003	Various	Hotop and Lineberger (1975)[c]
9	F	$2p^5$	$2p^6$	3.399	0.003	PLA,PLE	Hotop and Lineberger (1975)[c]
10	Ne	$2p^6$		Not stable		SEMI EMP	Hotop and Lineberger (1975)[c]
11	Na	3s	$3s^2$	0.547930	0.000025	LPT	Mead *et al.* (1983a)
12	Mg	$3s^2$		Not stable		e-SCATT	Burrow *et al.* (1976)
13	Al	$3s^23p$	$3s^23p^2$	0.442	0.010	LPES(O^-)	Feigerle *et al.* (1981a)
14	Si	$3p^2$	$3p^3$	1.385	0.005	LPES(K^-)	Kasdan *et al.* (1975)
15	P	$3p^3$	$3p^4$	0.7464	0.0003	LPT	Slater and Lineberger (1977)
16	S	$3p^4$	$3p^5$	2.0771197	0.000001	LPT(COAXIAL), LICR	Mead *et al.* (1983b); Stoneman and Larson (1983)
17	Cl	$3p^5$	$3p^6$	3.615	0.004	PLA,PLE	Hotop and Lineberger (1975)[c]
18	Ar	$3p^6$		Not stable		SEMI EMP	Zollweg (1969)
19	K	4s	$4s^2$	0.50147	0.0001	LPT	Slater *et al.* (1978)
20	Ca	$4s^2$		Not stable		SEMI EMP	Zollweg (1969)
21	Sc	$3d4s^2$	$3d4s^24p$	0.189	0.020	LPES(O^-)	Feigerle *et al.* (1981b)
22	Ti	$3d^24s^2$	$3d^34s^2$	0.080	0.014	LPES(O^-)	Feigerle *et al.* (1981c)
23	V	$3d^34s^2$	$3d^44s^2$	0.526	0.012	LPES(O^-)	Feigerle *et al.* (1981c)
24	Cr	$3d^54s$	$3d^54s^2$	0.667	0.012	LPES(O^-)	Feigerle *et al.* (1981c)
25	Mn	$3d^54s^2$		Not stable		SEMI EMP	Zollweg (1969)
26	Fe	$3d^64s^2$	$3d^74s^2$	0.164	0.035	LPES(O^-)	Engelking and Lineberger (1979)

27	Co	$3d^7 4s^2$	$3d^8 4s^2$	0.662	0.010	LPES(O^-)	Corderman *et al.* (1979)
28	Ni	$3d^8 4s^2$	$3d^9 4s^2$	1.157	0.010	LPES(O^-)	Corderman *et al.* (1979)
29	Cu	$3d^{10} 4s$	$3d^{10} 4s^2$	1.228	0.010	LPES(OH^-)	Hotop *et al.* (1973b)
30	Zn	$4s^2$		Not stable		e-SCATT	Burrow *et al.* (1976)
31	Ga	$4s^2 4p$	$4s^2 4p^2$	0.3	0.15	PT	Feldmann *et al.* (1973)
32	Ge	$4p^2$	$4p^3$	1.2	0.1	PT	Feldmann *et al.* (1977b)
33	As	$4p^3$	$4p^4$	0.81	0.03	PT	Feldmann *et al.* (1977b)
34	Se	$4p^4$	$4p^5$	2.0206	0.0003	LPT	Hotop *et al.* (1973a)
35	Br	$4p^5$	$4p^6$	3.364	0.004	PLA,PLE	Hotop and Lineberger (1975)[c]
36	Kr	$4p^6$		Not stable		SEMI EMP	Zollweg (1969)
37	Rb	$5s$	$5s^2$	0.48592	0.000020	LPT	Frey *et al.* (1978)
38	Sr	$5s^2$		Not stable		SEMI EMP	Zollweg (1969)
39	Y	$4d5s^2$	$4d5s^2 5p$	0.308	0.012	LPES(O^-)	Feigerle *et al.* (1981b)
40	Zr	$4d^2 5s^2$	$4d^3 5s^2$	0.427	0.014	LPES(O^-)	Feigerle *et al.* (1981c)
41	Nb	$4d^3 5s^2$	$4d^4 5s^2$	0.894	0.025	LPES(O^-)	Feigerle *et al.* (1981c)
42	Mo	$4d^5 5s$	$4d^5 5s^2$	0.747	0.010	LPES(O^-)	Feigerle *et al.* (1981c)
43	Tc	$4d^5 5s^2$	$4d^6 5s^2$	0.55	0.15	SEMI EMP	Zollweg (1969)
44	Ru	$4d^7 5s$	$4d^7 5s^2$	1.05	0.10	SEMI EMP	Zollweg (1969)
45	Rh	$4d^8 5s$	$4d^8 5s^2$	1.138	0.008	LPES(O^-)	Feigerle *et al.* (1981c)
46	Pd	$4d^{10}$	$4d^{10} 4s$	0.558	0.008	LPES(O^-)	Feigerle *et al.* (1981c)
47	Ag	$4d^{10} 5s$	$4d^{10} 5s^2$	1.303	0.009	LPES(O^-)	Hotop *et al.* (1973b)
48	Cd	$5s^2$		Not stable		e-SCATT	Burrow *et al.* (1976)
49	In	$5p$	$5p^2$	0.30	0.15	PT	Feldmann *et al.* (1973)
50	Sn	$5p^2$	$5p^3$	1.15	0.15	PT	Feldmann *et al.* (1977b)
51	Sb	$5p^3$	$5p^4$	1.07	0.05	PT	Feldmann *et al.* (1977b)
52	Te	$5p^4$	$5p^5$	1.9708	0.0003	LPT	Slater and Lineberger (1977)
53	I	$5p^5$	$5p^6$	3.0591	0.001	OGS	Webster *et al.* (1983)
54	Xe	$5p^6$		Not stable		SEMI EMP	Zollweg (1969)
55	Cs	$6s$	$6s^2$	0.471630	0.000025	LPT	Mead *et al.* (1983a)
56	Ba	$6s^2$		Not stable		SEMI EMP	Zollweg (1969)
57	La	$5d6s^2$	$5d^2 6s^2$	0.5	0.25	SEMI EMP	Angelov (1976)
72	Hf	$5d^2 6s^2$		Not stable		SEMI EMP	Zollweg (1969)

(*continued*)

TABLE II (*continued*)

Z	symbol	Configuration: Neutral	Configuration: Ion	Electron affinity[a] (eV)	Error (eV)	Method[b]	Reference
73	Ta	$5d^36s^2$	$5d^46s^2$	0.323	0.012	$LPES(O^-)$	Feigerle *et al.* (1981c)
74	W	$5d^46s^2$	$5d^56s^2$	0.816	0.008	$LPES(O^-)$	Feigerle *et al.* (1981c)
75	Re	$5d^56s^2$	$5d^66s^2$	0.12	0.10	SEMI EMP	Zollweg (1969)
76	Os	$5d^66s^2$	$5d^76s^2$	1.12	0.15	SEMI EMP	Zollweg (1969)
77	Ir	$5d^76s^2$	$5d^86s^2$	1.566	0.008	$LPES(O^-)$	Feigerle *et al.* (1981c)
78	Pt	$5d^96s$	$5d^96s^2$	2.128	0.002	LPT	Hotop and Lineberger (1973)
79	Au	$5d^{10}6s$	$5d^{10}6s^2$	2.308620	0.000025	LPT	Hendewerk *et al.* (1981)
80	Hg	$6s^2$		Not stable		e-SCATT	Burrow *et al.* (1976)
81	Tl	$6s^26p$	$6s^26p^2$	0.3	0.2	PT	Feldmann *et al.* (1973)
82	Pb	$6s^26p^2$	$6s^26p^3$	0.365	0.008	$LPES(O^-)$	Feigerle *et al.* (1981a)
83	Bi	$6s^25p^3$	$6s^26p^4$	0.947	0.010	$LPES(O^-)$	Feigerle *et al.* (1981a)
84	Po	$6p^4$	$6p^5$	1.9	0.3	SEMI EMP	Zollweg (1969)
85	At	$6p^5$	$6p^6$	2.8	0.2	SEMI EMP	Zollweg (1969)
86	Rn	$6p^6$		Not stable		SEMI EMP	Zollweg (1969)

[a] The statement "not stable" implies that any negative ion formed by simple electron addition to the indicated neutral configuration is not stable. Many of the ions so indicated possess metastable negative ion states with lifetimes sufficiently long (10^{-6} sec) to be detected in beam machines. The others have been observed as electron scattering resonances.

[b] Key to methods: Var Cal, variational calculation; SEMI EMP, semiempirical estimate; LICR, laser-ion-cyclotron reasonance; LPT, laser photodetachment threshold measurement; $LPES(X^-)$, laser photoelectron spectrometry (reference ion); OGS, optogalvanic spectroscopy; PLA, plasma absorption measurement; PLE, plasma emission; PT, photodetachment threshold (nonlaser); e SCATT, electron scattering from the neutral species; DIS ATT, dissociative attachment. LICR is discussed by Jopson and Larson (1981). OGS is discussed by Webster *et al.* (1983). See Hotop and Lineberger (1975) for detailed discussion of other methods.

[c] The primary data reference for this species is not in this reference. Rather, this recommended electron affinity (EA) arises from consideration of a variety of data sources; the reader is advised to consult this review for both the original sources and the arguments leading to the EA proposed here.

negative ion. In all cases where the ion has been studied by LPES or LPT methods, the indicated configuration has been confirmed, including, for example, the unusual configurations of Sc^- ($3d4s^24p$) and Y^- ($4d5s^25p$). The configurations indicated in other cases (e.g., Tl, Tc, Hf) are simply the plausible ones.

While many of these ions are known to possess bound electronically excited states and have spin–orbit splitting data available, these data are not presented here. Both these data and a detailed discussion of the choices given for the recommended electron affinity will be provided in a forthcoming revision of Hotop and Lineberger (1975).

VI. SUMMARY AND CONCLUSIONS

In this review, we have examined some typical applications of negative-ion-beam photodetachment devices to chemical problems. The four-year interval since the last review of this topic (Corderman and Lineberger, 1979) has seen a 3000-fold enhancement in the resolution of the tunable laser devices! This improvement is now affording ever more sophisticated tests of threshold behavior and of the autoionization process. It would appear that detailed probes of electron correlation via two-electron ejection from negative ions lie in the immediate future.

The photoelectron spectrometers are being applied to ever more complex, chemically relevant species. The work on HNO^- and metal carbonyls serves to illustrate this progress. Still, the need for higher resolution, shorter laser wavelengths, and better chemical control of the negative ion source continues to be a major problem.

The next few years, however, will show great progress in at least two of these areas. Ellis and Ellison (1983) now have in operation an instrument with significantly better resolution (20 meV) and are systematically studying a variety of organic anions. A new photoelectron spectrometer using wide-aperture electron optics, multichannel array detection of photoelectrons, and ion deceleration is being tested by Feigerle (1983). This device has demonstrated electron energy resolution of <4 meV and, with a resolution of 8 meV, a sensitivity over 1000-fold greater than present instruments. This development will make it possible to replace the present active discharge negative ion sources with a much less intense, but chemically well-characterized, flowing afterglow device. The experimenter will then have some control over the internal temperature and isomeric form of ions under study. The possibility then exists to determine the distribution of states populated in the ionic products of ion–molecule reactions. It seems safe to speculate that the next five years will see even more progress in the understanding of gas phase negative ions than have the past five.

ACKNOWLEDGMENT

It is a great pleasure to acknowledge the stimulating interactions with our "negative-ion colleagues"—C. S. Feigerle, S. M. Burnett, P. A. Schulz, U. Hefter, and H. B. Ellis. We have also benefited from many discussions with J. I. Brauman, P. C. Engelking, G. B. Ellison, S. R. Leone, and J. L. Hall. W. C. L. is pleased to acknowledge the receipt of a J. S. Guggenheim Fellowship and the award of a Faculty Fellowship by the Council on Research and Creative Work of the University of Colorado. Much of the work reported here was supported by the National Science Foundation through Grants CHE78-18424 and PHY79-04928.

REFERENCES

Albritton, D. L., Schmeltekopf, A. L., and Zare, R. N. (1976. *In* "Molecular Spectroscopy: Modern Research" (K. Narahari Rao, ed.), Vol. II, pp. 1–67. Academic Press, New York.

Angelov, B. M. (1976). *Chem. Phys. Lett.* **43**, 368–370.

Åslund, N., Neuhaus, H., Lagerqvist, A., and Andersén, E. (1964). *Ark. Fys.* **28**, 271–283.

Bancroft, J. L., Hollas, J. M., and Ramsay, D. A. (1962). *Can. J. Phys.* **40**, 322–347.

Bartmess, J. E., and McIver, R. T., Jr. (1979). *In* "Gas Phase Ion Chemistry" (M. T. Bowers, ed.), Vol. 2, pp. 87–121. Academic Press, New York.

Berry, R. S., and Leach, S. (1982). *Adv. Electron. Electron. Phys.* **57**, 1–144.

Blumberg, W. A. M., Itano, W. M., and Larson, D. J. (1979). *Phys. Rev. D* **19**, 139–148.

Branscomb, L. M. (1962). *In* "Atomic and Molecular Processes" (D. R. Bates, ed.), pp. 100–140. Academic Press, New York.

Breeze, P. A., Burdett, J. K., and Turner, J. J. (1981). *Inorg. Chem.* **20**, 3369–3378.

Breyer, F., Frey, P., and Hotop, H. (1978). *Z. Phys. A* **286**, 133–138.

Breyer, F., Frey, P., and Hotop, H. (1981). *Z. Phys. A* **300**, 7–24.

Bruna, P. J., and Marian, C. M. (1979). *Chem. Phys. Lett.* **67**, 109–114.

Burdett, J. K. (1973). *J. Chem. Soc. Chem. Commun.* 763–764.

Burrow, P. D., Michejda, J. A., and Comer, J. (1976). *J. Phys. B* **9**, 3225–3236.

Carroll, P. K., McCormack, P., and O'Connor, S. (1976) *Astrophys. J.* **208**, 903–913.

Compton, R. N., and Stockdale, J. A. D. (1976). *Int. J. Mass Spectrom. Ion Phys.* **22**, 47–55.

Corderman, R. R., and Lineberger, W. C. (1979). *Ann. Rev. Phys. Chem.* **30**, 347–378.

Corderman, R. R., Engelking, P. C., and Lineberger, W. C. (1979). *J. Chem. Phys.* **70**, 4474–4480.

Coxon, J. A. (1975). *J. Mol. Spectrosc.* **58**, 1–28.

Dalby, F. W. (1958). *Can. J. Phys.* **36**, 1336–1371.

Das, G. (1981). *J. Chem. Phys.* **74**, 5766–5774.

Davis, S. (1982). Private communication.

DeKock, R. L. (1971). *Inorg. Chem.* **10**, 1205–1211.

Dendramis, A., Van Zee, R. J., and Weltner, W., Jr. (1979). *Astrophys. J.* **231**, 632–636.

Dunlap, B. I., Yu, H. L., and Antoniewicz, P. R. (1982). *Phys. Rev. A* **25**, 7–13.

Engelking, P. C. (1982). *Phys. Rev. A* **26**, 740–745.

Engelking, P. C., and Lineberger, W. C. (1979). *Phys. Rev. A* **19**, 149–155.

Engelking, P. C., Corderman, R. R., Wendoloski, J. J., Ellison, G. B., ONeil, S. V., and Lineberger, W. C. (1981). *J. Chem. Phys.* **74**, 5460–5473.

Ellis, H. B., Jr., and Ellison, G. B. (1983). *J. Chem. Phys.* **78**, 6541–6558.

Feigerle, C. S. (1983). Ph.D. Thesis, University of Colorado, Boulder, Colorado.

Feigerle, C. S., Corderman, R. R., and Lineberger, W. C. (1981a). *J. Chem. Phys.* **74**, 1513–1515.

Feigerle, C. S., Herman, Z., and Lineberger, W. C. (1981b). *J. Elect. Spect. Related Phenom.* **23**, 441–450.

Feigerle, C. S., Corderman, R. R. Bobashev, S. V., and Lineberger, W. C. (1981c). *J. Chem. Phys.* **74**, 1580–1598.

Feigerle, C. S., Stevens, A. E., and Lineberger, W. C. (1984). To be submitted.
Feldmann, D. (1976). *Z. Phys. A* **277**, 19–25.
Feldmann, D. (1977). *Chem. Phys. Lett.* **47**, 338–340.
Feldmann, D., Rackwitz, R., Heinicke, E., and Kaiser, H. J. (1973). *Phys. Lett.* **45A**, 404.
Feldmann, D., Rackwitz, R., Heinicke, E., and Kaiser, H. J. (1977a). *Z. Phys. A* **282**, 143–148.
Feldmann, D., Rackwitz, R., Heinicke, E., and Kaiser, H. J. (1977b). *Z. Naturforsch.* **32a**, 302–306.
Frey, P., Breyer, F., and Hotop, H. (1978). *J. Phys. B* **11**, L589–L594.
Gailitis, M., and Damburg, R. (1963). *Proc. Phys. Soc.* **82**, 192.
Gottscho, R. A., Field, R. W., Dick, K. A. and Benesch, W. (1979). *J. Mol. Spectrosc.* **74**, 435–455.
Hayes, W., McCarvill, P. D., and Nevin, T. E. (1957). *Proc. Phys. Soc. A* **70**, 904–905.
Herzberg, G., and Lagerqvist, A. (1968). *Can. J. Phys.* **46**, 2363–2373.
Hefter, U., Mead, R. D., Schulz, P. A., and Lineberger, W. C. (1983). *Phys. Rev. A* **28**, 1429–1439.
Hendewerk, M., Mead, R. D., Schulz, P. A., and Lineberger, W. C. (1981). Unpublished.
Hotop, H., and Lineberger, W. C. (1973). *J. Chem. Phys.* **58**, 2379–2387.
Hotop, H., and Lineberger, W. C. (1975). *J. Phys. Chem. Ref. Data* **4**, 539–576.
Hotop, H., Patterson, T. A., and Lineberger, W. C. (1973a). *Phys. Rev. A* **8**, 762–774.
Hotop, H., Bennett, R. A., and Lineberger, W. C. (1973b). *J. Chem. Phys.* **58**, 2373–2378.
Hotop, H., Patterson, T. A., and Lineberger, W. C. (1974). *J. Chem. Phys.* **60**, 1806–1812.
Hougen, J. T. (1970). "The Calculation of Rotational Energy Levels and Rotational Line Intensities in Diatomic Molecules," Nat. Bur. of Std. Monograph 115. U. S. Government Printing Office, Washington, D. C.
Ishiwata, T., Tanaka, I., and Akimoto, H. (1978). *J. Phys. Chem.* **82**, 1336–1343, and references therein.
Jaffe, R. L. (1982). Private communication.
Janousek, B. K., and Brauman, J. I. (1979). *In* "Gas Phase Ion Chemistry" (M. T. Bowers, ed.), Vol. 2, pp. 53–86. Academic Press, New York.
Jones, L. H., McDowell, R. S., Goldblatt, M., and Swanson, B. I. (1972). *J. Chem. Phys.* **57**, 2050–2064.
Jones, P. L., Mead. R. D., Kohler, B. E., Rosner, S. D., and Lineberger, W. C. (1980). *J. Chem. Phys.* **78**, 4419–4432.
Jopson, R. M., and Larson, D. J. (1981). *Phys. Rev. Lett.* **47**, 789–792.
Kasdan, A., Herbst, E., and Lineberger, W. C. (1975). *J. Chem. Phys.* **62**, 541–548.
Klynning, L., and Kronekvist, M. (1972). *Physica Scripta* **6**, 61–65.
Lee, C. M. (1975). *Phys. Rev. A* **11**, 1692–1699.
Lineberger, W. C., and Woodward, B. W. (1970). *Phys. Rev. Lett.* **25**, 424–427.
Massey, H. S. W. (1976). "Negative Ions," 3rd ed. Cambridge Univ. Press, London and New York.
Mazeau, J., Gresteau, F., Hall, R. I., and Huetz, A. (1978). *J. Phys. B* **11**, L557–L560.
Mead, R. D., Schulz, P. A., and Lineberger, W. C. (1983a). Unpublished.
Mead, R. D., Lykke, K. R., and Lineberger, W. C. (1983b). Unpublished.
Mead, R. D., Hefter, U., Schulz, P. A., and Lineberger, W. C. (1984). *J. Chem. Phys.* (to be submitted).
Miller, T. M. (1981). *Adv. Electron. Electron Phys.* **55**, 119–188.
Moores, D. L., and Norcross, D. W. (1974). *Phys. Rev. A* **10**, 1646–1657.
Moseley, J. T. (1982). *In* "Applied Atomic Collision Physics" (H. S. W. Massey, E. W. McDaniel, and B, Bederson, eds.), Vol. 5, p. 269. Academic Press, New York.
Moseley, J. and Durup, J. (1981). *Ann. Rev. Phys. Chem.* **32**, 53–76.
Nevin, T. E., and Stephens, D. V. (1953). *Proc. R. Irish Acad.* **55**, 109–116.

Norcross, D. W., and Taylor, K. T. (1983). *Phys. Rev. A* (to be submitted).
Ogilvie, J. F. (1974). *Can. J. Spectrosc.* **19**, 171–177.
O'Malley, T. F. (1965). *Phys. Rev.* **137**, A1668–A1672.
Pekeris, C. L. (1958). *Phys. Rev.* **112**, 1649–1658.
Pekeris, C. L. (1962). *Phys. Rev.* **126**, 1470–1476.
Rau, A. R. P., and Fano, U. (1971). *Phys. Rev. A* **4**, 1751–1759.
Rives, A. B., and Fenske, R. F. (1981). *J. Chem. Phys.* **75**, 1293–1302.
Rouze, N., and Geballe, R. (1983). *Phys. Rev. A* **27**, 3071–3077.
Schulz, P. A., Mead, R. D., Jones, P. L., and Lineberger, W. C. (1982). *J. Chem. Phys.* **77**, 1153–1165.
Schulz, P. A., Mead, R. D., and Lineberger, W. C. (1983) *Phys. Rev. A* **27**, 2229–2232.
Scott, P. R., and Richards, W. G. (1975). *J. Chem. Phys.* **63**, 1690–1691.
Slater, J., and Lineberger, W. C. (1977). *Phys. Rev. A* **15**, 2277–2282.
Slater, J. Read, F. H., Novick, S. E., and Lineberger, W. C. (1978) *Phys. Rev. A* **17**, 201–213.
Smith, R. E. (1973). *Proc. R. Soc. London Ser. A* **332**, 113–127.
Stevens, A. E., Feigerle, C. S., and Lineberger, W. C. (1982). *J. Am. Chem. Soc.* **104**, 5026–5031.
Stevens, A. E., Feigerle, C. S., and Lineberger, W. C. (1983). *J. Chem. Phys.* **78**, 5420–5431.
Stoneman, R., and Larson, D. J. (1983). Private communication.
Van Zee, R. J., DeVore, T. C., and Weltner, W., Jr. (1978). *J. Chem. Phys.* **69**, 1869–1875.
Walch, S. P., and Baushlicher, C. W., Jr. (1983). *J. Chem. Phys.* **78**, 4597–4605.
Walker, T. E. H. (1973). *Chem. Phys. Lett.* **19**, 493–496.
Watanabe, S., and Greene, C. H. (1980). *Phys. Rev. A* **22**, 158–169.
Webster C. R., McDermid, I. S., and Rettner, C. T. (1983). *J. Chem. Phys.* **78**, 646.
Wigner, E. P. (1948). *Phys. Rev.* **73**, 1002–1009.
Wildt, R. (1939). *Astrophys. J.* **89**, 295–301.
Woo, S. B., Helmy, E. M., Mauk, P. H., and Paszek, A. P. (1981). *Phys. Rev. A* **24**, 1380–1390.
Zare, R. N., Schmeltekopf, A. L., Harrop, W. J., and Albritton, D. L. (1973). *J. Mol. Spectrosc.* **46**, 37–66.
Zeitz, M., Peyerimhoff, S. D., and Buenker, R. J. (1979). *Chem. Phys. Lett.* **64**, 243–249.
Zollweg, R. J. (1969). *J. Chem. Phys.* **50**, 4251–4261.

Chapter 23

Electronically chemiluminescent ion–molecule exchange reactions

CH. OTTINGER

MAX-PLANCK-INSTITUT FÜR STRÖMUNGSFORSCHUNG

GÖTTINGEN, FEDERAL REPUBLIC OF GERMANY

ISBN 0-12-120803-6

I. INTRODUCTION

A. Scope

In this chapter the investigation of ion–molecule exchange reactions by means of optical emission from the products will be reviewed. Thus, we will be concerned with processes of the type

$$\begin{array}{l} A^+ + BC \rightarrow AB^{+*} + C \\ \qquad\qquad\quad \downarrow \\ \qquad\qquad\quad AB^+ \end{array} \tag{1a}$$

or

$$\begin{array}{l} A^+ + BC \rightarrow AB^* + C^+ \\ \qquad\qquad\quad \downarrow \\ \qquad\qquad\quad AB, \end{array} \tag{1b}$$

where A, B, and C stand for atoms or groups of atoms. The asterisk indicates electronic excitation of the molecular reaction product, either charged [reaction (1a)] or neutral [reaction (1b)]. Only those states are of interest here which can undergo an allowed electronic transition into some lower state, indicated by AB^+ and AB, respectively. This will often, but not necessarily, be the electronic ground state of the product.

Reactions of the type (1a) and (1b) are called chemiluminescent because the luminescence originates from the products of true chemical exchange reactions, as characterized by the transfer of a heavy particle. Exchange reactions in this sense are to be distinguished from charge transfer reactions, in which merely an electron, but not an atom, is transferred. Such processes will also often lead to luminescence, e.g.,

$$A^+ + BC \rightarrow A + BC^{+*}. \tag{2}$$

They will be covered in Chapter 24 by J. Leventhal.

In most chemiluminescent ion–molecule reactions studied so far, the emphasis has been on observation of emission from the molecular rather than from the atomic product. Corresponding to reactions (1a) and (1b), there will also be reactions of the type

$$\begin{array}{l} A^+ + BC \rightarrow AB^+ + C^* \\ \qquad\qquad\qquad\qquad\quad \downarrow \\ \qquad\qquad\qquad\qquad\quad C \end{array} \tag{3a}$$

or

$$\begin{array}{l} A^+ + BC \rightarrow AB + C^{+*} \\ \qquad\qquad\qquad\qquad\ \downarrow \\ \qquad\qquad\qquad\qquad\ C^+ \end{array} \tag{3b}$$

However, the rich structure of a typical molecular emission band spectrum $AB^{+*} \rightarrow AB^{+}$ or $AB^{*} \rightarrow AB$ contains much more dynamical information on the exchange reaction itself than does a simple atomic emission line spectrum $C^{*} \rightarrow C$ or $C^{+*} \rightarrow C^{+}$, so that from the standpoint of molecular reaction dynamics the processes (1a) and (1b) are much more interesting than (3a) and (3b).

Only positive ions have so far been shown to lead to electronic chemiluminescence in exchange reactions. Indeed, a negative counterpart of reaction (1a) is unlikely, since electronically excited states of negative molecular ions will usually not be stable against autoionization (Massey, 1976; an exception is C_2^{-}). Negative counterparts of reaction (1b), however, could conceivably exist.

Consistent with the scope of this book, only gas phase ionic reactions will be covered. This is not to say that ion chemiluminescence in condensed phases does not occur—in fact, luminescence from radical ion reactions in solution is a field of study by itself (Cormier *et al.*, 1973). However, the emphasis here will be on the dynamical analysis of ion–molecule reactions as deduced from the observed spectra. In condensed media, fast secondary collisions will degrade the spectra so as to make them useless for an analysis of the primary dynamics.

Reactant ion collisional energies considered in this chapter will typically lie in the range of $\sim 1 - \sim 100$ eV. At very low energy, few ionic species will be capable of giving electronically excited products radiating in the visible or UV because this requires a high exothermicity, on the order of a few electron volts. At very high impact energy, on the other hand, exchange reactions like reaction (1) are very rare, the total energy of the system being so high that product dissociation takes over. Nevertheless, in one case, electronic luminescence from a thermal energy reaction has been observed, and there are several examples of ionic luminescence observed at collision energies of hundreds of electron volts.

B. *Advantages of Ionic versus Neutral Chemiluminescence*

Electronic chemiluminescence is, of course, not restricted to ion–molecule reactions. In fact, many more neutral than ionic exchange reactions giving rise to electronically excited products have been observed so far. These have recently been reviewed by Menzinger (1980). In many cases, a detailed dynamical analysis was not attempted, the emphasis of much of this work being on the overall luminescence light yield, in search for a chemical laser operating in the visible or ultraviolet(VIS/UV). Let us briefly list the principal advantages of using ionic reactants in chemiluminescence experiments.

1. Variable Collision Energy

Although in the neutral chemiluminescence studies a beam configuration was frequently employed, the dependence of the reaction characteristics on beam energy was only investigated in a few experiments, e.g., the work of Menzinger and his collaborators (Menzinger, 1980). However, a detailed understanding of the reaction dynamics must be based on measurements across a wide range of collision energies. Even in the much simpler case of elastic scattering, the two-body interaction potential can only be derived accurately from scattering data if the measurements are made at different velocities. In reactive scattering, the potential energy surface of the reacting system must remain quite ambiguous as long as experiments are done only at one collision energy, especially when only integral cross sections are measured, as is usually the case with chemiluminescence.

This is one reason why ionic chemiluminescence lends itself particularly well to fundamental studies of exchange reactions. The collision energy can very conveniently be tuned through an extremely wide range. It starts at about 1 eV_{Lab}, below which point the ion beam intensity, as a result of space charge effects, becomes too low to be useful, and reaches up to any energy of chemical interest. By contrast, the collision energy in neutral reactions can only be raised above the thermal range by means of special techniques such as beam seeding (e.g., Wren and Menzinger, 1975), charge transfer (e.g., Zehnle *et al.*, 1978), or even high-speed "paddle" rotors (Rettner and Simons, 1978). An easily adjustable collision energy is desirable for various reasons. Thresholds, excitation functions, energy-dependent branching ratios between different product channels, and the dependence of rotational–vibrational–electronic state populations on the collision energy can be measured and will, in an ideal analysis, collectively determine the potential energy surface. Moreover, it turns out that the majority of chemiluminescent ion–molecule reactions proceeds nonadiabatically, involving "jumps" from one potential energy surface to another. The probability of such a jump depends sensitively on the collision energy, typically increasing sharply in the electron volt region. Ion luminescence thus affords a means of studying these very interesting processes in detail. Finally, if one wishes to study chemiluminescence in the VIS/UV spectral range, i.e., from states with excitation energies in the electron volt range, it is obviously advantageous to be able to supply this energy from the reactants' kinetic energy. Then there are no thermochemical limitations in the choice of reactants. In contrast, in neutral chemiluminescence, the primary concern is always with finding a system that reacts with sufficient exothermicity, which explains the predominance of metal atom–oxygen or halogen compound reactions in that field.

2. *Versatile Reactant Preparation*

Reactant ions for luminescence studies are prepared and mass-selected by means of a primary mass spectrometer. Thus, one enjoys virtually complete freedom in choosing a particular reactant species. This is to be contrasted with the situation in neutral reaction studies, where either stable molecules or metal atoms have to be used or, in the case of radicals, interference from unwanted admixtures has to be judiciously avoided. The advantage of ions in this respect has so far not been fully exploited but will become obvious once a wider range of radical ions comes under study.

3. *Metastable Reactants*

Preparation of the reactant ions by electron impact leads, inevitably, to a certain proportion of ions in long-lived electronically excited states. Often these lie at excitation energies of several electron volts above the ground state and have usually an electronic symmetry, e.g., spin, different from that of the ground state. For both these reasons, they will react very differently from the same ion in its electronic ground state. If the lifetime of these ions is $\sim 10^{-5}$ sec or longer, they will reach the reaction zone and will contribute additional reaction entrance channels to the process being studied. Since it is possible to measure the proportion of metastable ions in the beam as well as to control it within limits, the observed overall reaction can be uniquely apportioned to the various reactant species present. Thus, with ions of a given chemical species, one can often study quite different reactions going on side by side. Chemiluminescence of metastable neutrals has also been observed (e.g., Brinkmann *et al.*, 1980; Rettner and Simons, 1979). In this case, a special electron bombardment source has to be installed in order to generate the metastable beam.

4. *Charge Transfer States*

A more subtle advantage of ionic over neutral systems stems from the doubling of the number of energy levels due to the possibility of charge transfer. Reactions (1a) and (1b) and (3a) and (3b) give examples of different charge distributions with otherwise the same arrangements of atoms. Charge-exchanged states constitute additional potential energy surfaces of the overall system. The examples (1a) and (1b) and (3a) and (3b) refer only to the product side, but on the reactant side, too, all electronic states of the system come in pairs: Close to the energy level of $A^+ + BC$ there will be the energy of $A + BC^+$, and the same is true for excited (metastable) reactant ion states. Thus, an ionic system will generally have twice as many electronic states per energy interval as the corresponding neutral system.

This enhances the possibility of nonadiabatic transitions; may, therefore, facilitate the population of a particular excited level; and generally causes ionic systems to have, a priori, a richer chemistry than their neutral counterparts.

To be sure, with some neutral reactions the possibility of charge transfer also exists. Along with any state A + BC, there will be, e.g., $A^+ + BC^-$ (and even $A^- + BC^+$), but as a result of the large difference between the ionization potential of A and the electron affinity of BC (or vice versa) these charge-exchanged states will usually lie at much higher energy than that of A + BC. Nevertheless, they may be important at close range as a result of energy lowering by Coulomb attraction, and this is the basis of the well-known harpooning mechanism.

5. *Spectroscopy*

Apart from dynamical information, one can expect new spectroscopic data from chemiluminescence experiments, both ionic and neutral. This is because products are formed and observed without secondary collisions. Hence, the initially populated electronic state can be observed, which under more conventional conditions often escapes detection due to collisional relaxation into other states. The spectroscopic value of chemiluminescence under beam conditions is especially great for ions because here, at fairly high impact energy, states can be reached which will not be populated under the conditions of discharges from which most spectroscopic information on ions has been derived. At the same time, there exists an unfortunate paucity of spectroscopic data of molecular ions. Among about 1000 species listed in the compilation of spectra by Huber and Herzberg (1979), only ~100 are ions.

C. *Advantages of Optical versus Mass Spectrometric Investigation of Ion–Molecule Reactions*

The study of reactive ion collisions using mass spectrometers for detection has reached a high level of perfection. Translational product energy distributions can be measured as a function of product scattering angle, giving important insight into the reaction mechanism. What, then, can luminescence studies contribute to this field? The most important specific advantages of the luminescence technique, as compared to beam scattering, are listed in the following.

1. *Resolution*

From the spectra, precise product internal energy distributions can be derived. In principle, these can also be obtained from the translational energy distributions measured in the conventional way, but the resolution is usually insufficient to resolve vibrational levels [for an example where, with special care, this has been achieved, see Jones *et al.*, 1977], and rotational levels have only been resolved in one case (Teloy, 1978). Spectroscopically, however, rotational lines are much easier to resolve, and correspondingly not only vibrational distributions $P(v')$ but also rotational distributions $P(J')$ have been obtained for most luminescent ion–molecule reactions.

2. Sensitivity

Employing photon counting techniques, very small cross sections for production of a particular electronically excited state can be measured. The low limit depends on the structure of the spectrum; cross sections of $\sim 10^{-3}$ Å^2 have been measured.

3. Discrimination

Frequently the formation of an electronically excited product state has a much smaller cross section than that of ground-state products. In these cases, it will be very difficult to detect mass spectrometrically the small proportion of excited ions against the background, in the translational energy spectrum, of ground-state ions. The luminescence technique, being "blind" to this background, does not have this problem. Therefore, minority reaction channels can often only be detected by their luminescence, even though the resolution of the translational energy measurement should have been sufficient. For example, electronically excited $CH^+(A\,^1\Pi)$ constitutes only a few percent of the total CH^+ production from $C^+ + H_2$ collisions, and this product is therefore not clearly distinguishable in translational energy spectra (Jones *et al.*, 1977).

4. Polarization

In addition to the internal (v', J') state distribution, the spatial distribution of the product angular momentum vectors can be inferred from measurements of the luminescence polarization (Rettner and Simons, 1978; Prisant *et al.*, 1981). By the use of a reactant ion beam, a reference axis is established relative to which the emitted light will usually be polarized. The degree of polarization gives indications about the reaction mechanism. For example, transition from complex formation to stripping with increasing collision energy will be accompanied by an increase in the degree of polarization. For a statistical reaction model, it can be shown that the rotational polarization is entirely analogous to the product angular distribution (Case and Herschbach, 1976). In the case of ion chemiluminescence, neither a luminescence polarization nor an angular distribution measurement has been made so far. It appears that the polarization would be easier to measure than the angular distribution, because the latter would require the detection of photons in coincidence with scattered ions, and would suffer from very low intensity.

In summary, then, ion chemiluminescence offers a number of significant advantages over other types of ion–molecule studies. On the other hand, it cannot be denied that a serious disadvantage of the method is its limitation to radiating states. However, the wealth of detail to be gleaned from the spectra offsets this drawback. Luminescence reactions should be considered as model studies which allow one to investigate dynamical features that may have quite general significance.

II. EXPERIMENT AND ANALYSIS

A. *Apparatus Configuration and Types of Experiments*

With the exception of one flow tube experiment (Tsuji *et al.*, 1982), all studies of VIS/UV exchange reaction ion luminescence have been done using the beam-gas configuration. A mass-selected ion beam, not necessarily well collimated but with well-defined energy, enters a collision cell filled with the reactant gas at low pressure. Differential pumping between the cell and the ion source and mass separator ensures that ion/gas collisions outside the cell are unimportant. Light issuing from the cell is focused onto an optical spectrometer and detected using a photomultiplier. Five main types of experiments are usually performed:

(a) Spectra are recorded as a function of the ion kinetic energy. Analysis of the spectra then gives the principal information sought, namely, the products' internal state distribution.

(b) The dependence of the light intensity on the collision gas pressure is measured in selected regions of the spectrum. This is done in order to establish that the light intensity increases linearly with the collision gas pressure, thus proving a simple bimolecular reaction mechanism between the selected ion and the admitted gas. Occasionally, nonlinear dependences have been observed, indicating more complex, secondary reaction mechanisms.

(c) The spectral distribution is measured at a few different pressures, e.g., slightly higher than normal and as low as is possible with the given intensity. This type of experiment, which is mostly done at the high optical resolution necessary to resolve rotational details, confirms that the spectra observed at "normal" pressure are unaffected by secondary relaxation, particularly rotational.

(d) Using low resolution, the relative light intensity is measured as a function of collision energy at a few selected wavelengths and normalized to the respective ion beam intensity. Since the latter depends on the beam energy (dropping rather sharply at the lowest energies), and the beam geometry may change due to space charge effects, it is essential that the intensity of that portion of the ion beam traversing the field of view of the spectrometer be measured. Combined with the results of (a), integration over the spectrum then gives the light yield (or the relative phenomenological luminescence cross section) as a function of energy. If necessary, the normalization to the energy-dependent ion beam intensity can be checked by comparison with some other ion impact-induced luminescence process for which the relative cross section is accurately known, such as the charge transfer reaction $He^+ + Ar \rightarrow Ar^{+*} + He$ (Lipeles *et al.*, 1970).

(e) Absolute light yields (and cross sections) are measured by calibration of the light detection system, either using a standard tungsten (or, in the UV, deuterium) lamp or by reference to some other luminescent reaction of known cross section, which can be studied in the same apparatus. This might be a luminescent charge-transfer reaction or electron impact-excited fluorescence [e.g., $e^- + N_2 \rightarrow N_2^{+*} + 2e^-$, Ajello 1971] if an electron beam is passed through the collision cell instead of the ion beam. Both the gas pressure and the exciting beam intensity have to be known absolutely for type (e) experiments. The latter, particularly, can present difficulties if the beam geometry is not well defined. This is less of a problem at the higher beam energies, and absolute cross sections at lower energies can then be obtained from the results of (d).

Experiments employing a secondary beam instead of the collision chamber have so far not been done. The light intensity would in most cases be prohibitively low, while the advantages would only be minor: For one, the spread of the relative [center-of-mass (CM)] collision energy would be somewhat reduced because the contribution from the thermal target motion would be eliminated. Except at very low collision energy and with light targets, however, this improvement would be only slight, compared to the additional CM energy spread resulting from the (laboratory system) ion beam energy distribution. Secondly, with a beam–beam configuration, the possibility of secondary reactions would be eliminated. However, with a cell these are usually not a problem and can be interesting in themselves.*

* For example, CN(B–X) emission in $C^+ + NO$ collisions was found to depend quadratically on the NO pressure. This was explained by the reaction sequence $C^+ + NO \rightarrow C + NO^+$, $C + NO \rightarrow CN^* + O$ (Brandt *et al.*, 1973). Thus, this is not an ionic, but a neutral, chemiluminescent reaction in the 2–10 eV collision energy range.

B. *Apparatus Description*

Figure 1 shows schematically the experimental arrangement used in Göttingen with which most luminescent reactions of the type (1) have been studied. The apparatus of the other group which has reported two such reactions is similar in principle (Harris and Leventhal, 1976). A description of the Göttingen apparatus has been given elsewhere (Brandt *et al.*, 1973; Ottinger and Simonis, 1978) and can be summarized as follows.

Ions are produced in a plasma-type ion source ("Colutron") operated at $\sim$0.1 torr, 300-mA discharge current. The voltage, which determines critically the percentage of metastable ions in the beam, was varied from 20–120 V, corresponding to low and high metastable fraction, respectively. The ion energy spread is on the order of $\sim$1 eV, although with fragment ions the energy distribution has a low-intensity tail extending several electron volts to the high-energy side. The C^+, B^+, and N^+ ions to be discussed here were produced from CO, BF_3, and N_2, respectively. Ions are extracted through an orifice in the anode by means of a conical electrode at -1000 eV across a 4-mm distance. The ion beam, amounting to $\sim 7\ \mu A$ at this point, is then focused by means of four electrodes and steered into the 60°, 5-cm radius magnetic mass separator, mounted inside the vacuum chamber, having a pole gap of 7 mm. A simple immersion lens, indicated by three electrodes in Fig. 1, decelerates the ions prior to their entry into the collision chamber to energies of $\gtrsim$1 eV. The entrance slit of the collision chamber serves at the same time as the mass spectrometer exit slit.

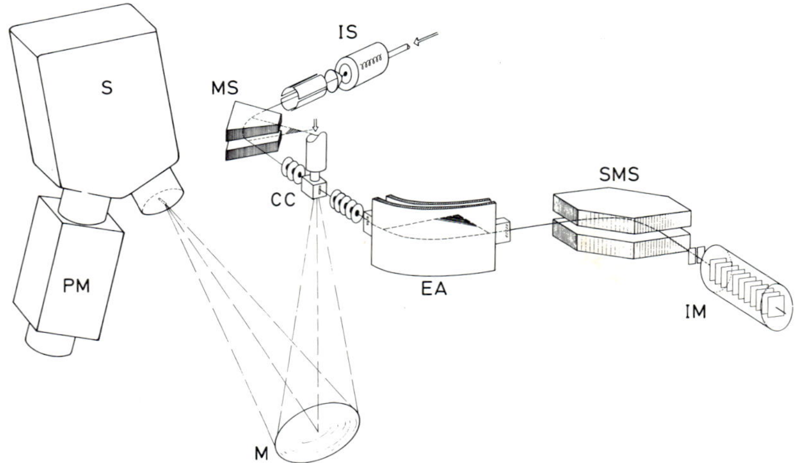

FIG. 1 Schematic of apparatus used in Göttingen for the study of luminescent ion–molecule reactions: IS, ion source with gas inlet; MS, mass separator magnet; CC, collision cell with gas inlet; EA, energy analyzer; SMS, secondary mass spectrometer: IM, ion multiplier; M, concave mirror; S, optical spectrometer; PM, photomultiplier.

The target gas is introduced into the collision chamber through a liquid nitrogen cold trap for purification. The pressure, as measured by a Barocel capacitance manometer and held constant by an automatic needle valve, was typically 10 mtorr for the recording of spectra. For measurements as a function of pressure, the range 0.1–20 mtorr was covered. The length of the collision cell is 24 mm. The ion beam intensity passing through the cell is typically $1 - 5 \times 10^{-8}$ A at the highest energy of 1000 eV, dropping by about a factor of 3 as the energy is lowered to 50 eV. The experiments were limited to $E_{\mathrm{Lab}} \gtrsim 1$ eV mainly on account of the ion energy spread and not by the ion beam intensity. At 1 eV, the intensity was typically $\sim 5 \times 10^{-10}$ A, while in favorable cases sufficient light intensity could still be obtained with ion currents well below 10^{-10} A.

Passing through the collision gas, the beam is attenuated, by between a few percent and about 50%. Only the central section of the collision chamber, 12 mm long, is viewed by the optical spectrometer. The "effective" ion beam intensity, as averaged over this portion, can be shown to be approximately

$$i_{\mathrm{eff}}(p) = i(p) \cdot \sqrt{i_0/i(p)}, \tag{4}$$

where $i(p)$ and i_0 are the currents passing through the cell at gas pressure p and without collision gas, respectively. This correction for the beam attenuation was applied both for measurements as a function of p and of beam energy.

As shown in Fig. 1, there is also provision for energy and mass analysis of ions leaving the collision cell in the forward direction.

Optical emission is observed through a MgF_2 window located in the bottom of the collision cell. A concave Al mirror (11-cm diameter, 50-cm radius of curvature) images the ion interaction region onto the entrance slit of an evacuable monochromator. It has a dispersion of 13.3 Å/mm. Using slit widths of 150 μm to 2 mm, depending on the requirements, the optical resolution was then between 2 and 26 Å full width at half minimum (FWHM).

A thermoelectrically cooled photomultiplier (usually EMI 6256 S) was used in the photon counting mode. The dark current was $\sim$0.5 counts/sec; typical luminescence count rates were between 1 and 10 counts/sec. Each spectrum was scanned repetitively, up to 100 times in succession in order to average over small drifts of the apparatus parameters. A stepping motor grating drive and synchronous data accumulation in a 1024-channel memory were used. Data accumulation was inhibited during each rapid (25 Å/sec) return of the grating. The total data acquisition time per spectrum was typically 2–20 hr.

As a precaution, with each reaction system a spectrum was usually taken at "negative ion energy," i.e., with the collision cell potential positive with respect to the ion source. Under these conditions, ions are prevented from entering the cell, and light can only be produced by electrons released from the immersion lens front electrode by fast ion impact and accelerated into the cell. This light would, with very similar intensity, also be produced at the small positive ion energies normally used. In all cases, this background was found to be negligible. In addition, it does not, of course, originate from the exchange reaction product but consists of spectra of the target gas or its ion itself and is, therefore, easily distinguishable. This background could, however, be a serious disturbance with experiments on luminescent charge transfer.

The vacuum system consists of five separately pumped chambers: one each for ion source, mass selector, collision cell, energy and secondary mass analyzer, and optical spectrometer plus mirror chamber. Except for the last, mercury diffusion pumps were used in order to avoid the formation of nonconducting surface layers on the electrodes surrounding the ion beam. For the same reason, all these electrodes, including the collision cell, but with the exception of the magnet pole faces, were constantly kept at $\sim 120°C$.

C. *Principle of Analysis*

Typically, the primary experimental result will consist of a measured spectrum, i.e., an intensity distribution as a function of wavelength. Each line in the spectrum can, ideally, be associated with product molecules in a particular initial vibrational–rotational level i with quantum numbers v', J'. [If the resolution is insufficient for complete separation of all lines, which is always the case, at least near the band heads, then techniques like computer simulation (to be described below for a specific case) will be used to assign the emission intensity in a certain wavelength interval to the contributing v', J' levels.] The emission intensity of molecules for the level i to some lower level k, I_{ik} is proportional to the number density n_i of the emitting molecules:

$$I_{ik} = n_i h\nu_{ik} A_{ik}. \tag{5}$$

Here, A_{ik} is the Einstein coefficient and ν_{ik} is the transition frequency; n_i is given by the balance between formation by reaction and depletion by radiation. We have

$$\text{rate of formation} = jaln\sigma_i, \tag{6}$$

where j is the incident ion current density, n the reactant gas number density, a the collision chamber entrance slit area, and l the length of ion path in the chamber. σ_i is the desired quantity, the state-specific cross section for formation of the product in the level i. Furthermore, we have

$$\text{rate of depletion} = n_i \sum_k A_{ik}, \tag{7}$$

so that

$$I_{ik} = \sigma_i h\nu_{ik} jaln \frac{A_{ik}}{\Sigma A_{ik}}. \tag{8}$$

Hence, σ_i is proportional to the measured spectral intensities, which is the basis of the method.

The Einstein coefficient is given by

$$A_{ik} = \frac{64\pi^4 \nu_{ik}^3}{3hc^3} \left| \int \psi_i(R) M(R) \psi_k(R)\, dR \right|^2 \frac{S(J_i, J_k)}{2J_i + 1}, \tag{9}$$

where the square of the integral over the initial and final state vibrational wave functions $\psi_i(R)$, $\psi_k(R)$ and the electronic transition moment $M(R)$ is called the band strength of the (v_i, v_k) band, and $S(J_i, J_k)$ is the London–Hönl factor. The band strength is often approximated by $|M(R_e)|^2 q(v_i, v_k)$, where R_e is the equlibrium internuclear distance of the emitting molecule and q is the Franck–Condon factor. Note that the intensity is independent of a scale factor of $M(R)$ because, under the assumptions made, the number of emitter molecules formed per second in the observation volume al equals the number of emitted photons per second. For a measurement of relative as well as absolute cross sections, it is therefore not necessary to know the absolute radiative lifetime. This is in sharp contrast to the infrared chemiluminescence experiments as, for example, done by Polanyi and co-workers [see, e.g., Brandt and Polanyi, (1978)], where the dominant depletion of emitter concentration is not by radiation but by wall deactivation, and A_{ik} has to be known absolutely for an analysis.

The quantity measured with this technique is, strictly speaking, the so-called emission cross section, which can only be identified with the cross section for formation of the excited species in a direct reaction (1), if (a) the emitting state is not populated by cascading from higher states; (b) nonradiative deactivation of the excited state does not occur; and (c) in the case of branching emission, all branches are either included in the measurement or allowed for. In addition, the excited-state lifetime has to be short enough so that all product ions emit within the observation region. These factors have been discussed by Thomas (1972), the last one also by Harris and Leventhal (1976). Conditions (a)–(c) are generally fulfilled in the cases studied: Higher states are not excited, or are excited only at high collision energy; the linear dependence of the emission intensity on the target gas pressure indicates absence of quenching; and emission branching does not occur in the cases studied. Loss of emitting products from the observation zone can be serious if the radiative lifetime is longer than ~ 1 μsec, corresponding to a flight distance of a few centimeters for a light ion at high energy.

Let this mean flight distance of an emitting species be λ_0 and let the observation region include a section of the ion beam from l_1 to l_2, measured from the collision chamber entrance slit. For example, in the Göttingen apparatus, $l_1 = 6$ mm, $l_2 = 18$ mm. By integration from l_2 to l_1, one then calculates a fractional light loss, due to the finite lifetime, of

$$\rho = \frac{\lambda_0}{l_2 - l_1}(e^{-l_1/\lambda_0} - e^{-l_2/\lambda_0}). \tag{10}$$

This is a generalization of the formula given by Harris and Leventhal (1976), which refers to the case $l_1 = 0$. It can easily be seen that ρ is smaller if $l_1 \approx \lambda_0$ than if $l_1 = 0$ (assuming the same $l_2 - l_1$ in each case), so that a geometry with the observation region well behind the entrance slit is to be preferred over one joining directly onto this slit.

D. *Metastable Ions*

As was pointed out above, it is important to be able to measure and control the proportion of metastable ions in the beam. The measurement is done by the differential attenuation technique introduced by Turner *et al.* (1968). It is based on the fact that the intensity i of a composite ion beam, containing the fractions f of metastable and $(1-f)$ of ground-state ions, will be attenuated, upon passage through a suitably chosen gas of pressure p, in a nonexponential fashion:

$$i(p) = i_0[fe^{-\alpha_1 p} + (1-f)e^{-\alpha_2 p}], \tag{11}$$

where $\alpha_{1,2}$ are proportional to the total cross sections for removal out of the beam of the two ion species. For $\alpha_1 > \alpha_2$, at large p the metastable component has been filtered out and the falloff is from there on exponential. Extrapolation of this section back to $p = 0$ gives the fraction $1-f$ as the ordinate intercept. The method is illustrated in Fig. 2 for the case of B^+ ions from BF_3 (Reichmuth, 1981). N_2O is used as the attenuating gas, following Lin *et al.* (1974a), because its ionization potential of 12.9 eV is close to the energy of the metastable $B^+(^3P)$ state, relative to $B(^2P)$(12.93 eV). Due to resonant charge transfer, the condition $\alpha_1 > \alpha_2$ is then fulfilled.

The principal parameter governing the metastable concentration is the ion source anode-to-cathode voltage U, to a lesser extent also the ion

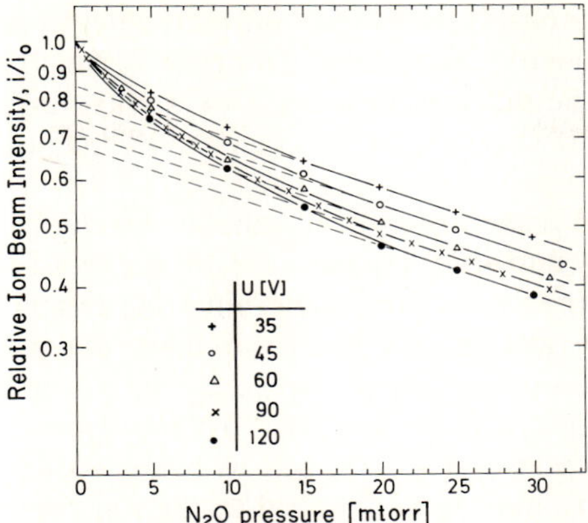

FIG. 2 Attenuation of a composite $B^+(^3P)/B^+(^1S)$ ion beam passing through the collision cell filled with N_2O, at 215 eV_{Lab}. Five different ion source operating voltages U (35–120 V) were used as indicated, corresponding to increasing fractions of metastable ions $B^+(^3P)$.

source pressure (Ottinger and Simonis, 1978). Figure 2 shows attenuation curves of B^+ in N_2O for five different values of U (Reichmuth, 1981). At the lowest U, f is the smallest, giving a reliable linear portion (in the semi-logarithmic plot) at large p. This aids the extrapolation to $p \rightarrow 0$ for the other curves because the limiting slope has to be the same independent of f.

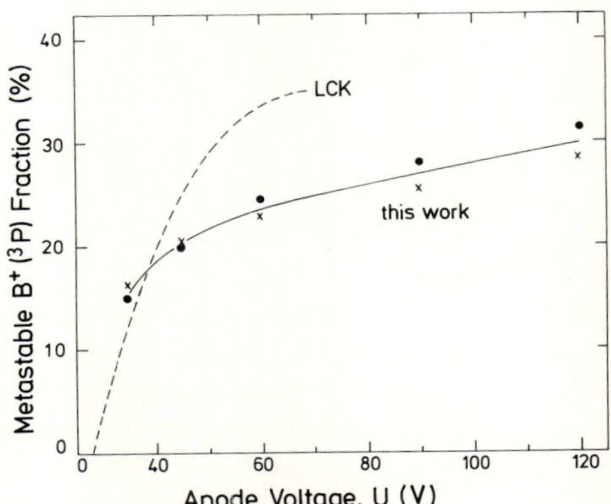

FIG. 3 Fractions of metastable ions $B^+(^3P)$ in the beam as a function of the ion source operating voltage U. ●, ion current as measured on the collision cell rear electrode (corresponds to Fig. 2); ×, current measurement on the energy analyzer entrance slit (see Fig. 1). Curve LCK shows analogous data from Lin *et al.* (1974a) for their ion source conditions.

The fractions f obtained from the five ordinate intercepts are plotted versus U in Fig. 3 [two slightly different experiments by Reichmuth (1981)]. Comparable data for the different apparatus of Koski's group are also shown (Lin *et al.*, 1974a). It can be seen that the $B^+(^3P)$ fraction can be varied between 15 and 30%.

A given optical emission will, in general, be due to reaction of both metastable and ground-state reactant ions, with cross sections σ_m and σ_g, respectively. It is then possible to determine the relative contribution $\gamma = \sigma_m/\sigma_g$ of these two species from a measurement of the light intensity as a function of the ion source voltage, $I(U)$. Let U_0 be some reference voltage, e.g., the highest voltage used, and $i(U)$ the total ion current through the collision cell (which as a rule increases with increasing U). Then, $Z(U) = I(U)/i(U)$, the light intensity normalized to the ion current, is a measure of the light yield per incident ion. We have

$$\begin{aligned}\frac{Z(U)}{Z(U_0)} &= \frac{\sigma_m f(U) + \sigma_g(1 - f(U))}{\sigma_m f(U_0) + \sigma_g(1 - f(U_0))} \\ &= \frac{1 + (\gamma - 1)f(U)}{1 + (\gamma - 1)f(U_0)} \\ &= F(U; \gamma). \end{aligned} \tag{12}$$

From the known dependence $f(U)$ characteristic for the ion source and ion species (Fig. 3), the right-hand side $F(U; \gamma)$ of Eq. (12) can then be plotted

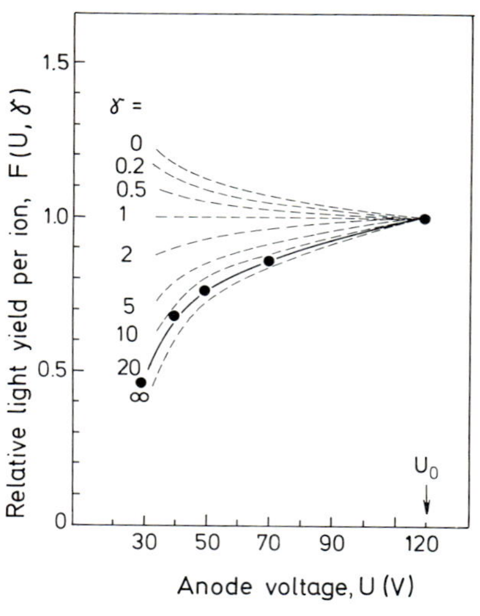

FIG. 4 Example of the determination of the cross-section ratio γ for luminescent reactions of ground-state and metastable ions. $B^+ + D_2 \rightarrow B(^2S) + D_2^+$(100 eV_{CM}, $\gamma = \sigma_m/\sigma_g$). Decreasing light yield with decreasing ion source voltage indicates predominant reaction of metastable ions.

for a variety of assumed cross-section ratios γ and the measured light yield ratios $Z(U)/Z(U_0)$ entered into this diagram. These points should then all lie on or close to one particular curve $F(U; \gamma)$, which gives the desired value of γ. Qualitatively speaking, if the light yield of some particular emission drops as the anode voltage is reduced, then this emission is primarily (or exclusively, $\gamma = \infty$) due to metastable ions; if the light yield increases, then it is primarily (or exclusively, $\gamma = 0$) due to ground-state ions. Figure 4 shows, as an example, results for excitation of the 2497-Å boron line by $B^+ + D_2$ charge transfer (Ottinger *et al.*, 1983). Here, the measured points lie close to the $\gamma = 20$ curve.

III. SELECTED RESULTS

The electronically chemiluminescent ion–molecule reactions studied up until 1981 have been compiled in two previous reviews (Menzinger, 1980; Tiernan and Lifshitz, 1981). An updated list is given in Table I. As can be seen, the number of reactions studied is not very large, compared to the numerous neutral chemiluminescent reactions (Menzinger, 1980). This is due to several factors. The low emission intensity is one of them, making data accumulation tedious. More importantly, however, in most cases studied, the specific virtues of ionic reactions, as outlined above, have been fully exploited, yielding a large body of data for each reacting system. For example, in most cases, spectra have been taken over a wide range of collision energies; careful analysis in terms of vibrational–rotational and electronic state population has been carried out; molecular constants have been derived for electronic states which were previously poorly known; new electronic states have been identified; relative and, in some cases, absolute cross sections have been measured; and the reactions have been assigned to ground-state or excited reactant ions. The data yield is such that a complete study of a typical reaction requires an effort on the order of 3–6 man-years.

In this chapter, the capabilities of the method will be illustrated by a discussion of four examples. These are

(a) the system $C^+ + H_2$, for which the internal product state distribution has been studied in particularly great detail;

(b) the system $B^+ + H_2$, with emphasis on the electronic properties;

(c) collisions of B^+ ions with hydrocarbons, showing the transition from type (1a) to type (1b) reactions; and

(d) reactions of N^+ ions with hydrocarbon molecules at comparatively high energy, with particular consideration of the reaction cross section as a function of energy.

TABLE I

Electronically Chemiluminescent Ion–Molecule Exchange Reactions

Reactants[a]	Luminescent product	References
$B^+(^3P) + H_2, D_2$	$BH^+, BD^+(A\,^2\Pi)$	Ottinger and Reichmuth (1981, 1984)
$B^+(^1S) + H_2, D_2$	$BH^+, BD^+(A\,^2\Pi)$; $BH^+, BD^+(B\,^2\Sigma^+)$	
$B^+(^3P) + CH_4$	$BH^+(A\,^2\Pi)$, $BH(A\,^1\Pi)$	
$B^+(^3P) + C_3H_8$	$BH(A\,^1\Pi)$; $BH(b\,^3\Sigma^-)$; $BH(C'\,^1\Delta)$ (?); $BH(B\,^1\Sigma^+)$ (?)	
$C^+(^2P) + H_2$	$CH^+(A\,^1\Pi)$	Appell *et al.* (1975); Harris *et al.* (1975); Kusunoki and Ottinger (1979a)
$C^+(^2P) + D_2$	$CD^+(A\,^1\Pi)$	Kusunoki and Ottinger (1979a)
$C^+(^4P) + H_2, D_2$	$CH^+, CD^+(b\,^3\Sigma^-)$	Kusunoki and Ottinger (1980, 1982)
$C^+(^2P) + H_2$	$CH^+(B\,^1\Delta)$	Appell *et al.* (1975); Harris *et al.* (1975)
$C^+(^2P) + H_2$	$CH(A\,^2\Delta)$	Kusunoki and Ottinger (1979a)
$C^+ + RH$ ($R = CH_3, C_2H_3, C_2H_5, C_3H_7$)	$CH(A\,^2\Delta)$, $CH(B\,^2\Sigma^-)$, $CH(C\,^2\Sigma^+)$, $H(n = 4)$, $C(^1P)$	Kusunoki (1978)
$C^+(^2P) + O_2$	$CO^+(A\,^2\Pi)$	Ottinger and Simonis (1975); Tsuji *et al.* (1982)
$C^+ + NO$[b]	$CN(B\,^2\Sigma^+)$	Brandt *et al.* (1973)
$N^+(^3P) + H_2$	$NH(A\,^3\Pi)$	Kusunoki *et al.* (1976); Kusunoki and Ottinger (1979b,c)
$N^+ + H_2, D_2$	$NH^+, ND^+(B\,^2\Delta)$	Kusunoki and Ottinger (1984)
$N^+(^3P) + RH$ ($R = CH_3, C_2H_3, C_2H_5, C_3H_7$)	$NH(A\,^3\Pi)$; also $CH(A\,^2\Delta, B\,^2\Sigma^-, C\,^2\Sigma^+)$, $H(n = 4)$, $C(^1P)$ from dissociation	Kusunoki *et al.* (1976); Kusunoki and Ottinger (1979b,c)
$N^+(^3P) + NO$	$N_2^+(B\,^2\Sigma_u^+)$; N_2^+ (quartet state)	Brandt and Ottinger (1973, 1975); Brandt (1976)
$O^+(^2D, ^2P$, also $^4S) + H_2$	$OH^+(A\,^3\Pi)$	Harris and Leventhal (1976)
$O^+(^2D, ^2P) + H_2$	$OH(A\,^2\Sigma^+)$	
$O^+(^2D) + CH_4$[c]	$OH(A\,^2\Sigma^+)$	Kusunoki *et al.* (1979)
$O^+(^2D$ (and ^2P?)) + RH ($R = C_2H_3, C_2H_5, C_3H_7$)	$OH(A\,^2\Sigma^+)$; also $CH(A\,^2\Delta, B\,^2\Sigma^-, C\,^2\Sigma^+)$, $H(n = 4)$, $C(^1P)$ from dissociation	

[a] The electronic state is given wherever it was determined directly or inferred from measured reaction threshold energies.

[b] Second order, probably via C as intermediate.

[c] Second order, probably via $OH^+(^1\Sigma^+)$ as intermediate.

A. $C^+ + H_2$

1. *Survey*

Chemiluminescence resulting from impact of low-energy carbon ions on hydrogen molecules was reported at about the same time by Harris *et al.* (1975) and Appell *et al.* (1975). This was one of the first well-characterized reactions of type (1) and has continued to be studied up until now (Kusunoki and Ottinger, 1979a, 1980, 1982), with much improved resolution, and including metastable C^+ ions. Interest in this reaction stems from earlier, very detailed, and thoroughly analyzed kinematic studies of the CH^+ product ion (Iden *et al.*, 1972; Mahan and Sloane, 1973; Jones *et al.*, 1976, 1977), from its inherent simplicity in terms of electronic structure, which has motivated numerous ab initio calculations (Sakai *et al.*, 1981 and references therein) and from the presumed importance of $C^+ + H_2$ collisions as the initial step in the formation of interstellar molecules, which in turn enable stars to evolve (Herbst *et al.*, 1977). Also, the spectroscopy of the possible products CH^+ and CH is reasonably well known, providing a good starting point for a chemiluminescence study.

Figure 5 gives the potential energy curves of the CH^+ ion, indicating two important electronic transitions (Huber and Herzberg, 1979). Figure 6 (Ottinger, 1978) shows low-resolution survey spectra observed with $C^+ + H_2$ collisions. Obviously, the $CH^+(A\,^1\Pi \rightarrow X\,^1\Sigma^+)$ emission, lying between 3500 and 4500 Å, is dominant at most energies. It disappears below 3-eV_{CM} collision energy, consistent with the threshold of 3.4 eV for the formation of excited product ions in the reaction

$$C^+(^2P) + H_2 \rightarrow CH^+(A\,^1\Pi) + H. \tag{13}$$

At 3 eV_{CM}, although nominally below the threshold energy, some $CH^+(A)$ is still formed as a result of the collision energy spread (see above).

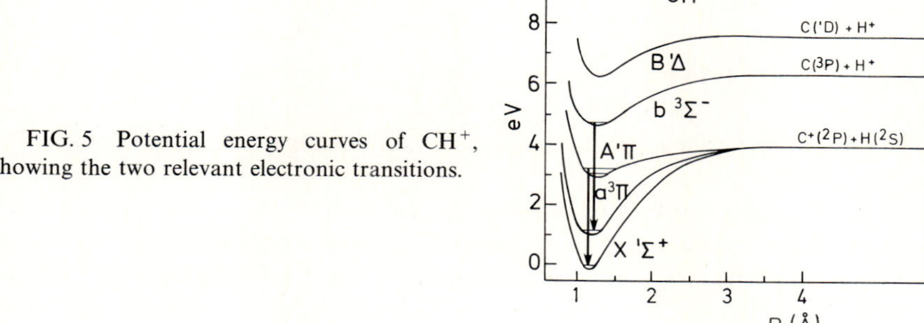

FIG. 5 Potential energy curves of CH^+, showing the two relevant electronic transitions.

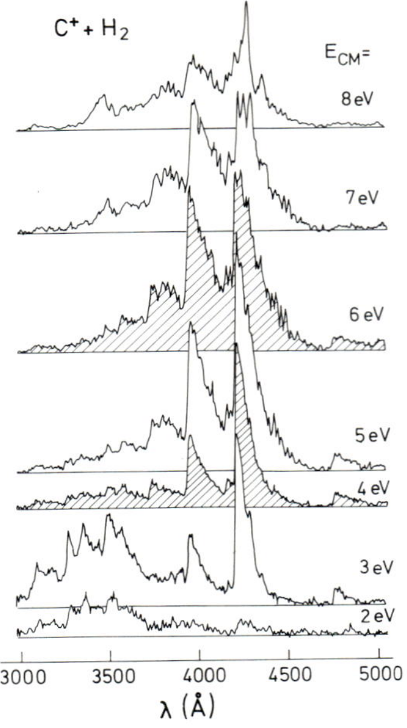

FIG. 6 Low-resolution (20-Å FWHM) spectra of chemiluminescence from $C^+ + H_2$ reactions. For clarity, not all spectra are shown on the same ordinate scale. [From Ottinger (1978).]

At the two lowest energies, emission is observed in the range of the known $CH^+(b\,^3\Sigma^- \rightarrow a\,^3\Pi)$ band system. However, $CH^+(b\,^3\Sigma^-)$, lying higher than $CH^+(A\,^1\Pi)$, can certainly not be populated by reaction of C^+ ions in their electronic ground state ^2P at 2 and 3 eV_{CM}. A careful analysis (Kusunoki and Ottinger, 1980) has proved that the emission in question is, in fact, due to the $CH^+(b \rightarrow a)$ transition, the emitting ion state being populated by reaction of metastable ions $C^+(^4P)$. These are typically present, in the Göttingen apparatus, in concentrations of only a few percent of the total C^+ beam (Ottinger and Simonis, 1978), but produce $CH^+(b)$ with a fairly large cross section. $C^+(^4P)$, with the electron configuration $2s2p^2$, has an energy 5.33 eV above the $C^+(^2P)$ ground state, making the reaction

$$C^+(^4P) + H_2 \rightarrow CH^+(b^3\Sigma^-) + H \tag{14}$$

exothermic by 0.36 eV, consistent with the observation of this emission down to 1 eV_{CM}. In addition, metastable reactant ions were unambiguously identified using the technique described above (Kusunoki and Ottinger, 1980).

Prior to these studies, only three bands of the $CH^+(b \rightarrow a)$ system were known, while obviously the chemiluminescence spectrum (Fig. 6) has a much richer structure in this region. This is even truer of the analogous CD^+ emission, for which similar survey spectra have been published

(Ottinger, 1978). This observed band structure is an example of the spectroscopic power of the ion impact luminescence method. Six new bands of $CH^+(b \rightarrow a)$ and twelve of $CD^+(b \rightarrow a)$ have been observed. From these data, vibrational constants of both the upper and the lower CH^+ triplet states were derived (Kusunoki and Ottinger, 1980) which formed the basis for a population analysis of $CH^+(b)$ from reaction (14), to be discussed below.

Harris *et al.* (1975) have also reported an emission from $C^+ + H_2$ collisions at short wavelengths (3250–3650 Å) but at higher energy (7.7 and 18.3 eV_{CM}). They ascribed it both to the $CH^+(b\,^3\Sigma^- \rightarrow a\,^3\Pi)$ and to the $CH^+(B\,^1\Delta \rightarrow A\,^1\Pi)$ transitions. Reasons have been given, however, why the triplet contribution, if any, must have been quite insignificant (Kusunoki and Ottinger, 1980). Interpretation of their spectra as predominantly CH^+ $(B \rightarrow A)$ appears to be much more likely. Appell *et al.* (1975) have also observed emission near 3470 Å at high collision energies (11 eV_{CM}) and tentatively ascribed it to $CH^+(B \rightarrow A)$. Both groups pointed out that, as a result of cascading transitions $B \rightarrow A \rightarrow X$, the $A \rightarrow X$ emission intensity should then be at least equal to the $B \rightarrow A$ intensity at all energies, while the observations show that the former becomes less than the latter at about $E_{CM} \geq 10$ eV. This question is still unresolved. Harris *et al.* (1975) consider loss of some $CH^+(A)$ emission due to instrumental effects (radiative lifetime comparable to the transit time through the observation region), while Appell *et al.* (1975) suggest the alternative interpretation of the 3470-Å feature as a new band system of CH^+, perhaps analogous to the $^1\Sigma^+ \rightarrow {}^1\Pi$ system of the isoelectronic molecule BH. More detailed studies of this feature of the $C^+ + H_2$ chemiluminescence, at better resolution, will be necessary to settle this problem.

2. *Cross Sections*

The relative cross sections for excitation of the CH^+ singlet $(A \rightarrow X)$ and triplet $(b \rightarrow a)$ systems are shown in Figs. 7 and 8, respectively (Appell *et al.*, 1975; Kusunoki and Ottinger, 1980). The shape of these curves is uncorrected for the energy-dependent light loss due to the finite lifetime. At 5 eV_{CM}, corresponding to 35 eV_{Lab}, the CH^+ laboratory system velocity would be 2.1×10^6 cm/sec for a stripping-like mechanism. For $CH^+(A \rightarrow X)$, radiative lifetimes between 280 and 815 nsec have been reported. Assuming the latest result (Grieman *et al.*, 1981), 815 nsec, to be correct, this corresponds to a mean ion flight distance of $\lambda_0 = 1.7$ cm before radiation occurs or, from Eq. (10), to a ~50% light loss. At 10 eV_{CM} the loss would be ~60%. Thus, relative cross sections such as shown in Figs. 7 and 8 are much less affected by the finite radiative lifetime than are absolute cross sections.

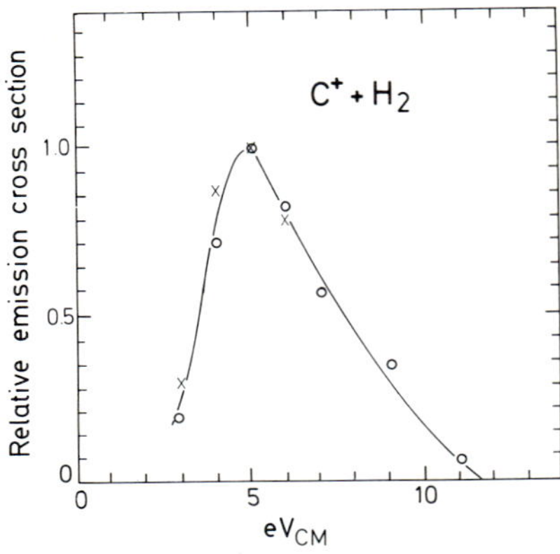

FIG. 7 Cross section for CH^+(A–X) emission from $C^+ + H_2$ reactions, as a function of the collision energy. At the low energies, the spectrum was integrated (×). At higher energies, data are from the normalized (1, 0) band intensity only, in order to eliminate contributions from CH emission. [From Appell *et al.* (1975).]

Figure 7 shows a steep rise of the cross section above threshold and, past the maximum, a surprisingly sharp decline. This drop is probably due to competition of nonradiative channels, setting in at high energy, where the density of electronic energy surfaces increases rapidly. Product dissociation will also contribute to the sharp falloff. In contrast to the peaked cross section curve for the endothermic CH^+(A) formation, the cross section for CH^+(b) (Fig. 8) shows the monotonic falloff typical for an exothermic reaction. The curves for 3360 and 3280 Å also demonstrate that $CH^+(b\,^3\Sigma^-)$ is apparently not formed from $C^+(^2P)$, the major component of the ion

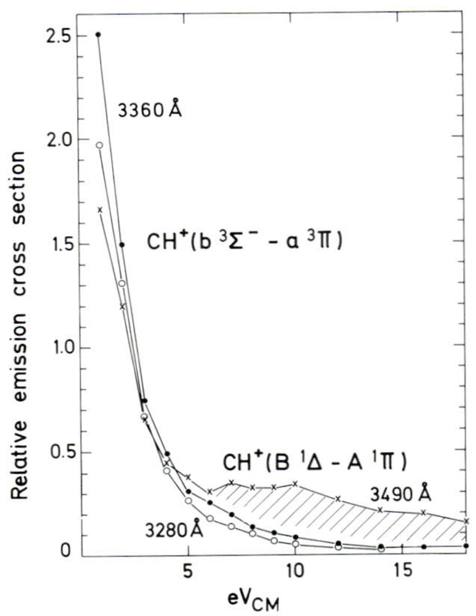

FIG. 8 Cross section for $CH^+(b - a)$ emission from $C^+ + H_2$ reactions, as a function of the collision energy. [From Kusunoki and Ottinger (1980).]

beam. This reaction would be endothermic by 5 eV. The light yield (Fig. 8) gives no indication for an additional channel opening up at that energy.

An absolute cross section of 0.04 Å^2 for the $CH^+(A)$ formation at 5 eV_{CM} was given by Appell *et al.* (1975), which would represent only ~2% of the total CH^+ cross section as measured mass spectrometrically by Lindemann *et al.* (1972). However, due to the difficulties of such measurements indicated above, this result is uncertain within a factor of 3. Also, the ~50% correction from the lifetime has not been applied. Jones *et al.* (1977) have seen evidence of $CD^+(A)$ formation, from $C^+ + D_2$ at 28 eV_{Lab} (=7 eV_{CM}), from the CD^+ translational energy distribution measured in the forward direction, and feel that the fraction of CD^+ in the A state must be considerably greater than 2% in order to be detectable in this way. However, their technique does not discriminate between $CD^+(A)$ and highly vibrationally excited CD^+(X or a).

The cross section for $CH^+(b)$ formation by reaction (14) can be related to that for $CH^+(A)$ formation [Eq. (13)], based on the known ratio 3:100 of $C^+(^4P)$ to $C^+(^2P)$ in the ion beam. One then finds a cross section for reaction (14) of at least 2 Å^2 at 1 eV_{CM} (Kusunoki and Ottinger, 1980). This is a lower limit because of the long lifetime of $CH^+(b)$. Calculated oscillator strengths f (Kusunoki *et al.*, 1980) for the b → a transition are about a factor of 10 smaller than for A → X, making the lifetime $\tau \sim \lambda^2/f$ longer by a factor of ~6.5, or ~5 μsec. At 1 eV_{CM}, the mean flight distance before radiation is then $\lambda_0 \sim 5.4$ cm, which means that only ~20–25% of the light will be emitted in the observation region. The total cross section for reaction (14) may therefore be as large as ~10 Å^2.

3. *Product Vibrational Distributions*

Figure 9 shows $CH^+(A \rightarrow X)$ spectra measured with much higher resolution. Similar spectra were also measured for $C^+ + D_2$ collisions (Kusunoki and Ottinger, 1979a). Figure 9 shows the (1, 0) and the (0, 0) bands, at ~3950 and ~4220 Å, respectively. Even qualitative inspection reveals a distinctly increasing excitation of $v' = 1$, relative to $v' = 0$, with increasing collision energy. At the same time, the rotational excitation also increases strongly, as can be seen by the appearance of more and more rotational lines on the long-wavelength side of each band. Strong rotational excitation of the $CH^+(A)$ product was also noted by Harris *et al.* (1975). In the detailed work of Kusunoki and Ottinger (1979a) both vibrational and rotational distributions were extracted from the spectra by means of a computer simulation. Based on the known molecular constants and on level populations $P(v', J')$ which were adjusted by trial and error, synthetic spectrum contours were calculated which matched the observed spectrum very

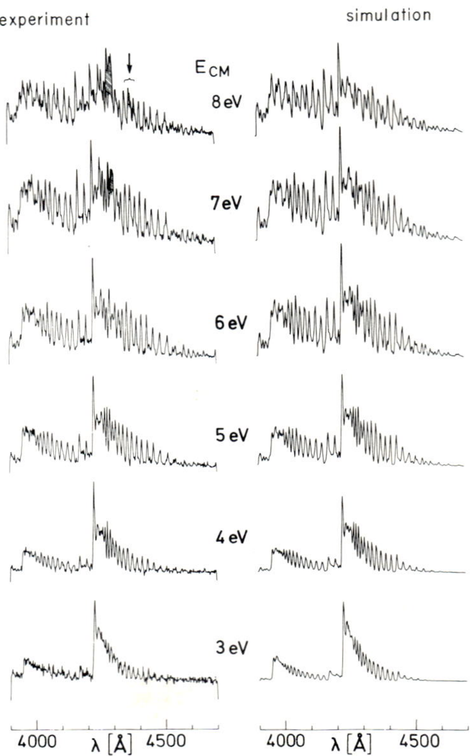

FIG. 9 High-resolution (6-Å FWHM) spectra of chemiluminescence from $C^+ + H_2$ reactions (left). The computer simulation (right) is based on vibrational and rotational distributions of the $CH^+(A\,^1\Pi)$ emitter, as shown in Figs. 10 and 12. The shaded parts of the 7- and 8-eV spectra are due to CH emission and are, therefore, not reproduced by the simulation. [From Kusunoki and Ottinger (1979a).]

well. They are shown in Fig. 9 on the right-hand side. In this calculation, account was taken of the dependence of the transition moment on the internuclear distance, known from ab initio theory (Yoshimine *et al.*, 1973). It affects strongly the relative band intensities. Furthermore, the effect of centrifugal distortion of the emitting CH^+ ion on the intensities was allowed for, by calculating J-dependent band oscillator strengths. Rotation can change the intensities by a factor of 3, for the highest J of ~20, as a result of the quite shallow (~1-eV-deep) A $^1\Pi$ potential well.

Comparing the observed and the simulated spectra shown in Fig. 9, one notices some excess intensity in the former, near 4300–4400 Å, at 7 and 8 eV_{CM}. Considering the otherwise very successful simulation, these extra fea-

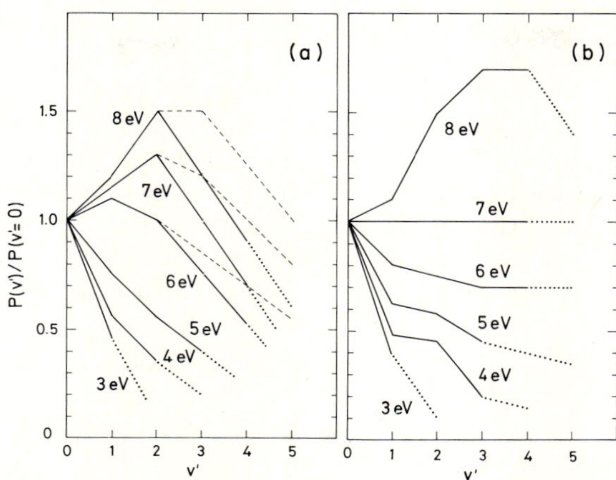

FIG. 10 Vibrational level population of (a) $CH^+(A\,^1\Pi)$ and (b) $CD^+(A\,^1\Pi)$ from $C^+(^2P) + H_2(D_2)$ collisions, obtained from chemiluminescence spectrum analysis by simulation. [From Kusunoki and Ottinger (1979a).]

tures must be due to some band system other than $CH^+(A \rightarrow X)$. It was identified by Appell *et al.* (1975) and by Kusunoki and Ottinger (1979a) as $CH(A\,^2\Delta \rightarrow X\,^2\Pi)$ emission from the reaction

$$C^+ + H_2 \rightarrow CH(A\,^2\Delta) + H^+ \tag{15}$$

which is endothermic by 6.3 eV. Above this threshold this very minor ($\sigma \sim 10^{-3}$ Å^2) reaction channel opens up, parallel to reaction (13). It has not yet been studied in detail.

The vibrational distributions of $CH^+(A)$ and $CD^+(A)$ are shown in Fig. 10. They were obtained through the simulation technique as applied to both the high-resolution spectra (Fig. 9) and spectra such as shown in Fig. 6, taken at lower resolution, but covering a wider wavelength range and therefore encompassing more bands. Both for the $C^+ + H_2$ and the $C^+ + D_2$ reaction, the population of the higher v' levels relative to $v' = 0$ falls off monotonically at low collision energy, while at high energy a vibrational inversion builds up.

Two problems arise for the higher v'. First, the excited-state lifetime, in general, depends on v'. If light loss due to a long radiative lifetime is not negligible, then bands from different v' levels may be differentially discriminated against in the spectrum. In the case of $CH^+(A \rightarrow X)$, the radiative lifetime increases with v'. Calculated band oscillator strengths (Kusunoki and Ottinger, 1979a) show that, e.g., $\tau(v' = 2) = 1.8 \times \tau(v' = 0)$. If the long lifetime $\tau(v' = 0) = 815$ nsec (Grieman *et al.*, 1981) is correct, this would, for example, require the population ratio $P(v' = 2)/P(v' = 0)$, at 5 eV_{CM}, to be corrected upwards by ~60%. In Fig. 10, adapted from Kusunoki and Ottinger (1979a), no such corrections were applied because literature values of τ available at the time of these experiments were considerably smaller.

The second problem concerns predissociation. For $v' = 3, 4, 5$, the highest rotational levels lie above the crest of the effective potential and are therefore unstable. They do not contribute then to the emission cross section, but should be included in the reaction cross section. Their contribution, as estimated from the extrapolated rotational distributions, is shown by the dashed lines in Fig. 10.

The most striking feature of the vibrational distributions is the change from a monotonically declining to an inverted shape as the collision energy is increased. The 3-, 4-, and 5-eV curves can, very roughly, be described by Boltzmann distributions with temperatures $T_{vib} = 3000$, 4000, and 8200 K, respectively. It is known from the kinematic studies (Iden *et al.*, 1972; Mahan and Sloane, 1973) that up to about 4-eV collision energy reaction proceeds largely through an intermediate $CH_2{}^+$ complex, while a direct (or impulsive) mechanism becomes dominant as the energy is increased up to ~7 eV_{CM}. Mahan and Sloane (1973), however, stress the evidence that at an energy as low as 2.8 eV, there are signs of a transition from the complex mechanism to the direct reaction, or else a separate, direct type reaction path is contributing even at this energy. In the light of these studies, it is very interesting to note that the reaction branch to $CH^+(A)$ shows the same trend. At low energy, this reaction, too, may be going through a $CH_2{}^+$ complex which is sufficiently long-lived to permit random distribution among the internal degrees of freedom, as is indicated by the Boltzmann-like $P(v')$. At high energy, on the other hand, the measured inverted distribution $P(v')$ proves that a direct reaction mechanism takes over because theory shows that breakup of a long-lived complex will never give an inverted $P(v')$ in the products (Light, 1979).

For the $CH^+(CD^+)$ triplet formation [reaction (14)], the vibrational distributions derived from spectrum simulations are shown in Fig. 11. Measurements were made only at two energies in each case. It is very striking that for this reaction the relative population of most excited vibrational levels, except the highest ones, is smaller at the higher collision energy. This trend is opposite to that documented in Fig. 10 for the singlet CH^+ reaction and is contrary to the intuitive expectation. A fundamental diffference between reactions (13) and (14) is obviously that the former is endothermic while the latter is exothermic. For exothermic reactions, Polanyi and coworkers (Ding *et al.*, 1973) have formulated the rule of "translational adiabaticity," stating that increased reagent translation is in these cases channeled preferentially into additional product translation (and, to a lesser

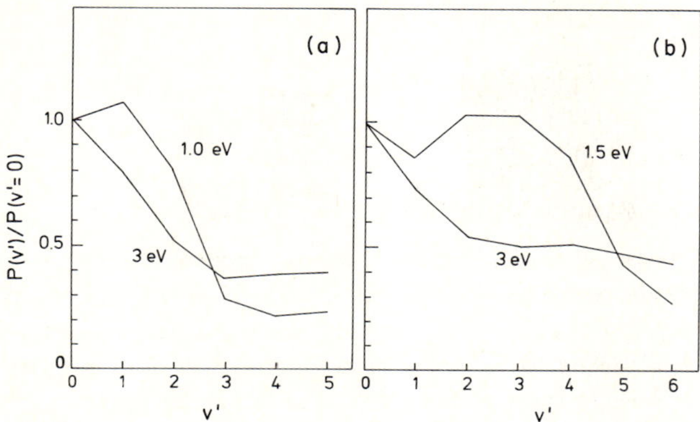

FIG. 11 Vibrational level population of (a) CH^+(b) [$CH^+(^3\Sigma^-)$] and (b) CD^+(b) [$CD^+(^3\Sigma^-)$] from $C^+(^4P) + H_2(D_2)$ collisions. [After Kusunoki and Ottinger (1982).]

extent, rotation) while the mean product vibration remains constant (or was, in fact, observed to decrease slightly). The measured $P(v')$ for reaction (14) appears to be another example of this rule.

However, the mere fact that reaction (14) is exothermic can certainly not be the only reason for the "anomalous" trend of $P(v')$ with E_{CM}. The reaction

$$C^+(^2P) + O_2 \rightarrow CO^+(A\,^2\Pi) + O \tag{16}$$

is also exothermic (by 0.69 eV). It was recently observed in a flow tube by Tsuji *et al.*, (1982) and is the only electronically chemiluminescent ion–molecule reaction to have been studied at thermal energy. From the spectrum, vibrational distributions $P(v')$ were derived in the pressure range 0.65–3 torr, and collisional relaxation was accounted for by extrapolation to $p \rightarrow 0$. The final $P(v')$ was found to follow very roughly a Boltzmann distribution having $T_{vib} = 2700 \pm 400$ K. This reaction was earlier also studied under beam conditions (Ottinger and Simonis, 1975). $P(v')$, as obtained from a spectrum simulation (Ottinger, 1976), was here clearly nonthermal, but the closest Boltzmann distribution, at 3.6 eV_{CM}, had $T_{vib} = 17{,}000$ K. Comparison with the flow tube result shows that, for this exothermic reaction, the vibrational population increases with increasing collision energy. Independently of the flow experiments, which might be affected by some residual relaxation, this trend is confirmed by additional beam experiments (Ottinger, 1976) at 0.7 to 5.6 eV_{CM}. (At higher energies, complications from luminescent charge transfer set in.) Although an exact simulation was done only at 3.6 eV_{CM}, the trend in the spectra is clearly toward higher vibrational excitation with increasing collision energy.

The conclusion from a comparison of $P(v')$ for reactions (13), (14), and (16) is, therefore, that the sign of the energy balance of a reaction is certainly not the only factor in determining the dependence of $P(v')$ on the collision energy.

4. *Product Rotational Distributions*

Returning to reaction (13), the rotational distributions $P(J')$ of $CH^+(A)$, obtained by the spectrum simulation, are shown in Fig. 12 (from Kusunoki and Ottinger, 1979a). The best simulation of the particularly critical and well-resolved bands (0, 0) and (1, 0) was achieved using the same $P(J')$ for $v' = 0$ and $v' = 1$. Spectra of bands from other v', at lower resolution, were also successfully simulated with this $P(J')$, which was thus taken to be independent of v'. Only with regard to predissociation was v' taken into account: $P(J')$ was truncated at $J' = 22$, 19, and 16 for $v' = 3$, 4, and 5, respectively (arrows in Fig. 12). Note the very high rotational excitation of $CH^+(A)$ produced in this way, giving spectroscopic information on high rotational levels which are difficult to observe otherwise. CH^+ rotational levels with similarly high J have recently been observed by laser-induced fluorescence $CH^+(A \leftrightarrow X)$ of trapped fragment ions produced by electron impact (Grieman *et al.*, 1981). Conventional light sources, on the other hand, produce only lines up to $J = 14$ (Carrington and Ramsay, 1982). Note also the absence of rotational relaxation: A Boltzmann $P(J')$ corresponding to the collision cell temperature ($\sim$400 K) would peak at $J' = 3$

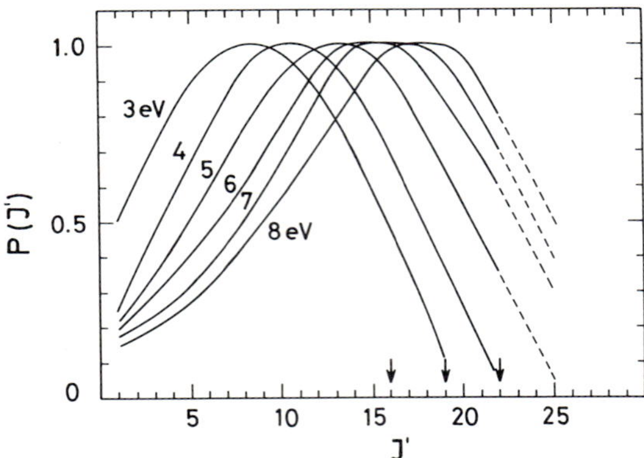

FIG. 12 Rotational level population of $CH^+(A)$ from $C^+(^2P) + H_2$ collisions. The same distributions apply, within the error limits, for $CH^+(A)$ in different vibrational levels, except for cutoffs for $v' = 5$, 4, and 3 (arrows, from left to right) due to rotational predissociation. [From Kusunoki and Ottinger (1979a).]

and, according to a well-known model of rotational energy transfer (Polanyi and Woodall, 1972), a partly relaxed $P(J')$ should be bimodal, with a secondary peak at these low J'. As a direct check on this point, it was verified that the measured high-resolution spectrum contour was independent of the H_2 target gas pressure.

Although the $P(J')$ distributions are clearly nonthermal, they can be approximated, in the region of their maxima, by Boltzmann distributions corresponding to $T_{rot} = 2500$ and 4000 K for 3 and 4 eV_{CM}, respectively. These "temperatures" are similar to the T_{vib} values given above at the corresponding energies, thus lending support to the idea that at the lowest impact energy some kind of "complex" mechanism, involving energy exchange between all three atoms, is operative. At higher collision energy the rotational distributions become totally nonthermal, just like the vibrational distributions.

For $CH^+(b)$ from reaction (14), even higher rotational excitation was observed (Kusunoki and Ottinger, 1982). For this state of CH^+, which is more strongly bound than $CH^+(A)$, rotational predissociation is less critical. The average $CH^+(b)$ angular momentum, at 3 eV_{CM}, was measured to be $\langle J' \rangle = 24\ \hbar$, compared to $\langle J' \rangle = 9\ \hbar$ for $CH^+(A)$ at the same CM energy. A simple straight-line trajectory model for the partitioning of angular momentum between relative motion and molecular rotation was used by Kusunoki and Ottinger (1979a, 1982) to translate these $\langle J' \rangle$ values into average impact parameters $\langle b \rangle$ for the two reactions. The results were $\langle b \rangle = 0.17$ Å and 2.7 Å for reactions (13) and (14), respectively, at 3 eV_{CM}. This large difference corresponds plausibly to the very different reaction cross sections. In fact, the effective target sizes, $\pi \langle b \rangle^2$ and the measured cross sections given above are both just in the same ratio of about 250. This very good agreement is probably coincidental, since the impact parameter is very likely not the only factor determining the relative reactivity, but it is gratifying that the total cross-section measurements and the analysis of the spectra are so nicely consistent.

The details of the model from which the $\langle b \rangle$ values were derived are not critical for the general conclusion. A primitive and a more refined straight-line model were compared by Kusunoki and Ottinger (1979a) and gave similar results for $\langle b \rangle$. If, in addition, the assumption of straight-line trajectories were dropped, the $\langle b \rangle$ values would still not change greatly. This is especially true for reaction (13) at threshold, where the relative motion of the products is zero and $\langle b \rangle$ is simply found from equating the observed $\langle J \rangle$ with the reactants' relative angular momentum. Thus, if at low energy a complex-like mechanism obtains, this would certainly invalidate the straight-line model adopted for convenience by Kusunoki and Ottinger (1979a), but not the $\langle b \rangle$ values obtained.

Of the wealth of dynamical information contained in the $P(v')$, $P(J')$ distributions, only very little has been exploited up until now. Features such as the width of the rotational distribution, the trend of $\langle J' \rangle$ with impact energy, the onset of vibrational inversion, will have to await an interpretation until the reactive scattering has been theoretically modeled in detail. The only sufficiently general, though not rigorous, method available is that of classical trajectories (Truhlar and Muckerman, 1979). It presupposes knowledge of the relevant potential energy hypersurface(s) over a wide range of internuclear distances, covering this range fairly evenly and with "chemical accuracy" (often thought to be on the order of 1 kcal/mole). Such comprehensive quantum chemical calculations are now becoming available for simple systems like $(C—H—H)^+$ and $(B—H—H)^+$. For application to the present problem of electronic chemiluminescence, obviously at least one excited potential energy surface will have to be known. Electronic-state correlation diagrams can serve to classify the various surfaces. For the $C^+ + H_2$ reactions (13) and (14), such adiabatic correlations have been given by Appell *et al.* (1975) and by Kusunoki and Ottinger (1980) for doublet and quartet states, respectively.

It appears from the experience with the reactions listed in Table I that the rule of conservation of overall spin multiplicity is generally obeyed at the energies studied. On this basis, $C^+(^2P) + H_2$ collisions could lead to both $CH^+(A\,^1\Pi) + H(^2S)$ and to $CH^+(b\,^3\Sigma^-) + H(^2S)$ (although the latter reaction does not appear to occur, see above and Fig. 8). $C^+(^4P) + H_2$, on the other hand, can, with spin conservation, only give $CH^+(b\,^3\Sigma^-)$ and not $CH^+(A\,^1\Pi)$; indeed, Fig. 6 shows that below the threshold for reaction (13), triplet CH^+ formation appears to be the only reaction channel possible.

Inspection of the adiabatic state correlations (Appell *et al.*, 1975; Kusunoki and Ottinger, 1980) shows immediately that neither reaction (13) nor (14) can proceed along a single adiabatic surface. Both must involve a nonadiabatic transition somewhere along the respective reaction paths from a surface, originating from the reactants and correlating with nonradiating product states, to the nearest surface which correlates with the observed product. This surface will usually be one of the same symmetry, A′ or A″. For example, for reaction (13) the most likely path will involve a "surface hopping" from the second $^2A'$ surface, repulsive in the region of $C^+ + H_2$ approach, to the third $^2A'$ surface, which then correlates with $CH^+(A) + H$. An alternative route would be from the lowest to the first excited $^2A''$ surface (Sakai *et al.*, 1981). For the exothermic reaction (14), the surface hop would be from the third $^4A''$ surface down to the second. More general nonadiabatic transitions between surfaces of the $CH_2{}^+$ system have been discussed in the literature, namely, those between states of A′ and A″ symmetry, by means of a "dynamic Renner effect" (Herbst and

Delos, 1976; Herbst *et al.*, 1977), the analog of Π–Σ transitions brought about by angular coupling in diatomic systems.

Any modeling of the $C^+ + H_2$ reaction by means of trajectory calculations would necessarily have to take the nonadiabatic transitions into account, e.g., by means of the surface-hopping trajectory (SHT) method (Truhlar and Dixon, 1979). In a first step, one would only consider transitions between surfaces of like symmetry. The critical areas on these surfaces as pertaining to reaction (13) have already been identified in the ab initio work by Sakai *et al.* (1981). Trajectory calculations along these lines are in progress. It will be interesting to see if they will be able to explain the observations or if symmetry-changing nonadiabatic transitions will have to be considered as well.

B. $B^+ + H_2$

1. Survey

The system $(B—H—H)^+$ possesses only four valence electrons and is therefore much more tractable by quantum chemical methods than the system $(C—H—H)^+$. Several groups are engaged in calculating the potential energy hypersurfaces for this system. First results using the diatomics in molecules (DIM) method have already been obtained (Muckerman, 1981; Schneider and Zülicke, 1981) and other work is in progress (Rosmus, 1982). Experimental studies of this system, particularly of its excited states by means of luminescence from $B^+ + H_2$ collisions, are, therefore, of great interest. Furthermore, the reaction product BH^+ is spectroscopically not very well known, so that the technique of ion chemiluminescence can be expected to make contributions to its spectroscopy. In the only earlier spectroscopic study (Almy and Horsfall, 1937), no more than two bands of the $BH^+(A\,^2\Pi \rightarrow X\,^2\Sigma^+)$ system were reported, of which only one was rotationally analyzed. The chemiluminescence studies (Ottinger and Reichmuth, 1981) have prompted recent research on BH^+, both experimental (Ramsay and Sarre, 1982) and theoretical (Guest and Hirst, 1981; Klein *et al.*, 1982; Gerratt, 1981; Vázquez, 1981; Kusunoki, 1982). Finally, two independent measurements of product angular and energy distributions of the ion–molecule reactions $B^+ + D_2$ (Sondergaard *et al.*, 1979, 1981) and $B^+ + H_2$ (Friedrich and Herman, 1982) have been done, which both indicated large yields of electronically excited product. Unlike the $C^+ + H_2$ case, the product translational energy measurements showed here two clearly separated groups, the slower of which was identified as $BD^+(BH^+)$ in the $A(^2\Pi)$ state by kinematic considerations. At low kinetic energy, the angular distributions of the energy-resolved product ions were also measured so that, by integration over the angle, the total yield of $BH^+(A)$ relative to $BH^+(X)$ could be obtained. This ratio amounts to 0.13 at 2.25 eV_{CM} collision energy (Herman, 1982). Note that product translational energy distributions in the lab system forward direction only, such as reported for BD^+ by Sondergaard *et al.* (1979) in their Fig. 1, give visually a wrong impression of the relative yield, overemphasizing the excited product. After transformation to the velocity scale and application of the lab → CM Jacobian factor, one finds from the data of Sondergaard *et al.* (1979) about 0.25 for the ratio $BD^+(A)/BD^+(X)$ at 2.93 eV_{CM}. (The widths of the cm angular distributions of ground-state and excited products, which in principle also enter here, happen to be about the same in the present case.) Alternatively, one can extrapolate from Herman's data at 2.25 eV_{CM} to 3 eV_{CM}, using the measured relative $BH^+(A)$ cross section (see Fig. 14 below) and the relative overall cross section (ground-state plus excited product) as measured mass spectrometrically (Lin *et al.*, 1974b). In this way, one finds a $(A\,^2\Pi)/(X\,^2\Sigma^+)$ ratio of 0.24 at 3 eV_{CM}.

$B^+ + H_2$ chemiluminescence experiments were done with B^+ ions prepared from BF_3, as in the kinematic studies. It is known that large percentages of metastable $B^+(^3P)$ are obtained in this way; see Fig. 2 and Lin *et al.* (1974a). $B^+(^3P)$ has an energy 4.63 eV above the $B^+(^1S)$ ground state. The reactions

$$B^+(^1S) + H_2 \rightarrow BH^+(A\,^2\Pi) + H \tag{17}$$

$$B^+(^3P) + H_2 \rightarrow BH^+(A\,^2\Pi) + H \tag{18}$$

are endothermic by 5.80 and 1.17 eV, respectively (Friedrich and Herman, 1982).

2. *$BH^+(A)$ Spectra and Cross Sections*

Figure 13 shows the observed spectra in the 3300–4100-A region (Ottinger and Reichmuth, 1981). They consist of many bands, much more than were previously known, and, in particular, include the (1, 0) band. This band is valuable for a determination of vibrational constants of $BH^+(A)$ and $BH^+(X)$, which cannot be derived on the basis of the (0, 0) and (1, 1) bands alone, the only bands previously reported. Special efforts were therefore made to measure the wavelengths of the heads of the new bands as accurately as possible. Part of the $\Delta v = 0$ sequence was measured at 2-Å FWHM resolution, the highest so far employed in these studies, and the (1,0) band head was measured at 4-Å resolution. It was found to lie at 3491.5 Å. In a recent conventional spectroscopic study of the (0,0), (1, 1), and (1,0) bands (Ramsay

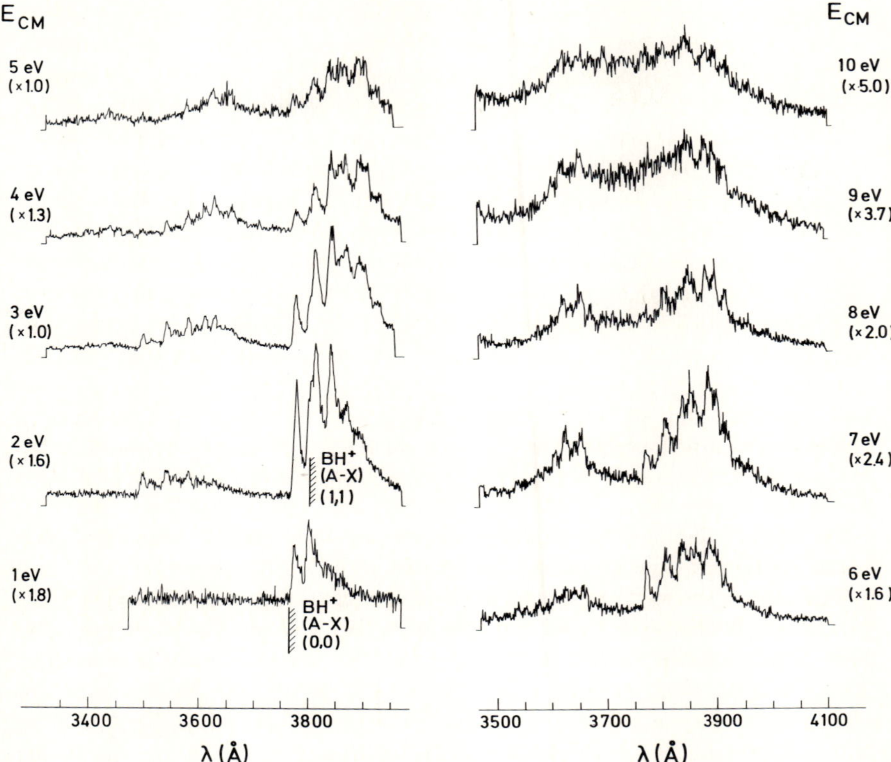

FIG. 13 Spectra of chemiluminescence from $B^+ + H_2$ reactions for ten collision energies as indicated. The ordinate scale of each spectrum has been expanded by the factors given in parentheses. [From Ottinger and Reichmuth (1981).]

and Sarre, 1982), the head of the latter was measured at 3492.33 Å, confirming the chemiluminescence results.

From the numerous band heads measured in the chemiluminescence spectra, and using the very accurate rotational constants from Almy and Horsfall (1937) and Ramsay and Sarre (1982), vibrational BH^+(A) and BH^+(X) constants were derived (Ottinger and Reichmuth, 1982). Although not nearly of the accuracy of conventional data where available, they are at present the only experimentally determined vibrational constants. On this basis, computer simulations of the spectra shown in Fig. 13 were performed which reproduced the observed structure very well, thus confirming the overall set of constants. In a similar way, vibrational constants were determined, by spectrum simulation, for the CH^+ triplet transition (Kusunoki

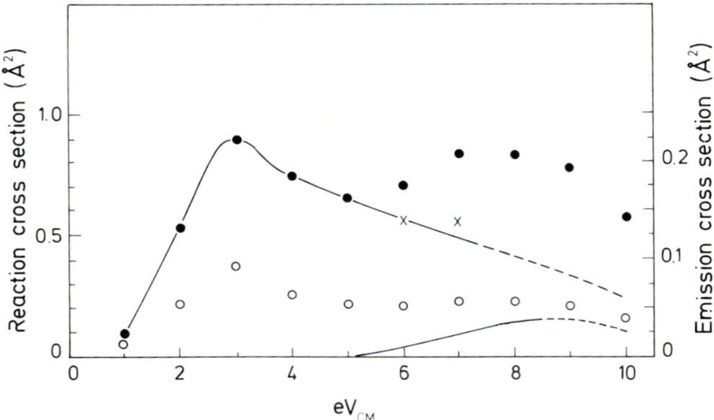

FIG. 14 Emission cross sections (right-hand ordinate scale) and reaction cross sections (left-hand ordinate scale) for $B^+ + H_2$ chemiluminescence. ○, raw data; ●, data after correction for light loss due to finite emitter lifetime. Lower curve, reaction (17); upper curve, reaction (18).

and Ottinger, 1980, 1982). These reactions are thus good examples of how ion chemiluminescence studies can contribute to spectroscopy.

Figure 14 shows the dependence of the cross section for $BH^+(A\,^2\Pi)$ formation, in absolute units, on the $B^+ + H_2$ CM collision energy. The open points are raw data obtained by integration over the measured spectra, taking into account, of course, the wavelength dependence of the detection system sensitivity as calibrated using a tungsten ribbon lamp. The absolute cross section (right-hand ordinate scale of Fig. 14) was found from the known geometry and from the manufacturer's data on the photomultiplier sensitivity. The alternative calibration method of using a luminescent collision with known cross section as a standard gives results which agree within the expected accuracy of about a factor of 2.

The raw data were then corrected for the light loss due to the finite $BH^+(A, v')$ lifetime as described below. The resulting emission cross section is marked by dots in Fig. 14 and is also referred to the right-hand ordinate scale.

The BH^+ emission exhibits a threshold close to 1 eV_{CM}, which identifies the reaction at low energies uniquely as formation of $BH^+(A)$ from metastable $B^+(^3P)$ impact [reaction (18)]. The absolute cross section for this reaction, in the low-energy regime, is then four times the emission cross section because of the 25% concentration of $B^+(^3P)$ ions in the B^+ beam. This reaction cross section is given by the upper curve in Fig. 14 and is referred to the left-hand ordinate scale.

Some light intensity is observed slightly below the nominal threshold energy of 1.17 eV_{CM} of reaction (18). This can be explained by the projectile ion cm energy spread and (after Chantry, 1971) by the thermal target motion ($\sim$0.15 and $\sim$0.6 eV_{CM}, respectively), similar to the $C^+ + H_2$ case (Fig. 7). At its maximum at 3 eV_{CM}, the cross section amounts to about 0.9 $Å^2$. This figure is based on the measured ion current leaving the collision cell. The effective ion current may have been greater and is difficult to assess because at low energies the beam spreads considerably inside the cell. The true reaction cross section is, therefore, possibly somewhat smaller than 0.9 $Å^2$ at its maximum. This is consistent with earlier mass spectrometric measurements of the total yield of BD^+ from reactions of $B^+(^3P)$ with D_2 and the kinematically determined proportion of electronically excited product. At 3 eV_{CM}, Lin *et al.* (1974b) measured a 2-$Å^2$ total reaction cross section, of which, from the above, 20%, or 0.4 $Å^2$, should be associated with product ions in the A $^2\Pi$ state.

The sharp cross-section maximum is similar to that of the $C^+ + H_2$ cross section (Fig. 7). At higher energies, the overall BH^+ emission cross section (Fig. 14) exhibits a second broad peak. This can be associated with the contribution of ground-state $B^+(^1S)$ reaction (17), to the BH^+(A) production, setting in at 5.8 eV_{CM}. Above that energy, a rough partitioning of the overall BH^+(A) emission cross section between reactions (17) and (18) is possible. At 6 and 7 eV, some guidance to this division was obtained from the spectrum simulation discussed below (Reichmuth, 1981), which permitted a reasonably safe determination of the cross section for reaction (18) beyond the onset of reaction (17) (× in Fig. 14). Above 7 eV_{CM}, the spectrum became too unstructured for this method to be useful, and a short extrapolation of the cross section (upper curve in Fig. 14) had to be used.

The (apparent) emission cross section for the ground-state $B^+(^1S)$ reaction (17) is then given by the excess of the overall emission cross section (●) over the extrapolated upper curve, both as referred to the right-hand scale. From this, the true reaction cross section for reaction (17) (lower curve, referred to the left-hand scale) is obtained by multiplication by 4/3 to allow for the 75% concentration of $B^+(^1S)$.

3. *BH^+(A) Internal State Distributions*

BH^+ has a longer radiative lifetime than CH^+ so that, in this case, the loss of light signal as a result of emission outside the observation region is more serious. Moreover, the lifetime depends rather strongly on the vibrational level v'. In the case of $B^+ + H_2$, a wider range of v' levels was included in the simulation (up to $v' = 8$) than with $C^+ + H_2$, and the kinetic energy varied by a larger factor. For these reasons, all relative and absolute BH^+(A) cross-section data have been corrected for the light loss. $BH^+(A \rightarrow X)$ lifetime measurements do not appear to have been made, but

Klein *et al.* (1982) have calculated transition probabilities. For the (0, 0) band, they give an emission Einstein coefficient of $A_{00} = 0.926 \times 10^6$ $\sec^{-1}$. Furthermore, they calculated the dependence of the electronic transition moment $M(R)$ on the internuclear distance R, and from this function one can calculate relative transition probabilities for all other bands. In the present case, this was done using the R-centroid approximation:

$$A \sim [M(R^{c}_{v',v''})]^2 \cdot q_{v',v''} \cdot \nu^3_{v',v''}, \tag{19}$$

where

$$R^{c} = \int \psi_{v'}(R) \cdot R \cdot \psi_{v''}(R)\, dR \bigg/ \int \psi_{v'}\, \psi_{v''}\, dR, \tag{20}$$

and $q_{v',v''}$, $\nu_{v',v''}$ are the Franck–Condon factor and transition frequency of the band. After normalization to A_{00}, Einstein coefficients were thus obtained for all other contributing v', v'' transitions, and absolute lifetimes of the BH^+(A) vibrational levels were obtained by summing over v'':

$$\tau(v') = 1 \bigg/ \sum_{v''} A_{v'v''}. \tag{21}$$

Lifetimes $\tau(v')$ range between 1.08 μsec for $v' = 0$ and 2.44 μsec for $v' = 8$.

A spectrum simulation yielded in a first step a raw vibrational distribution $P(v')$ (Reichmuth, 1981), which was then corrected using these $\tau(v')$ and Eq. (10). The final vibrational distributions are shown in Fig. 15. In the correction of the total $BH^+(A \to X)$ cross section as a function of energy (Fig. 14), the light loss was calculated from $\tau(v')$ for each contributing v' and weighted with $P(v')$ from Fig. 15.

Owing to the complex band structure, the rotational distributions of BH^+(A) could not be determined in the same detail as those of CH^+(A), although their dependence on v' was in this case taken into account. For simplicity, the distributions were assumed to be Boltzmann, so that for each band, at each energy, only one parameter T_{rot} was determined from the simulation. Any dependence of the band intensity on the (mean) rotation was neglected in this case; with the deep ($\sim$3-eV) BH^+(A) potential well, the centrifugal distortion is much less than for CH^+(A).* The best-fit values of T_{rot} obtained from the $BH^+(A \to X)$ spectrum simulation are plotted in Fig. 16.

The internal state distributions of BH^+(A) can now be compared to the kinematic results. In Fig. 17, the translational energy distribution of BD^+

* For example, at $T_{rot} = 5500$ K, a rather high value, the most probable J is 13, and this changes the Franck–Condon factors $q_{v'v''}$ by at most a few percent of the dominant $q_{v'v''}$.

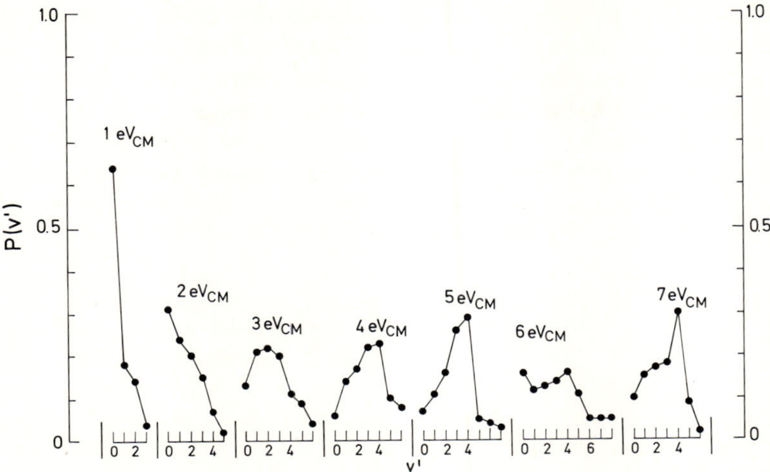

FIG. 15 Vibrational level population of $BH^+(A\,^2\Pi)$ from $B^+ + H_2$ collisions, derived from computer simulation of the spectra shown in Fig. 13 and corrected for effect of finite lifetime. At each energy, $P(v')$ is normalized such that $\Sigma P(v') = 1.0$.

is shown, as measured by Koski's group (Sondergaard *et al.*, 1979) in the forward direction at 3 eV_{CM}. From the experience with the CH^+/CD^+ systems, it is expected that there is no isotope effect of the internal energy distributions [although the distributions over the quantum levels, $P(v')$ and $P(J')$, shift toward higher v' and J' for the heavier isotope]. The bars in Fig. 17 show the vibrational level population of $BH^+(A)$ derived from the

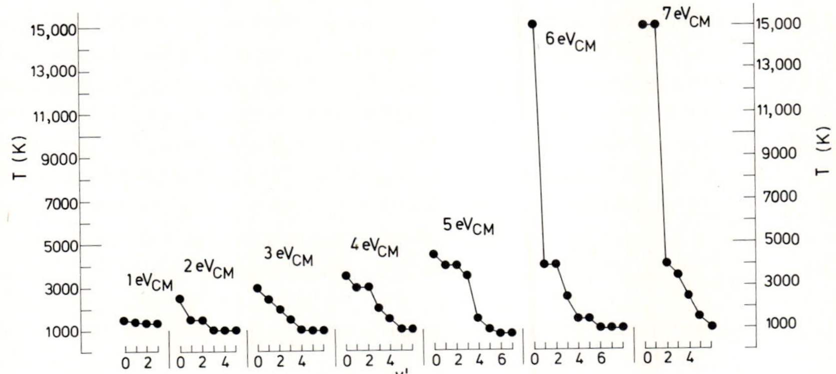

FIG. 16 Rotational temperatures of $BH^+(A\,^2\Pi)$, from $B^+ + H_2$ collisions, for seven collision energies and vibrational levels as indicated.

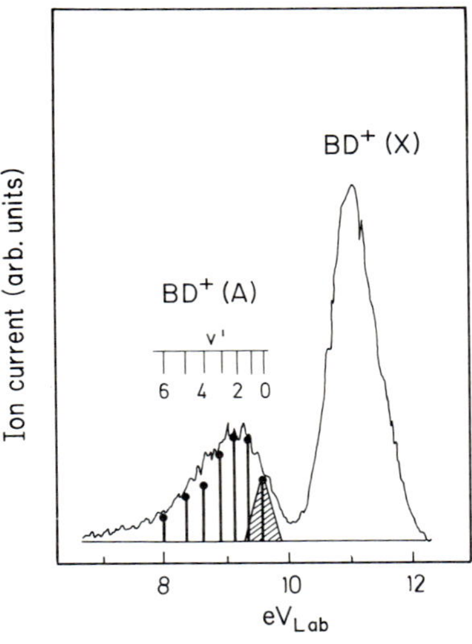

FIG. 17 Comparison of BH^+ translational energy distributions as measured for 3 eV_{CM} $B^+ + D_2$ collisions by Sondergaard *et al.* (1979) with that expected from the vibrational level population as derived from $BH^+(A)$ chemiluminescence (Fig. 15). Shaded triangle indicates the resolution of the translational energy measurement.

optical spectra, represented on the translational energy scale of Koski's experiment. The higher v' levels, having more internal energy, appear at a correspondingly smaller translational energy. The location of the bars as shown takes into account the mean rotational energy for each v' level. This shifts the pattern to the left by roughly one vibrational level spacing. For $v' = 0$, the apparatus resolution function of Koski's experiment is also shown. As can be seen, the agreement between results from these two totally different experiments is very good. It has to be borne in mind, however, that the optical experiment averages over all scattering angles while Koski's data refer to the forward direction only. The internal state distribution may well depend on the scattering angle, although—especially in the analysis of neutral beam reactive scattering—the assumption of separable cm angular and velocity distributions has often proved to be a reasonable approximation. Friedrich and Herman (1982) have given angle-averaged product kinetic energy distributions for the $B^+ + H_2$ reaction. Unfortunately, the averaging, due to various instrumental factors, blurs the separation between the $BH^+(X)$ and $BH^+(A)$ parts of the distribution. Also, these workers report data only up to 2.25 eV_{CM}. At this energy, their angle-averaged results indicate a vibrational level population nicely consistent with the optical results.

From the analysis of the forward-scattered $BD^+(A)$, Sondergaard *et al.* (1979) conclude that the reaction proceeds by a stripping-like mechanism up to 8 eV_{CM} (although at very low energy, 1.57 eV, they note deviations from the spectator stripping limit). This can again be checked from the optical spectra. According to the stripping model, the kinetic energy of B^+ relative to H is converted into BH^+ excitation. For the present case of a 1.17-eV endothermicity, we specify that, of the excess cm energy, the fraction prescribed by the stripping model should appear as "internal" (vibration plus rotation) BH^+ excitation, i.e.,

$$E_{\text{int}} = \frac{13}{24}(E_{\text{CM}} - 1.17\,\text{eV}). \tag{22}$$

In Fig. 18, this stripping dependence of E_{int} is shown, together with the average rotational and vibrational $BH^+(A)$ energies, $\langle E_{\text{rot}} \rangle$ and $\langle E_{\text{vib}} \rangle$, as derived from the optical results shown in Figs. 15 and 16. At threshold, the measured $\langle E_{\text{int}} \rangle$ is not zero as it should be, due to energy resolution effects discussed above. At 2 and 3 eV, $\langle E_{\text{int}} \rangle$ is in agreement with the stripping model, but above this point it clearly falls short of the model prediction. This need not be in conflict with the results from Koski's group, which was based on forward scattering only. The discrepancy would indicate that at larger scattering angles, which are all included in the optical results, internal product excitation is much less than at smaller angles, where the stripping approximation may still apply. Indeed, by definition a stripping reaction will always give a forward-scattered product. Any substantial sideways scattering must be caused by interaction with the third atom, thus violating the stripping assumption and channeling energy into

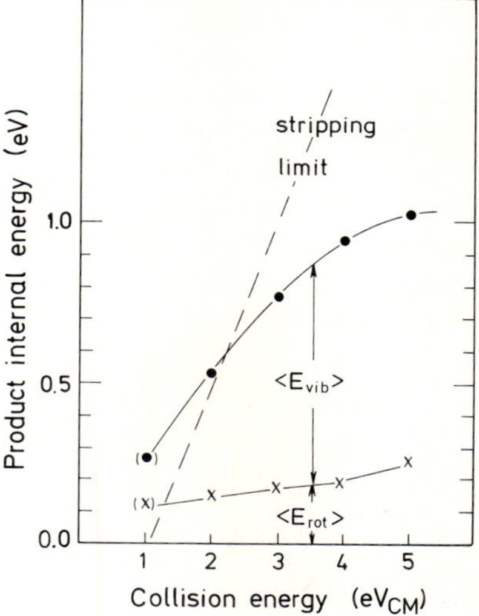

FIG. 18 Comparison of the mean internal energy $\langle E_{\text{vib}} \rangle + \langle E_{\text{rot}} \rangle$ of $BH^+(A)$, derived from the chemiluminescence spectra, with the prediction of the spectator stripping model.

relative product translation, leaving less for internal excitation. This point illustrates nicely how the optical experiments, besides giving more detail, can also complement kinematic experiments where comprehensive intensity contour maps are lacking.

A comparison of the vibrational distribution with the rotational distribution is also interesting, particularly in the light of the kinematic results. Much like in the $CH^+(A)$ case, $P(v')$ is not very different from a Boltzmann distribution at low collision energy. Figure 19 shows the $P(v')$ data at 2 eV_{CM} on a semilogarithmic scale plotted versus the vibrational energy $G_{v'}$. The first four points lie close to a line giving a 12,200 K Boltzmann distribution. (For $v' = 4$ and 5, $G_{v'}$ exceeds the nominally available energy, as is reflected by the much lower population of these levels.) At the same collision energy, the best-fit rotational temperatures are only 1000–2500 K (see Fig. 16). Obviously, the rotational and vibrational degrees of freedom of the $BH_2{}^+$ system are quite decoupled at this energy. This is in sharp contrast to the $CH_2{}^+$ system. In that case, at a comparable energy, namely, 1 eV above threshold—i.e., at $\sim$4 eV_{CM}—T_{vib} and T_{rot} of $CH^+(A)$ were found to be about the same, namely, 4000 K (Kusunoki and Ottinger, 1979a).

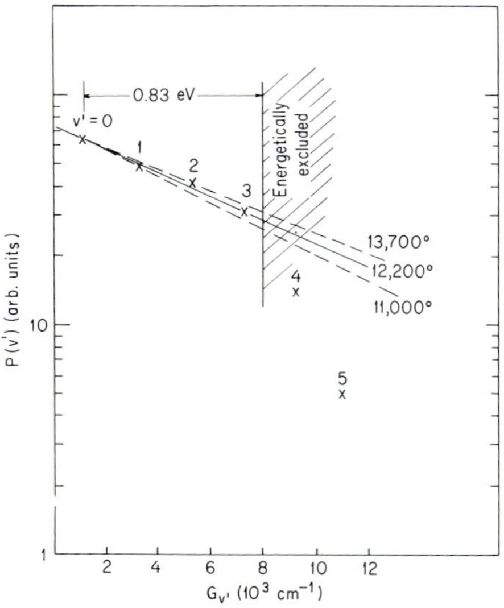

FIG. 19 Semilogarithm plot of the $BH^+(A)$ vibrational level population at 2 eV_{CM}, showing a vibrational temperature much higher than the rotational temperature at the same energy (Fig. 16).

The scattering contour diagrams [for $C^+ + H_2/D_2$ (Mahan and Sloane, 1973; Iden *et al.*, 1972) and for $B^+ + H_2/D_2$ (Friedrich and Herman, 1982; Sondergaard *et al.*, 1979)] exhibit, for both reactions, a somewhat asymmetric forward-backward peaking of product intensity, indicative of a short-lived ("osculating") complex. The different groups do not agree very well as to the extent of the asymmetry, which demonstrates the difficulties of the low-energy ion scattering experiments. Only in the BH^+ case is the electronically excited component kinematically distinguishable from the ground-state product. For this component, Sondergaard *et al.* (1979) do not report any intensity in the cm backward direction at 2.24 eV_{CM}, while Friedrich and Herman (1982), at the same energy, find a 1.9:1.0 forward-backward intensity ratio. From this, they calculate that the mean lifetime of the excited $BH_2{}^+$ complex is about 20% shorter than the period of its rotation.* The intensity asymmetry increases rapidly with the collision energy, reflecting rapid shortening of the complex lifetime. The optical spectra show how the internal degrees of freedom respond to the shortening of the lifetime. In particular, the vibrational distribution of $BH^+(A)$ with a high vibrational temperature at 2 eV_{CM}, changing to an inverted distribution at 3 eV_{CM}, clearly points to a rapid transition from partial energy equilibration to a direct reaction mechanism.

This comparison of the optical and the beam scattering results for $B^+ + H_2/D_2$ can be summarized as follows. All studies agree that close to threshold of reaction (18) the three particles $(B—H—H)^+$ interact strongly, while the reaction mechanism becomes more "direct" with increasing energy. According to the scattering contours of Sondergaard *et al.* (1979), this transition occurs at 2–3 eV_{CM}; Friedrich and Herman (1982) find a rather abrupt drop of the complex lifetime above 2 eV_{CM}, and the optical results show a sudden change of the vibrational distribution again between 2 and 3 eV_{CM}, from very "hot," but quasi-thermal, to inverted. The average internal product excitation $\langle E_{int} \rangle$ as derived from the spectra shows that even at 5 eV_{CM} the ideal spectator stripping conditions are not met. At 2 eV, however, $\langle E_{int} \rangle$ is consistent with the stripping model. This finding, which is apparently misleading, may simply indicate that an average is a less reliable criterion than the complete distribution of the quantity in question.

The rotational distributions of $BH^+(A)$ (shown in Fig. 16) exhibit a striking anomaly. The rotational temperature, which from 1 to 5 eV_{CM} increases steadily up to 4500 K, jumps at 6 eV_{CM} to 15,000 K for $v' = 0$, and

* The importance of complex formation in this reaction has also been verified theoretically by C. Zuhrt, using classical trajectory calculations (ref. 25 in Friedrich and Herman, 1982).

for $v' = 1$ as well at 7 eV_{CM}. It was not possible to simulate the 6- and 7-eV spectra without this anomalous T_{rot} distribution. The principal feature forcing $T_{rot}(v' = 0)$ to the extremely high value was the sharp (0, 0) band R-branch head at 3766 Å, which, as can be seen in Fig. 13, decreases and broadens continuously from 1 to 5 eV, but reappears suddenly at 6 eV_{CM}. The same effect was found for the $B^+ + D_2$ reaction (Reichmuth, 1981). It is very likely that this behavior is associated with the onset of reaction (17) at 6 eV. A rationalization in terms of the electronic structure will now be given.

Figure 20 shows the relevant adiabatic electronic-state correlations for the collinear reaction $B^+ + H_2$. The energy profiles along the $^1\Sigma^+$ path [reaction (17)] and the $^3\Pi$ path [reaction (18)] are, although schematic, based on DIM calculations by Muckerman (1981) and Schneider and Zülicke (1981), respectively. The former indicates the presence of a ridge in the ground-state (singlet) surface, the latter a deep basin in the excited (triplet) surface. This basin, incidentally, is responsible for the complex formation discussed above. Reaction (18) can proceed along this path adiabatically, both in collinear arrangement, as shown in Fig. 20, and in bent conformations of the triatomic system, where the $^3\Pi$ surface simply splits into two surfaces: $^3A'$ and $^3A''$. Reaction (17), on the other hand, requires a nonadiabatic transition between the lowest and the second singlet surface. In $C_{\infty v}$ geometry (Fig. 20), these have different symmetry, $^1\Sigma^+$ and $^1\Pi$, and do not interact. Reaction (17) becomes allowed only in bent conformations and involves a transition from the $^1A'$ ground state to the $^1A'$ component of the $^1\Pi$ surface (dashed arrow in Fig. 20). The ground-state potential ridge will facilitate this transition, although the shape of the ridge will depend on the conformation angle. In summary, then, reaction (17) is restricted to the approach of B^+ at some angle relative to the H_2 molecule axis, while reaction (18) can proceed at any orientation. The preferred bent reaction geometry of reaction (17) will favor high rotational product excitation, such as is observed at the onset of this reaction channel.

This reasoning is an example of how observed microscopic reaction characteristics can be derived from stereochemical properties which, in turn, are based on the electronic symmetry of the system. Thus, it comes close to the ultimate goal of studies of this nature, which is an understanding, and maybe

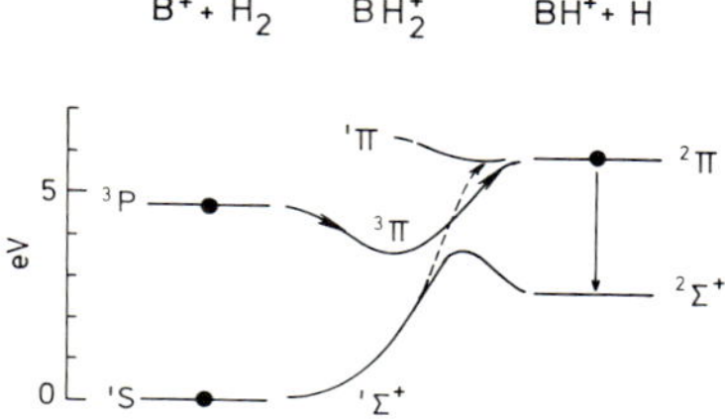

FIG. 20 Electronic-state correlations for the linear $(B—H—H)^+$ system. $B^+(^3P)$ reacts adiabatically; $B^+(^1S)$ reacts nonadiabatically (and only in nonlinear conformations) to give $BH^+(A\ ^2\Pi)$.

prediction, of the reaction attributes founded on the electronic properties of the reactants. Whether, in fact, in this particular reaction electronic symmetry is the only factor causing the exceedingly high product rotational excitation from reaction (17) remains to be explored by classical trajectory studies of this system. With quantum chemical calculations of the energy surfaces becoming available at the present time, this is an extremely interesting and promising system for further experimental and theoretical studies.

4. $BH^+(B\,^2\Sigma^+ \rightarrow X\,^2\Sigma^+)$ *Emission*

At collision energies $E \geqq 8\ eV_{CM}$, a previously unknown band system of BH^+ has been observed (Ottinger and Reichmuth, 1981). It was interpreted, by analogy with the isoelectronic molecule BeH, as a $B\,^2\Sigma^+ \rightarrow X\,^2\Sigma^+$ transition. From the observed threshold, $B^+(^1S)$ was identified as the reactant ion, and this was confirmed by a direct measurement as summarized in Eq. (12). The reaction then reads

$$B^+(^1S) + H_2 \rightarrow BH^+(B\,^2\Sigma^+) + H. \tag{23}$$

Interestingly, $B^+(^3P)$ ions do not contribute significantly to $BH^+(B)$ formation, while $BH^+(A)$ is produced from both species, although with far greater cross section from $B^+(^3P)$. There appears to be a general preponderance for a reactant ion in a given electronic state to give a specific electronic state on the product side. A similar case was found in the $C^+ + H_2$ system, where $C^+(^2P)$ forms only $CH^+(A\,^1\Pi)$ but not $CH^+(b\,^3\Sigma^-)$ (although this would be spin-allowed).

The newly discovered $BH^+(B\,^2\Sigma^+)$ state has the same asymptote $B(^3P) + H(^2S)$ as the $A\,^2\Pi$ state but a shallower minimum lying at much larger internuclear distances. These potential energy curves have recently been calculated ab initio by a number of workers (Guest and Hirst, 1981; Klein *et al.*, 1982; Vázquez, 1981; Kusunoki, 1982), confirming the qualitative predictions about the location of the minimum. Moreover, Guest and Hirst (1981) as well as Klein *et al.* (1982) have calculated spectroscopic constants for the $BH^+(B \rightarrow X)$ transition, and the latter workers also give the $BH^+(B \rightarrow X)$ transition moment function and from it absolute Einstein emission coefficients for 31 bands. In Fig. 21, a calculated bar diagram of this transition is compared with a measured low-resolution survey spectrum. The bar diagram is adapted from Klein *et al.* (1982). At the wavelengths of each Q-branch band head, the calculated relative emission intensity is plotted, weighted with the apparatus sensitivity function which is also shown.

The very good overall agreement between the bar diagram and the measured spectrum contour confirms the earlier interpretation of this part of the spectrum as $BH^+(B \rightarrow X)$. Two superimposed atomic lines, at 2497 Å and

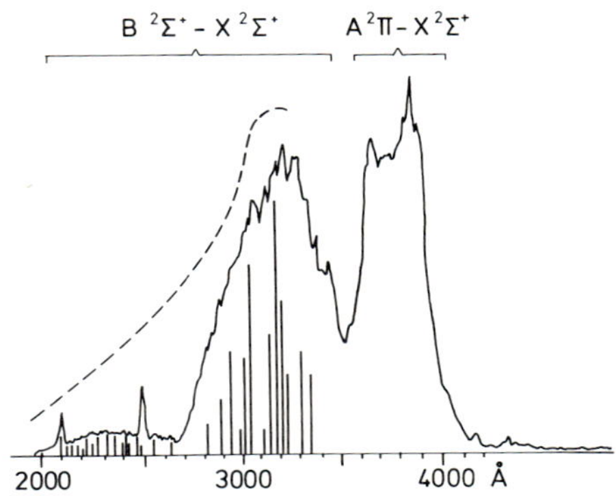

FIG. 21 Low-resolution spectrum of $B^+ + H_2$ chemiluminescence (10 eV_{CM}; 26-Å FWHM), showing two band systems. In the region of the newly discovered B–X system, the relative detection sensitivity is shown by the dashed curve. Bars give B–X band intensities as calculated by Klein *et al.* (1982).

2089 Å, are the two boron atom resonance lines $B(^2S, ^2D \to ^2P)$ originating from luminescent charge transfer (Reichmuth, 1981):

$$B^+(^1S, ^3P) + H_2 \to B(^2S, ^2D) + H_2{}^+. \tag{24}$$

In the 3500–4000-Å region, the $BH^+(A \to X)$ system dominates. However, continuum $BH^+(B \to X)$ emission may contribute here, too. Klein *et al.* (1982) have concluded from their calculated Einstein coefficients that, because summation over the bound X-state levels does not lead to convergence, significant $BH^+(B)$ emission into X-state continuum levels is to be expected. This emission would underlie the $A \to X$ system and may account for the very unstructured appearance of this system at high energies. In fact, Fig. 13 shows that the $A \to X$ bands start to become blurred at just about the "appearance potential" (8 eV_{CM}) of the $B \to X$ system.

The $B \to X$ bands fall into two broad groups: 2000–2600 Å and 2800–3400 Å. The v', v'' numbering given by Klein *et al.* (1982) shows that the former group is characterized by low v'' ($v'' = 0$ to 3), the latter by high v'' (6 to 9). v' ranges from low to high numbers in both groups, the shorter wavelengths in each group, of course, generally belonging to higher v'($v' \sim 5$ to 8). Figure 22 shows a series of $BH^+(B \to X)$ spectra measured at higher resolution (8-Å FWHM) (Reichmuth, 1981). The $B \to X$ system appears now clearly resolved into bands. A detailed analysis has not yet been made, but, for a general orientation, the v' numbering from Klein *et al.* (1982) is again shown. The 2497-Å boron line is included and indicates the resolution. For intensity reasons, only the group of bands at longer wavelengths could be

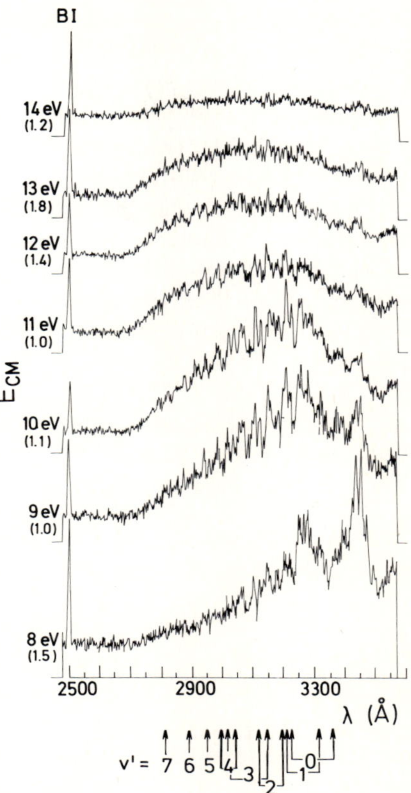

FIG. 22 High-resolution $BH^+(B–X)$ chemiluminescence spectrum. Wavelengths of predominant emission from some v' levels as calculated by Klein *et al.* (1982) are indicated. The atomic boron line at 2497 Å results from luminescent charge transfer.

measured at this resolution. The 8-eV spectrum is at long wavelengths overlapped by $BH^+(A \rightarrow X)$ emission, while at higher energies this contribution is much less. The trends of the relative intensity distribution within the $BH^+(B \rightarrow X)$ system with increasing energy are very instructive. The regions at shorter wavelengths (~ 2800 Å), corresponding to emission from high v', become progressively accentuated in going from 8 to 13 eV. This indicates the "normal" pattern of increasing $P(v')$ with increasing collision energy. Note, however, that the emission of this group of bands appears to break off sharply at ~ 2675 Å. It is easy to see why this limit is never exceeded toward the blue, even at the highest energies: 2675 Å corresponds to a transition energy of 4.63 eV, which is exactly the separation of the asymptotes $B^+(^3P) + H$ and $B^+(^1S) + H$ of the BH^+ $B\,^2\Sigma^+$ and $X\,^2\Sigma^+$ states, respectively. 2675-Å emission then represents transitions between the extreme long-range limbs of the upper and the lower BH^+ $^2\Sigma^+$ states, which is consistent with the observed dense vibrational structure in this region. According to the spectra shown in Fig. 22, the BH^+ B-state is apparently populated up to its highest vibrational levels, even at low collision energy ($\sim 9\ eV_{CM}$). Increasing the energy then cannot populate additional levels, such as was observed with $CH^+(A)$ and $BH^+(A)$; instead, the relative weight of the high v' levels increases with energy.

Figure 23 illustrates the situation pictorially. The vertical arrows represent the B → X bands as calculated by Klein *et al.* (1982). They are drawn, in accordance with the Franck–Condon principle, near the classical v', v'' turning points. The separation into a group at longer and one at shorter wavelengths is clearly evident. They consist, respectively, of transitions from the right and from the left limb of the $B\,^2\Sigma^+$ curve, into the high and the low parts of the $X\,^2\Sigma^+$ curve. The curved arrows symbolize the trend of the B-state vibrational excitation with increasing energy, pushing, on the right limb, toward the 2675-Å blue limit of the long wavelength group. There is an analogous blue limit of the short wavelength group at 1921 Å, corresponding to the 6.45-eV energy separation between the highest B-state

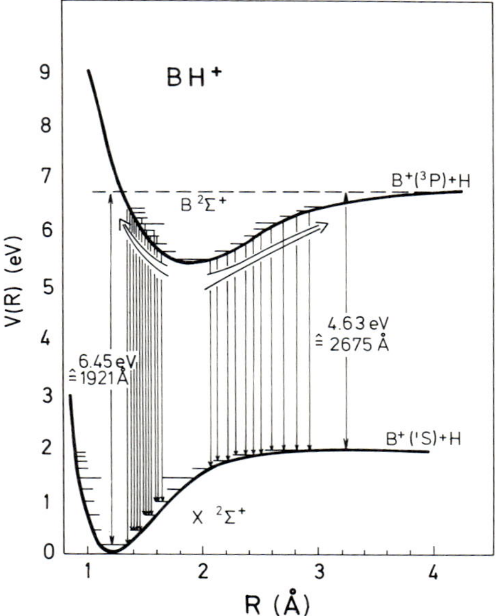

FIG. 23 $B\,^2\Sigma^+$ and $X\,^2\Sigma^+$ potential energy curves of BH^+ and v', v'' transitions as calculated by Klein *et al.* (1982). These fall into two groups, as seen in Fig. 21; the longer wavelength group is also shown in Fig. 22. The curved arrows emphasize the trend toward increasing population of high v' levels with increasing collision energy, evident from Fig. 22. 2675 Å and 1921 Å are the short wavelength limits of the spectra shown in Figs. 22 and 21, respectively.

vibrational levels and X, $v'' = 0$. For 10 eV_{CM}, this limit can in fact be seen in Fig. 21, although it could not be determined over a whole range of collision energies.

C. B^+ + Hydrocarbons

Particularly rich chemiluminescence spectra have been obtained by bombarding hydrocarbon molecules with B^+ ions (Reichmuth, 1981). Some preliminary results are shown in Fig. 24. With CH_4 as target gas, BH^+ ($A \rightarrow X$) emission was observed at an energy as high as 20 eV_{CM}, exhibiting a spectrum contour which resembles that measured with $B^+ + H_2$ at 6 eV_{CM} (see Fig. 13). In addition, in the 4250–4400-Å region, emission from neutral BH in the $A\,^1\Pi \rightarrow X\,^1\Sigma^+$ system appears, an example of a reaction according to Eq. (lb). With C_3H_8 as a target, no BH^+ emission is seen. Instead, the BH($A \rightarrow X$) system dominates, and, in addition, BH($b\,^3\Sigma^- \rightarrow a\,^3\Pi$) appears. Note that these two band systems are analogous to their isoelectronic counterparts $CH^+(A \rightarrow X)$ and $CH^+(b \rightarrow a)$ discussed above. In addition, BH($C'\,^1\Delta \rightarrow A\,^1\Pi$) emission may underlie the $A \rightarrow X$ bands, and there is evidence of some emission near 3450 Å, perhaps due to BH($B\,^1\Sigma^+ \rightarrow A\,^1\Pi$). The atomic line at 4177 Å is B($^2D \rightarrow {}^2P$) in second grating order.

It is interesting that, in going from H_2 to CH_4 to larger hydrocarbons, the emissions observed shift from the ionic to the neutral product, i.e., from reaction (1a) to (1b). A similar trend has also been found with other projectile ions. C^+, while giving predominantly CH^+ emission with H_2, gives

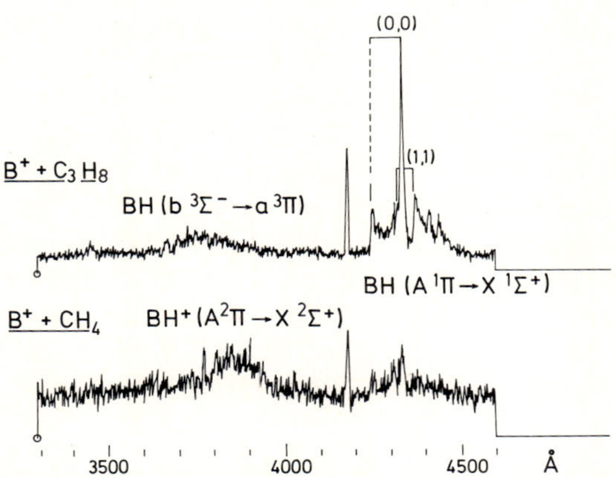

FIG. 24 Chemiluminescence spectra of $B^+ + CH_4$ and $B^+ + C_3H_8$ reactions (20 eV_{CM}), showing transition from ionic to neutral hydride emitters.

TABLE II

Predominance of Ionic (i) versus Neutral (n) Hydride Reaction Product from B^+, C^+, N^+, O^+ RH Collisions

	R IP (eV)	H 13.6	CH_3 9.8	C_3H_7 8.1
BH	9.8	i	$i + n$	n
CH	10.7	i	n	n
NH	13.8	$i + n$	n	n
OH	13.2	$i + n$	n	n

strong CH(A → X), (B → X), and (C → X) emission with various hydrocarbon molecules (Kusunoki, 1978). $O^+ + H_2$ collisions have been studied by Harris and Leventhal (1976), who observed overlapping spectra of both OH^+(A) and OH(A). $O^+ + C_2H_4$, C_2H_6, and C_3H_8, on the other hand, resulted in OH(A) emission only (Kusunoki *et al.*, 1979).* With N^+, the situation parallels that of O^+:$N^+ + H_2$ gives both NH(A) (Kusunoki *et al.*, 1976; Kusunoki and Ottinger, 1979b,c) and NH^+ ($B\,^2\Delta$) (Kusunoki and Ottinger, 1984), while with CH_4 through C_3H_8 only NH(A) is obtained.

It appears that all these observations can be rationalized in terms of a simple rule. Let us consider the reacting system at the state just prior to the final separation of the products, ignoring the electronic excitation. It appears reasonable to assume that of the two possible charge distributions, $AB^+ + C$ or $AB + C^+$, the energetically lower one will be favored. In other words, the most loosely bound electron will normally attach itself to the species with the higher ionization potential. This rule is somewhat analogous to the so-called "Stevenson's rule" in mass spectrometry. In Table II, this prediction is specified for the examples given above, i.e., with atom A = boron, carbon, nitrogen, oxygen; atom B = hydrogen; species C = hydrogen, methyl, propyl. Based on the balance of the ionization potentials given, the expected charge arrangements "i" ("n") for "hydride product formed as ion (neutral)" are shown in the table. "$i + n$" has been entered whenever the ionization potentials of AB and C are very similar, so that comparable amounts of ionic and neutral hydride are expected. It turns out that the prediction is in all cases in agreement with the observation. The electronic excitation energy of the luminescent products can be disregarded for this qualitative argument because it will disfavor both "i" and "n" by

* CH_4 was found to be an exception in this experiment, giving OH(A) predominantly in a two-step process, $O^+ + CH_4 \rightarrow OH^+ + CH_3$, $OH^+ + CH_4 \rightarrow OH(A) + CH_3{}^+$, with a peculiarly low rotational excitation of OH(A).

approximately the same amount, since the observed excitation energies of ionic and neutral hydrides are of a similar magnitude.

D. N^+ + Hydrocarbons

Luminescent collisions of N^+ ions with CH_4, C_2H_4, C_2H_6, and C_3H_8, as well as with H_2, have been extensively studied in Göttingen (Kusunoki *et al.*, 1976; Kusunoki and Ottinger, 1979b, c). The observed product was in all cases NH(A $^3\Pi$), radiating into the NH(X $^3\Sigma^-$) ground state:

$$\begin{aligned} &N^+ + RH \rightarrow NH(A\,^3\Pi) + R^+, \\ &R = H, CH_3, C_2H_3\ C_2H_5, C_3H_7. \end{aligned} \tag{25}$$

The reaction with H_2 is $\sim$4 eV endothermic, with CH_4 and C_2H_4 about thermoneutral, and with C_2H_6, C_3H_7 $\sim$1.5 eV exothermic.

1. Internal State Distributions

The results of the spectrum analysis (Kusunoki and Ottinger, 1979b) can be summarized as follows. NH(A) was predominantly observed in the vibrational level $v' = 0$. The population ratio of $v' = 1$ relative to $v' = 0$ increases from 1.6 to $\sim$10 eV_{CM} and then levels off at about $\sim$0.5 (except for H_2, where it appears to reach 1.0). The $v' = 2$ population is very small throughout. Interpretation of the measured vibrational excitation is difficult because NH(A) is strongly predissociated at about $v' = 2$. The predissociation was very clearly observed in the rotational distributions. Their peak lies with CH_4 at low rotational quantum numbers ($N' \sim 7$) at 1.6 eV_{CM}, shifting up to the predissociation limit ($N' = 25$–31) at 8 eV_{CM}. Increasing the collision energy further fills in $P(N')$ at low N', making the distribution almost rectangular with a sharp cutoff at the predissociation limit. For $v' = 1$, predissociation occurs above $N' = 15$ and hence suppresses the overall $v' = 1$ population severely at energies as low as 5 eV_{CM}. This is a good illustration of a general problem with the chemiluminescence method: Predissociation may preclude a measurement of the "original" internal product excitation, such as would pertain if only one unperturbed electronically excited state were contributing. The dynamical analysis was therefore in the present case limited to low collision energies ($\lesssim 5$–10 eV_{CM}). It revealed that for reaction with CH_4 the stripping model is a very good approximation. The internal product energy as derived directly from the spectrum was found to increase linearly with the collision energy in just the proportion which the stripping model prescribes. For C_2H_4, C_2H_6, and C_3H_8, on the other hand, a product excitation higher than that calculated for the spectator stripping case was measured.

2. Cross Sections

In the rest of this section, we will concentrate on the NH luminescence cross section (Kusunoki *et al.*, 1976; Kusunoki and Ottinger, 1979c), particularly for reactions of N^+ with CH_4, C_2H_4, and C_2H_6. At the lowest energies the cross section is quite large, on the order of 1 Å^2. The energy dependence of the cross section is shown in Fig. 25. There is a sudden drop, by almost one order of magnitude, between 10 and 15 eV_{CM}, followed by a shoulder and finally a nearly exponential falloff up to ~ 150 eV_{CM}.

At the highest energies, where the cross section is only ~ 0.01 Å^2, a secondary process became noticeable and was corrected for (see Fig. 25). It is ascribed to the reaction sequence

$$\begin{aligned} N^+ + RH &\rightarrow N + RH^+ \\ N + RH &\rightarrow NH(A) + R. \end{aligned} \tag{26}$$

The neutralization step may, at least partly, take place in the gas effusing from the collision chamber entrance slit. The resulting beam of fast atoms enters the collision cell and can give rise to interfering emissions whenever the ionic reaction under study is very weak. This has also occasionally been observed in the Göttingen apparatus at very low ion kinetic energy because the neutral atom beam, not being affected by the ion retarding field, can then become comparatively strong relative to the ion beam. These interfering processes are, however, always recognizable on account of the quadratic dependence on the target gas pressure and can be subtracted.

The solid lines in Fig. 25, giving the corrected cross section, show the remarkable fact that the hydrocarbon reactions (25) can proceed up to very high collision energies, probably the highest ever observed for any ion–molecule reaction. The overall shape of the energy dependence curve suggests clearly two distinct mechanisms.

At low energies, a direct N^+–H interaction occurs, which, in the case of CH_4, assumes the limiting form of spectator stripping. For any such direct mechanism there is a certain maximum kinetic energy, beyond which product dissociation sets in. In the clearcut case of CH_4, the calculated spectator stripping limit agrees well with the observed drop of the $\sigma(E)$ function, and for the other target gases the sudden decrease of $\sigma(E)$ is certainly also associated with termination of the direct mechanism.

The high-energy portion of the $\sigma(E)$ curves has been analyzed in detail by Kusunoki and Ottinger (1979c). Reaction in this regime was explained

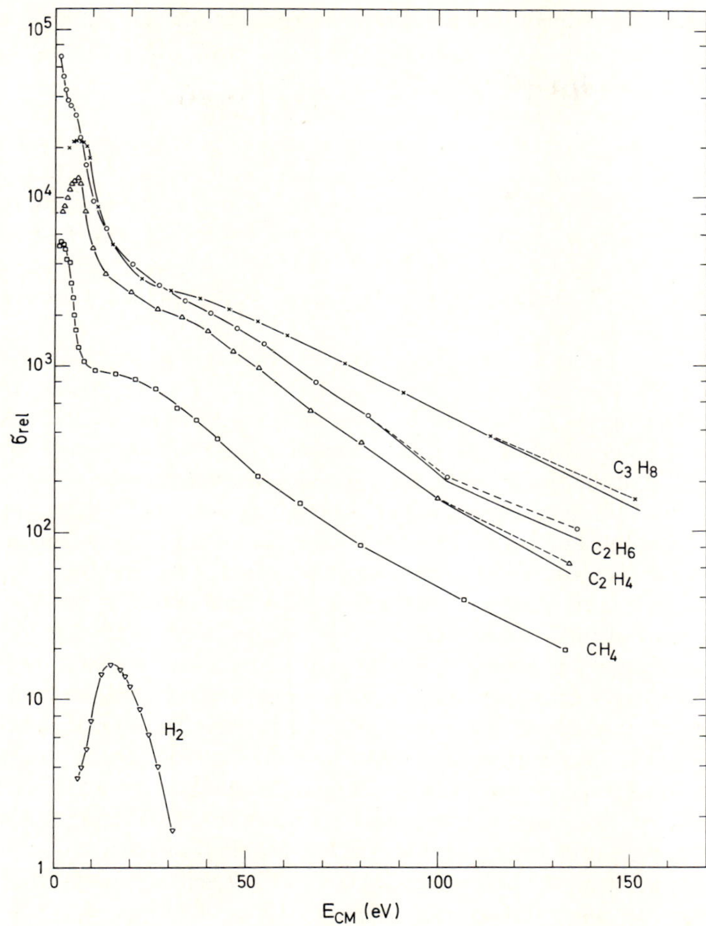

FIG. 25 Relative cross section for $NH(A\,^3\Pi)$ formation from reactions $N^+ + RH$ [Eq. (25)] as a function of collision energy. A low-energy stripping-like and a high-energy knockout mechanism can be distinguished with the four hydrocarbons, the latter exhibiting a nearly exponential falloff of cross section with energy. [From Kusunoki and Ottinger (1979c).]

as an intramolecular two-step process. The primary interaction was assumed to be between the projectile ion and a carbon atom. If this collision is sufficiently central, then the struck carbon atom will take over most of the projectile's kinetic energy, thus slowing the projectile down to such an extent that it can subsequently capture one of the hydrogen atoms of the same molecule to form a stable NH(A) product. The captured hydrogen atom, for simplicity, is assumed to remain at rest, unaffected by the impact on the carbon atom.

This mechanism hinges on the fact that N^+ and C have similar masses, enabling an efficient energy transfer. The model is thus supported by the observation of analogous high-energy luminescent reactions of B^+ and O^+ ions, also having a similar mass as C, with the same hydrocarbon molecules. BH emission as shown in Fig. 24 was found to persist up to very high kinetic energy (Reichmuth, 1981), and the energy dependence of OH emission from O^+ impact (Kusunoki *et al.*, 1979) showed two separable contributions similar to the N^+ case.

Further evidence for the model is the observation of molecular fragments $C(^1P)$, $H(n = 4)$, and $CH(A\,^2\Delta,\ B\,^2\Sigma^-,\ C\,^2\Sigma^+)$ by means of their emission (Kusunoki and Ottinger, 1979c; Kusunoki *et al.*, 1979), which has an onset at about 20–50 eV_{CM} and thereafter a fairly flat energy dependence. Dissociation of the target molecule is a necessary consequence of the model assumption. Taking the formation of excited fragments as an indicator of the overall fragmentation implied in the model, the energy dependence of fragmentation is then quite consistent with the assumed mechanism.

Final confirmation of the high-energy reaction model comes from trajectory calculations, as described in detail by Kusunoki and Ottinger (1979c). This work is here summarized and somewhat extended. In general terms, the quasi-exponential falloff of the reaction cross section reflects the requirement of a sufficiently central N^+–C collision. In the calculation, any such collision was considered to lead to reaction if it resulted in slowing down the projectile to less than a specified critical energy, which is related to the NH(A) dissociation energy and is a constant. This requirement leads to a decreasing effective target area with increasing impact energy. Assuming a realistic, exponentially repulsive N^+ –C interaction potential, the projectile energy loss at a given impact parameter and energy was determined from computer trajectory calculations. The resulting reactive cross section, as a function of energy, is compared in Fig. 26 with the high-energy portion of the measured $N^+ + CH_4$ cross section. It can be seen that the agreement is quite good from the normalization point at 50 eV_{CM} up to 250 eV_{CM}.

Having thus confirmed the model for CH_4, the measured cross sections for C_2H_6 and C_2H_4 can be discussed. A comparison between the two latter ones is straightforward. The two molecules differ only in the number of H atoms available; therefore, the $N^+ + C_2H_6$ cross section should be $6/4 = 1.5$ times that of the $N^+ + C_2H_4$ cross section, independent of energy. This expectation is borne out by the experiment, as the cross-section ratio plotted in Fig. 27 shows.

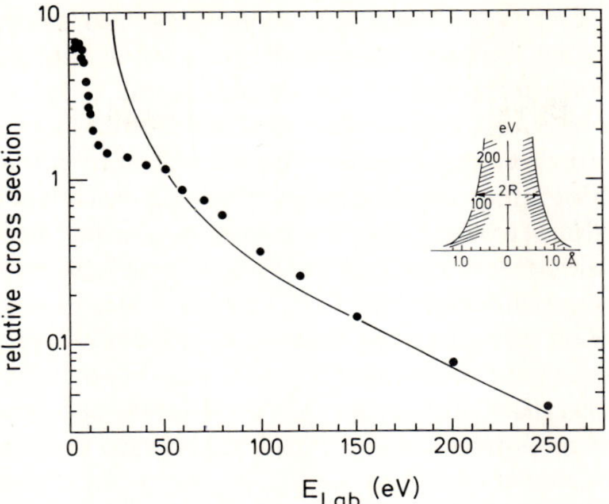

FIG. 26 Cross section for $N^+ + CH_4$ chemiluminescent reaction. (Fig. 25), (●) compared with results from a model calculation (curve), based on trajectories in a repulsive potential (see text). The exponential potential is shown in the inset.

A comparison between C_2H_4 and CH_4 is more difficult. One might expect a ratio of 2:1, based on the number of C atoms. However, this would overestimate the ratio since the second C atom will, for certain orientations of the C_2H_4 molecule, be "hidden" behind the first—in its "shadow," so to speak. On the other hand, the four H atoms will also not be equally efficient in the two molecules. For CH_4, they are all equivalent, being directly bound to the struck C atom. For C_2H_4, however, this is only true for two

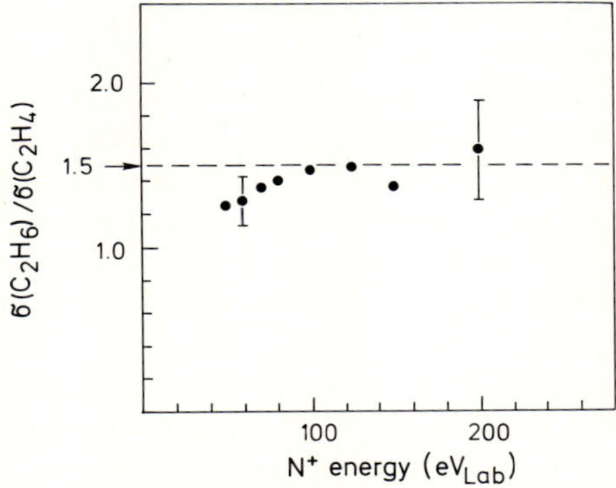

FIG. 27 Measured cross-section ratio for NH(A) production from $N^+ + C_2H_6$ and C_2H_4, showing proportionality with the number of hydrogen atoms.

H atoms. The other two are removed from the site of the primary impact. The assumption made above, that the H atoms stay approximately at rest during the dissociative collision, will only be fulfilled for a fraction of all collisions. This fraction will be relatively higher for the two H atoms not bound to the struck C atom because the C—C bond will absorb some of the shock resulting from the N^+ impact. Owing to this "buffer" action, two of the four H atoms of C_2H_4 will, therefore, have a better chance of being captured than the other two, which for their part will have the same chance as all four H atoms in CH_4.

The buffer effect will, therefore, tend to increase the C_2H_4-to-CH_4 cross-section ratio above the value of 2:1 while the shadow effect will tend to decrease it below this value. Figure 28 shows the measured cross-section ratios. They are, in contrast to Fig. 25, plotted versus the lab system energy because the reaction model as stated above refers to impact of N^+ on a stationary target and reaction of the decelerated projectile with a stationary H atom. Figure 28 shows that, at low energies, the buffer effect is considerable and more than compensates the shadow effect. At higher energies, however, the buffer effect appears to become unimportant. This is plausible because very violent shocks will be transmitted through the C—C bond.

Beyond these qualitative statements, the buffer effect has so far not been checked systematically by model calculations. It is doubtful if a realistic quantitative description is possible within this simple framework. The shadow effect, however, is calculable. Kusunoki and Ottinger (1979c) have given results for two different approaches. One is simply based on geometrical considerations, assigning to each C atom a cross section which

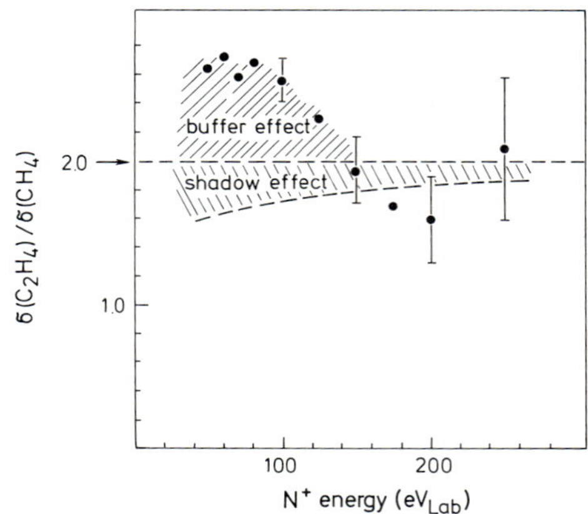

FIG. 28 Measured cross-section ratio for NH(A) production from $N^+ + C_2H_4$ and CH_4, showing deviations (explained in the text) from the value of 2.0 expected on the basis of the number of carbon atoms.

decreases with energy in the way shown in Fig. 26. The result was a C_2H_4/CH_4 cross-section ratio which increases from 1.57 at 50 eV_{lab} to 1.83 at 250 eV_{lab}. Thus, the shadow effect becomes less important at high energy because the target sizes are then smaller and overlap correspondingly less. A more refined calculation gave very similar results. In this approach, the trajectory of the projectile was actually followed through the entire scattering sequence, including events where first one and then the second carbon atom is hit. Hard-sphere scattering was assumed. Computer-generated "reactivity maps" can serve to classify various scattering sequences. An example is shown in Fig. 29. In a plane perpendicular to the direction of flight of the projectile, the effective C_2H_4 target areas are shown for ten orientation angles of the molecular axis. 0° refers to orientation of the molecule parallel to the projectile velocity vector. The outer circles indicate the hard-sphere target atom radius, decreasing with energy. Crosshatched areas show the range of impact parameters which lead to reaction after collision with only one carbon atom. In addition, two other types of collision

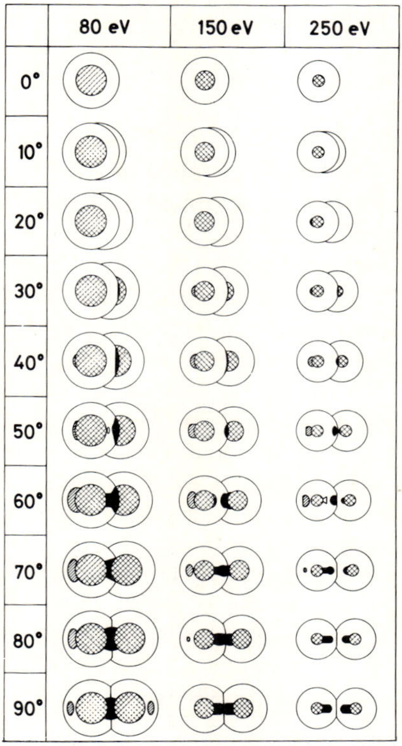

FIG. 29 Reactive areas of a C_2H_4 (or C_2H_6) molecule, viewed along the direction of the incident N^+, for ten different angles between this direction and the molecular axis. Crosshatching indicates direct reaction and its shadow effect. Black areas indicate "double" reaction, and simple hatching indicates recoil reactive collisions (see text). [From Kusunoki and Ottinger (1979c).]

make significant contributions. One, termed "double collision," requires an impact occurring close to the center of the molecule (black areas in Fig. 29). Then, scattering off the first atom and subsequently off the second leads to an overall energy loss sufficient for reaction. Finally, glancing impact (banana-shaped hatched areas) can also lead to reaction; although the projectile energy loss is small, the struck C atom recoils off the other C atom into a direction close to that of the scattered projectile. The relative energy between H atoms attached to this C atom and the projectile can then be small enough to enable reaction ("recoil collision").

After averaging over the orientation angles, it is found that the effect of the double and recoil collisions is most pronounced at high energies. At 250 eV_{lab}, a cross-section ratio $\sigma_{C_2H_4}/\sigma_{CH_4} = 2.45$ is calculated for the combination of all three scattering mechanisms, compared to 1.85 for the direct mechanism alone. The apparent rise of the measured cross-section ratio (Fig. 28) at the highest energies may thus be due to the two more involved types of scattering illustrated by Fig. 29. While quantitative agreement with the experiment would require much more elaborate model calculations, the interpretative value of the rather crude reactivity maps is considerable.

IV. CONCLUSIONS

The study of ion–molecule reactions by means of spontaneous optical emission of the products affords unique possibilities in regard to measurement of internal product excitation. The method is to be considered as complementary to the more traditional ion beam scattering studies. Using the reactions of carbon, boron, and nitrogen ions with hydrogen and hydrocarbons as examples, the scope as well as specific problems of this method have been outlined.

In the $C^+ + H_2$ case, the product (CH^+) rotational-vibrational excitation distribution has been analyzed in great detail, probably more than for any other reaction. Conclusions regarding the mechanism derive from the results, although a full interpretation of the wealth of data will be possible only by means of a specific theoretical treatment such as classical trajectory calculations.

With the $B^+ + H_2$ example, the emphasis has been on the electronic structure of the reacting system. In this very simple case, it appears possible to trace certain product characteristics directly to fundamental properties of the reactants. In particular, an anomalously high rotational energy seems to derive from the overall electronic symmetry. The $B^+ + H_2$ reaction has also been discussed as a case where new spectroscopic information on a molecular ion (BH^+) has been obtained through the technique of ion chemiluminescence. Vibrational molecular constants have been derived, and a new band system has been found.

The example of N^+ + hydrocarbon reactions has been discussed with emphasis on the relative cross section at high collision energies. This contrasts with the other, low-energy reactions in that here the details of the potential energy hypersurface are unimportant, and the reactions can be described in terms of hard- or soft-sphere collisions. Simple and more elaborate collision models have been discussed.

The remaining chemiluminescent ion–molecule reactions listed in Table I, like the ones selected here, give emission in the visible or near-UV range. It is rather striking that no vacuum–UV (VUV) emitter is included, although, with ions, the available energy is not a limiting factor. Attempts at observing VUV chemiluminescence have so far been unsuccesful. Perhaps the competition of other channels in the regime of high electronic excitation suppresses the emission. This should be checked systematically in future work. Chances for VUV emission are probably best if the reactant ions are already in a high-lying excited state. This is supported by the observation of the $H_2(a\,^3\Sigma_g{}^+ \rightarrow b\,^3\Sigma_u{}^+)$ VUV continuum emission as a result of energy transfer from highly excited metastable Ar^+ ions colliding with ground state H_2 (Brandt and Ottinger, 1979).

In addition to an extension of the wavelength range, expansion of the ion chemiluminescence work can be expected in two main directions: On the one hand, the chemical scope will broaden. Metal ions and metal atoms should be included, and polyatomic target molecules have hardly been studied. It would also be interesting to bombard simple atoms like N or O with diatomic ions and study the formation of the same product from different reactants, e.g., to compare $N^+ + NO \rightarrow N_2{}^+ + O$ (Brandt and Ottinger, 1973, 1975) with $NO^+ + N \rightarrow N_2{}^+ + O$.

The second direction of development could be in the area of more precise definition of the physical parameters of a collision, such as orientation of the angular momentum, and scattering angle. The former could be obtained from polarization measurements; the latter would involve photon/ion coincidences. These very refined experiments would be especially valuable with systems for which the quantum chemical description is well advanced. Quite generally, as has been emphasized throughout this article, chemiluminescence experiments should always be conducted in close contact with the theoretical research because it is only by the combined efforts of experiment and theory that a complete understanding of ionic reactions, at least in simple cases, will ultimately be achieved.

ACKNOWLEDGMENTS

It is a pleasure to acknowledge the very substantial contributions of my co-workers. Most of the results on C^+ and N^+ collisions are due to the untiring efforts of Prof. I. Kusunoki during his stay in Göttingen as a visiting scientist. The $B^+ + H_2$ experiments, largely unpublished as yet, form the main part of the thesis of Mr. J. Reichmuth. I am indebted to him and to Dr. Z. Herman, Prague, for numerous discussions on the $B^+ + H_2$ system. Thanks are also due to Dr. D. Brandt and Dr. J. Simonis, who competently assembled the apparatus and perfected it in earlier experiments, as well as to Dr. S. Zimmermann and Mr. D. Neuschäfer, who contributed at later stages. I would like to thank Dr. P. Rosmus, Frankfurt, for making his work on BH^+ available to me prior to publication. Financial support from the Deutsche Forschungsgemeinschaft and the Alexander-von-Humboldt Stiftung is gratefully acknowledged.

REFERENCES

Ajello, J. M. (1971). *J. Chem. Phys.* **55**, 3158–3168.

Almy, G. M., and Horsfall, R. B. (1937). *Phys. Rev.* **51**, 491–500.

Appell, J., Brandt, D., and Ottinger, Ch. (1975). *Chem. Phys. Lett.* **33**, 131–135.

Brandt, D. (1976). Thesis, Bonn, published as Bericht 4/1979, Max-Planck-Institut für Strömungsforschung, Göttingen.

Brandt, D., and Ottinger, Ch. (1973). *Chem. Phys. Lett.* **23**, 257–261.

Brandt, D., and Ottinger, Ch. (1975). *In* "Electronic and Atomic Collisions" (J. S. Risley and R. Geballe, eds.), pp. 580–581. Univ. of Washington Press, Seattle, Washington.

Brandt, D., and Ottinger, Ch. (1979). *Phys. Rev. A* **19**, 219–230.

Brandt, D., and Polanyi, J. C. (1978). *Chem. Phys.* **35**, 23–34.

Brandt, D., Ottinger, Ch., and Simonis, J. (1973). *Ber. Bunsenges.* **77**, 648–655.

Brinkmann, U., Schmidt, V. H., and Telle, H. (1980). *Chem. Phys. Lett.* **73**, 530–535.

Carrington, A., and Ramsay, D. A. (1982). *Physica Scripta* **25**, 272–274.

Case. D. A., and Herschbach, D. R. (1976). *J. Chem. Phys.* **64**, 4212–4222.

Chantry, P. J. (1971). *J. Chem. Phys.* **55**, 2746–2759.

Cormier, M. J., Hercules, D. M., and Lee, J. (eds.) (1973). "Chemiluminescence and Bioluminescence," pp. 143–208. Plenum Press, New York.

Ding, A. M. G., Kirsch, L. J., Perry, D. S., Polanyi, J. C., and Schreiber, J. L. (1973). *Faraday Discuss. Chem. Soc.* **55**, 252–276.

Friedrich, B., and Herman, Z. (1982). *Chem. Phys.* **69**, 433–442.

Gerratt, J. (1981). Private communication.

Grieman, F. J., Mahan, B. H., O'Keefe, A., and Winn, J. S. (1981). *Discuss. Faraday Soc.* **71**, 191–203.

Guest, M. F., and Hirst, D. M. (1981). *Chem. Phys. Lett.* **80**, 131–134.

Harris, H. H., and Leventhal, J. J. (1976). *J. Chem. Phys.* **64**, 3185–3191.

Harris, H. H., Crowley, M. G., and Leventhal, J. J. (1975). *Phys. Rev. Lett.* **34**, 67–70.

Herbst, E., and Delos, J. B. (1976). *Chem. Phys. Lett.* **42**, 54–57.

Herbst, E., Schubert, J. G., and Certain, P. R. (1977). *Astrophys. J.* **213**, 696–704.

Herman, Z. (1982). Private communication.

Huber, K. P., and Herzberg, G. (1979). "Molecular Spectra and Molecular Structure," Vol. IV. Van Nostrand-Reinhold, New York.

Iden, C. R., Liardon, R., and Koski, W. S. (1972). *J. Chem. Phys.* **56**, 851–854.

Jones, C. A., Wendell, K. L., Kaufman, J. J., and Koski, W. S. (1976). *J. Chem. Phys.* **65**, 2345–2352.

Jones, C. A., Wendell, K. L., and Koski, W. S. (1977). *J. Chem. Phys.* **66**, 5325–5329.
Klein, R., Rosmus, P., and Werner, H. J. (1982). *J. Chem. Phys.* **77**, 3559–3570.
Kusunoki, I. (1978). Unpublished results.
Kusunoki, I. (1982). Private communication.
Kusunoki, I., and Ottinger, Ch. (1979a). *J. Chem. Phys.* **71**, 4227–4241.
Kusunoki, I., and Ottinger, Ch. (1979b). *J. Chem. Phys.* **70**, 699–709.
Kusunoki, I., and Ottinger, Ch. (1979c). *J. Chem. Phys.* **70**, 710–721.
Kusunoki, I., and Ottinger, Ch. (1980). *J. Chem. Phys.* **73**, 2069–2075.
Kusunoki, I., and Ottinger, Ch. (1982). *J. Chem. Phys.* **76**, 1845–1852.
Kusunoki, I., and Ottinger, Ch. (1984). *J. Chem. Phys.* (in press).
Kusunoki, I., Ottinger, Ch., and Simonis, J. (1976). *Chem. Phys. Lett.* **41**, 601–605.
Kusunoki, I., Ottinger, Ch., and Zimmermann, S. (1979). *J. Chem. Phys.* **71**, 894–908.
Kusunoki, I., Sakai, S., Kato, S., and Morokuma, K. (1980). *J. Chem. Phys.* **72**, 6813–6815.
Light, J. C. (1979). *In* "Atom-Molecule Collision Theory" (R. B. Bernstein, ed.), pp. 647–653. Plenum Press, New York.
Lin, K., Cotter, R. J., and Koski, W. S. (1974a). *J. Chem. Phys.* **60**, 3412–3416.
Lin, K., Watkins, H. P., Cotter, R. J., and Koski, W. S. (1974b). *J. Chem. Phys.* **60**, 5134–5135.
Lindemann, E., Frees, L. C., Rozett, R. W., and Koski, W. S. (1972). *J. Chem. Phys.* **56**, 1003–1004.
Lipeles, M., Swift, R. D., Longmire, M. S., and Weinreb, M. P. (1970). *Phys. Rev. Lett.* **24**, 799–801.
Mahan, B. H., and Sloane, T. M. (1973). *J. Chem. Phys.* **59**, 5661–5675.
Massey, H. (1976). "Negative Ions," 3rd ed., p. 222. Cambridge Univ. Press, London and New York.
Menzinger, M. (1980). *Adv. Chem. Phys.* **42**, 1–61.
Muckerman, J. T. (1981). Private communication.
Ottinger, Ch. (1976). "Chemolumineszente Ionen-Molekül-Reaktionen," Bericht 18/1976. Max-Planck-Institut für Strömungsforschung, Göttingen.
Ottinger, Ch. (1978). *In* "Electronic and Atomic Collisions" (G. Watel, ed.), pp. 639–662. North Holland. Publ. Amsterdam.
Ottinger, Ch., and Reichmuth, J. (1981). *J. Chem. Phys.* **74**, 928–933.
Ottinger, Ch., and Reichmuth, J. (1984). To be published.
Ottinger, Ch., and Simonis, J. (1975). *Phys. Rev. Lett.* **35**, 924–927.
Ottinger, Ch., and Simonis, J. (1978). *Chem. Phys.* **28**, 97–112.
Ottinger, Ch., Reichmuth, J., and Yang, M. (1983). *Chem Phys.* **76**, 61–69.
Polanyi, J. C., and Woodall, K. B. (1972). *J. Chem. Phys.* **56**, 1563–1572.
Prisant M. G., Rettner, C. T., and Zare, R. N. (1981). *J. Chem. Phys.* **75**, 2222–2230.
Ramsay, D. A., and Sarre, P. J. (1982). *J. Chem. Soc. Faraday Trans. 2* **78**, 1331–1338.
Reichmuth, J. (1981). Diplomarbeit, Max-Planck-Institut für Strömungsforschung, Göttingen.
Rettner, C. T., and Simons, J. P. (1978). *Chem. Phys. Lett.* **59**, 178–182.
Rettner, C. T., and Simons, J. P. (1979). *Faraday Discuss. Chem. Soc.* **67**, 329–342.
Rosmus, P. (1982). Private communication.
Sakai, S., Kato, S., Morokuma, K., and Kusunoki, I. (1981). *J. Chem. Phys.* **75**, 5398–5409.
Schneider, F., and Zülicke, L. (1981). *I. Eur. Conf. At. Phys., Heidelberg, Book of Abstracts* (J. Kowalski, G. zu Putlitz, and H. G. Weber, eds.), pp. 1108–1109. European Physical Society, Heidelberg, 1981.
Sondergaard, N. A., Sauers, I., Jones, A. C., Kaufman, J. J., and Koski, W. S. (1979). *J. Chem Phys.* **71**, 2229–2234.
Sondergaard, N. A., Sauers, I., Kaufman, J. J., and Koski, W. S. (1981). *J. Phys. Chem.* **86**, 533–537.

Teloy, E. (1978). *In* "Electronic and Atomic Collisions" (G. Watel, ed.), pp. 591–603. North Holland Publ., Amsterdam.

Thomas, E. W. (1972). "Excitation in Heavy-Particle Collisions." Wiley (Interscience), New York.

Tiernan, T. O., and Lifshitz, C. (1981). *Adv. Chem. Phys.* **45**, 81–224.

Turner, B. R., Rutherford, J. A., and Compton, D. M. J. (1968). *J. Chem. Phys.* **48**, 1602–1608.

Truhlar, D. G., and Dixon, D. A. (1979). *In* "Atom-Molecule Collision Theory" (R. B. Bernstein, ed.), pp. 595–636. Plenum Press, New York.

Truhlar, D. G., and Muckerman, J. T. (1979). *In* "Atom-Molecule Collision Theory" (R. B. Bernstein, ed.), pp. 505–561. Plenum Press, New York.

Tsuji, M., Susuki, T., Endoh, M., and Nishimura, Y. (1982). *Chem. Phys. Lett.* **86**, 411–414.

Vázquez, G. J. (1981). Private communication.

Wren, D. J., and Menzinger, M. (1975). *J. Chem. Phys.* **63**, 4557–5559.

Yoshimine, M., Green, S., and Thaddeus, P. (1973). *Astrophys. J.* **183**, 899–902.

Zehnle, L., Clemens, E., Martin, P. J., Schäuble, W., and Kempter, V. (1978). *J. Phys. B* **11**, 2133–2141.

Chapter 24
The emission of light from excited products of charge exchange reactions

J. J. LEVENTHAL
DEPARTMENT OF PHYSICS
UNIVERSITY OF MISSOURI—ST. LOUIS
ST. LOUIS, MISSOURI

I. INTRODUCTION

One of the most powerful techniques for studying inelastic collisions between ions and neutral atoms or molecules is that in which analysis of radiation from the decay of excited reaction products is used to determine the energy partitioning among the available final states. This technique has

ISBN 0-12-120803-6

been particularly useful in ion beam experiments designed to study electron exchange in ion/atom or ion/molecule collisions. Although luminescence experiments of this type have been used for many years, primarily at kinetic energies in the kiloelectron volt range, it is only within the past 8 to 10 years that experiments at energies as low as 1 eV have been performed. The discussion in this chapter will, therefore, be restricted to these more recent experiments and will generally focus on work in which the relative kinetic energy was below ~ 1000 eV.

Much of the information on final state distributions from low-energy inelastic collisions that had been obtained before luminescence was utilized for this purpose was obtained from experiments that depended upon kinematic analysis of the collision products. Such experiments require velocity and, usually mass, analysis of the product ions. While photons cannot be detected as efficiently as ions, nor can they be as easily steered to a detector as ions, the ability to analyze light from the decay of excited products offers the advantage that the data are of extremely high precision. These data then lead to very accurate determinations of the final state distributions. The precision to which the product internal energy may be specified depends upon the accuracy to which the photon energy may be measured. The Einstein relation

$$E = hc/\lambda, \tag{1}$$

where E is the energy of a photon of wavelength λ, h is Planck's constant, and c is the speed of light, may be written

$$E(\text{in eV}) \cong 12{,}400/\lambda(\text{in Å}). \tag{2}$$

The uncertainty in the photon energy corresponding to the uncertainty in the wavelength measurement is

$$\Delta E \simeq (12{,}400/\lambda^2)\Delta\lambda, \tag{3}$$

so that, for quite modest optical resolution, say 50 Å, the excitation energy may be specified, for yellow light (~ 6000 Å), to ± 0.017 eV. Thus, if the light signal is of sufficient magnitude to permit spectral analysis, an extremely accurate determination of the product internal energy may be made.

As a simple, but important, example of the information that experiments of this type can yield, consider charge transfer in He^+/He collisions. Because this is *symmetric* charge transfer, i.e., the reactant ion is the same species as the once ionized neutral reactant, it has been customary to assume that electron transfer proceeds exclusively by the resonant process for which the product He^+ and He are in their respective ground states. If indeed this is the case, then of course no radiation would be observed. Figure 1 is a luminescence spectrum from He^+/He collision at 875 eV, and Fig. 2 is an energy level diagram of HeI showing the transitions that are observed (in

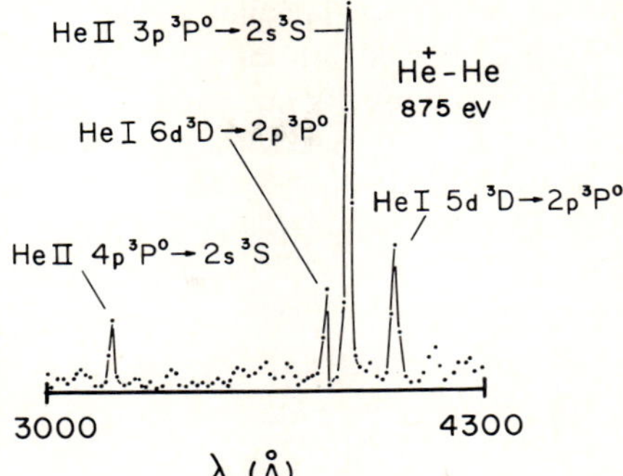

FIG. 1 Portion of the spectrum from 875-eV (lab) He^+/He collisions. The resolution is 10 Å full width half maximum (FWHM); data are not corrected for relative spectral efficiency.

the near UV–visible–near IR) in 1000-eV He^+/He collisions (Bearman and Leventhal, 1978). The spectral lines all result from the decay of excited product He atoms. The observation of these lines is a clear demonstration that some He^+/He collisions are taking place with the conversion of as much as

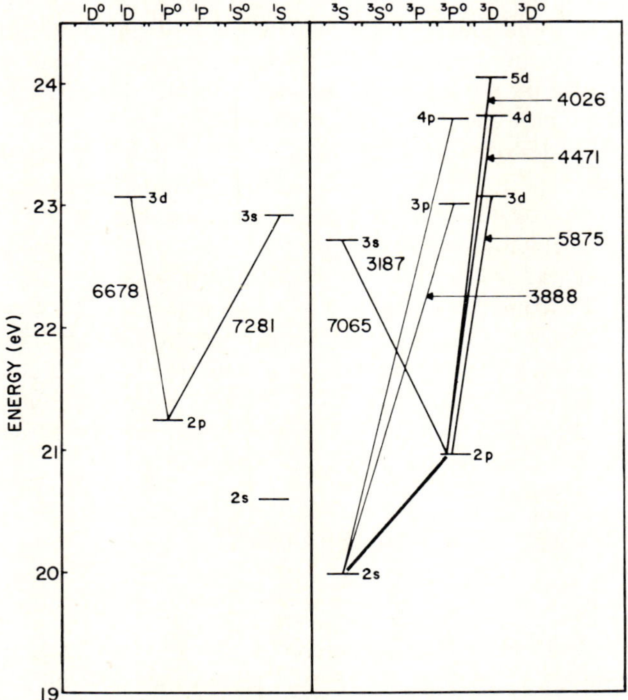

FIG. 2 Partial term diagram of HeI showing the near-uv visible, near-ir transitions observed in 875-eV (lab) He^+/He collisions. Wavelengths are in angstrom. [From Bearman and Leventhal (1978).]

23 eV from translational to internal energy. In fact, based on the strength of the observed luminescence signal, these data show that, assuming only charge exchange is occurring, roughly 1% of the total charge transfer cross section is taken up by these highly endothermic processes, a somewhat surprising conclusion. Therefore, by performing what is now a relatively simple luminescence experiment, a long-standing presumption was unequivocally tested.

While these experiments, which combine low-energy ion beam techniques with those of emission spectroscopy, can provide a wealth of information, the restrictions of this method should also be pointed out. Obviously, the combined ion beam/luminescence method cannot be used when the products are exclusively in either ground or metastable (nonradiating) states. This discrimination also makes it impossible to include these states when tabulating final state distributions from luminescence data. Therefore, for this purpose, the apparatus must either be expanded to include components for kinematic analysis, or data from other experiments must be invoked. This restriction is usually not serious because, in many cases, either the nonradiating states are unimportant or the required data already exist. However, an advantage of the discrimination against products in nonradiating quantum states is that it is possible in some instances to study minor reaction channels without interference from the predominant channel. This last is a decided advantage when it is the minor channel that is of interest.

In the remainder of this chapter some of the experimental considerations that are important for these studies will be discussed. Emphasis will be placed on those features that are unique to the acquisition of luminescence data and the treatment of these data. In this context the definitions of "light" and "luminescence" will be expanded beyond their usual meanings to include all wavelengths from the so-called soft x-ray region, ~250 Å, to the near IR, ~9000 Å. The major part of this chapter will, however, be taken up with a discussion of specific experiments, the information from which could only be obtained by analyzing the light from decay of excited reaction products. Emphasis will be placed on the pertinence of these experiments to the general subject of inelastic atomic and molecular processes.

II. EXPERIMENTAL CONSIDERATIONS

A. *Apparatus*

The determination of state-specific cross sections in charge exchanging collisions requires a marriage of the techniques of ion and molecular beam experiments with those of emission spectroscopy. A schematic diagram of one of the apparatuses in use for such studies is shown in Fig. 3. In this ap-

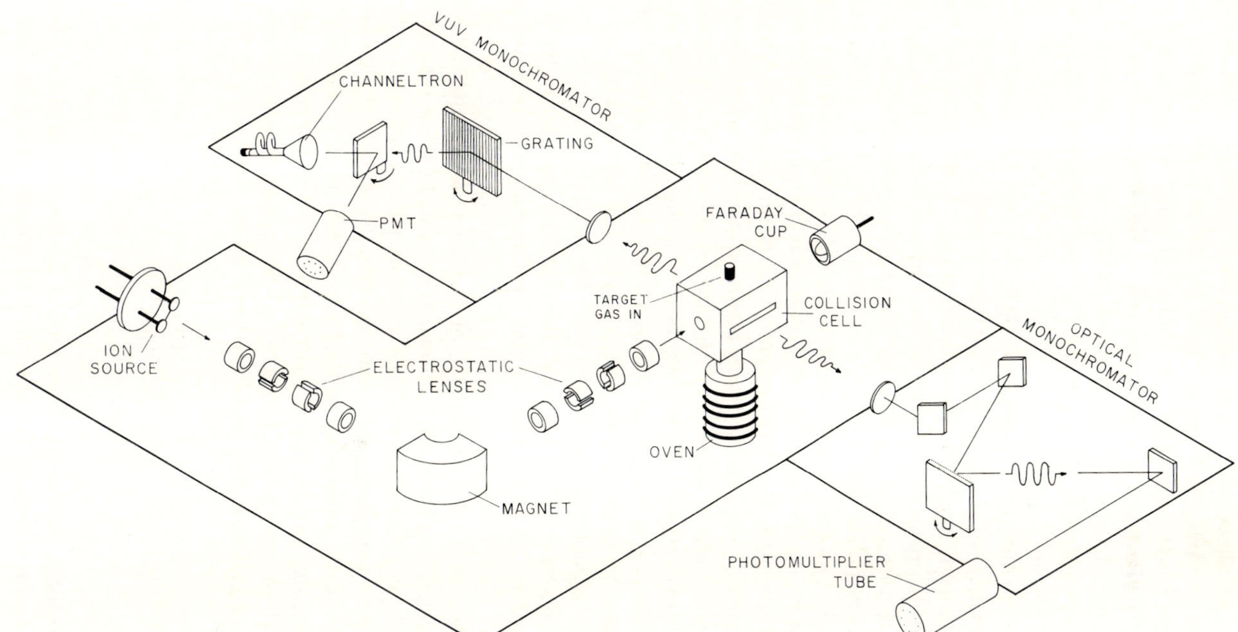

FIG. 3 Schematic diagram of an apparatus used to observe radiation from decay of excited products of ion/neutral collisions. This particular apparatus is equipped with two monochromators, making it possible to obtain two collision-produced spectra simultaneously.

paratus, ions are produced in an electron impact source and focused into a beam with a set of cylindrical electrostatic lenses. After magnetic mass selection, the beam is accelerated or retarded with a second set of cylindrical electrostatic lenses and enters a collision cell at the desired kinetic energy; ions that pass through the cell are then collected with a Faraday cup and the current monitored with an electrometer. The neutral reactants may be either gaseous molecules or metal atoms. If a gas is used, molecules are admitted to the collision cell through a variable leak valve and the pressure is maintained at some constant value below 1 μm, which corresponds to about 3×10^{13} atoms per cubic centimeter. At these low densities, usually only single-collision processes are important, although cases have occurred in which the pressure had to be lowered further to maintain this condition (Harris and Leventhal, 1976).

If the neutral reactant is to be a metal atom, it is necessary to use the atomic beam oven located beneath the collision cell, in which case a crossed-beam experiment is performed. Atom densities in the beam are roughly one order of magnitude lower than those employed in the beam-gas experiments so that single-collision conditions again prevail. Although the literature is replete with atomic beam source designs, the special requirements of the optical experiments impose additional constraints on this source. Since single photon counting is employed in this work, it is necessary to eliminate as much of the blackbody radiation from the oven as possible. Alternatively, the ion beam can be electrically modulated and photons detected synchronously, but this procedure more than doubles the data acquisition time.

Radiation resulting from the decay of excited reaction products is dispersed and detected at right angles to the plane defined by the ion and atomic beams. In this particular apparatus, two computer-controlled monochromators, which may be used simultaneously, are employed—one on each side of the plane. One monochromator, an f/3.5 Ebert mount instrument, together with focusing lenses and a cooled photomultiplier tube (PMT), comprises the "optical system" and is used for analysis of light with wavelength in the approximate range 1800–9000 Å. The other monochromator is a normal incidence f/4.5 vacuum ultraviolet (VUV) device which is useful for wavelengths longer than about 400 Å. Various detectors have been used with this "VUV system," depending upon the wavelength region of interest. Below about 1200 Å, a channeltron is used, but above this wavelength any of several solar blind PMTs have been employed. In other experiments, the normal incidence VUV spectrometer has been replaced with an f/7.8 Seya–Namioka grazing incidence instrument for dispersal of radiation in the 100–400 Å range.

In one mode of operation the monochromator setting is fixed and the collision energy varied to obtain the energy dependence of a given emission, thus obtaining "emission cross sections." In the other mode of operation the ion beam energy is fixed and a collision-produced spectrum assembled by scanning the monochromator stepwise through the wavelength range of interest. The wavelength steps $\Delta\lambda$ are chosen to be consistent with the resolution (slit size) employed in that particular experiment. Photon counts are recorded for a preset time at each wavelength setting, yielding a count rate versus wavelength collision-produced spectrum. Examination of these spectra and the changes that occur as the ion kinetic energy is varied reveals changes in the energy partitioning among final states of the products, as well as information on the conversion of kinetic energy to internal energy.

Because photon signals are frequently quite low, it is sometimes necessary to signal average by repetitively scanning the wavelength range of interest. Figure 4 is an example of signal averaging used to recover emissions in the

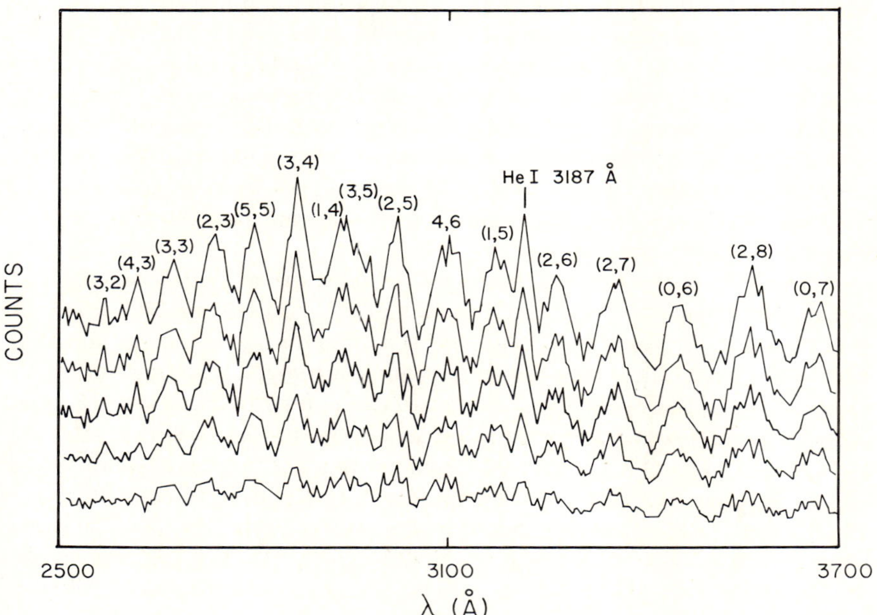

FIG. 4 Portion of the spectrum from 200-eV (lab) He^+/O_2 collisions, illustrating the signal averaging technique employed in the luminescence experiments described in the text. The (v', v'') designation identifies bands from the $O_2{}^+(A \rightarrow X)$ second negative emission system. Although impossible to identify after the first scan, these bands are clearly distinguishable after the fifth scan.

second negative system of O_2^+ resulting from 200-eV He^+/O_2 collisions. These emissions are very weak compared to those from the first negative system and from some atomic oxygen lines (Harris *et al.*, 1974), but are shown to illustrate the signal averaging technique. Each scan in Fig. 4 took approximately 1 hr, so that after 5 hrs a clearly identifiable spectrum was obtained.

B. *Treatment of the Data*

After having acquired both collision-produced spectra and the kinetic energy dependences of particular emissions, there are several corrections to the raw data that may be necessary. The most obvious of these is that of taking proper account of the spectral efficiency of the optical or VUV systems. This problem is, of course, not peculiar to experiments of this type but is an important consideration in any experiment in which radiation spectra are the principal data. The procedures to be followed and the necessary equipment for proper calibration are described in a number of texts on optical and vacuum ultraviolet spectroscopy (Samson, 1967). Standard lamps, which may be used for calibration of the optical system, are available commercially. By combining the calibration data using a deuterium lamp with those obtained using a tungsten halide lamp, the entire range of the optical system may be covered and an *absolute* calibration obtained. This absolute calibration, together with knowledge of the ion beam current, neutral atom, or molecule density, and scattering path length lead to a determination of absolute emission cross sections.

Calibration of the VUV system is less straightforward than for the optical system. Frequently the only practical method of obtaining a relative calibration is by folding together the grating and detector efficiencies as supplied by the manufacturers. One of the important advantages of having the simultaneous use of the optical and VUV spectrometers is that, using a known branching ratio from a given excited state and the absolute calibration of the the optical system, it may be possible to determine the absolute calibration of the VUV system at a few wavelengths. For example (Barrett and Leventhal, 1980), if $He(3p\,^1P^0)$ can be produced in situ, as it is in He^+/H_2 collisions, the known branching ratio of this state to 537 Å and 5015 Å, together with the absolute efficiency of the optical system, leads to an absolute efficiency of the VUV system at 537 Å.

It was emphasized above that when the kinetic energy dependence of the radiation at a given wavelength is determined, the resulting data are *emission* cross sections. The quantities of interest are usually excitation cross sections, i.e., the cross sections for production of a given state in a particular reaction. The transformation from emission cross section to excitation cross section necessitates consideration of both branching from and cascading into the state in question. For example, referring to Fig. 2, it can be seen that production of He(3d ^{3}D) results in 3d ^{3}D → 2p ^{3}P^0 5875 Å radiation. However, in order to convert the observed emission cross section for 5875 Å into an excitation cross section for production of He(3d ^{3}D), it is necessary to either detect the 3d ^{3}D → 3p ^{3}P^0 18.6-μ radiation or have knowledge of the branching ratio from the upper state. In this experiment, the infrared line cannot be detected, but the branching ratio is known (Wiese *et al.*, 1966).

As noted above, it is also necessary to consider cascading when converting emission cross sections to excitation cross sections. It is possible that the upper level of an observed transition may be populated by cascading from some higher-lying level rather than directly in the collision process. If the radiation from the higher-lying level can be detected, this effect may be accounted for and the appropriate corrections made. In the example above, this is not possible for *all* transitions that terminate on He(3d ^{3}D). For example, the 1.9-μm 4f ^{3}F^0 → 3d ^{3}D line is beyond the range of the optical system so that inferences about the possible population of such levels must be drawn from either the remaining data or from theory.

In addition to the considerations outlined above, there are circumstances under which an additional correction of the data may be necessary. This correction is peculiar to beam experiments and is a consequence of the fact that the emitting species may be moving with nearly uniform velocity. Such a case can occur in charge transfer when the (fast) neutralized ions are excited. In this case the momentum of the emitters is virtually the same as that of the reactant ions so that these excited species may move out of the field of view of the monochromator before emitting radiation. The extent to which this effect is important in any given experiment depends on both the velocity of the emitters and the lifetime of the state.

Different residence times in the detection region as a function of kinetic energy may be approximately accounted for by assuming that emitting products are formed uniformly in the collision cell and integrating across the dimension of the slit (Harris and Leventhal, 1976). This procedure yields an expression for N', the number of emitters that are observed, in terms of the number formed N_0, the velocity v, the lifetime of the excited state τ, and the slit length l:

$$N'/N_0 = [1 + (v\tau/l)\exp(-l/v\tau) - (v\tau/l)]. \tag{4}$$

Figure 5 is a graph of the fraction observed, N'/N_0, versus laboratory kinetic energy, assuming that the emitting species are hydrogen atoms in the 2p state. Thus, if L_α emissions from, say, 2000 eV-H^+/X charge transfer are detected, roughly 40% of the H(2p) escape the field of view of the VUV system before emitting at 1216 Å.

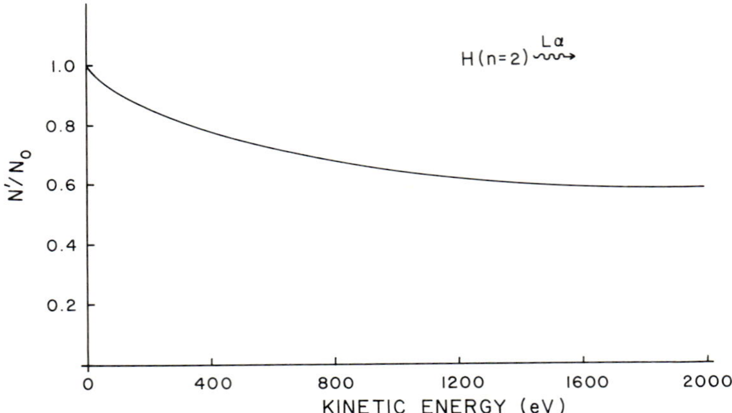

FIG. 5 Graph of Eq. (4) showing the ratio of the number of emitters observed N', to the number formed N_0, if the emitting species is H(2p). The lifetime τ is 1.6×10^{-9} sec, and the path length l is taken to be 2 cm.

III. ION–ATOM SYSTEMS

The interpretation of experiments in which both the projectile ion and the target species are atoms is usually much simpler than for even triatomic systems because the molecular nature of the collisions may be easily visualized in two dimensions. In other words, potential energy curves describe the various electronic states rather than potential energy surfaces or hypersurfaces. Of course, the apparent simplicity of the diatomic systems is usually somewhat nullified by either imprecise or nonexistent knowledge of the curves. Nevertheless, even when the potential energy curves are not known, it is frequently possible to estimate their character by consideration of appropriate isoelectronic analogs, thus facilitating interpretation of the data.

The most studied diatomic collision systems are those in which rare gas ions interact with rare gas atoms, primarily because a crossed-beam configuration is not required. A number of both homonuclear and heteronuclear rare gas systems have been studied, principally in the laboratories of Tiernan and of Isler, and the results are described in a recent review (Tiernan and Lifshitz, 1981).

Because of the importance of rare gas ion/metal atom charge transfer as laser pumping mechanisms, several systems of this type have been investigated. Although, in many cases, the search for new laser transitions in some specified wavelength range was the motivation for these experiments, a great deal of information of a fundamental nature has been extracted from the data. Helium ions are especially important in charge transfer lasers because the high recombination energy (RE) of He^+ ions makes it energetically possible to form *excited* metal ions in the charge transfer process. In fact, laser action has been produced on many transitions between energy levels of metal ions using charge transfer excitation (CTE) as a means of producing the requisite population inversion (Willett, 1974; Davis and King, 1975).

An example of the type of fundamental information that can be obtained by observing light from ion/atom collisions is provided by the results of He^+–Mg experiments (Myers and Leventhal, 1978). Because the sum of the first two ionization potentials of Mg is lower than the He^+ RE(24.6 eV) the charge transfer reaction,

$$He^+(1s) + Mg(3s^2\,{}^1S) \rightarrow (Mg^+)^* + He(1s^2\,{}^1S), \qquad (5)$$

where the asterisk signifies electronic excitation, is exothermic for production of *any* excited state of Mg^+. Exothermic processes are those for which the kinetic energy of the products is greater than the kinetic energy of the reactants. The opposite is true for endothermic reactions, which require conversion of reactant kinetic energy to internal energy. Therefore, cross sections for endothermic processes must have a kinetic energy threshold while cross sections for exothermic processes need not have such a threshold. In fact, cross sections for exothermic processes usually do not have thresholds. However, in the present case, as shown in Fig. 6, the cross sections for production of several excited states of Mg^+ were observed to behave in a fashion

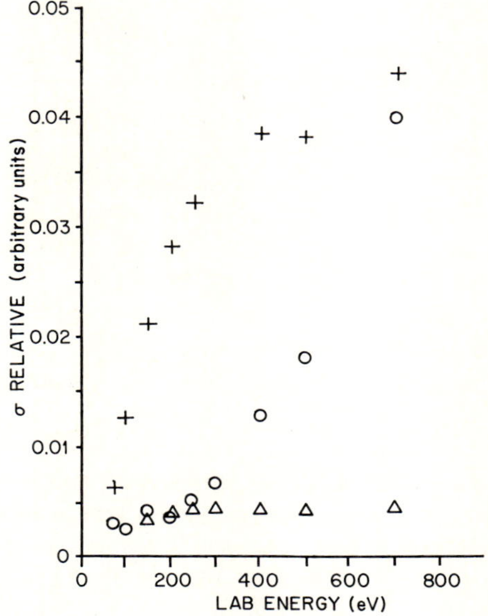

FIG. 6 Emission cross sections for the indicated lines resulting from He^+/Mg collisions as functions of He^+ laboratory kinetic energy [+, 279 nm (MgII); ○, 285 nm (MgI); △, 293 nm (MgII)]. [From Myers and Leventhal (1978).]

that is typical for endothermic processes, and a threshold was observed. This unusual phenomenon was interpreted as being a consequence of the infinite number of continuum (He + Mg^{++} + e) states available to the system. Figure 7 is an energy level diagram that includes schematic molecular potential energy curves for the He^{+}–Mg entrance channel and He–Mg^{++} exit channel. These potential energy curves are not known precisely; however, Kulander and Dahler (1975) have shown that the He^{+}–Mg entrance channel is repulsive; the presumption that the He–Mg^{++} curve is repulsive is based on the inert configuration of these two closed-shell species. That is, the He–Mg^{++} curve should be very similar to that for He–Ne; this curve is the boundary for the (He–Mg^{++}–e) continuum.

At the lowest energies it is likely that one of the infinity of Mg^{++} exit channels will be favored. In fact Mg^{++} has been observed from thermal energy He^{+}/Mg collisions (Arrathoon *et al.*, 1973; Latush and Sem, 1972; Zhukov *et al.*, 1977). As the kinetic energy is increased the system can pass through the continuum and access one of the exothermic $(Mg^{+})^{*}$ exit chan-

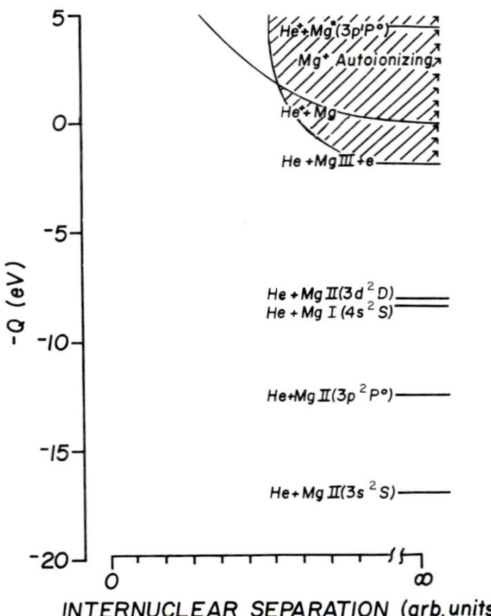

FIG. 7 Energy level diagram for various states of $[HeMg]^{+}$ at infinite separation. The ordinate shows the difference between initial and final relative kinetic energies, Q, for production of these final states from ground-state He^{+}/Mg collisions; $Q > 0$ corresponds to an exothermic collision. Also shown are schematic potential energy curves for ground-state He^{+}–Mg and He–Mg^{++} – e. The crosshatched portion of the diagram represents the continuum of states available to the He–Mg^{++} – e system. [From Myers and Leventhal (1978).]

nels. In addition, the relative kinetic energy must be sufficient for the system to "climb" the repulsive He^+–Mg curve to a crossing point, with one of the curves representing an exothermic exit channel. Thus, while production of Mg^+ is exothermic from a thermochemical standpoint, the picture outlined above shows that there is an activation energy for formation of excited magnesium ions. The results of this study illustrate the necessity of considering the *molecular* nature of the collision process.

It is interesting to note that, in contrast to the situation for a large number of the elemental metals, there are only a few observed Mg^+ laser transitions (Wilett, 1974; Davis and King, 1975), none of which are pumped by He^+/Mg charge transfer. The results of the crossed-beam He^+–Mg experiments strongly suggest that the reason for the failure to effect a population inversion by charge transfer is the propensity of this system for the (He–Mg^{++}–e) exit channel. This example also illustrates a situation in which higher energy data aid in the interpretation of thermal energy (discharge) data.

Another illustration of the necessity of considering the molecular nature of the collision is provided by the study of He^{++}/Li collisions. Shipsey *et al.* (1978) investigated this system theoretically and, from consideration of the molecular potential energy curves for the $[HeLi]^{++}$ diatom, concluded that He^{++}/Li collisions would lead almost exclusively to production of $He^+(n = 3)$. Since the spacing of the energy levels in the one-electron He^+ ion is four times that in the hydrogen atom, $He^+(n = 3)$ radiates L_β and L_α (from cascading) at one-fourth the wavelengths of the analogous Lyman lines in hydrogen, 256 Å and 304 Å, respectively. Thus, preferential population of $He^+(n = 3)$ in this charge transfer reaction could be useful in the quest to develop a superradiant device at soft x-ray wavelengths.

In order to experimentally investigate the theoretical prediction of Shipsey *et al.* (1978), it was necessary to construct a grazing incidence spectrometer and incorporate this device in a crossed-beam apparatus. This was successfully accomplished and collision-produced spectra obtained (Barrett and Leventhal, 1980, 1981). Figure 8 shows one such spectrum taken at 2-keV laboratory energy. While the L_α and L_β lines are of comparable magnitude, because the H_α line at 1640 Å cannot be detected in this experiment, it is not possible to directly determine how much, if any, of the L_α results from cascading and how much is the result of direct formation of $He^+(n = 2)$ in the He^{++}/Li collision. This is illustrated in Fig. 9, which shows the allowed transitions between all states, *nl*, for $n = 1$, 2, and 3.

In the absence of information on the population of specific angular momentum states, the data were compared to a synthetic spectrum generated by assuming the relative populations of the angular momentum states. This was possible because the transition probabilities for one-electron atoms are

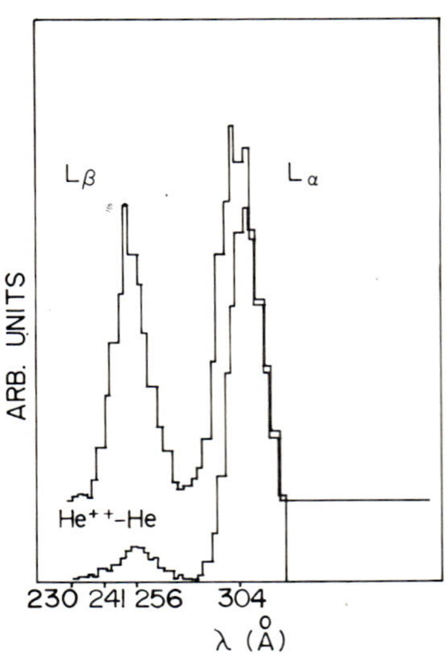

FIG. 8 Emission spectrum between 230 and 312 Å for He^{++}/Li collisions at 2 keV (lab). The spectral resolution is 18 Å FWHM; data are not corrected for relative spectral responses. Also shown for comparison is a spectrum from He^{++}/He collisions at 3 keV. [Barrett and Leventhal (1980).]

accurately known (Bethe and Salpeter, 1977). If only $He^+(n = 3)$ is formed, and if the s, p, and d states are statistically populated, 1:3:5, then the ratio of L_β/L_α would be 0.44. Since the ratio is observed to be larger than 0.44, the angular momentum levels cannot be statistically populated, and it is likely that $He^+(n = 3)$ is indeed the exclusive product.

Actually the L_β/L_α ratio was observed to depend on the collision energy. At low energies, around 100 eV, the L_β/L_α ratio was about one-half and increased to nearly unity at 2 keV before decreasing again to one-half, after which it remained constant to the upper limit of the experiment, 7 keV. While the theoretical treatment of Shipsey *et al.* (1978) did not provide information on the angular momentum states of the product $He^+(n = 3)$, it was possible to construct a crude model based on the intraconversion of angular momen-

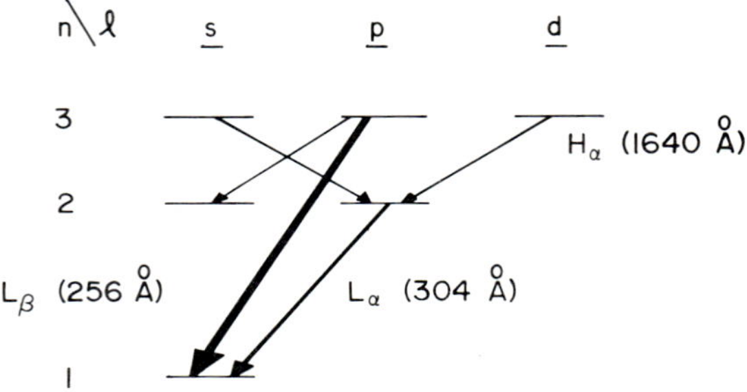

FIG. 9 Term diagram for $n = 1$, 2, and 3 of He^+ showing the electric dipole allowed L_α, L_β, and H_α transitions, respectively. [From Barrett and Leventhal (1980).]

tum, collisional (ion–atom), electronic orbital, and electronic spin, to qualitatively interpret the data (Barrett and Leventhal, 1980, 1981). Of course, the vector sum of these angular momenta must remain constant during the collision. If straight-line trajectories are assumed, it can be shown that the change in collisional angular momentum is given by

$$\Delta L = \{14.4b(\text{in Å})/[E_L(\text{in eV})]^{1/2}\}[h/2\pi], \tag{6}$$

where b is the impact parameter and E_L is the laboratory kinetic energy. Insertion of appropriate values of b and E_L yield values of ΔL of order $(h/2\pi)$. For example, if $b = 3$ Å and $E_L = 2$ keV, $\Delta L \simeq (h/2\pi)$. Thus, according to the model, a few quanta of angular momentum may be exchanged, but only below $\sim$2 keV. Noting that the reactants have zero electronic orbital angular momentum, there should then be a preference for $He^+(3s)$, which cannot yield L_β (see Fig. 9), above $\sim$2 keV. Thus, the ratio L_β/L_α is expected to be smaller at higher energies than at intermediate energies, where $He^+(3p)$, which does yield L_β, is more likely than $He^+(3s)$. Finally, at the lowest energies, the model predicts that $He^+(3d)$, which leads only to L_α, is preferred and the L_β/L_α ratio is expected to decrease again. The behavior of the L_β/L_α ratio that is observed in the experiment is consistent with this picture so that, while the model is crude, it does successfully rationalize the data.

The formation of specific angular momentum states in one-electron atoms produced by charge transfer has also been studied in the beam-gas experiments of Van Zyl *et al.* (1980). In this work, Balmer emissions from decay of excited hydrogen atoms formed in H^+/Ar collisions were analyzed using optical filters and a cooled PMT. This optical system could be moved along the proton beam axis to permit the detection of light at various penetration depths into the collision cell. Because the 3s and 4s states of hydrogen are relatively long-lived (Bethe and Salpeter, 1977), H(3s) and H(4s) atoms formed by charge transfer travel further than other $H(n = 3)$ and $H(n = 4)$ atoms before decaying. Therefore, a partial separation of the radiation from these states can be effected by observing Hα and Hβ as functions of penetration depth into the collision cell. Figure 10 shows the emission cross sections as functions of kinetic energy. Also shown for comparison are earlier data for Hα.

Although the formation of $H(n = 3)$ and $H(n = 4)$ in H^+/Ar collisions is highly endothermic (greater than 10 eV), these products are formed with non-negligible cross sections. Under normal circumstances it does not seem likely that the requisite curve crossings would occur if the asymptotic entrance and exit channels differ in energy by such a substantial amount. It has, however, been suggested that in this case the crossing is accessed through an intermediate $H^-–Ar^{++}$ potential energy curve, as shown in Fig. 11. Here again the molecular nature of the collision is shown to be important.

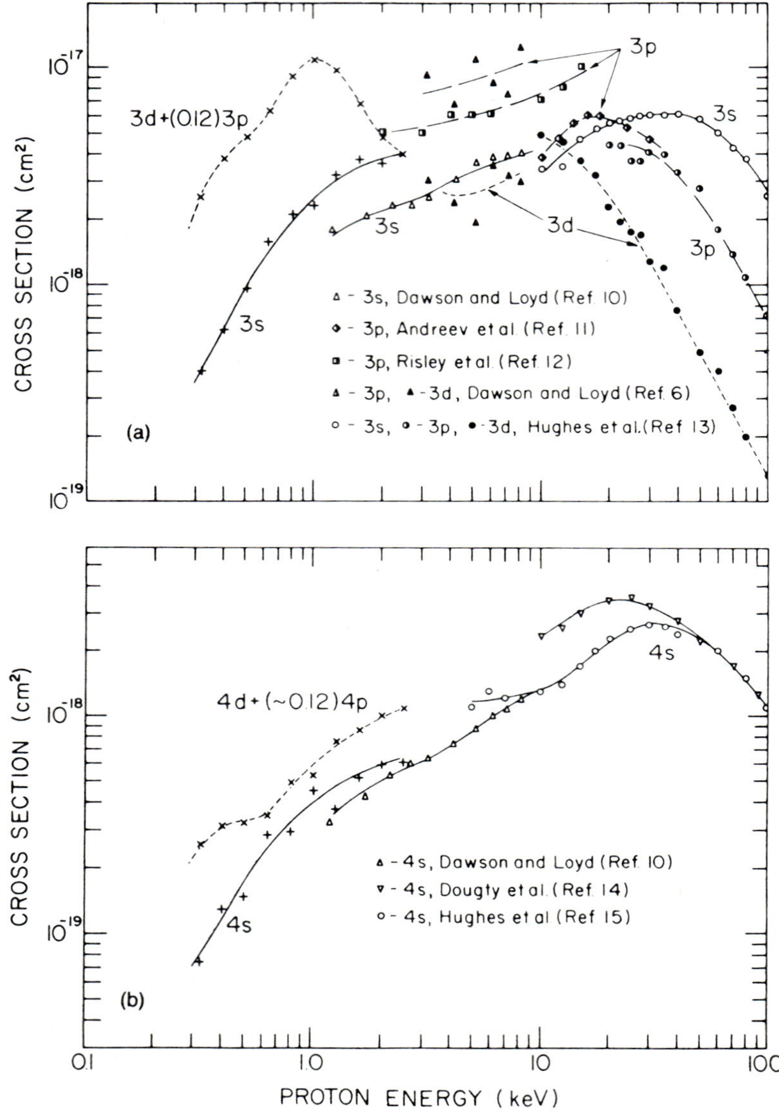

FIG. 10 Comparison of emission cross sections for formation of the (a) $n = 3$ and (b) $n = 4$ levels of hydrogen in H^+/Ar collision measured by Van Zyl *et al.* (1980) with previous measurements (Dawson and Loyd, 1974; Andreev *et al.*, 1967; Risley *et al.*, 1978; Dawson and Loyd, 1977; Hughes *et al.*, 1970; Doughty *et al.*, 1978; Hughes *et al.*, 1967). [From Van Zyl *et al.* (1980).]

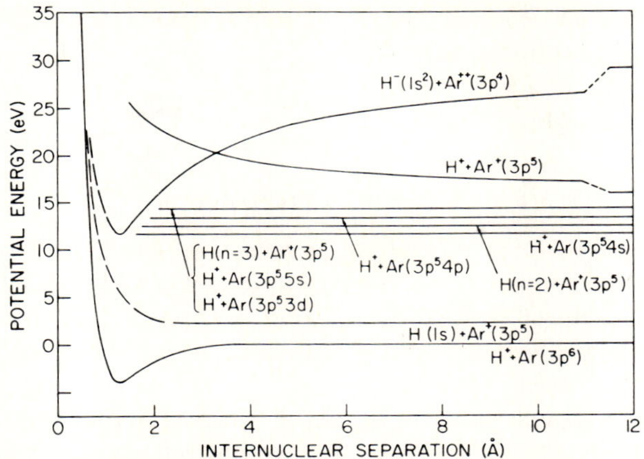

FIG. 11 Estimated potential energy curves for the ArH^+ molecule. [From Van Zyl *et al.* (1980).]

Another illustration of the detail that can be obtained by analyzing the light emitted from excited reaction products is provided by studies of spin conservation in double-electron transfer in collisions involving reactant He^{++} ions (Myers *et al.*, 1978). The total spin angular momentum quantum number of the reactants is given by

$$S = (s_1 + s_2), (s_1 + s_2 - 1), \ldots, |s_1 - s_2|, \tag{7}$$

where s_1 and s_2 are the spin quantum numbers of each reactant. These reactants may be ions, atoms, molecules, or any combination of these. The Wigner spin rule (Massey *et al.*, 1971; Moore, 1973) states that, if the coupling between electron spin and electron orbital angular momenta is weak, total electronic spin will be conserved. Thus, if s_3 and s_4 are the spin quantum numbers of the products, then one of the set of numbers $(s_3 + s_4)$, $(s_3 + s_4 - 1), \ldots, |s_3 - s_4|$ must be included in the set given in Eq. (7).

For most ion/atom collision systems, the data are usually inconclusive with respect to the spin rule. For example, consider He^{++}/Ne double-charge transfer in which emissions from neutral helium are observed:

$$He^{++} + Ne(2p^6\,{}^1S) \rightarrow He(1snl) + Ne^{++}(2p^4). \tag{8}$$

In this case, $s_1 + s_2 = 0$, but since each product may be either a singlet or a triplet, no definitive information would be provided by such an experiment. Note that, as a consequence of the Laporte rule (Herzberg, 1944), transitions between two terms of the same electron configuration are prohibited, so that radiation from excited $Ne^{++}(2p^4)$ cannot be observed. If, however, the neutral electron donor has only two electrons, He or H_2, the spins of which are antiparallel, then a definitive test of the spin rule may be conducted. In the double-electron capture processes,

$$He^{++} + He(1s^2\,{}^1S) \rightarrow He(1snl) + He^{++} \tag{9}$$

and

$$He^{++} + H_2(X\,{}^1\Sigma_g{}^+) \rightarrow He(1snl) + 2H^+, \tag{10}$$

only singlet states of the excited He product are permitted if spin is to be conserved. Thus, comparison of the signal from the 5875 Å, 3d $^3D \rightarrow 2p\,{}^1P^0$, line with that of the 6678 Å, 3d $^1D \rightarrow 2p\,{}^1P^0$, line (see Fig. 2) provides a measure of the extent to which spin is conserved in He^{++}/He and He^{++}/H_2 double-charge transfer collisions.

It is true that double-electron capture is a rather special inelastic process, however, because the spin rule could be definitively tested, experiments on these reactions were undertaken. Furthermore, the conclusions drawn from the data should be of general validity and not restricted to double-electron capture, so that a better understanding of the coupling between collisional, electronic orbital, and electronic spin angular momenta was sought. To a large extent, this goal was attained.

The data are summarized in Fig. 12, which contains collision-produced spectra for several He^{++}–X systems. The fact that the triplet line is larger than the singlet line in He^{++}/Ne and He^{++}/O_2 collisions [panels (a) and (b)] is not surprising since spin may be conserved by formation of either state of the neutral helium product. The observed ratio is roughly that observed in the conventional discharge spectrum of helium (Striganov and Sventitaskii, 1968). When the electron donor is H_2, the triplet line is almost, but not entirely, absent [panels (c) and (d)], indicating a high degree of spin conservation. The He^{++}–H_2 data exhibit two other interesting features. As the He^{++} kinetic energy is increased, the triplet/singlet ratio increases. This increase is accompanied by a decrease in the total cross section for production of He(1s 3d) by about a factor of 10 over the kinetic energy range 10–600 eV. However, for He^{++}/He collisions, the triplet/singlet ratio is comparable to that observed for He^{++}/Ne collisions [panel (e)], but the cross section for this highly endothermic ($\sim$23-eV) double-capture process is only about 10^{-2} of that for He^{++}/H_2 at 10 eV.

Attempts to form a coherent picture of these observations provide insights into spin conservation during atomic and molecular collisions. It must be remembered that the quantity "electronic spin" represents the magnetic dipole moment of the electron. As such, it may be "flipped" only by magnetic forces, which are caused by moving charges such as an impinging ion or the orbital motion of electrons in one of the interacting species. For the case under consideration here, it may be assumed that the required magnetic field

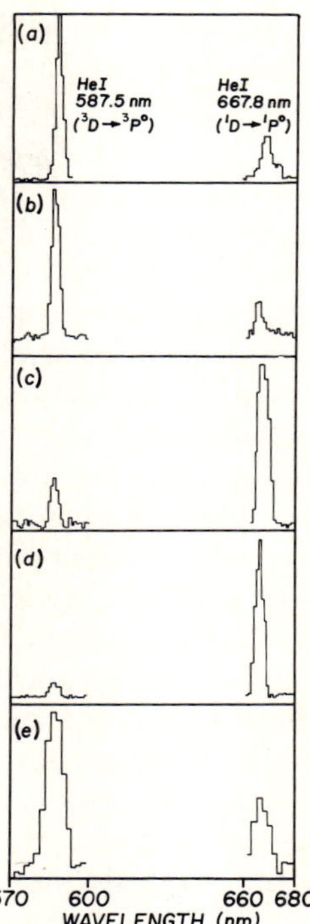

FIG. 12 Spectra for He^{++} collisions with various neutral species at the indicated laboratory kinetic energies: (a) O_2 (4 eV), (b) Ne (250 eV), (c) H_2 (600 eV), (d) H_2 (10 eV), and (e) He (500 eV). The resolution for the spectra in panels (a)–(d) is 5 nm FWHM. Because of low photon yields from He^{++}/He collisions, the resolution in (e) is 10 nm FWHM; background counts have been subtracted. All data are corrected for spectral efficiency of the optical system. The vertical scales are adjusted to make the largest peak full scale and do not, therefore, represent absolute photon count rates. The wavelength region between 600 and 660 nm is omitted here, although, in some cases, there were emissions from species other than HeI in that region; e.g., Hα is produced in He^{++}/H_2 collisions. [From Myers *et al.* (1978).]

is provided by the moving He^{++} ion and, assuming straight-line trajectories, the torque on the electron magnetic moment may be calculated. When integrated from $t = -\infty$ to $t = +\infty$, the torque impulse on the electron magnetic moment is obtained; this torque impulse must be equal to the change in angular momentum, ΔL. The result is

$$|\Delta L| \approx 10^{-4} \frac{a_0}{b_e} \left(\frac{h}{2\pi}\right), \qquad (11)$$

where a_0 is the Bohr radius and b_e is the "impact parameter" for the ion/electron collision. It is important to note the distinction between b_e and the conventional impact parameter b. According to Eq. (11), a spin flip can occur for very close ion/electron encounters, $\sim 10^{-4}\, a_0$. This suggests penetration of the electron cloud surrounding the target atom or molecule, as would be the case for close ion/neutral encounters. If such close encounters are the only collisions for which reaction occurs, as in a very small cross-section process, then spin need not be conserved for that particular process. This is precisely the observation for He(1s 3d) production in He^{++}/He collisions.

For He^{++}/H_2 collisions, the cross section for formation of He(1s 3d) is considerably larger, so that those *close* ion/electron encounters that result in He(1s 3d) are only a small fraction of the total number of events that lead to this product. Thus spin will be conserved in the overwhelming majority of cases for which double capture to He(1s 3d) occurs. The relatively weak triplet line that is observed, 5875 Å, may be regarded as arising from those small impact parameter ion/electron collisions that form He(1s 3d).

The changing triplet/singlet ratio in He^{++}/H_2 collisions is also qualitatively in accord with the above considerations. The decrease in the total reaction cross section as the kinetic energy is increased may be viewed as an increase in the fraction of encounters that lead to He(3d ^{3}D) in spin-flip double-electron capture. That is, while the effective impact parameter for He^{++}–H_2 production of He(1s 3d) decreases, the number of events in which He(1s 3d) is formed, but for which spin is not conserved, remains roughly constant. In other words, the target molecule is the same "size" regardless of the He^{++} kinetic energy.

The general conclusion that emerges from this study is that processes for which the cross section is relatively large exhibit a high degree of spin conservation. Processes for which the reaction cross section is small do not necessarily conserve spin. Earlier in this section interchange of electronic spin, electronic orbital, and collisional angular momenta was discussed as it pertained to the He^{++}–Li experiments, and the change in spin angular momentum was neglected. Since the cross section for selective production of $He^{+}(n = 3)$ was observed to be large, the assumption that spin was conserved was clearly justified. Furthermore, the conclusions about spin conservation are not restricted to ion/neutral collisions. The magnetic field required to flip the electronic spin can be provided by the interpenetration of the electron clouds in neutral/neutral collisions. It is the necessity of close encounters for spin flip, and the concomitant small cross sections, that is responsible for the general validity of the Wigner rule.

In the experiments on spin conservation described above, both He(3d ^{1}D) and He(3d ^{3}D) were observed in those collision systems, such as He^{++}/Ne, for which no spin restrictions apply. This may not, however, be the case if some other factor, such as the interaction potential, the curve or surface, dictates otherwise. An example of such a situation exists for the case of He^{+}/Li charge transfer (Myers *et al.*, 1979). For this doublet/doublet entrance channel, charge transfer leading to either a singlet/singlet or a triplet/singlet exit channel is permitted by the spin rule. However, the results of these experiments show that He(3d ^{3}D) is formed but that He(3d ^{1}D) is not formed.

An explanation of these observations may be found by examining the character of the potential energy curves for the $[HeLi]^+$ molecule. The He^{+}–Li system is isoelectronic with H/Li so that the well-known LiH curves

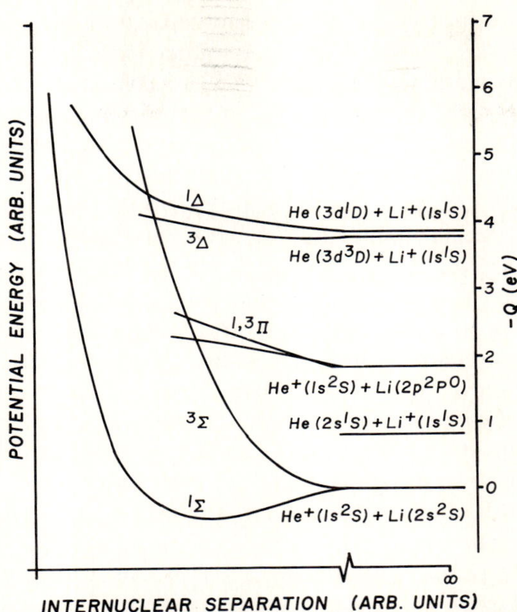

FIG. 13 Energy level diagram for various states of $[HeLi]^+$ at infinite separation. The right-hand ordinate shows the difference between the initial and final relative kinetic energies, Q, for production of these final states from ground-state He^+/Li collisions. The energy separation between the He(3d ^{3}D) and He(3d ^{1}D) states has been exaggerated for clarity. Schematic potential energy curves for several states of the $[HeLi]^+$ molecule are indicated. [From Myers *et al.* (1979).]

(Herzberg, 1950) can serve as a guide when constructing schematic $[HeLi]^+$ curves. Figure 13 shows such a representation. As the collision occurs and the internuclear separation decreases, the system may evolve along either the attractive $^1\Sigma$ or the repulsive $^3\Sigma$ curves, in accordance with the relative orientations of the electronic spins. Note that, while $[HeLi]^+$ is heteronuclear, the situation is analogous to that in the simplest of all neutral molecules, H_2. If the collision occurs along the $^3\Sigma$ curve, then, because this state is repulsive, crossings with other triplet states—such as those associated with the He(3d ^{3}D)–Li^+(1s^2 ^{1}S) separated atoms—occur at internuclear separation of the order of molecular dimensions, ~1–2 Å. Of course, crossings of the $^3\Sigma$ state with singlet states also occur, but the transition from reactants to products requires spin–orbit coupling in this case. The collision may also evolve along the $^1\Sigma$ state of $[HeLi]^+$, in which case the crossings with other singlet states occur at very small internuclear separations, i.e., along the repulsive wall. Thus, since it is reasonable to assume that those reactions that occur at larger internuclear separation have larger cross sections, the preference for triplet products rather than singlets is rationalized. Of course the formation of any given set of reaction products depends on the transition probability from the curve representing the entrance channel to that representing the exit channel. This point is also illustrated by the data of these experiments. While 5875-Å radiation from He(3d ^{3}D) products is observed, 3889-Å radiation from He(3p ^{3}P) is not. This is the case, in spite of the fact that, because the energy of the latter state is slightly lower than that of the former, the He(3p ^{3}P)–Li^+(1s^2 ^{1}S) manifold of curves will be encountered before the manifold corresponding to He(3d ^{3}D)–Li^+(1s^2 ^{1}S).

The discussion so far has been concerned exclusively with excited states of the *products* of charge transfer. It is also possible to study the effects of *reactant* excitation by observing the light from excited state formation.

It has been shown (Hagstrum, 1956) that above about 30 eV, e/Ar and e/Kr collisions produce both ground-state ions and ions in high-lying metastable states. Because the lifetimes of the metastable states are sufficiently long to survive the flight from the ion source to the collision region, the results of beam experiments utilizing 50-eV electron impact-produced Ar^+ should reflect the presence of these energy-rich ions. If the electron energy in the source is adjustable below and above the appearance potentials of these metastable states, a comparison of the effects of ground-state ions with the effects of excited ions may be made. This technique has been employed, and the differences in reactivity of ground state Ar^+ and metastable Ar^+ have been observed.

Rothwell *et al.* (1978) examined the formation of excited atoms in Ar^+/H_2 collisions as functions of the electron impact energy used to form the reactant Ar^+. Figure 14 is a summary of their results. The top panel shows that, at 6-eV center-of-mass kinetic energy, the production of radiating argon atoms with 3p^{5}4s configuration, as evidenced by observation of the 1048- and 1067-Å radiation, and the production of excited hydrogen atoms, as evidenced by the 1216-Å L_α radiation, require at least 32-eV electron impact energy. The cross section for formation of these species increases above this electron energy and becomes constant in the vicinity of 40–45 eV.

Almost identical behavior is observed for the production of metastable states of Ar in Ar^+/H_2 charge transfer, as shown in the middle panel of Fig. 14. Two experiments were performed to detect these nonradiating states. In the first, the neutralized argon atoms emerging from the Ar^+/H_2 collision region were detected with a channel particle multiplier, the presumption being that ground-state argon atoms could not energize the detector. These data are represented by the open triangles in Fig. 14b.

In the second experiment the neutralized argon atoms from 13-eV Ar^+/H_2 collisions entered a second collision cell, this one containing neutral argon atoms. All ultraviolet wavelengths in the range 1040–1300 Å emerging from

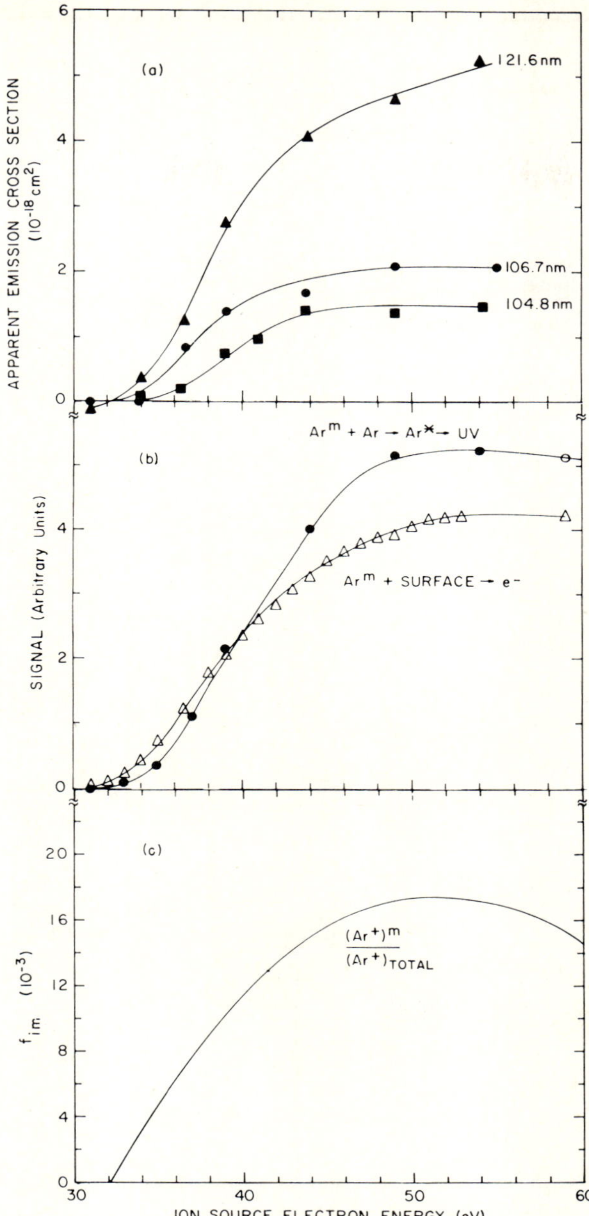

FIG. 14 Ionizing electron energy dependence of the following quantities: (a) apparent emission cross sections for the 104.8- and 106.7-nm ArI lines and L_α from HI (1 nm = 10 Å); (b) total UV for Ar/Ar collisions and detector surface response to neutral argon atoms leaving the charge transfer region; (c) fraction of metastable Ar^+ as reported by Hagstrum (1956). [From Rothwell *et al.* (1978).]

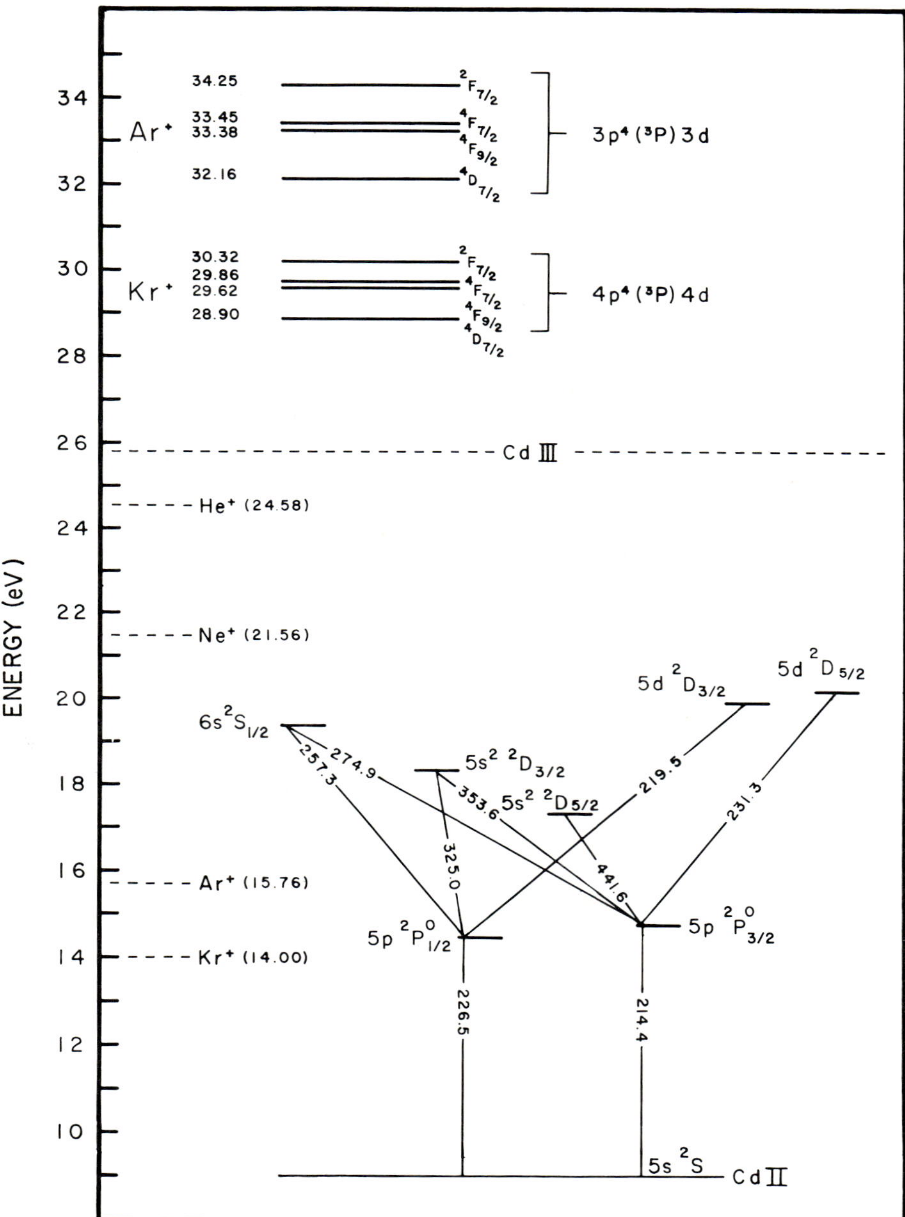

FIG. 15 Partial term diagram for CdII showing the transitions observed from 2-eV rare gas ion/Cd collisions. Energies are relative to the ground state of CdI, and wavelengths are in nanometers. Also shown are pertinent states of some rare gas ions. [From Barrett *et al.* (1981).]

this second cell were detected and assumed to be the result of quenching of the metastable component of the neutralized Ar atom beam. These data are shown as closed circles in Fig. 14b. Clearly the total ultraviolet radiation and channel particle multiplier signals exhibit the same behavior as the 1048-, 1067- and 1216-Å signals in Fig. 14a.

In the earlier work of Hagstrum (1956), the fraction f_{im} of metastable Ar^+ produced by electron impact as a function of electron energy was estimated; the data are shown in Fig. 14c. The close correspondence between the dependences of f_{im} and the excited-state production in Ar^+/H_2 collisions as functions of electron energy strongly suggest that the metastable component of the Ar^+ beam is responsible for the formation of the excited collision products. Furthermore, if the "apparent" emission cross sections shown in Figure 14a are scaled in accordance with f_{im} in Fig. 14c, the true cross section for excited state production in collisions with metastable Ar^+ are of the order of 1–5 Å^2. Thus, although no estimate of the cross section for production of the metastable argon atoms is given, it is clear that the total charge transfer cross section for collisions between metastable Ar^+ and H_2 exceeds 10 Å^2.

The difference in the reactivities of ground-state and metastable ions was also convincingly demonstrated in crossed-beam experiments on Ar^+/Cd charge transfer (Barrett *et al.*, 1981). Figure 15 is a partial term diagram for Cd^+; also shown on the diagram are the REs of ground-state rare gas ions, i.e., the ionization potential of the rare gas atoms, and the locations of the manifolds of Ar^+ and Kr^+ metastable states that are presumed to be present when the ions are formed by 50-eV electron impact. If there is no conversion of translational energy to internal energy, only the 5p $^2P^0$ states of Cd^+ can be formed in Ar^+/Cd collisions; $(Ar^+)^*$/Cd collisions, on the other hand, are capable of producing any state of singly charged cadmium as well as the ground state of Cd^{++}. Figure 16 shows several rare gas ion/Cd spectra, including one from $(Ar^+)^*$/Cd collisions at 1.5-eV center-of-mass kinetic energy (2-eV laboratory energy). The asterisk indicates that the argon ions were produced by electrons with 50-eV kinetic energy, above the threshold for production of the metastable states. These data show that the 3d ^2D, 6s ^2S, and $4d^95s^2$ ^2D states are formed. (Higher resolution spectra of the blue end of this scan have shown that all four lines originating in the 6s ^2S and 5d ^2D states are present.) Since, at 1.5-eV relative energy, these states are inaccessible with ground-state ions, the presence of metastable states in the $(Ar^+)^*$ beam is verified. This conclusion is strengthened by the absence of any collision-produced radiation from 1.5-eV Ar^+/Cd collisions when the ionizing electron energy is below the metastable state threshold.

Furthermore, the emission cross sections for each line observed in $(Ar^+)^*$/Cd collisions is quite large. For example, based on the total measured $(Ar^+)^*$ ion current, the emission cross section for the 2749-Å line, at 1.5-eV relative

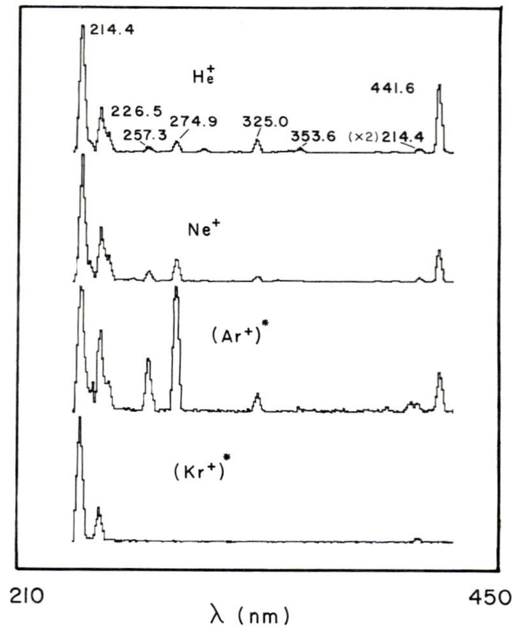

FIG. 16 Emission spectra from 2-eV (cm) rare gas ion/Cd collisions. The emissions are from excited states of CdII. The spectra have been corrected for relative spectral response; resolution is 1 nm FWHM. The spectra are normalized to the same amplitude. The argon and krypton spectra were taken with source conditions that produce metastable states in the beam. [From Barrett *et al.* (1981).]

energy, is determined to be 1.3 Å^2. However, this is a lower limit because the metastable population is unknown. Since the 1.5-eV Ar^+–Cd data clearly show that ground-state ions do not react to form any excited cadmium ions, the metastable component of the $(Ar^+)^*$ beam is solely responsible for the observed excited Cd^+ ions. Hagstrum (1956) estimated that the metastable component comprises only about 1% of the total number of argon ions formed by 50-eV electron impact (see Fig. 14). If this estimate is correct, then the emission cross section for the 2749-Å Cd^+ line produced in collisions between metastable ions and cadmium atoms is ~130 Å^2. The *total* CTE cross section for charge transfer with metastable argon ions is then the order of 1000 Å^2, a very large cross section indeed!

Because of the similarities between argon and krypton that had been observed in previous experiments on electron impact production of metastable ions, it might be expected that the results of Kr^+–Cd and $(Kr^+)^*$–Cd experiments would be similar to the analogous argon data. The asterisk again refers to ions produced by electrons with 50-eV kinetic energy. However, the krypton and argon experiments yielded widely divergent results. The Cd^+ term diagram in Fig. 15 shows that, with transfer of a fraction of an electron volt from kinetic to internal energy, Kr^+/Cd charge transfer can form either of the states in the first excited multiplet, the 5p $^2P^0$ states, of Cd^+; all other Cd^+ states are prohibited. However, as in the case of $(Ar^+)^*$/Cd, $(Kr^+)^*$/Cd collisions can form any Cd^+ excited state as well as Cd^{++}. Nevertheless, the collision-produced spectra from (Kr^+)/Cd and $(Kr^+)^*$/Cd are identical. One such spectrum, taken at 1.1-eV center-of-mass kinetic energy (2-eV laboratory energy), is shown in Fig. 16; the only features are the 2144- and 2265-Å lines from the 5p $^2P^0 \rightarrow$ 5s ^2S transitions. Additional experiments with $(Kr^+)^*$ beams were performed, using higher electron impact energies, but no difference in the Kr^+/Cd and $(Kr^+)^*$/Cd spectra could be detected. Thus, while all excited Cd^+ are allowed in $(Kr^+)^*$/Cd charge transfer, only the two lowest lying excited states are formed, the same excited states as in Kr^+/Cd collision.

Although it is tempting to infer that the metastable levels of the krypton ion are not a major component of the beam, even when relatively high ionizing electron energies are used, such a conclusion would be at variance with previous experiments. Therefore, since there is no reason for the metastable krypton ions to be absent, it must be concluded that these ions are simply not as reactive with cadmium atoms as are metastable argon ions.

The most plausible explanation for the different reactivities of $(Ar^+)^*$ and $(Kr^+)^*$ with Cd atoms is a density-of-states argument. Figure 17 is a combined energy level diagram for Ar, Ar^+, Kr, Kr^+, and Cd. The scale is adjusted so that the mean energies of the metastable levels of Ar^+ and Kr^+ coincide with each other and with the Cd^+(6s ^2S) state. Recall that this state of Cd^+ is preferentially populated in $(Ar^+)^*$/Cd collisions but not in $(Kr^+)^*$/Cd collisions. Formation of Cd^+(6s ^2S), accompanied by formation of either ground-state Ar or Kr in each of these collision systems, is highly exothermic. Thus, the required curve crossings probably do not occur, and it is unlikely that this reaction proceeds. However, nearly thermoneutral charge transfer, with product Cd^+(6s ^2S), can occur, provided it is accompanied by formation of an *excited* neutral Ar or Kr. For the krypton case, there are only two possible states—the $5s[\frac{3}{2}]^0$ and the $5s[\frac{1}{2}]^0$—that can lead to a nearly thermoneutral reaction, but there is a dense manifold of such states in the argon case. Thus, it may be argued that the probability of curve crossing leading to formation of Cd^+(6s ^2S) in the argon system is many times greater than for the krypton system simply because there are many more potential energy curves in the vicinity of the potential energy curves corresponding to the reactants. Here again we have the situation in which results from an atom/atom collision experiment are interpreted on the basis of the molecular nature of the collision process. This is true even though

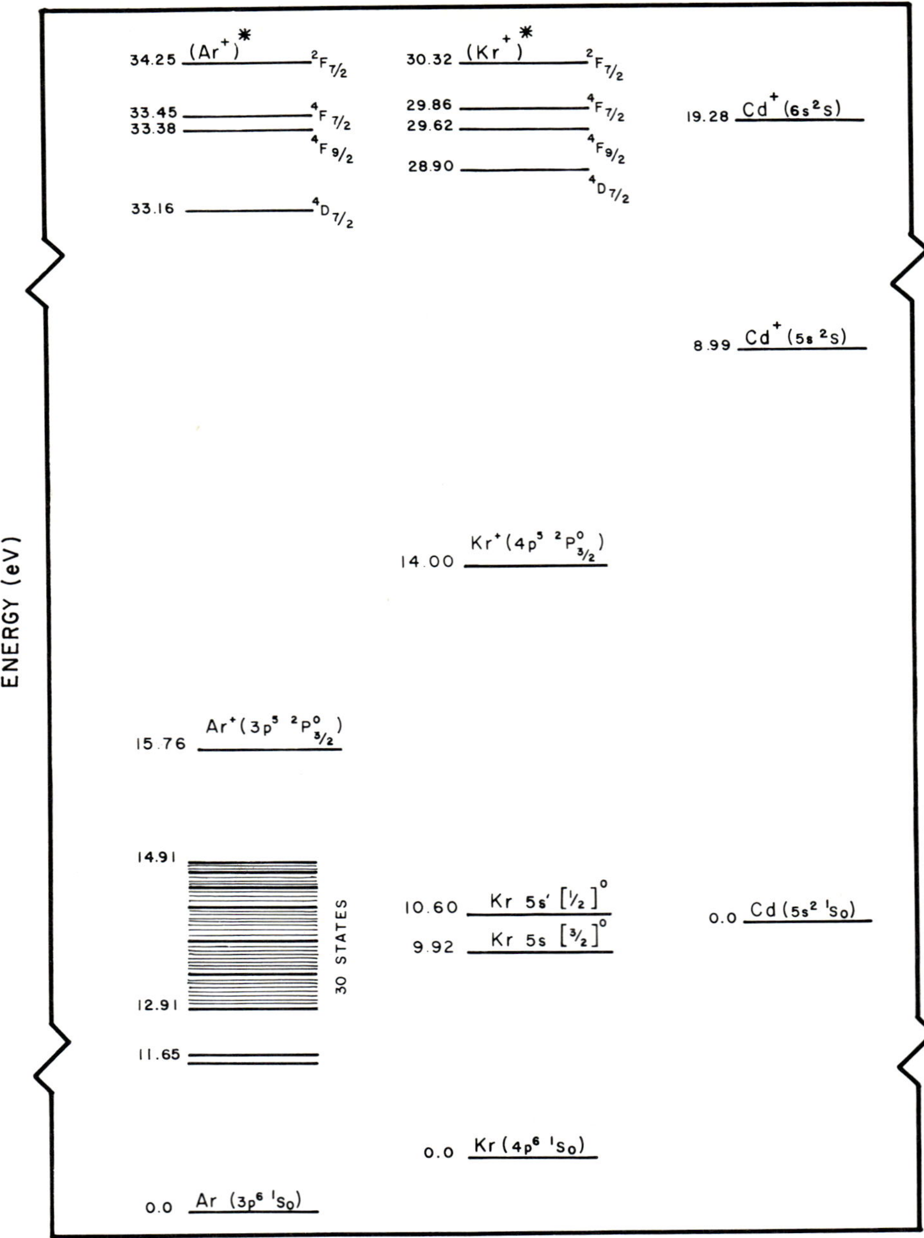

FIG. 17 Energy level diagram for argon, krypton, and cadmium. The 6s ^{2}S state of CdII is aligned with the energy centers of the metastable states of Ar^+ and Kr^+. [From Barrett *et al.* (1981).]

virtually nothing is known about any of the $[ArCd]^+$ potential energy curves. The deductions were made solely on the basis of the asymptotic atom-atom energies.

A further refinement of the technique of using radiation from decay of excited products to analyze inelastic collisions has been employed recently to study reactions involving *state-selected* excited reactants (Kushawaha *et al.*, 1980). State selection was accomplished by using a tunable dye laser to excite ground-state sodium atoms in studies of charge transfer in H^+/Na collisions. The purpose of the investigation was to compare the cross section for

$$H^+ + Na(3s) \rightarrow H(n = 2) + Na^+ \tag{12}$$

with that for

$$H^+ + Na(3p) \rightarrow H(n = 2) + Na^+. \tag{13}$$

Reaction (12) is endothermic by 1.74 eV, but reaction (13) is exothermic by 0.34 eV. Since the potential energy curves for $H(n = 2)$–Na^+ lie closer to the H^+–Na(3p) curve than the H^+–Na(3s) curve, it was anticipated that the cross section for L_α production from the H^+–Na(3p) entrance channel would be much larger than that for H^+–Na(3s).

In order to perform these experiments, it was necessary to incorporate a tunable laser into the apparatus. A cw dye laser was used, and the laser beam mechanically modulated so that L_α photons could be detected in synchronization with the laser beam, thus yielding comparative data for H^+/Na(3s) and H^+/Na(3p) collisions.

The results showed that, contrary to prior expectations, no enhancement in L_α production occurred at kinetic energies above 35 eV. A relatively small enhancement was observed, less than a factor of 2, below 35 eV. The failure to detect any significant difference between H^+/Na(3s) and H^+/Na(3p) charge exchange was investigated theoretically by Kimura *et al.* (1982), who used pseudo-potential molecular structure calculations to obtain the low-lying potential energy curves for the NaH^+ molecule. They showed that, indeed, the experimental result is consistent with the nature of the curves and the couplings between them. Using the method of perturbed stationary states, they calculated cross sections for charge transfer in H^+/Na(3s) and H^+/Na(3p) collisions. These calculations, which were carried out for kinetic energies in the range 0.1 to 10 keV, showed that the cross section for production of $H(n = 2)$ in H^+/Na(3s) collisions was nearly identical to that for H^+/Na(3p) collisions. They attributed this surprising result to the extent of the coupling between the various entrance and exit channels. Figure 18 shows the calculated potential energy curves. Two molecular states arise from the H^+–Na(3p) entrance channel, the F $^2\Sigma$ and E $^2\pi$ states, which have

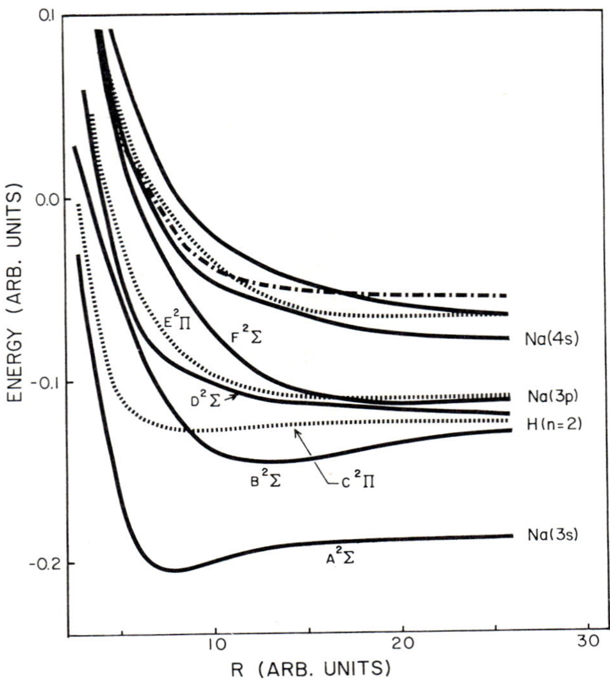

FIG. 18 Calculated potential energy curves for NaH^+. The ${}^2\Sigma^+$ molecular states are denoted by solid lines, the ${}^2\Pi$ states by dotted lines, and the ${}^2\Delta$ states by dash-dotted lines. [From Kimura *et al.* (1982).]

statistical weights $\frac{1}{3}$ and $\frac{2}{3}$, respectively. While the F state couples strongly to the states arising from the $H(n = 2)$–Na^+ exit channel, the E state does not. Thus roughly two-thirds of the incident flux is precluded from reaction.

IV. ION–MOLECULE SYSTEMS

Of all the charge transfer excitation experiments that have been conducted by observing light from the decay of excited reaction products, there probably exist more data for atomic ion/diatomic molecule systems than for any other. Most of the early experiments were undertaken in an effort to acquire laboratory data that simulated reactions that occur in the earth's upper atmosphere (McNeal and Birely, 1973). In addition to the measurement of cross sections for various CTE processes, emphasis was placed on the determination of vibrational state distributions within the excited electronic state of the product diatomic ion. In an effort to better understand the nature of these collisions, the observed vibrational state distributions have been compared to those obtained from models.

The most commonly employed model, and the one to which all vibrational state distributions are compared as a first approximation, is the Franck–Condon, or vertical transition model. The steps that take place under the assumptions of this model are shown schematically in Fig. 19. The neutral diatomic reactant AB is certainly in the electronic ground state at 300 K, and almost always exclusively in the ground vibrational level. Removal of the transferred electron in the CTE reaction results in an electronically excited molecular ion, $AB^+(b)$. This transition is represented by three straight arrows, each originating on the ground vibrational level of the ground electronic state of the AB molecule but each terminating on a different vibrational level of $AB^+(b)$. In the Franck–Condon model, it is these transitions that are assumed to occur vertically, i.e., they occur with no change in the internuclear separation of the AB molecule. The extent to which each of the vibrational levels of the excited electronic state of the AB^+ molecule is populated depends on the square of the overlap integrals between these levels and $v = 0$ of the ground electronic state of AB—the Franck–Condon factors (Herzberg, 1950).

Data are acquired by observing the radiation, represented by the wavy arrows in Fig. 19, when $AB^+(b)$ decays to a lower state $AB^+(a)$, which may or may not be the electronic ground state of the molecular ion. In

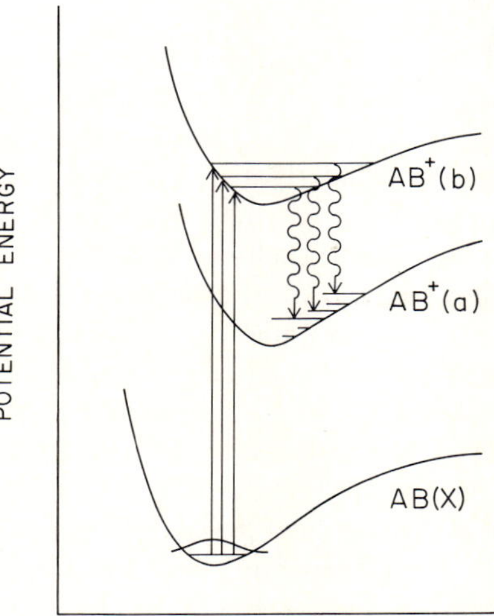

FIG. 19 Schematic potential energy diagram illustrating the vertical transition, or Franck–Condon, model for ionization of a diatomic molecule AB. At room temperature, AB will usually be in the ground vibrational level of the ground electronic state, the X state. Ionization to various vibrational levels (three are shown) of electronically excited AB, the b state, is assumed to occur with no change in the A–B internuclear separation. This assumption of vertical ionization in the collision process, indicated by three straight arrows in the diagram, is the only assumption of the model since the subsequent radiative decay, indicated by three wavy arrows, certainly occurs vertically, i.e., with no change in the A–B internuclear separation.

order to compare the data to the model, synthetic spectra may be constructed from knowledge of the Franck–Condon factors for the AB(X, $v = 0) \rightarrow AB^+$(b, v') and the AB^+(b, v') $\rightarrow AB^+$(a, v'') transitions. The light signal, in quanta per second, for a particular band, (v', v''), is given by (Herzberg, 1950)

$$\xi v',v'' = [K f_{v'} \bar{R}_e{}^2 q_{v',v''} E(\lambda_{v',v''})]/[\lambda_{v',v''}]^3, \tag{14}$$

where K is a constant that includes geometrical factors, $f_{v'}$ is the fraction of molecules in the level v', $\bar{R}_e$ is the average electronic transition moment, $q_{v',v''}$ is the Franck–Condon factor for the (v', v'') transition, $\lambda_{v',v''}$ is the wavelength of the band, and $E(\lambda_{v',v''})$ is the spectral efficiency of the detection system at $\lambda_{v',v''}$. The only assumption of the vertical transition model is that when the electron is transferred during the collision, the product diatomic ion has the same internuclear separation as the neutral reactant diatomic molecule. Thus, $f_{v'}$ in Eq. (14) is simply the Franck–Condon factor for the transition from AB(X, $v = 0$) to AB^+(b, v'). That is, it is assumed that the vibrational state distribution within AB^+(b) is the same as it would be if the AB(X, $v = 0$) had been photoionized by a radiation continuum having wavelength limits ranging from that corresponding to the AB(X, $v = 0) \rightarrow AB^+$(b, $v = 0$) energy to the analogous energy for ionization to the highest vibrational level of AB^+(b). Since the emission of radiation in the AB^+(b, v') $\rightarrow AB^+$(a, v'') transition is surely vertical, the inclusion of $q_{v',v''}$ in Eq. (14) is not an assumption of the model.

Since K is a constant and $\bar{R}_e$ may be assumed constant for a given band system, knowledge of $E(\lambda)$ and the respective $\lambda_{v',v''}$, together with the necessary Franck–Condon factors, permits generation of a synthetic spectrum based on the assumption of vertical ionization. In practice, however, the synthetic spectra are usually calculated under the assumption that $E(\lambda)$ is a constant for all λ, and the experimentally determined spectra are corrected using the known $E(\lambda)$.

Although useful as a starting point, this model represents an oversimplified picture of the collision since it treats the diatomic target and the product diatomic ion as isolated molecules. This view ignores the chemical nature of the reactant species. Figures 20 and 21 show portions of the collision-produced spectra for 800-eV H_2^+/CO, H_3^+/CO, H_2^+/N_2, and H_3^+/N_2 collisions, and illustrate the importance of chemistry in determining the partitioning of energy among the final states of the products. The red-shaded bands in Fig. 20 are from the CO^+(A $\rightarrow$ X) comet tail system (Krupenie, 1966; Rosen, 1970). It is clear that, aside from the H_γ and H_δ line from atomic hydrogen, the two spectra differ substantially. The (4, 0) and (3, 0) peaks are stronger relative to (2, 0) for H_3^+/CO than for

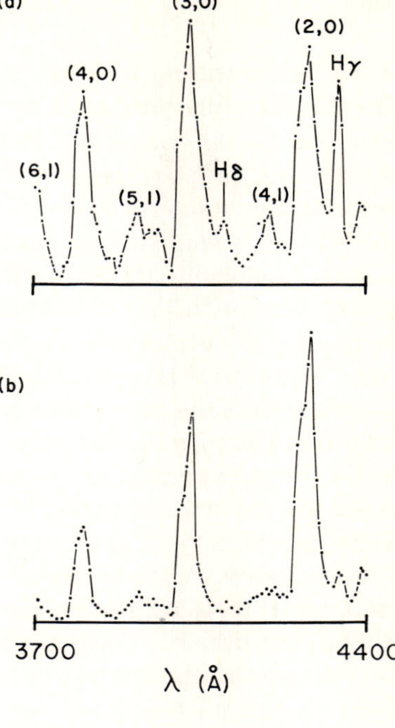

FIG. 20 Partial emission spectra from 800-eV (lab) (a) H_3^+/CO and (b) H_2^+/CO collisions. The spectra have not been corrected for relative spectral response; the resolution is 10 Å FWHM. The (v', v'') designation identifies bands from the $CO^+(A \rightarrow X)$ comet tail emission system. H_γ and H_δ from excited atomic hydrogen are also identified.

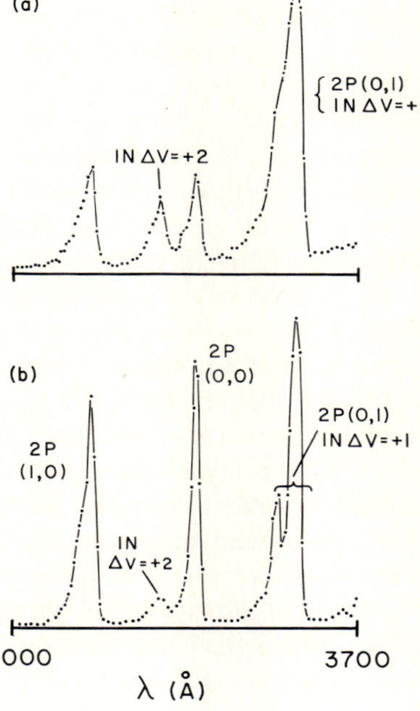

FIG. 21 Emission spectra from 800-eV (lab) (a) H_3^+/N_2 and (b) H_2^+/N_2 collisions. The spectra have not been corrected for relative spectral response; the resolution is 10 Å FWHM. The (v', v'') designation identifies bands from the $N_2(C \rightarrow B)$ second positive emission system; the Δv designation identifies sequences of bands from the $N_2^+(B \rightarrow X)$ first negative emission system.

H_2^+/CO, indicating a higher degree of vibrational excitation with H_3^+. The relative strengths of the (6, 1) and (5, 1) bands in the two spectra support this conclusion. Thus, in spite of the fact that the relative kinetic energies are roughly the same, the degree of vibrational excitation in the two collision systems is quite different. Spectra acquired at different kinetic energies show that this conclusion is true for collisions in which the H_3^+ and H_2^+ velocities are the same.

The features of the spectra in Fig. 21 show that two different inelastic processes are occurring in both H_2^+/N_2 and H_3^+/N_2 collisions. The peaks labeled 1N are sequences of bands from $N_2^+(B\,^2\Sigma_u^+ \rightarrow X_2\,^+\Sigma_g)$ first negative emissions and are indicative of CTE processes. The peaks labeled $2P$ are bands from the $N_2(C\,^3\Pi_u \rightarrow B\,^3\Pi_g)$ second positive system; the presence of these emissions in the spectrum shows that collision-induced excitation (CIE) is competing with charge transfer. Comparison of the signal from the $1N(\Delta v = +2)$ sequence with that from the $2P(0, 0)$ band shows that CIE competes more effectively with charge transfer in H_2^+/N_2 collisions than in H_3^+/N_2 collisions. This is consistent with the Wigner spin rule, since formation of a triplet/singlet exit channel from the singlet/singlet $H_3^+/\text{–}N_2$ entrance channel violates the rule (see Section III), while formation of the triplet/doublet exit channel from the doublet/singlet $H_2^+\text{–}N_2$ entrance conserves spin.

A great many of the studies of vibrational state distributions have employed reactant N_2 molecules. This is because of the importance of this molecule in atmospheric studies and due to the fact that the well-known 1N emission system of N_2^+ transitions yields radiation in an easily detectable range, ~2800–5000Å (Lofthus and Krupenie, 1977; Rosen, 1970). Furthermore, N_2 gas is relatively inexpensive. CO has also been used for studies of this type because radiation from the $A\,^2\Pi \rightarrow X\,^2\Sigma$ comet tail system of CO^+ is also well known, in a desirable region of the spectrum, and similarly inexpensive.

In 1969 Moore and Doering reported the results of a systematic study of vibrational distributions within $N_2^+(B)$ resulting from CTE in reactions between some simple ions, such as H^+ and H_2^+, and N_2 molecules. They concluded that the observed vibrational distributions depended heavily on the laboratory velocity of the reactant ion, and, contrary to the data discussed above, very little on the chemical nature of the ions. They observed that at the lowest velocities the vibrational distributions were not in accord with the vertical transition model, but tended toward the model result at velocities in excess of $\sim 10^8$ cm/sec. They attributed this phenomenon to distortion of the target molecule electron cloud by the impinging ion at low velocities and the failure of the molecule to distort during the short collision times characteristic of higher velocities.

Subsequently, Lipeles (1969) studied the vibrational distribution within $CO^+(A^2\Pi)$ resulting from Ar^+/CO collisions and found that the data were not in accord with the vertical transition model at $v = 7 \times 10^6$ cm/sec. He calculated "distorted Franck–Condon factors" by assuming that the incoming ion polarizes the diatomic reactant molecule and that vertical transitions from CO(X, $v = 0$) to CO^+(A) take place from the deformed CO(X). Using this polarization distortion model, Lipeles accounted for the observed vibrational distributions.

Any model based on target molecule distortion caused by the incoming projectile ion is, in fact, accounting, however crudely, for the molecular nature of the collision. Although adjustment of the isolated CO and CO^+ potential energy curves could rationalize the data, the fact that the electron would have to be transferred at an Ar^+–CO separation of only 1.38 Å (Lipeles, 1969) to fit the data makes it unlikely that ion-induced polarization is the dominant effect. At such small separations the ion-induced dipole potential is ~10 eV, which is unrealistically large.

In an effort to clarify the situation, a new series of experiments was undertaken (Kelley *et al.*, 1977a, 1977b, 1978). In addition to CTE, CIE was also studied so that vibrational distributions resulting from different types of inelastic processes could be compared. Figure 22 shows a spectrum resulting from 50-eV He^+/N_2 collisions (Kelley *et al.*, 1977b). The emissions are primarily from the $N_2{}^+$(1N) system from the CTE reaction

$$He^+ + N_2(B, v = 0) \rightarrow N_2{}^+(B, v') + He. \tag{15}$$

The bar graph represents a synthetic spectrum based on the vertical transition model. The bars have been adjusted so that the height of the bar representing the (0, 0) band, the strongest band in the synthetic spectrum, matches the data. As expected, there is considerable disparity between the observed spectra and those that would be observed if vertical collisional processes had occurred.

The second spectrum shown in Fig. 22 is also one of $N_2{}^+$(1N) emissions. In this case, however, the data are from $N_2{}^+$/He CIE, i.e.,

$$N_2{}^+(X, v = 0) + He \rightarrow N_2{}^+(B, v') + He. \tag{16}$$

The synthetic spectrum, which is of course different from the synthetic spectrum from He^+/N_2 CTE, based on the vertical transition model is also shown in the figure. It is clear that the vibrational state distribution within the $N_2{}^+$(B) product of $N_2{}^+$/He CIE is also not in accord with the model. In this case, the charge resides with the diatom so that ion-induced distortion cannot be responsible for the deviation from vertical excitation in the collision.

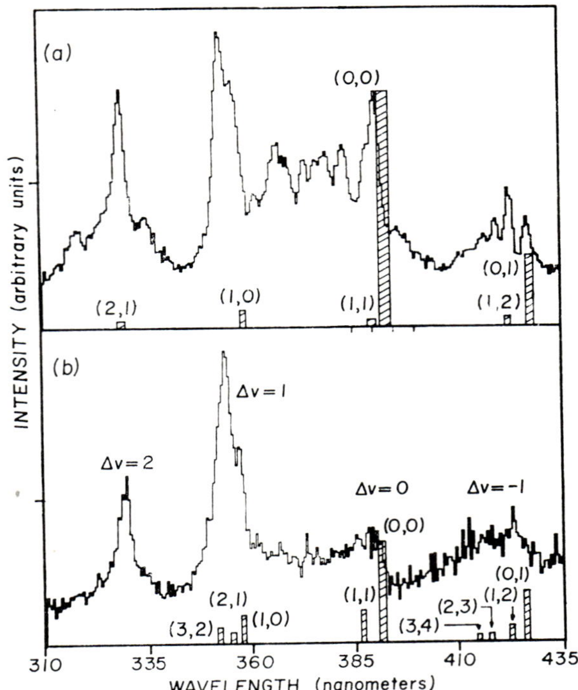

FIG. 22 Emission spectra from (a) He^+/N_2 and (b) N_2^+/He collisions at 50- and 100-eV laboratory kinetic energy, respectively. The resolution is 2 nm FWHM; data are corrected for relative spectral response. Also shown in each panel is a bar graph representing a model spectrum calculated under the assumption of vertical excitation and vertical decay. In each panel, the height of the (0, 0) bar has been adjusted to match the data. The Δv designation identifies sequences of bands from the $N_2^+(B \rightarrow X)$ first negative emission system. [From Kelley *et al.* (1977b).]

A great many additional experiments were performed, and these experiments led to the same general conclusions: Nonvertical transitions are the rule at low kinetic energy, but the vibrational distributions tend toward those that would be produced in vertical collisional transitions as the kinetic energy is increased. An example of this trend is shown in Fig. 23 for H^+/O_2 collisions (Kelley *et al.*, 1978). The emissions are from the $O_2^+(b \rightarrow a)$ first negative system (Krupenie, 1972). The bar graph represents the spectrum that would result if vertical ionizing transitions occurred. It is clear that the 50-eV spectrum is not in accord with the model but that the model result is approached as the kinetic energy is increased.

On the basis of the above discussion it is clear that more general considerations than distortion by ion-induced polarization are required. A complete treatment of either CTE or CIE would require detailed knowledge

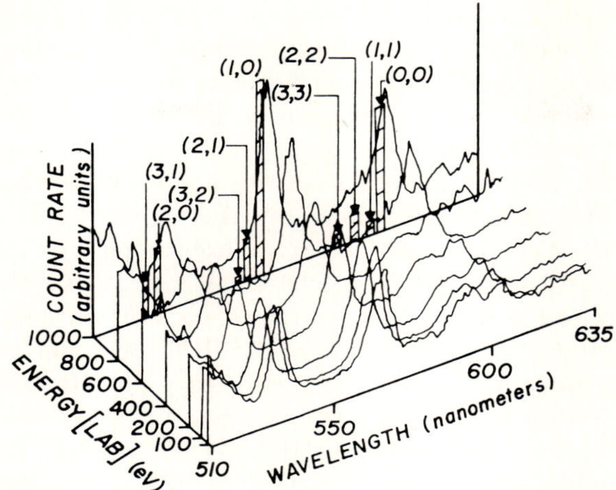

FIG. 23 Emission spectra from H^+/O_2 collisions at several collision energies. The resolution is 2 nm FWHM; data are corrected for relative spectral response. The locations of the heads of several bands of the $O_2^+(b \rightarrow a)$ first negative emission system are shown. A bar graph representing a model spectrum calculated under the assumption of vertical excitation and vertical decay is also shown; the height of the bar representing the (1, 0) band, the largest in the model spectrum, has been adjusted to match the height of the (1, 0) peak in the 1000-eV data, the largest peak in all the spectra shown. [From Kelley *et al.* (1978).]

of the two potential energy surfaces involved and the coupling between them. In the absence of this detailed knowledge, it is necessary to construct a model which is applicable to either CTE or CIE. In fact, such a model should also be capable of adequately describing neutral/neutral collisions since ionizing collisions between hydrogen atoms and nitrogen molecules have been shown to yield $N_2^+(B)$ with non-Franck–Condon vibrational distributions (Birely, 1974). The model that was developed (Kelley *et al.*, 1977a, 1978) to explain the observations is based on the assumption that deviations from vertical transitions during the collision are caused by short-range interactions between the reactants, and these short-range interactions result in direct translational–vibrational (T–V) excitation in both the initial and final electronic states of the diatomic molecule.

This model was applied to N_2^+/He CIE, and the experimental observations were qualitatively reproduced (Kelley *et al.*, 1978). It was also shown that the distributions and trends that were observed in other collision systems, for both CTE and CIE, would also be in qualitative agreement with the model. Although a specific mechanism, T–V transfer, was invoked as part of the model in order to perform calculations, the fact that short-range forces are involved means that the reaction occurs at such small impact parameters that the diatomic molecule cannot be regarded as being isolated.

During the reaction, as for example when an electron is transferred in an atomic ion/diatomic molecule collision, the reactants comprise an ephemeral triatomic molecule. Indeed, the target diatom is distorted, but it is distorted by the chemical forces that are operative in any short-range encounter between two atoms or molecules.

In spite of the fact that the overwhelming body of accumulated data for low-energy collisions show vibrational distributions that are not in accord with the Franck–Condon model, there are exceptions. For example, CTE with He_2^+ and N_2, O_2, or CO reactants has been shown to yield vibrational distributions that are consistent with vertical ionization at kinetic energies as low as 1 eV (Leventhal *et al.*, 1975; Bearman *et al.*, 1976a). Similar observations have been made with Ne_2^+ reactant ions (Bearman *et al.*, 1976b). Figure 24 illustrates this situation for 11-eV He_2^+/CO collisions. The emissions are from the $CO^+(B \rightarrow X)$ system and, when compared to the bar graph, clearly indicate agreement with the vertical transition model. Although these results seem to be inconsistent with the data and the model described above, they, in fact, contribute to form a coherent picture.

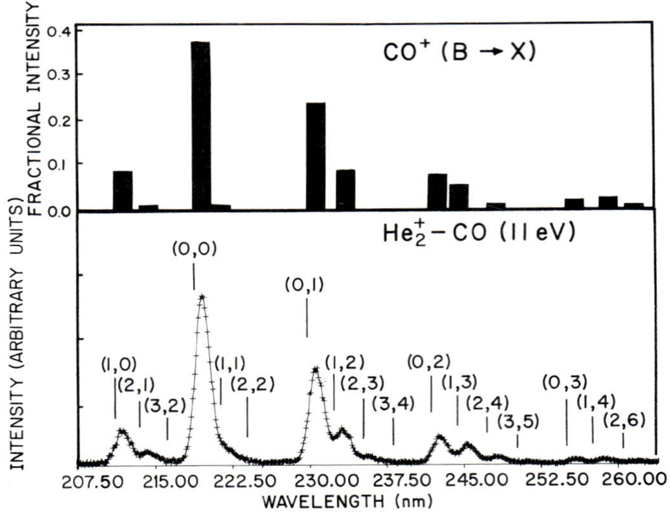

FIG. 24 Portion of the spectrum from 11-eV He_2^+/CO collisions. The resolution is 0.4 nm FWHM; data are corrected for relative spectral response. The locations of the heads of several bands of the $CO^+(B \rightarrow X)$ first negative emission system are shown. A bar graph representing a model spectrum calculated under the assumption of vertical excitation and vertical decay is also shown; the height of the bar representing the (0, 0) band has been adjusted to match the height of the (0, 0) band in the data. [From Bearman *et al.* (1976a).]

A major difference between these data is that the cross sections for CTE in collisions of $He_2{}^+$ with N_2, CO, or O_2 are about two orders of magnitude larger than the cross sections for the processes that produce vibrational distributions that are not in accord with the vertical transition model. These large cross sections suggest long-range encounters so that the electron is transferred at large reactant separation. In terms of the model, these long-range interactions do not produce T–V transfer so that no deviation from the vertical transition model should occur. In terms of the $[He_2CO]^+$, $[HeN_2]^+$, or $[HeO_2]^+$ potential hypersurfaces, the reactants are far enough away from each other when ionization of the diatomic target occurs that the reactants are essentially isolated from each other.

One of the most interesting aspects of these experiments is the unusually large cross section that is observed for many CTE processes involving $He_2{}^+$ and $Ne_2{}^+$. Why should CTE cross sections with protons, $H_2{}^+$, Ar^+, and most of the commonly used ionic species be $\sim 10^{-2}$ Å^2, while some cross sections for processes involving $He_2{}^+$ and $Ne_2{}^+$ are orders of magnitude larger? The explanation probably lies in the unique character of these two ions. For most "common" ions such as $H_2{}^+$, charge transfer yields a stable neutral "parent." For example, the CTE reaction

$$H_2{}^+ + X \rightarrow (X^+)^* + H_2 \tag{17}$$

has a two-body, $(X^+)^*$ and H_2, final state. Because the ground states of both He_2 and Ne_2 are unbound, the analogous reactions such as

$$He_2{}^+ + X \rightarrow (X^+)^* + He + He \tag{18}$$

yield three-body final states. Figure 25 shows the ground-state potential energy curves for the helium and neon dimer systems. It is clear that the RE of each ion in the ground electronic and vibrational levels depends upon the internuclear separation of the two atoms when the electron is transferred. Furthermore, because the repulsive potentials support no bound states, the RE may be continuously variable. Unlike the other rare gas dimer ions, the probable ranges of REs for $He_2{}^+$ and $Ne_2{}^+$, shown as the crosshatched columns in Fig. 25, are sufficient to both ionize and excite most molecules without conversion of kinetic energy to internal energy in the collision. This is another way of saying that a CTE reaction may be energy resonant, or that ΔE, the heat of reaction, may be nearly zero. Such processes are considered to be the most likely because, as stated previously, the potential energy surfaces representing the entrance and exit channels lie very close in the separated reactant limit, thus making a crossing of the hypersurfaces likely. The unique feature of the $He_2{}^+$ and $Ne_2{}^+$ CTE processes is that, from this point of view, there are an infinite number of exit channel hypersurfaces which may be accessed by the system so that the probability of

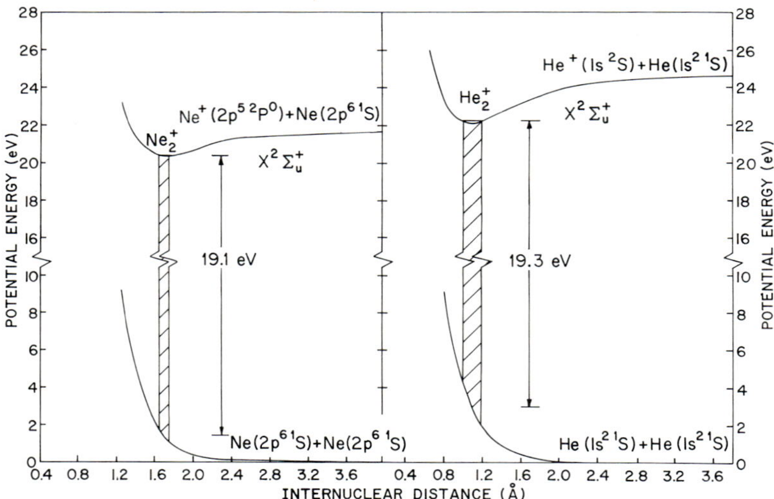

FIG. 25 Potential energy curves for helium and neon dimers. The curves for neon are taken from the calculations of Cohen and Schneider (1974) and Schneider and Cohen (1974). The repulsive He–He potential is taken from Foreman *et al.* (1974), and the $He_2{}^+(X\,{}^2\Sigma_u{}^+)$ curve is a Morse function with molecular parameters given by Ginter and Ginter (1968). The cross-hatched portions show the vertical electronic recombination energies of $Ne_2{}^+$ and $He_2{}^+$.

reaction is quite high. Each of the infinity of hypersurfaces corresponds to a different kinetic energy of the product helium atoms. This kinetic energy is, of course, not quantized for a three-body final state.

The above discussion may be cast in other (equivalent) terms. For example, the transferred electron in a charge exchange process may be viewed as oscillating between the projectile ion and the once ionized target. In the $H_2{}^+$–X case, the electron "alternates" between the $H_2{}^+$ and the X^+ during the course of the collision. Whether or not charge transfer occurs depends upon whether it ends up on the $H_2{}^+$ or the X^+. For reactions involving $He_2{}^+$ or $Ne_2{}^+$, no such oscillation can occur. When the electron resides with the rare gas dimer, ion dissociation immediately occurs so that the electron oscillation cannot continue. Thus the $He_2{}^+$ or $Ne_2{}^+$ may be viewed as a sink for electrons. As was remarked above, all of these arguments are basically the same and may be summarized by saying that the three-body final state has more phase space available than the two-body final state, so that, other things being equal, the process having a three-body final state is more likely.

Vibrational distributions of electronically excited products of CTE have also been studied in the laboratory of Ottinger (Brandt *et al.*, 1973; Ottinger and Simonis, 1978). Their findings are consistent with those described above and showed that O^+/N_2 CTE gave a broad vibrational distribution within $N_2{}^+(B)$ when the ions in the O^+ beam were essentially all in the $2p^3\,{}^4S^0$ ground state (Ottinger and Simonis, 1978). In this case, the excitation cross section for production of $N_2{}^+(B)$ was measured to be 4.6×10^{-2} Å^2. However, when the ion source conditions were adjusted so that the population of the metastable $O^+(2p^3\,{}^2P^0)$ component of the beam was increased, the vibrational distribution changed drastically in favor of that predicted by the vertical transition model. The cross section for production of $N_2{}^+(B)$ in $O^+(2p^3\,{}^2P^0)/N_2$ collisions was measured to be ~ 4 Å^2, two orders of magnitude larger than that for $O^+(2p^3\,{}^4S^0)/N_2$ CTE. Furthermore, the energy defect for production of $N_2{}^+/(B, v = 0)$ in collisions with the metastable is only -0.10 eV, but is -5.12 eV for collisions with ground-state O^+. Thus, the larger cross section is attributed to the near resonance for $O^+(2p^3\,{}^2P^0)/N_2$ CTE. In terms of potential energy surfaces, this entrance channel lies very close to the exit channel, but the $O^+(2p^3\,{}^4S^0)$–N_2 entrance channel does not. These results show that, here again, large cross sections correspond to Franck–Condon distributions, but small cross sections do not.

In the same set of experiments, Ottinger and Simonis (1978) also studied N^+/N_2 collisions. While production of $N_2{}^+(B)$ in collisions with ground-state ions $N^+(2p^2\,{}^3P)$ is endothermic by 4.19 eV, production of $N_2{}^+(B)$ in collisions with ions in the $2p^2\,{}^1S$ metastable state is very nearly thermoneutral. Nevertheless, unlike the nearly resonant $O^+(2p^3\,{}^2P^0)$–N_2 case, $N^+(2p^2\,{}^1S)$ was found to be essentially unreactive. This observation makes it clear that factors other than energetics must be considered. These authors point out that the reaction

$$O^+(2p^3\,{}^2P^0) + N_2(X\,{}^1\Sigma_g{}^+) \rightarrow O(2p^4\,{}^3P) + N_2{}^+(B\,{}^2\Sigma_u{}^+) \tag{19}$$

is allowed according to the Wigner spin rule, but

$$N^+(2p^2\,{}^1S) + N_2(X\,{}^1\Sigma_g{}^+) \rightarrow N(2p^3\,{}^4S^0) + N_2{}^+(B\,{}^2\Sigma_u{}^+) \tag{20}$$

is not. Thus, even though the potential energy surfaces for the entrance and exit channels in reaction (20) lie very close to each other, the spin–orbit coupling required to effect the transition from reactants to products is not sufficient for reaction to occur.

In addition to the studies of vibrational distributions described above, a number of other charge transfer processes have been investigated by analyzing collision-produced light. For example, vibrational distributions of $N_2(C)$, resulting from $N_2{}^+$/alkali atom collisions, have been measured (Barrett and Leventhal, 1979). The results of these experiments are consistent with the general conclusions stated above. The study of charge transfer has also been extended to systems involving triatomic targets. Because their ground states have linear configurations, and because they are both nontoxic and inexpensive, the CO_2 and N_2O molecules have received a good deal of attention. It has been shown that, at kinetic energies in the range 20–1000 eV, the only radiating product of charge transfer between most singly charged ions and N_2O is $N_2O^+(\tilde{A})$, and that the apparent vibrational distributions are independent of both kinetic energy and the ion (Bearman *et al.*, 1977). The exceptions are He^+ and Ne^+, which upon impact on N_2O, yield both $N_2O^+(\tilde{A})$ and $N_2{}^+(B)$ fragments. Here, again, the spectra are independent of kinetic energy except for the appearance of some atomic lines from highly endothermic charge transfer processes at the higher energies. The higher degree of fragmentation observed in the He^+–N_2O and Ne^+–N_2O systems has been attributed to the high RE for each of these ions. These are the only two (singly charged) ions for which formation of $N_2O^+(\tilde{C})$ by CTE is an exothermic process. Using the N_2O^+ correlation diagram (Lorquet and Cadet, 1971) as a guide, it was proposed that the incipient $N_2O^+(\tilde{C})$ then predissociates to $N_2{}^+(B)$ and $O(2p^4\,{}^3P)$, so that the internal energy is of major importance. Therefore, only He^+/N_2O and Ne^+/N_2O electron exchange leads to excited fragments $N_2{}^+(B)$, and this conclusion is supported by additional experiments performed with incident doubly charged argon ions (Bearman *et al.*, 1977). Since the RE of Ar^{++} to ground-state Ar^+ is 27.6 eV, higher than that of either He^+ and Ne^+, it might be expected that the Ar^{++}/N_2O collision-produced spectrum would closely resemble those of He^+/N_2O and Ne^+/N_2O. Indeed, the collision-produced spectra for each of the three collision systems were virtually identical. In fact, except for atomic lines of He, Ne, and Ar^+, the spectra changed very little as a function of kinetic energy.

It was remarked above that for impact of most (all but two) singly charged ions, the *apparent* vibrational distributions within $N_2O^+(\tilde{A})$ are independent of both the ion and the kinetic energy. This observation seems at variance with the strong dependencies on both kinetic energy and ion chemistry observed with diatomic targets. Closer examination shows that this is not necessarily the case. First, as usual, it is desirable to compare the observed spectrum with one that is known to result from vertical exciting transitions. Such a spectrum may be obtained as before, by using known Franck–Condon factors for the appropriate molecular transitions and generating a synthetic spectrum. An alternate method is to compare the collision-produced spectrum with one that is observed from either high-energy electron impact or photoionization, both of which should produce vertical transitions. This latter approach was taken, and the comparison showed that ion impact on N_2O and photoionization of N_2O (Lee and Judge, 1974) produced identical spectra. This observation would, in most cases, lead to the conclusion that ion impact resulted in vertical ionizing collisions. However, in this case it has been shown that the higher vibrational levels of $N_2O^+(\tilde{A})$ predissociate (Eland, 1973) so that different vibrational distributions of nascent production $N_2O^+(\tilde{A})$ would not yield spectra that differed appreciably from the photoionization spectrum.

An interesting aspect of the N_2O work is that fluorescence experiments have been performed in other laboratories at both higher and lower energies than the 20–1000-eV range discussed above. Parker and Chandra (1981) studied charge transfer between N_2O and a number of positive ions in the kinetic energy range 600–5000 eV. They found that, although the fluorescence from He^+ impact was considerably weaker than from $H_2{}^+$ impact, the spectra were similar. Gerard *et al.* (1979), using an ion cyclotron resonance apparatus, found that thermal energy He^+/N_2O collisions produced both $N_2O^+(\tilde{A})$ and $N_2{}^+(B)$, as is the case for higher kinetic energies, but they did not detect any emissions from Ne^+/N_2O. They concluded that, in spite of the nominally exothermic nature of the charge transfer leading to either excited product, the conversion of kinetic energy to internal energy must play a role in Ne^+/N_2O charge transfer.

Subsequent beam experiments (Myers and Leventhal, 1979) designed to bridge the gap between the thermal energy experiments and energies previously studied using beams were undertaken. Collision-produced spectra were acquired at laboratory kinetic energies down to 1 eV. As the kinetic energy is lowered, the $N_2{}^+(B \rightarrow X)$ fluorescence decreased relative to that from $N_2O^+(\tilde{A} \rightarrow \tilde{X})$, but the cross sections for each of these emissions increases. Thus, while these results demonstrate that conversion of kinetic energy to internal energy is important in this system, further work is required to couple the thermal and superthermal energy data.

Several fluorescence studies of charge transfer between various ions and CO_2 molecules have also been carried out. Haugh and Birely (1974) found that the principal emissions are from the $\tilde{B} \rightarrow \tilde{X}$ and $\tilde{A} \rightarrow \tilde{X}$ systems of $CO_2{}^+$ (Samson and Gardner, 1973). They observed significant changes in the Ne^+/CO_2 collision-produced spectra as a function of kinetic energy. These changes show that the vibrational distribution of the $\tilde{A}$ state changes as the kinetic energy is varied; they did not report any changes in the vibrational distribution from impact of other ions, including He^+. However, Bregman-Reisler and Doering (1974) also investigated charge transfer He^+/CO_2 collisions as well as $H_2{}^+/CO_2$ collisions and noted that, for both systems, the $CO_2{}^+(\tilde{A})$ vibrational distribution changed as a function of kinetic energy for kinetic energies below $\sim$2 keV. They also found that, above 2 keV, the spectra from $\tilde{A} \rightarrow \tilde{X}$ emissions did not change as a function of kinetic energy and were very similar to those arising from impact of high-energy protons and hydrogen atoms. Of more importance from the standpoint of comparison with models, the high-energy ion impact spectra were identical to those resulting from electron impact (Coplan and Mentall, 1973) on CO_2, indicating that as the ion energy is raised, the vibrational distributions, as usual, tended toward those produced by vertical ionizing collisions.

Because charge transfer in collisions between He_2^+ and N_2 and between He_2^+ and CO had provided a great deal of information of a fundamental nature, in addition to the practical application of these charge transfer systems in producing both laser action (Collins *et al.*, 1974; Collins and Cunningham, 1975; Laudenslager *et al.*, 1976; Rothe and Tan, 1977) and stimulated emission (Waller *et al.*, 1975), a comprehensive study of He_2^+/CO_2 collisions was undertaken (Bearman *et al.*, 1976c). These experiments were motivated by the fact that $CO_2^+ \tilde{B} \rightarrow \tilde{X}$ and $\tilde{A} \rightarrow \tilde{X}$ emissions occur in the near ultraviolet so that, if this reaction had sufficiently high cross section and led to a population inversion, it might be possible to produce laser action at one or more of these wavelengths. Indeed, emission cross sections comparable to the He_2^+–N_2 and He_2^+–CO values were measured for He_2^+/CO_2 CTE. To date, however, no reports of laser action from these transitions have appeared in the literature. A collision-produced spectrum, taken at 11 eV, is shown in Fig. 26. Emission cross sections for several

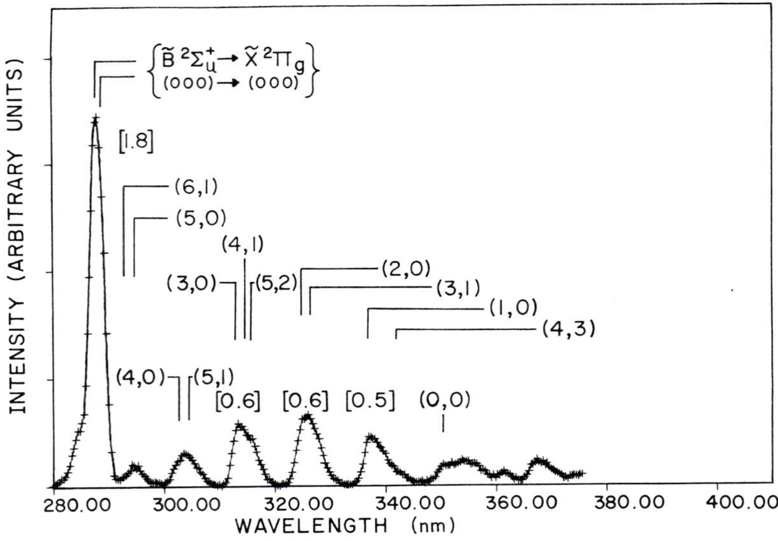

FIG. 26 Emission spectrum from 11-eV He_2^+/CO_2 collisions. The resolution is 1 nm FWHM; data are corrected for relative spectral response. The locations of the origins of several bands of the CO_2^+ ($\tilde{A}\,^2\Pi_u \rightarrow \tilde{X}\,^2\Pi_g$) system are shown; these transitions are $(v'00) \rightarrow (v''00)$, with only (v', v'') being given. Deslandres tables from Judge *et al.* (1969) have been used to identify the vibrational bands. Also shown are the locations of the heads of two bands of the CO_2^+ ($\tilde{B} \rightarrow \tilde{X}$) system. Emission cross sections in angstroms squared are indicated in square brackets for some of the prominent spectral features.

of the observed bands are indicated in square brackets. These results are consistent with those discussed earlier for large cross-section processes inasmuch as vertical ionizing transitions are occurring even at 11 eV. This conclusion is predicated on the correspondence between the He_2^+/CO_2 collision-produced spectrum and that resulting from e/CO_2 ionization (Coplan and Mentall, 1973).

V. SUMMARY

The discussion in this chapter has been intended to illustrate the high precision to which the energy partitioning among quantum states of the products of charge transfer can be determined by dispersal and detection of radiation from decay of excited collision products. The focus has been on aspects of energy transfer in electron-exchanging collisions that could only be, or at least best be, studied by using these luminescence techniques. An attempt was made to emphasize the applicability of the conclusions drawn from these experiments to energy disposal in atomic and molecular interactions in general.

While the observation of collision-produced "light" does indeed yield very precise information, interpretation of these data for even simple atom/atom collision systems is frequently limited by imprecise knowledge of the potential functions. Nevertheless, in many cases, deductions based on comparisons with isoelectronic species or molecules having constituent atoms with similar chemical properties can be of considerable aid. On the other hand, for cases in which nothing about the potential energy curves or surfaces is known, inferences about the nature of these functions may be drawn from the data.

Perhaps the best characterized of all product state distributions studied using luminescence techniques are vibrational distributions of electronically excited product diatomic ions or neutral molecules. The large quantity of data that have been acquired over a wide range of ion kinetic energies, ~ 1 eV to ~ 1 MeV, leads to the conclusion that a proper description of the interaction requires consideration of the molecular aspect of the collision.

There are several extensions and refinements of luminescence experiments that will no doubt be undertaken in the near future. To date, only one ion/atom experiment has been performed in which luminescence from interactions involving one state-selected, laser-excited reactant has been observed (Kushawaha *et al.*, 1980). As laser technology develops, it is expected that many more such experiments will be performed.

It would also be desirable to add the capability of performing experiments involving free radical neutral reactants. Of course, the quintessential experiments would be those employing reactant hydrogen atoms. Obviously, a great deal of fundamental information could be obtained if this were possible.

ACKNOWLEDGMENTS

I wish to express my thanks for support from the Department of Energy, Division of Chemical Sciences (Contract No. DE-AS02-76-ER02718), and from the Office of Naval Research (Contract No. N00014-76-C-0760). I would also like to acknowledge the many contributions of my collaborators to the work described in this chapter. Special thanks are due to P. L. DeVries and C. E. Burkhardt for critical readings of the manuscript.

REFERENCES

Andreev, E. P., Ankudinov, V. A., Bobashev, S. V., and Matveev, V. B. (1967). *Sov. Phys.-JETP* **25**, 232–235.

Arrathoon, R., Littlewood, I. M., and Webb, C. E. (1973). *Phys. Rev. Lett.* **31**, 1168–1170.

Barrett, J. L., and Leventhal, J. J. (1979). *J. Chem. Phys.* **71**, 4015–4020.

Barrett, J. L., and Leventhal, J. J. (1980). *Appl. Phys. Lett.* **36**, 869–871.

Barrett, J. L., and Leventhal, J. J. (1981). *Phys. Rev. A* **23**, 485–490.

Barrett, J. L., Mlynczak, M. G., and Leventhal, J. J. (1981). *J. Chem. Phys.* **75**, 2705–2710.

Bearman, G. H., and Leventhal, J. J. (1978). *Phys. Rev. A* **17**, 80–89.

Bearman, G. H., Earl, J. D., Harris, H. H., and Leventhal, J. J. (1976a). *Phys. Rev. A* **13**, 1734–1742.

Bearman, G. H., Earl, J. D., Pieper, R. J., Harris, H. H., and Leventhal, J. J. (1976b). *Appl. Phys. Lett.* **29**, 108–110.

Bearman, G. H., Harris, H. H., and Leventhal, J. J. (1976c). *Appl. Phys. Lett*, **28**, 345–347.

Bearman, G. H., Harris, H. H., and Leventhal, J. J. (1977). *J. Chem. Phys.* **66**, 4111–4115.

Bethe, H. A., and Salpeter, E. E. (1977). "Quantum Mechanics of One- and Two-Electron Atoms." Plenum Press, New York.

Birely, J. H. (1974). *Phys. Rev. A* **10**, 550–562.

Brandt, D., Ottinger, Ch., and Simonis, J. (1973). *Ber. Bunsenges, Phys. Chem.* **77**, 748–655.

Bregman-Reisler, H., and Doering, J. P. (1974). *Chem. Phys. Lett.* **27**, 199–202.

Cohen, J. S., and Schneider, B. (1974). *J. Chem. Phys.* **61**, 3230–3239.

Collins, C. B., and Cunningham, A. J. (1975). *Appl. Phys. Lett.* **27**, 127–128.

Collins, C. B., Cunningham, A. J., and Stockton, M. (1974). *Appl. Phys. Lett.* **25**, 344–345.

Coplan, M. A., and Mentall, J. E. (1973). *J. Chem. Phys.* **58**, 4912–4915.

Davis, C. C., and King, T. A. (1975). *Adv. Quantum Electron.* **3**, 169–454.

Dawson, H. R., and Loyd, D. H. (1974). *Phys. Rev. A* **9**, 166–170.

Dawson, H. R., and Loyd, D. H. (1977). *Phys. Rev. A* **15**, 43–47.

Doughty, B. M., Goad, M. L., and Cernosek, R. W. (1978). *Phys. Rev. A* **18**, 29–35.

Eland, J. H. D. (1973). *Int. J. Mass. Spectrom. Ion Phys.* **12**, 395–398.

Foreman, P. B., Rol, P. K., and Coffin, K. P. (1974). *J. Chem. Phys.* **61**, 1658–1665.

Gerard, M., Govers, T. R., and Marx, R. (1979). *Chem. Phys.* **36**, 247–257.

Ginter, M. L., and Ginter, D. S. (1968). *J. Chem. Phys.* **48**, 2284–2291.

Hagstrum, H. D. (1956). *Phys. Rev.* **104**, 309–316.

Harris, H. H., and Leventhal, J. J. (1976). *J. Chem. Phys.* **64**, 3185–3191.

Harris, H. H., Crowley, M. G., and Leventhal, J. J. (1974). *Chem. Phys. Lett.* **29**, 540–546.

Haugh, M. J., and Birely, J. H. (1974). *J. Chem. Phys.* **60**, 264–269.

Herzberg, G. (1944). "Atomic Spectra and Atomic Structure." Dover, New York.

Herzberg, G. (1950). "Spectra of Diatomic Molecules." Van Nostrand-Reinhold, New York.

Hughes, R. H., Dawson, H. R., and Doughty, B. M. (1967). *Phys. Rev.* **164**, 166–170.

Hughes, R. H., Stigers, C. A., Doughty, B. M., and Stokes, E. D. (1970). *Phys. Rev. A* **1**, 1424–1432.

Judge, D. L., Bloom, G. S., and Morse, A. L. (1969). *Can. J. Phys.* **47**, 489–497.

Kelley, J. D., Bearman, G. H., Harris, H. H., and Leventhal, J. J. (1977a). *In* "State to State Chemistry" (P. R. Brooks and E. F. Hayes, eds.), ACS Symp. Ser., pp. 171–174. American Chemical Society, Washington, D.C.

Kelley, J. D., Bearman, G. H., Harris, H. H., and Leventhal, J. J. (1977b). *Chem. Phys. Lett.* **50**, 295–300.

Kelley, J. D., Bearman, G. H., Harris, H. H., and Leventhal, J. J. (1978). *J. Chem. Phys.* **68**, 3345–3351.

Kimura, M., Olson, R. E., and Pascale, J. (1982). *Phys. Rev. A* **26**, 1138–1141.

Krupenie, P. H. (1966). The Band Spectrum of Carbon Monoxide. Nat. Bur. St. Ref. Data Ser., U.S. National Bureau of Standards, Washington, D. C.

Krupenie, P. H. (1972). *J. Phys. Chem. Ref. Data* **1**, 423–534.

Kulander, K. C., and Dahler, J. S. (1975). *J. Phys. B* **8**, 2679–2696.

Kushawaha, V. S., Burkhardt, C. E., and Leventhal, J. J. (1980). *Phys. Rev. Lett.* **45**, 1686–1688.

Latush, E. L., and Sem, M. F. (1972). *JETP Lett.* **15**, 457–459.

Laudenslager, J. B., Pacala, T. J., and Wittig, C. (1976). *Appl. Phys. Lett.* **29**, 580–582.

Lee, L. C., and Judge, D. L. (1974). *J. Phys. B* **7**, 626–629.

Leventhal, J. J., Earl, J. D., Harris, H. H. (1975). *Phys. Rev. Lett.* **35**, 719–722.

Lipeles, M. (1969). *J. Chem. Phys.* **51**, 1252–1253.

Lofthus, A., and Krupenie, P. H. (1977). *J. Phys. Chem. Ref. Data* **6**, 113–307.

Lorquet, J. C., and Cadet, C. (1971). *Int. J. Mass. Spectrom. Ion Phys.* **7**, 245–254.

Massey, H. S. W., Burhop, E. H. S., and Gilbody, H. B. (1971). "Electronic and Ionic Impact Phenomena," Vol. III. Oxford Univ. Press (Clarendon), London and New York.

McNeal, R. J., and Birely, J. H. (1973). *Rev. Geophys. Space Phys.* **11**, 633–692.

Moore, J. H. (1973). *Phys. Rev. A* **8**, 2359–2362.

Moore, J. H., and Doering, J. P. (1969). *Phys. Rev.* **182**, 176–180.

Myers, G. D., and Leventhal, J. J. (1978). *Phys. Rev. A* **18**, 434–440.

Myers, G. D., and Leventhal, J. J. (1979). *J. Chem. Phys.* **70**, 2030–2031.

Myers, G. D., Ambrose, J. G., James., P. B., and Leventhal, J. J. (1978). *Phys. Rev. A* **18**, 85–88.

Myers, G. D., Barrett, J. L., and Leventhal, J. J. (1979). *Phys. Rev. A* **20**, 797–803.

Ottinger, Ch., and Simonis, J. (1978). *Chem. Phys.* **28**, 97–112.

Parker, J. E., and Chandra, H. (1981). *Int. J. Mass Spectrom. Ion Phys.* **39**, 15–25.

Risley, J. S., de Heer, F. J., and Kerkdijk, J. (1978). *J. Phys. B* **11**, 1759–1781.

Rosen, B. (1970). "Spectroscopic Data Relative to Diatomic Molecules." Pergamon, Oxford.

Rothe, D. E., and Tan, K. O. (1977). *Appl. Phys. Lett.* **30**, 152–154.

Rothwell, H. L., Amme, R. C., and Van Zyl, B. (1978). *J. Chem. Phys.* **68**, 4326–4327.

Samson, J. A. R. (1967). "Techniques of Vacuum Ultraviolet Spectroscopy." Wiley, New York.

Samson, J. A. R., and Gardner, J. L. (1973). *J. Geophys. Res.* **78**, 3663–3667.

Schneider, B., and Cohen, J. S. (1974). *J. Chem. Phys.* **61**, 3240–3243.

Shipsey, E. J., Redmon, L. T., Browne, J. C., and Olson, R. E. (1978). *Phys. Rev. A* **18**, 1961–1966.

Striganov, A. R., and Sventitaskii, N. S. (1968). "Tables of Spectral Lines of Neutral and Ionized Atoms." Plenum Press, New York.

Tiernan, T. O., and Lifshitz, C. (1981). *In* "The Excited State in Chemical Physics" (J. W. McGowan, ed.), Part 2, pp. 81–224. Wiley, New York.

Van Zyl, B., Rothwell, H. L., and Neumann, H. (1980). *Phys. Rev. A* **21**, 730–737.

Waller, R. A., Collins, C. B., and Cunningham, A. J. (1975). *Appl. Phys. Lett.* **27**, 323–325.

Wiese, W. L., Smith, M. W., and Glennon, B. M. (1966). "Atomic Transition Probabilities," Vol. I, Nat. Bur. St. Ref. Data Ser., U.S. National Bureau of Standards, Washington, D.C.

Willett, C. S. (1974). "Introduction to Gas Lasers: Population Inversion Mechanisms." Pergamon, Oxford.

Zhukov, V. V., Latush, E. L., and Sem, M. F. (1977). *Sov. Phys. J.* **20**, 957–959.

Chapter 25

Relaxation dynamics of open-shell cations studied by photoelectron–photon coincidence spectroscopy

JOHN P. MAIER
FRITZ THOMMEN
PHYSIKALISCH-CHEMISCHES INSTITUT
UNIVERSITÄT BASEL,
BASEL, SWITZERLAND

ISBN 0-12-120803-6

I. INTRODUCTION

A. General

The detection of the radiative decay of a variety of electronically excited organic open-shell cations in the gas phase (Maier, 1980, 1982) provided an opportunity to investigate not only their spectroscopic structure but radiationless processes in general. Such studies on open-shell cations present certain advantages with respect to closed-shell molecules. These are, on the one hand, due to the favorable disposition of the electronic states of the open-shell organic cations and, on the other hand, due to experimental artifacts.

The former aspect, the structural advantages, are illustrated by the scheme shown in Fig. 1. In open-shell organic cations, the energy separation between the first excited doublet state $^2\tilde{A}$ and the ground-state $^2\tilde{X}$ is less than between the corresponding states of the same multiplicity (i.e., S_1 and S_0) in the parent molecules. Thus, for the polyunsaturated cations which decay radiatively, the wavelength of the photons to be detected lies in a favorable region, 300–800 nm. Furthermore, a closed-shell species in the first excited singlet state S_1 can couple to levels of at least two lower lying states, namely, to the first excited triplet state T_1 and to the electronic ground-state S_0. In open-shell cations, the decay pattern of the first excited doublet state $^2\tilde{A}$ is simplified because the lowest state of higher multiplicity

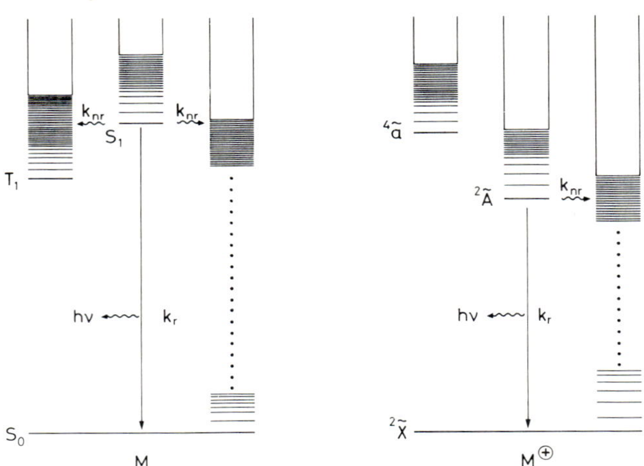

FIG. 1 Schematic comparison of the relative positions of the lowest electronic states of closed-shell (M) and open-shell (M^+) species and the relaxation pathways for the first excited electronic states with the same multiplicity as their ground states. The symbols k_{nr} and k_r refer to the nonradiative and radiative transitions, respectively.

$^4\tilde{a}$ is expected to lie to higher energy for a similar geometry (Carsky and Zahradnik, 1973). Thus, the excited radical cation can decay either radiatively, $^2\tilde{A} \rightarrow {}^2\tilde{X}$, or nonradiatively, $^2\tilde{A} \rightsquigarrow {}^2\tilde{X}$, to the ground state. Experimental advantages arise in quantitative studies of radiative and nonradiative transitions (e.g., internal conversion, fragmentation) in open-shell cations (Leach *et al.*, 1980) because the number of particles (i.e., photons, ions, or electrons) can be individually counted. This is achieved by coincidence techniques whereby two of the particles are concurrently detected. The now well-established technique using photoelectrons and ions yields the branching ratios for state-selected cations (Baer, 1979). In this chapter, the photoelectron–photon coincidence technique is the focus of attention since this allows the absolute fluorescence quantum yields, in addition to the lifetimes, for open-shell cations in selected vibrational levels of the excited state to be measured. These data can then be converted into the rate constants of the radiative and nonradiative channels depleting the respective level. In contrast, for neutral species, the determinations of the fluorescence quantum yields are by and large indirect (Avouris *et al.*, 1977).

B. A Brief History

An electron-photon coincidence technique was initially applied to determine lifetimes of excited atoms and molecules (Imhof and Read, 1971). The excited species were formed using monochromatic electrons, and their states of interest were defined by the energy loss of inelastically scattered electrons which were measured in coincidence with photons emitted by the species. Later on, this approach was extended to lifetime determinations of $N_2^+\ \tilde{B}\,^2\Sigma_u^+$, $CO^+\ \tilde{B}\,^2\Sigma^+$ and $CO_2^+\ \tilde{A}\,^2\Pi_u$, $\tilde{B}\,^2\Sigma_2^+$ (Smith *et al.*, 1975). The lifetimes for the ions are, however, averaged values for the populated levels of the electronic state.

The energy within the excited electronic state is, on the other hand, defined using the photoelectron–photon coincidence technique (Bloch and Turner, 1975). The cations are generated by He(Iα) photoionization of the molecular species M in its ground state ^{1}X, and energy-selected photoelectrons e_{KE} are detected in coincidence with wavelength undispersed photons $h\nu$ emitted by the excited cations $M^+\ ^2\tilde{A}(v')$, i.e.,

$$M\,^1X \xrightarrow{He(I\alpha)} M^{\oplus}\,^2\tilde{A}(v') + \boxed{e_{K.E.}}$$

$$M^{\oplus}\,^2\tilde{A}(v') \rightsquigarrow \boxed{h\nu} \quad (\text{in coincidence with } e_{K.E.})$$

$$M^{\oplus}\,^2\tilde{A}(v') \downarrow M^{\oplus}\,^2\tilde{X}$$

This was used to determine the lifetimes of some diatomic (N_2^+ $\tilde{B}\,^2\Sigma_u^+$, CO^+ $\tilde{B}\,^2\Sigma^+$) and triatomic (N_2O^+ $\tilde{A}\,^2\Sigma^+$, CO_2^+ $\tilde{A}\,^2\Pi_u$, $\tilde{B}\,^2\Sigma_u^+$) cations in selected vibrational levels and, subsequently, of fluoro-substituted benzene cations (Ames *et al.*, 1976).

The direct determinations of the fluorescence quantum yields were carried out, first, for some vibrational levels of CO_2^+ $\tilde{A}\,^2\Pi_u$, $\tilde{B}\,^2\Sigma_u^+$ (Schlag *et al.*, 1977) and N_2O^+ $\tilde{A}\,^2\Sigma^+$ (Frey *et al.*, 1978), using dispersed synchrotron radiation to generate the cations. The threshold electrons ($e_{KE} = 0$) were measured in coincidence with undispersed photons.

Another approach has been to detect mass-analyzed cations in coincidence with the wavelength-undispersed photons using a photoionization source (Eland *et al.*, 1976). Lifetimes and fluorescence quantum yields have been obtained by this method for the excited electronic states of the cations of carbon dioxide (Leach *et al.*, 1978), of the fluoro-substituted benzenes (Dujardin *et al.*, 1980a), and of chloro- and dichloroacetylene (Dujardin *et al.*, 1980b). The respective values are again averaged values for the populated levels of the excited electronic state.

We built a He(Iα) photoelectron–photon coincidence apparatus a few years ago (Maier and Thommen, 1980a) in order to study systematically and quantitatively the intramolecular pathways of open-shell organic cations whose emission spectra we had observed previously (Maier, 1980). This entailed the determination of their fluorescence quantum yields and lifetimes in defined vibrational levels of their excited doublet electronic states. In this review chapter, we first describe the experiment and associated features; then, the information and trends forthcoming from these investigations are illustrated by examples, as well as their relationship to the theories of radiationless transitions (Jortner and Mukamel, 1973; Freed, 1976).

II. PHOTOELECTRON–PHOTON COINCIDENCE EXPERIMENT

A. *Principle*

The principle of the photoelectron–photon coincidence technique is to measure the time delay between the ejection of photoelectrons with specific energy and any subsequently emitted photons. The necessary experimental arrangement is schematically depicted in Fig. 2. Open-shell cations are generated using the He(Iα) photon resonance line, and their internal energy is defined by the energy of the ejected photoelectron, e_{KE}, which also provides a time reference for the creation of the cation. These energy-selected electrons are then measured in coincidence with wavelength-undispersed photons. The events are stored in a multichannel analyzer, and the result is a

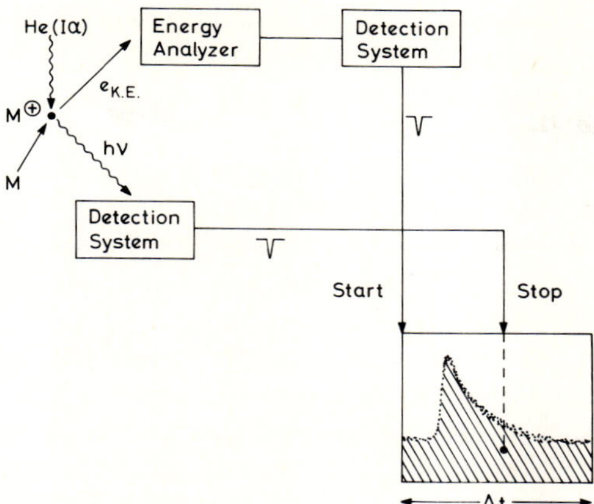

FIG. 2 Principal features of the photoelectron–photon coincidence technique.

curve of true coincidences, provided that the excited cation decays radiatively, superimposed on a uniform background of random coincidences. True coincidences arise only if the electron and the photon stem from the same cation of the specified internal energy.

B. Apparatus

The apparatus, which is depicted schematically in Fig. 3, is constructed within and around a stainless steel tank which houses the electron analyzer (Maier and Thommen, 1980a). It is surrounded by three perpendicular pairs of Helmholtz coils to eliminate magnetic fields. The apparatus consists of two parts: one for the electron energy analysis and the other for the detection of undispersed photons. The sample is introduced through an externally adjustable capillary at pressures less than 10^{-3} torr to ensure collision-free conditions. The base pressure in the vacuum tank during the measurement is $\leqslant 5 \times 10^{-5}$ torr. A collimated photon beam perpendicular to the gas jet ionizes the sample and also defines the ionization region. The ejected photoelectrons are energy-selected by a 55-mm radius hemispherical analyzer, constructed from copper, with a distance of 15 mm between the hemispheres. The 1-mm wide, 10-mm high, curved entrance slits yield a half-width of 55 meV for argon. Because of the differences in the flight times of photoelectrons through the analyzer, there is a time broadening which is apparent in the rise time of the coincidence curves. This is found to be, for example, ≈ 10 nsec for 10-eV electrons. The electrons are detected by a channeltron.

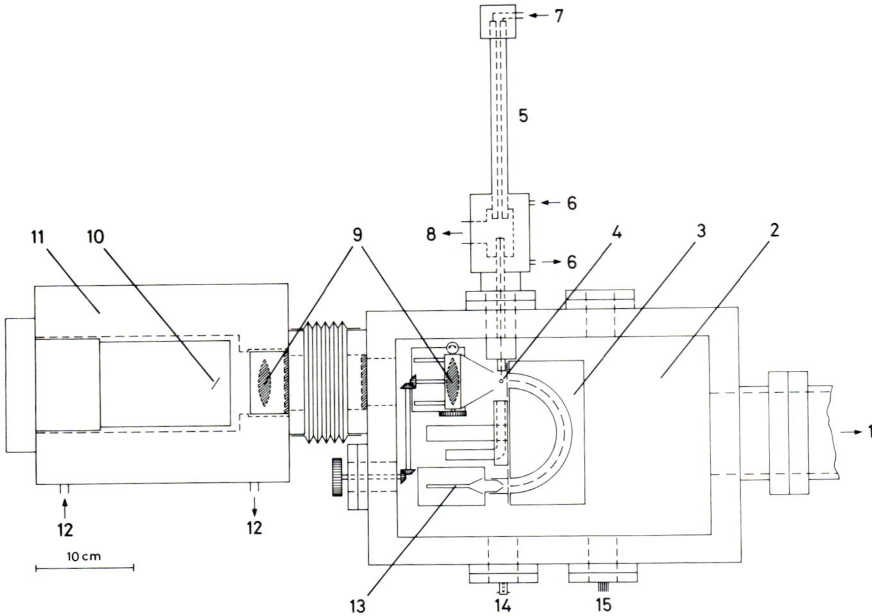

FIG. 3 Arrangement of the photoelectron–photon coincidence apparatus used for the measurements of fluorescence quantum yields and lifetimes of state-selected cations: 1, to the diffusion pump; 2, stainless steel vacuum tank; 3, hemispherical electron analyzer; 4, ionization region; 5, discharge lamp; 6, water cooling system; 7, rare gas inlet; 8, to the rotary pump; 9, lens system; 10, photocathode; 11, cooled photomultiplier housing; 12, $N_2(l)$ cooling system; 13, channeltron; 14, high voltage feedthrough; and 15, low voltage feedthroughs.

On the photon side, the ionization region is focused by means of two f/1, 50-mm biconvex lenses onto the GaAs photocathode of a cooled (233-K) photomultiplier (RCA C31034 A). This type was chosen because it has a uniform and high ($\approx 30\%$ at 500-nm) quantum efficiency over a large wavelength region. Thus, there is only a few percent variation in the photon detection probability over the typical range (≈ 100 nm) of an emission system of an organic cation. The relative quantum efficiency was established by using a calibrated standard lamp and a monochromator system. The absolute determination of the detection probability of photons as a function of wavelength was accomplished by coincidence measurements for cations with unity fluorescence quantum yield whose emission systems lie at different wavelengths (e.g., $N_2^+\ \tilde{B}\,^2\Sigma_u^+$, $CO_2^+\ \tilde{A}\,^2\Pi_u$, hexafluorobenzene cation $\tilde{B}\,^2A_{2u}$).

Both parts of the apparatus were empirically optimized to minimize the background rates. The electron analyzer is coated with soot, yielding a rate

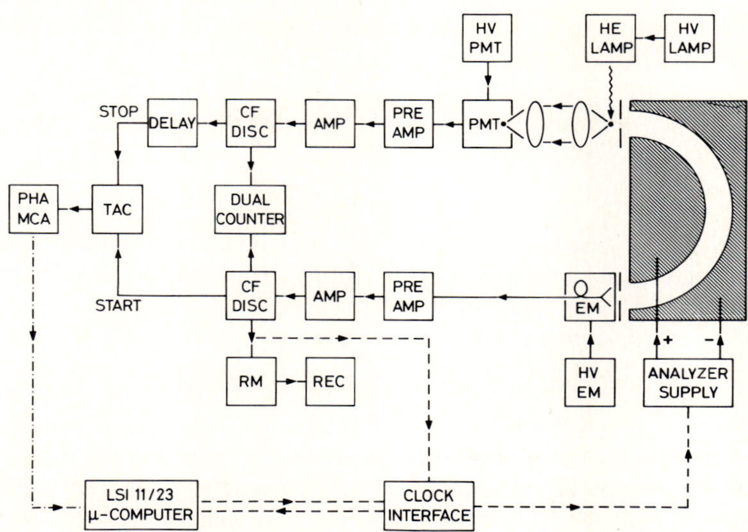

FIG. 4 Block diagram of the electronics of the coincidence apparatus: →, accumulation of coincidences; –·→·–, transfer of the coincidence data; – → –, recording of photoelectron spectra.

of scattered electrons of a few hertz, while a light trap together with an aligned baffle system reduces the rate of stray photons to about 500 Hz.

The block diagram in Fig. 4 describes the electronics which are used to determine the time distribution of emitted photons following the creation of cations. Because the photon rate usually exceeds the electron rate by an order of magnitude, the electron events are used to start the measurements of the timing sequences. The photon events are artificially delayed to compensate for the flight time of the photoelectrons through the analyzer. The time differences between the electron and photon events are then measured by using a time-to-amplitude converter and a pulse–height analyzer of a multichannel storage device. Finally, the coincidence curves are transferred via a serial line interface to an LSI 11/23 microcomputer, where the data are evaluated.

C. *Information*

Fluorescence Quantum Yields

In order to deduce the fluorescence quantum yield $\phi_F(v')$ of the cation in the state-selected level v' from the coincidence curve, the following six equations are defined. The rate of detected electrons N_e' consists of two parts, namely, of the electrons which define the internal energy of the cation N_e and any scattered electrons $N_{e,s}$.

$$N_e' = N_e + N_{e,s}. \tag{1}$$

The rate of true electrons N_e is given by

$$N_e = N_i g(v') f_e, \tag{2}$$

where N_i is the ionization rate, $g(v')$ the branching ratio of the ionization process, and f_e the collection efficiency of electrons within the selected energy bandpass.

The rate of detecting photons $N_{h\nu}$ is also comprised of two terms:

$$N_{h\nu} = N_i f_{h\nu}(\lambda)\, \Sigma_{v'} \phi_F(v') g(v') + N_{h\nu,s}. \tag{3}$$

In Eq. (3), $N_{h\nu,s}$ represents the rate of detection of photons resulting from the visible emission lines of the rare gas photon source and the dark count rate of the photomultiplier. $f_{h\nu}(\lambda)$ is the wavelength-dependent collection efficiency of photons. The rate of detecting true coincidences N_t is defined by

$$N_t = N_i g(v') f_e f_{h\nu}(\lambda) \phi_F(v'). \tag{4}$$

Equations (2) and (4) yield

$$N_t = N_e f_{h\nu}(\lambda) \phi_F(v'). \tag{5}$$

The ability to calibrate the wavelength-dependent factor $f_{h\nu}(\lambda)$ by coincidence measurements with cations which have unity fluorescence quantum yields (vide supra), and the determination of N_e and N_t, allows one to evaluate the absolute fluorescence quantum yields $\phi_F(v')$ of open-shell cations which emit photons in the 200–850-nm range. In practice, N_e' is measured during the accumulation period, and the $N_{e,s}$ correction is merely 1–2%. N_t is evaluated from the coincidence curve by subtracting the uniform background of false coincidences. The rate of the latter, N_f, within the time interval Δt is given by

$$N_f = N_e' N_{h\nu} \Delta t, \tag{6}$$

provided that the rates N_e' and $N_{h\nu}$ are small compared with $1/\Delta t$. This, of course, limits the rate of ionization which can be employed.

Typical conditions during the experiments are $N_e = 0.1$–1 kHz and $N_{h\nu} = 1$–10 kHz. As the collection efficiencies are $f_{h\nu} \approx f_e \approx 3 \times 10^{-3}$, accumulation times of 10–20 hr are customary and, as a consequence, $\phi_F(v') \approx 10^{-3}$ is the lower detection limit. Absolute $\phi_F(v')$ values are otherwise obtained with an uncertainty of 5–10%.

Lifetimes

The lifetimes $\tau(v')$, corresponding to the $\phi_F(v')$ values, are extracted from a semilogarithmic plot of the coincidence curves (after subtraction of background) by a least squares linear fit or by fitting a model consisting of sums of exponentials to the decay data using a nonlinear least squares algorithm.

The lifetimes and the fluorescence quantum yields often drop off simultaneously and, as a consequence, the lower limit for the lifetime measurement is given by the rise time of the coincidence curve (usually between 10 and 15 nsec).

Systematic Errors

The generated cations leave the coincidence region after a few microseconds due to their thermal velocities. Therefore, the measured lifetimes τ_{exp} and the associated fluorescence quantum yields $\phi_{F,exp}$ may be too small. The values can be roughly corrected using Eqs. (7) and (8):

$$\frac{1}{\tau_{exp}} = \frac{1}{\tau} + \frac{1}{\tau_{app}} \tag{7}$$

$$\frac{\phi_F}{\phi_{F,exp}} = 1 + \frac{\tau_{exp}}{\tau_{app}}. \tag{8}$$

ϕ_F and τ are the corrected values, and τ_{app} is the dwell time of the cations in the observation region, which depends on their mass. Lifetime measurements for a long-lived ion, $CO^+(\tilde{A}\,^2\Pi, v' = 1)$, $\tau \approx 3.6$ μsec (Möhlmann and de Heer, 1976), yield an estimate for τ_{app} of ≈ 8 μsec; as a result, experimental values for cations with $\tau > 200$ nsec have to be corrected. Thus, the practical time scale of this coincidence experiment lies between 10^{-10} and 10^{-6} sec. The lower limit is determined by the detection limit for true coincidences, $\phi_F(v') > 10^{-3}$, and the upper one by the losses due to the escape of the cations from the observation region. Furthermore, no pressure dependence is apparent in the 10^{-4}–10^{-3} torr range used for the coincidence experiments.

III. INTRAMOLECULAR RELAXATION OF OPEN-SHELL CATIONS: MODEL THEORIES

The relaxation processes of electronically excited radical cations in the gas phase which are considered in view of the time scale of the photoelectron–photon coincidence experiment (10^{-6}–10^{-10} sec), and the pressure conditions, are radiative decay and the radiationless transitions of internal

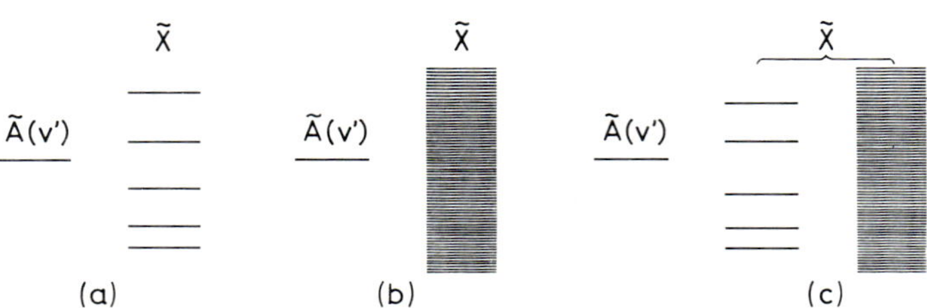

FIG. 5 Model energy level schemes for radiationless transition theory: (a) resonance limit, (b) statistical limit, and (c) intermediate case.

conversion, fragmentation, and/or isomerization. Intramolecular vibrational relaxation and collisional deactivation are expected to have an influence in the range $>10^{-6}$ sec (Leach *et al.*, 1980). Also, in cases when fluorescence and fragmentation are both detected, it could be shown that the latter step is slower than internal conversion (Forster *et al.*, 1981; Maier and Thommen, 1981b). Thus, the relaxation of open-shell cations in their lowest excited states under collision-free conditions can be discussed as competition between radiative decay $^2\tilde{A} \rightarrow {}^2\tilde{X}$ and internal conversion $^2\tilde{A} \rightsquigarrow {}^2\tilde{X}$. In this chapter, we outline the main features of radiationless transition theories of particular relevance to the decay behavior of excited open-shell cations.

Radiationless transition models have been applied almost exclusively to neutral molecules (Avouris *et al.*, 1977). The three cases which are commonly distinguished are the "resonance limit," the "statistical limit," and the "intermediate case," which are characterized by the structure of the vibronic levels of a lower electronic state which are coupled to isoenergetic vibronic levels of a higher electronic state (Jortner and Mukamel, 1973; Freed 1976).

This is a typical situation for open-shell cations because mostly only two electronic states are involved in the relaxation process (cf. Fig. 1). The excited state $\tilde{A}$ is coupled radiatively to the ground-state $\tilde{X}$ and nonradiatively via matrix elements of the nuclear kinetic energy $v(\tilde{A}-\tilde{X})$ to one or more of a set of isoenergetic levels of $\tilde{X}$ (Leach *et al.*, 1980). In Fig. 5, the energy level schemes for the three cases are drawn. In the following sections, the results of these model cases and their consequences on observed fluorescence quantum yields and lifetimes are briefly discussed.

A. *Resonance Limit*

In the case of small cations, the density of vibrationally excited ground-state levels $\rho(\tilde{X})$ is small, and thus one has a system of interacting discrete $\tilde{A}$ and $\tilde{X}$ levels which form resonant states (cf. Fig. 5a). Because the recurrence time $\tau_{rec} = \hbar\rho(\tilde{X})$ between $\tilde{A}$ and $\tilde{X}$ levels is short compared with the relaxation for $\tilde{X}$ levels, exponential decay is expected. In absence of fragmentation or collisional relaxation, the excited system decays only radiatively and the measured fluorescence quantum yield is equal to unity, provided the observation period is long enough. On the other hand, $\tilde{A}$–$\tilde{X}$ coupling induces a redistribution of $\tilde{A}$-level oscillator strength, which implies a lengthening of the fluorescence lifetime of the resulting mixed levels compared with that of a pure $\tilde{A}$ level.

B. *Statistical Limit*

In this case, the density of isoenergetic vibrational levels of $\tilde{X}$ is so high that a quasi-continuum is formed [Fig. 5(b)]. Moreover, the $\tilde{A}$–$\tilde{X}$ recurrence time is much greater than the maximum time that the excited ion is isolated. Thus, irreversible radiationless transition occurs in large polyatomic cations. The nonradiative rate constant $k_{nr}(v')$ is related to the average matrix elements $\langle v(\tilde{A}-\tilde{X})\rangle$ and to the density of states $\langle\rho(\tilde{X})\rangle$ by the Fermi–Wentzel golden rule [Eq. (9)]:

$$k_{nr}(v') = \frac{2\pi}{\hbar}\langle v(\tilde{A}-\tilde{X})\rangle^2\langle\rho(\tilde{X})\rangle. \tag{9}$$

The average lifetime $\tau(v')$ and fluorescence quantum yield $\phi_F(v')$ of a level v' of the excited state $\tilde{A}$ are then determined by the radiative $k_r(v')$ and nonradiative $k_{nr}(v')$ rate constants:

$$\tau(v') = \frac{1}{k_r(v') + k_{nr}(v')}, \tag{10}$$

$$\phi_F(v') = \frac{k_r(v')}{k_r(v') + k_{nr}(v')}. \tag{11}$$

The decay of an excited cation in the statistical limit will be monoexponential, and the fluorescence lifetime $\tau(v')$ may be shorter than the radiative lifetime, $\tau_r(v') = k_r(v')^{-1}$, with a fluorescence quantum yield, $\phi_F(v') \leqslant 1$.

When the energy difference $\Delta E(\tilde{A}-\tilde{X})$ between the vibrationless levels of the excited and the ground state is large compared with the vibrational quantum $\hbar\omega_i$ of the acceptor mode i, which is responsible for the nonradiative transition, the rate constant can be approximated by (Freed, 1976)

$$k_{nr}(0) \approx \left(\frac{2\pi\hbar\omega_i}{\Delta E(\tilde{A}-\tilde{X})}\right)^{1/2}\frac{\beta(\tilde{A}-\tilde{X})^2}{\hbar^2\omega_i}\exp(-x_i)\exp\left(-\Omega\frac{\Delta E(\tilde{A}-\tilde{X})}{\hbar\omega_i}\right), \tag{12}$$

where $\beta(\tilde{A}-\tilde{X})$ is an electronic coupling factor and

$$x_i = \frac{M_i\omega_i}{2\hbar}\,[Q_{eq}(\tilde{A}) - Q_{eq}(\tilde{X})]^2 \tag{13}$$

is a dimensionless expression for the displacement from the equilibrium positions for normal mode i. M_i is the corresponding reduced mass, and $Q_{eq}(\tilde{A})$ and $Q_{eq}(\tilde{X})$ are the normal coordinates. In general, $x_i \ll 1$, and

$$\Omega = \ln\frac{\Delta E(\tilde{A}-\tilde{X})}{x_i\hbar\omega_i} - 1 \tag{14}$$

is a slowly varying function of $\Delta E(\tilde{A}-\tilde{X})$. Hence, the radiationless transition rate $k_{nr}(0)$ is an exponentially decreasing function of the energy gap.

In cases where ω_i is the frequency of the CH stretching mode, which acts as the acceptor, $k_{nr}(0)$ will be affected by deuteration and there is a decrease in the nonradiative rate [due to the exponential factor in Eq. (12)]. In other words, more quanta are required to absorb the energy $\Delta E(\tilde{A}-\tilde{X})$.

C. *Intermediate Case*

Radiationless transition processes in the statistical limit are the primarily encountered situation for open-shell organic cations. However, there are a handful of cations which show features characteristic of the intermediate case description of intramolecular decay. The manifold of vibronic levels of the $\tilde{X}$ state can be divided into two classes [Fig. 5(c)]: those which form a quasi-continuum, as in the statistical limit, and others which are strongly coupled, as in the resonance limit (Lahmani *et al.*, 1974). For these intermediate case open-shell cations, quasi-biexponential decay curves with a short τ_1 and a long τ_2 component are observed. These values for the decay times and the total fluorescence quantum yield, $\phi_{F,tot}(v')$, can be described by simple analytical expressions (Tramer and Voltz, 1979):

$$\frac{1}{\tau_1} \approx k(\tilde{A}) + \Delta/\hbar, \tag{15}$$

$$\frac{1}{\tau_2} \approx k(\tilde{A})/N + k(\tilde{A}, \tilde{X}), \tag{16}$$

$$\phi_{F,tot}(v') \approx \frac{k_r(v')}{\Delta/\hbar + k(\tilde{A})} + \frac{k_r(v')}{Nk(\tilde{A}, \tilde{X})}. \tag{17}$$

The short time behavior of the system is similar to the statistical limit and consists of the rates $k(\tilde{A}) = k_r(v') + k_{nr}(v')$ and $\Delta/\hbar$. The latter describes the radiationless coupling of the initially populated level v' of $\tilde{A}$ with isoenergetic discrete levels of $\tilde{X}$. The resultant levels whose $\tilde{A}$ character is not irreversibly lost then decay with the rate $1/\tau_2$. $k(\tilde{A}, \tilde{X})$ is the sum of the rates for vibrational relaxation, collisional deactivation, and the radiationless transition to the dense manifold of $\tilde{X}$ levels. The number of levels with $\tilde{A}$ and $\tilde{X}$ character N is proportional to the amplitude ratio A_1/A_2 of the prompt and long decay components. N grows with increasing level density $\rho(\tilde{X})$ and width Δ of distribution of oscillator strength of the initially populated level of the $\tilde{A}$ state. Thus, with increasing vibrational excess energy, τ_2 increases and the amplitude A_2 decreases while τ_1 becomes shorter and, as a consequence, the experimental detection of τ_1 and τ_2 becomes more and more difficult. In addition to quasi-biexponential decay curves, fluorescence quantum yields $\phi_{F,tot}(v') \leqslant 1$ are expected because irreversible nonradiative decay from the initially populated level of the excited state or from resonant levels to the quasi-continuum of isoenergetic $\tilde{X}$ levels is also possible.

IV. EXAMPLES AND APPLICATIONS

A. *Small Open-Shell Cations*

Photoelectron–photon coincidence measurements on diatomic and triatomic open-shell cations have mainly been used to calibrate the apparatus (Maier and Thommen, 1980a) because they have unity fluorescence quantum yields, e.g., N_2^+, CO_2^+. In addition, many of the triatomic cations which are known to decay radiatively have lifetimes in the microsecond range (Maier, 1979), i.e., CS_2^+ $\tilde{A}\ ^2\Pi_u$, H_2O^+ $\tilde{B}\ ^2A_1$, H_2S^+ $\tilde{B}\ ^2A_1$, and XCN^+ $\tilde{A}\ ^2\Sigma^+$, where, X = Cl, Br, I, and thus are not suitable for this coincidence technique (cf. Section II.C.). Therefore, measurements for systems in which radiative decay and fragmentation are both detected, thus leading to $\phi_F(v')$ values of less than unity, have been of special interest.

This situation occurs for N_2O^+ $\tilde{A}\ ^2\Sigma^+$ and COS^+ $\tilde{A}\ ^2\Pi$. In both cases, the $\tilde{A}$ state is predissociated by a $^4\Sigma^-$ state yielding $NO^+ + N$ or $S^+ + CO$ products (Brehm *et al.*, 1974). As a result, photodissociation techniques have also been applied to study the spectroscopic structure (Larzilliere *et al.*, 1980). Recently, lifetimes (cf. Table I) of selected vibrational levels of these two cations in their $\tilde{A}$ electronic states have been measured by means of optical selection of emitted photons following electron impact excitation (Klapstein and Maier, 1981). On the other hand, measurements of fluorescence quantum yields $\phi_F(v')$ (Maier and Thommen, 1980a) are limited by the low resolution of the coincidence experiment as seen in their photoelectron

TABLE I

Fluorescence Quantum Yields $\phi_F(v')$ and Lifetimes $\tau(v')$ Obtained from Photoelectron–Photon Coincidence and Electron Impact Measurements on N_2O^+ and COS^+ in Their First Excited States

State		Photoelectron–photon coincidence[a] $\phi_F(v')$	Photoelectron–photon coincidence[a] $\tau(v')$/nsec	Electron excitation[b] $\tau(v')$/nsec
N_2O^+ $\tilde{A}\,^2\Sigma^+$	0^0	1.00 ± 0.05	243 ± 12	240 ± 24
	1^1	0.71 ± 0.05	198 ± 19	180 ± 18
	1^2			39 ± 4
	2^1			190 ± 19
	2^2			56 ± 6
	3^1	0.59 ± 0.06	181 ± 18	180 ± 18
COS^+ $\tilde{A}\,^2\Pi_{3/2}$	0^0	0.17 ± 0.04 (both levels)	93 ± 9 (both levels)	107 ± 11
$^2\Pi_{1/2}$	0^0			73 ± 7

[a] Maier and Thommen (1980a).
[b] Klapstein and Maier (1981).

spectra (Figs. 6 and 7). Nevertheless, the data for N_2O^+ show that the radiative rates, $k_r(v') = \phi_F(v')/\tau(v')$, remain constant ($\approx 4.1 \times 10^6$ sec^{-1}) (Table I). Together with the lifetime data, the predissociation rates $k_{pred}(v')$ for various vibrational levels can be evaluated using the relationship $1/\tau(v') = k_r(v') + k_{pred}(v')$. The inferred rates for the two cations are summarized in Fig. 8.

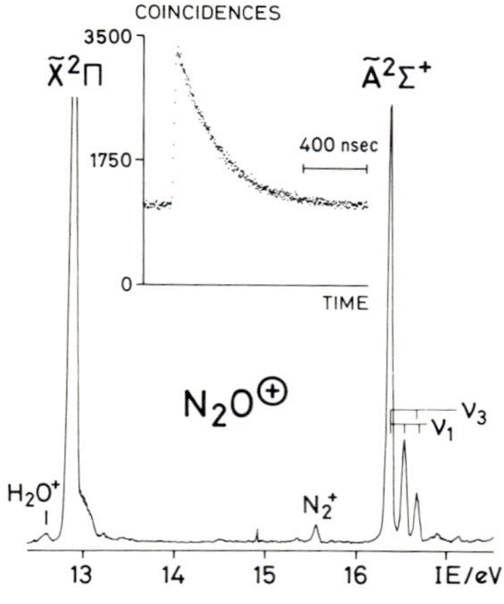

FIG. 6 He(Iα) photoelectron spectrum of N_2O recorded under coincidence conditions. The inset shows a coincidence curve obtained for the vibrationless level of N_2O^+ $\tilde{A}\,^2\Sigma^+$, $N_e = 600$ Hz, $N_{h\nu} = 6.6$ kHz, $N_t = 1.93$ Hz, accumulation time 25 hr. [From Maier and Thommen (1980a).]

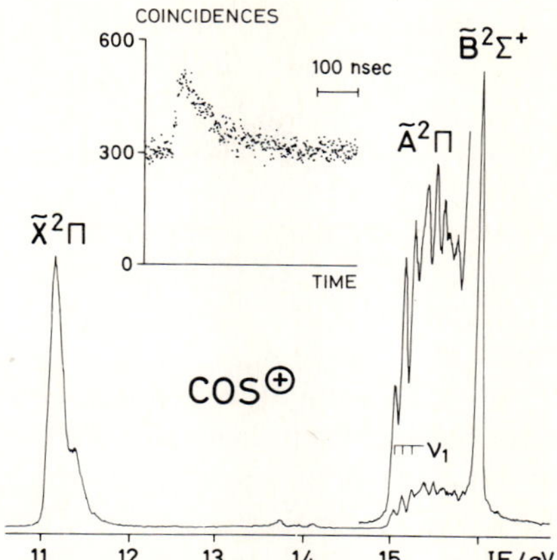

FIG. 7 Photoelectron–photon coincidence curve for the 0° level of the $\tilde{A}\,^2\Pi$ state of COS^+; $N_e = 280$ Hz, $N_{h\nu} = 5$ kHz, $N_t = 0.10$ Hz accumulated in 39 hr and the He(Iα) photoelectron spectrum showing the levels studied. [From Maier and Thommen (1980a).]

For the vibrationless level of N_2O^+ $\tilde{A}\,^2\Sigma^+$, fragment ions are not detected, although this lies above the thermodynamic threshold. The coincidence data (Table I) imply a predissociation rate $<2 \times 10^5$ sec^{-1}. Higher vibrational levels are, however, depleted by both radiative and predissociation pathways, as is evidenced by the fluorescence quantum yield and the ion branching ratio measurements (Brehm *et al.*, 1974).

On the other hand, photoelectron–photon coincidences were registered only for the 0^0 levels of COS^+ in its $\tilde{A}\,^2\Pi_\Omega$ state (Maier and Thommen, 1980a). The two spin-orbit components, $\Omega = \frac{3}{2}$ and $\frac{1}{2}$, were not resolved in

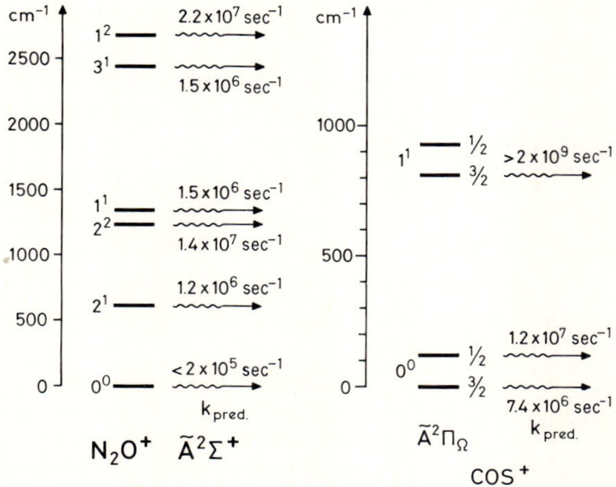

FIG. 8 Evaluated predissociation rates for various vibrational levels of N_2O^+ and COS^+ in their first excited electronic states. [From Klapstein and Maier (1981).]

the coincidence experiment. Thus, the given fluorescence quantum yield is an average value for the two components, but the individual $\phi_F(0^0, \Omega)$ values can be estimated from the measured lifetimes (Erman and Larsson, 1981) for the $\Omega = \frac{3}{2}$ and $\frac{1}{2}$ levels assuming a constant radiative rate. This yields $\phi_F(0^0, \frac{3}{2}) \approx 0.21$ and $\phi_F(0^0, \frac{1}{2}) \approx 0.14$. The evaluated predissociation rates increase noticeably, although the $0^0, \Omega = \frac{1}{2}$ level lies only ≈ 120 cm^{-1} above the $0^0, \Omega = \frac{3}{2}$ level. This trend can be followed further because the lack of detection of true coincidences for the 3^1 level implies a predissociation rate $> 2 \times 10^9$ sec^{-1}.

In conclusion, the combination of photoelectron–photon coincidence data, together with lifetime measurements at higher resolution, yields predissociation rates for the various vibrational levels. In the case of N_2O^+ $\tilde{A}\ ^2\Sigma^+$, these are in a rather good agreement with data from theoretical model calculations (Beswick and Horani, 1981).

B. *Large Open-Shell Cations*

The characteristic features of the decay of excited open-shell cations in the statistical limit are exponential decay and $\phi_F(0^0)$ values usually less than unity. In Table II, the cations which have been studied by the photoelectron–photon coincidence technique and the transitions showing these features are listed. For all these cations, the excited state, $^2\tilde{A}$ (or $^2\tilde{B}$), lies 1.6–3.6 eV above the ground state $^2\tilde{X}$, and the radiative rate constants $k_r(0^0)$ for the ions where the excited states are generated by a π^{-1} ionization process are found to be $\approx 1 - 3 \times 10^7$ sec^{-1}. For the cations in which the excited state is formed by a σ^{-1} ionization process, the $k_r(0^0)$ values are at least an order of magnitude smaller.

The nonradiative rates $k_{nr}(0^0)$ refer to the internal conversion process from the vibrationless level of the excited state to the quasi-continuum provided by the isoenergetic vibrational levels of the ground state. The $k_{nr}(0^0)$ values were estimated for the cations in which the lifetimes $\tau(0^0)$ were not available, assuming similar radiative rates $k_r(0^0)$ for a homologous series. In addition, for all the species studied, the nonradiative rate increases monotonously (linearly or exponentially) with vibrational excess energy. In the following discourse, some open-shell cations are used as examples of this relaxation behavior.

Substituted Diacetylene Cations

Diacetylene and dihalo-substituted diacetylene cations belong to the category of large polyatomic open-shell cations for which fragmentation channels are not accessible for their first excited doublet state (Allan *et al.*, 1978).

TABLE II

Polyatomic Cations Which Relax in Their Excited Electronic States According to the Statistical Limit Model and for Which Photoelectron–Photon Coincidence Measurements Have Been Carried Out[a].

Cation	Transition	$\Delta E(0_0^0)^b/\text{cm}^{-1}$	$\phi_F(0^0)$	$\tau(0^0)$/nsec	$k_r(0^0)/\text{sec}^{-1}$	$k_{nr}(0^0)/\text{sec}^{-1}$	Reference
$Cl{-}C{\equiv}C{-}C{\equiv}N^+$	$\tilde{A}\,^2\Sigma^+ \rightarrow \tilde{X}\,^2\Pi$	18,990	0.73	376	1.9×10^6	7.2×10^5	c
$Br{-}C{\equiv}C{-}C{\equiv}N^+$	$\tilde{A}\,^2\Pi \rightarrow \tilde{X}\,^2\Pi$	18,601	0.30	17	1.8×10^7	4.1×10^7	
$I{-}C{\equiv}C{-}C{\equiv}N^+$		15,567	0.03	—	—	$\simeq 5 \times 10^8$	
$H{-}C{\equiv}C{-}C{\equiv}C{-}H^+$	$\tilde{A}\,^2\Pi_u \rightarrow \tilde{X}\,^2\Pi_g$	19,762	0.72	72	1.0×10^7	3.9×10^6	d
$D{-}C{\equiv}C{-}C{\equiv}C{-}D^+$		19,762	0.80	79	1.0×10^7	2.5×10^6	
$H{-}C{\equiv}C{-}C{\equiv}C{-}Cl^+$	$\tilde{A}\,^2\Pi \rightarrow \tilde{X}\,^2\Pi$	19,710	0.79	41	1.9×10^7	5.1×10^6	e, f
$H{-}C{\equiv}C{-}C{\equiv}C{-}Br^+$		17,940	0.59	23	2.6×10^7	1.8×10^7	f
$Cl{-}C{\equiv}C{-}C{\equiv}C{-}Cl^+$	$\tilde{A}\,^2\Pi_u \rightarrow \tilde{X}\,^2\Pi_g$	19,090	0.47	21	2.2×10^7	2.5×10^7	e, f
$Br{-}C{\equiv}C{-}C{\equiv}C{-}Br^+$		16,585	0.19	10	1.9×10^7	8.1×10^7	f
$N{\equiv}C{-}C{\equiv}C{-}C{\equiv}N^+$		16,780	0.12	13	9.2×10^6	6.8×10^7	g
cis-1,2-difluoroethylene$^+$	$\tilde{A}\,^2A_1 \rightarrow \tilde{X}\,^2B_1$	28,880	0.07	320	2.2×10^5	2.9×10^6	h
$CH_3{-}C{\equiv}C{-}Cl^+$	$\tilde{A}\,^2E \rightarrow \tilde{X}\,^2E$	28,150	0.33	18	1.8×10^7	3.7×10^7	f
$CH_3{-}C{\equiv}C{-}Br^+$		21,780	0.25	15	1.7×10^7	5.0×10^7	
$CH_3{-}C{\equiv}C{-}C{\equiv}N^+$	$\tilde{A}\,^2A_1 \rightarrow \tilde{X}\,^2E$	18,390	0.26	210	1.2×10^6	3.5×10^6	c
$CH_3{-}C{\equiv}C{-}C{\equiv}C{-}H^+$	$\tilde{A}\,^2E \rightarrow \tilde{X}\,^2E$	20,375	0.94	50	1.9×10^7	1.2×10^6	i
$CH_3{-}C{\equiv}C{-}C{\equiv}C{-}D^+$		20,396	0.92	46	2.0×10^7	1.7×10^6	
$CD_3{-}C{\equiv}C{-}C{\equiv}C{-}H^+$		20,375	0.94	51	1.8×10^7	1.2×10^6	
$CD_3{-}C{\equiv}C{-}C{\equiv}C{-}D^+$		20,392	0.98	53	1.8×10^7	$\leqslant 1 \times 10^6$	
$CH_3{-}C{\equiv}C{-}C{\equiv}C{-}Cl^+$		19,840	0.55	22	2.5×10^7	2.0×10^7	f
$CH_3{-}C{\equiv}C{-}C{\equiv}C{-}Br^+$		18,170	0.23	10	2.3×10^7	7.7×10^7	
$CH_3{-}C{\equiv}C{-}C{\equiv}C{-}C{\equiv}N^+$		17,720	0.10	7	1.4×10^7	1.3×10^8	c
$CH_3{-}C{\equiv}C{-}C{\equiv}C{-}CH_3{}^+$	$\tilde{A}\,^2E_u \rightarrow \tilde{X}\,^2E_g$	20,562	0.74	25	3.0×10^7	1.0×10^7	i
$CD_3{-}C{\equiv}C{-}C{\equiv}C{-}CD_3{}^+$		20,560	0.92	32	2.9×10^7	2.5×10^6	
$CF_3{-}C{\equiv}C{-}C{\equiv}C{-}CF_3{}^+$		19,510	0.88	46	1.9×10^7	2.6×10^6	j
$C_2H_5{-}C{\equiv}C{-}C{\equiv}C{-}H^+$	$\tilde{A}\,^2A'' \rightarrow \tilde{X}\,^2A''$	20,004	0.004	—	—	$\simeq 7 \times 10^9$	i

(continued)

TABLE II (*continued*)

Cation	Transition	$\Delta E(0^0_0)^b/\mathrm{cm}^{-1}$	$\phi_F(0^0)$	$\tau(0^0)$/nsec	$k_r(0^0)/\mathrm{sec}^{-1}$	$k_{nr}(0^0)/\mathrm{sec}^{-1}$	Reference
1,3-difluorobenzene$^+$		18,632	0.006	—	—	$\simeq 1 \times 10^9$	*k*
1,3-dichlorobenzene$^+$		18,770	0.004	—	—	$\simeq 5 \times 10^9$	*l*
1,4-dichlorobenzene$^+$		19,620	0.005	—	—	$\simeq 4 \times 10^9$	
1,2,3-trifluorobenzene$^+$		22,288	0.74	48	1.5×10^7	5.4×10^6	*k*
1,2,4-trifluorobenzene$^+$		24,277	0.14	10	1.4×10^7	8.6×10^7	
1,3,5-trifluorobenzene$^+$		21,867	1.00	59	1.7×10^7	$\leqslant 1 \times 10^6$	
d_3-1,3,5-trifluorobenzene$^+$		21,867	1.00	58	1.7×10^7	$\leqslant 1 \times 10^6$	
1,3,5-trichlorobenzene$^+$	$\tilde{B}(\pi^{-1}) \rightarrow$	15,410	0.41	22	1.9×10^7	2.7×10^7	*l*
1,3,5-trifluorotrichlorobenzene$^+$	$\tilde{A}(\pi^{-1}), \tilde{X}(\pi^{-1})$	16,950	0.66	35	1.9×10^7	9.7×10^6	
1,3,5-trifluorotribromobenzene$^+$		13,510	0.05	—	—	$\simeq 3 \times 10^8$	
1,2,3,4-tetrafluorobenzene$^+$		23,277	0.90	50	1.8×10^7	2.0×10^6	*k*
1,2,3,5-tetrafluorobenzene$^+$		23,320	0.99	50	2.0×10^7	$\leqslant 1 \times 10^6$	
1,2,4,5-tetrafluorobenzene$^+$		24,437	0.61	32	1.9×10^7	1.2×10^7	
d_2-1,2,4,5-tetrafluorobenzene$^+$		24,437	0.74	38	1.9×10^7	6.8×10^6	
Pentafluorobenzene$^+$		23,137	0.98	48	2.0×10^7	$\leqslant 1 \times 10^6$	
Hexafluorobenzene$^+$		21,645	1.00	49	2.0×10^7	$\leqslant 1 \times 10^6$	
C_2H_5—C≡C—C≡C—$C_2H_5^+$	$\tilde{A}\,^2A'' \rightarrow \tilde{X}\,^2A''$	20,076	0.19	8	2.4×10^7	1.0×10^8	*m*

[a] The determined data (with uncertainty of 5–10%) for 0^0 level of the excited state are presented.
[b] The energy gap $\Delta E(0^0_0)$ refers to the indicated radiative transition.
[c] Unpublished results.
[d] Maier and Thommen (1980b).
[e] Maier *et al.* (1981a).
[f] Maier and Thommen (1982c).
[g] Maier *et al.* (1982).
[h] Maier and Thommen (1981b).
[i] Forster *et al.* (1981).
[j] Klapstein *et al.* (1982).
[k] Maier and Thommen (1981a).
[l] Maier and Thommen (1982b).
[m] Maier *et al.* (1981b).

Thus, radiative decay and internal conversion are the two competing relaxation pathways which can be quantitatively followed from the coincidence data. Photoelectron–photon coincidence measurements have been carried out for different levels of the substituted diacetylene cations, $X{-}(C{\equiv}C)_2{-}X^+$, where X = H, D, Cl, Br within their $\tilde{A}\ ^2\Pi_u$ states (Maier and Thommen, 1980b, 1982c). The He(Iα) photoelectron spectra recorded under coincidence conditions are presented in Fig. 9, and the selected internal energies are marked by the arrows. The measured fluorescence quantum yields $\phi_F(v')$ and lifetimes $\tau(v')$, as well as the evaluated radiative $k_r(v')$ and nonradiative $k_{nr}(v')$ rate constants are given in Table III.

True coincidences were also detected for excited vibrational levels within their first excited electronic states. The $k_r(v')$ values remain constant, but the $k_{nr}(v')$ rates increase with excess energy. This is typical for the irreversible radiationless transitions to the dense manifold of isoenergetic levels of the ground state. In addition, the $k_{nr}(0^0)$ values increase exponentially with decreasing difference between the vibrationless levels of the excited and the ground-state $\Delta E(\tilde{A}, \tilde{X})$. These are 2.45 eV for X = H, D; 2.32 eV for X = Cl; and 2.04 eV for X = Br in the series of the substituted diacetylene cations, $X{-}C{\equiv}C{-}C{\equiv}C{-}X^+$. This is in accord with the energy gap law [cf. Equation (12)]. Analogous observations have been found in other series of cations, e.g., 1,3,5-substituted halobenzene cations (Maier and Thommen 1981a; 1982b).

In addition, a deuterium effect is predicted by the energy gap law for the diacetylene cations X = H and X = D. There is a decrease in the nonradiative rate because more quanta of the CD stretching mode are required to absorb the energy, $\Delta E(\tilde{A}, \tilde{X}) = 2.45$ eV. The determined ratio k_{nr}^H/k_{nr}^D is ≈ 1.5. This effect is, of course, well known for aromatic hydrocarbons (Avouris *et al.*, 1977), where the CH stretching vibrations are the accepting modes. This seems also to be the case for the diacetylene cations.

Hexafluorobenzene Cation

A further aspect of relaxation behavior of electronically excited cations revealed by the photoelectron–photon coincidence technique is illustrated by hexafluorobenzene cation.

The first coincidence experiments (Ames *et al.*, 1976) suggested that the radiative decay depletes also higher excited electronic states, which is indeed the case (Maier and Thommen, 1981a) (cf. Fig. 10 and Table IV). This was not known from emission spectroscopy (Allan and Maier, 1975) because, on formation of hexafluorobenzene cations in the $\tilde{C}\ ^2B_{2u}$ state, the emitted photons fall in the same wavelength region (400–600 nm) as

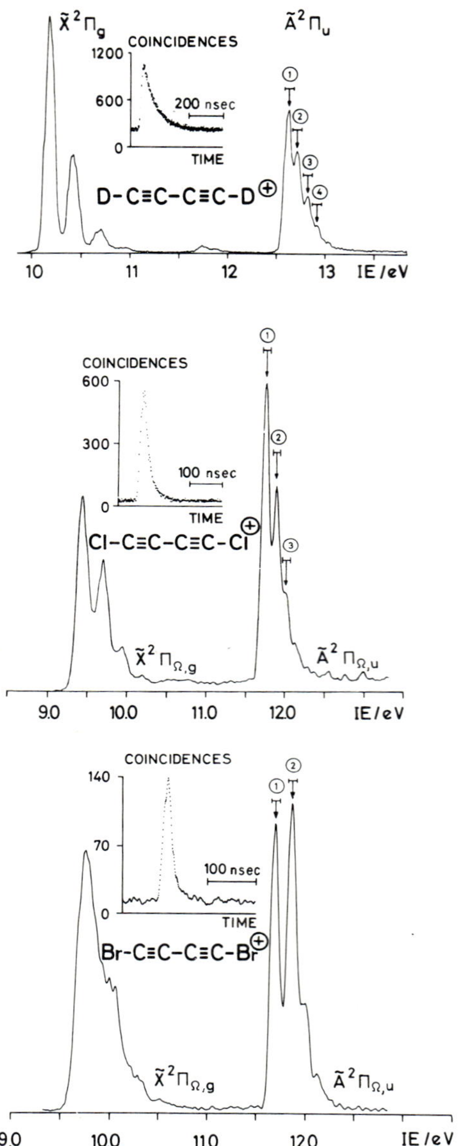

FIG. 9 He(Iα) photoelectron spectra of substituted diacetylene cations recorded under coincidence conditions. The insets show photoelectron–photon coincidence curves for the levels indicated by ①.

TABLE III

Fluorescence Quantum Yields $\phi_F(v')$, Lifetimes $\tau(v')$, and Evaluated Radiative $k_r(v')$ and Nonradiative $k_{nr}(v')$ Rate Constants at Different Ionization Energies (IE) Within the $\tilde{A}\,^2\Pi_u$ States of Substituted Diacetylene Cations

Cation	State	IE/eV	$\phi_F(v')$	$\tau(v')$/nsec	$k_r(v')/10^7\,sec^{-1}$	$k_{nr}(v')/10^7\,sec^{-1}$
H—C≡C—C≡C—H$^+$	$\tilde{A}\,^2\Pi_u$	12.62 ①[a]	0.72 ± 0.04	72 ± 3	1.0 ± 0.1	0.4 ± 0.1
		12.72 ②	0.62 ± 0.04	62 ± 3	1.0 ± 0.1	0.6 ± 0.1
		12.82 ③	0.55 ± 0.04	59 ± 3	0.9 ± 0.1	0.8 ± 0.1
		12.92 ④	0.48 ± 0.05	56 ± 3	0.9 ± 0.1	0.9 ± 0.1
D—C≡C—C≡C—D$^+$	$\tilde{A}\,^2\Pi_u$	12.62 ①	0.80 ± 0.04	79 ± 4	1.0 ± 0.1	0.3 ± 0.1
		12.72 ②	0.72 ± 0.04	72 ± 4	1.0 ± 0.1	0.4 ± 0.1
		12.82 ③	0.65 ± 0.04	66 ± 3	1.0 ± 0.1	0.5 ± 0.1
		12.92 ④	0.58 ± 0.05	63 ± 3	0.9 ± 0.1	0.7 ± 0.1
Cl—C≡C—C≡C—Cl$^+$	$\tilde{A}\,^2\Pi_u$	11.74 ①	0.47 ± 0.05	21 ± 2	2.2 ± 0.3	2.5 ± 0.6
		11.90 ②	0.14 ± 0.02	17 ± 3	0.8 ± 0.2	13.5 ± 2.9[b]
		12.06 ③	0.035 ± 0.007	—	—	60.7 ± 15.0[b]
Br—C≡C—C≡C—Br$^+$	$\tilde{A}\,^2\Pi_{3/2,u}$	11.22 ①	0.19 ± 0.02	10 ± 2	1.9 ± 0.4	8.1 ± 2.0
	$^2\Pi_{1/2,u}$	11.40 ②	0.16 ± 0.02	10 ± 2	1.6 ± 0.4	10.0 ± 2.6[b]

[a] These number correspond to the numbering in Fig. 9.
[b] Calculated, using $k_r(0°)$ value.

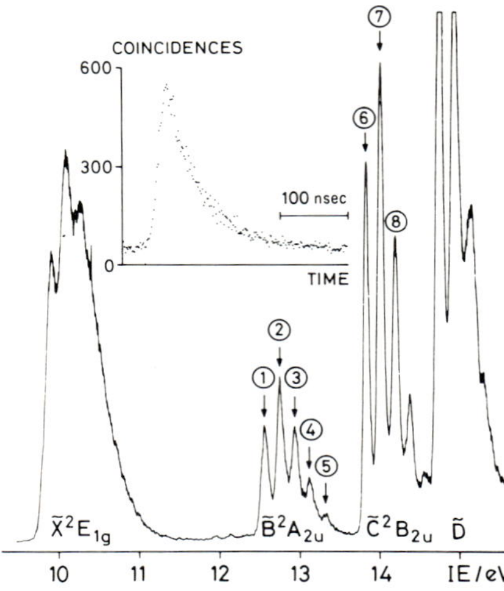

FIG. 10 Photoelectron–photon coincidence curve for the 0° level of the $\tilde{B}\,^2A_{2u}$ state of hexafluorobenzene cation; $N_e = 150$ Hz, $N_{h\nu} = 4.5$ kHz, $N_t = 0.22$ Hz accumulated in 24 hr, and the He(Iα) photoelectron spectrum showing the levels studied. [From Maier and Thommen (1981a).]

from the $\tilde{B}\ ^2A_{2u}$ state. This was established by means of coincidence measurements, using cutoff filters (Maier and Thommen, 1981a), and true coincidences were detected over a range of 1.7-eV excess energy above the lowest level of the $\tilde{B}\ ^2A_{2u}$ state, as indicated by the arrows above the photoelectron bands (Fig. 10).

In Table IV, the measured $\phi_F(v')$ and $\tau(v')$ data and the evaluated rate constants for hexafluorobenzene cation are given. The radiative rates $k_r(v')$ remain constant within the error limits whereas the nonradiative rates

TABLE IV

Radiative $k_r(v')$ and Nonradiative $k_{nr}(v')$ Rate Constants for the Hexafluorobenzene Cation at Selected IE, Evaluated from the Coincidence Data $\phi_F(v')$ and $\tau(v')$.

State	IE/eV	$\phi_F(v')$	$\tau(v')$/nsec	$k_r(v')/10^7\ \text{sec}^{-1}$	$k_{nr}(v')/10^7\ \text{sec}^{-1}$
$\tilde{B}\,^2A_{2u}$	12.56 ①[a]	1.00 ± 0.05	49 ± 3	2.0 ± 0.2	⩽0.1
	12.75 ②	1.00 ± 0.05	49 ± 3	2.0 ± 0.2	⩽0.1
	12.93 ③	0.88 ± 0.05	48 ± 3	1.8 ± 0.2	0.3 ± 0.1
	13.12 ②	0.65 ± 0.05	36 ± 4	1.8 ± 0.2	1.1 ± 0.3
	13.32 ⑤	0.25 ± 0.10	20 ± 4		6.1 ± 3.3[b]
$\tilde{C}\,^2B_{2u}$	13.88 ⑥	0.01 ± 0.002	<15		202 ± 44[b]
	14.08 ⑦	0.007 ± 0.002			290 ± 86[b]
	14.28 ⑧	≈0.001			≈2000[b]

[a] These numbers correspond to the locations in Fig. 10.
[b] Calculated, using $k_r(v') = 2.0 \times 10^7\ \text{sec}^{-1}$.

$k_{nr}(v')$ increase exponentially with excess energy without drastic irregularities on passing from the $\tilde{B}$ to the $\tilde{C}$ state. This has also been found for other halobenzene cations and is in accord with the statistical limit scheme. As fragmentation channels for the $\tilde{B}\ {}^2A_{2u}$ or $\tilde{C}\ {}^2B_{2u}$ states are not accessible, internal conversion is the presumed nonradiative decay pathway. For the levels for which unity fluorescence quantum yield has been measured, $k_{nr}(v') \leq 10^6\ sec^{-1}$ is an upper limit.

The detection of true coincidences for the $\tilde{C}$ state shows that either this state or another isoenergetic state is radiatively depleted. The radiative transition $\tilde{C} \rightarrow \tilde{B}$ can be energetically excluded because the observed emission spectrum (Allan and Maier, 1975) lies in the 400–600-nm wavelength region. Thus, either a radiative transition takes place from the $\tilde{C}$ state to highly excited vibrational levels of the ground state, i.e., $\tilde{C} \rightarrow \tilde{X}^*$, or a radiationless step $\tilde{C} \rightsquigarrow \tilde{B}^*$ is followed by the $\tilde{B}^* \rightarrow \tilde{X}^*$ radiative transition between the excited levels of these two states. The latter explanation is favored because the adiabatic ionizing transition to the $\tilde{C}$ state is dominant and, therefore, a corresponding 0_0^0 band of the $\tilde{C} \rightarrow \tilde{X}$ transition should be detected around 310 nm.

From all the available information on the decay behavior of the hexafluorobenzene cation in its lowest excited electronic states, the decay scheme presented in Fig. 11 has been constructed. The upper limit for the radiationless transition $\tilde{C} \rightsquigarrow \tilde{B}^*$ is $k_{nr} \leqslant 10^{12}\ sec^{-1}$ because the photoelectron spectrum still shows distinct vibrational structure.

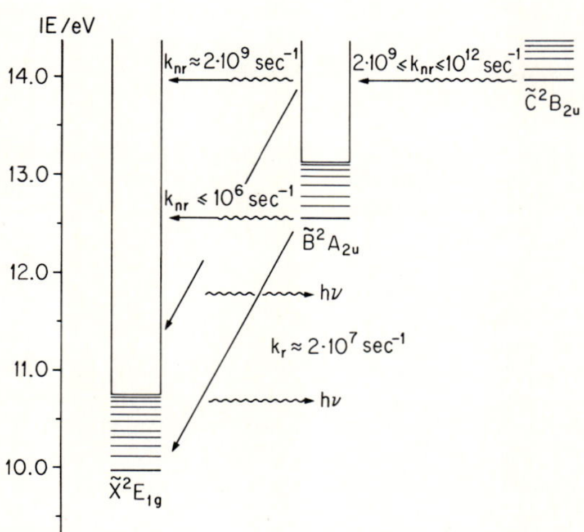

FIG. 11 Scheme of the inferred relaxation behavior of electronically excited hexafluorobenzene cations.

Competition Between Radiative and Fragmentation Decay

In a relatively small category of open-shell radical cations, fragmentation is observed in addition to the radiative decay from state-selected levels (Maier, 1979; 1981). These are listed in Table V. The example to be discussed is *cis*-1,2-difluoroethylene cation.

In this case, the vibrationless level of the $\tilde{A}\,^2A_1$ state lies already ≈ 1 eV above the thermodynamical threshold for the loss of HF from the parent ion. Photoelectron-photoion coincidence measurements show that on formation of *cis*-1,2-difluoroethylene cations near the threshold of the $\tilde{A}\,^2A_1$ state, parent ions are detected but their branching ratio falls off rapidly on

TABLE V

Organic Open-Shell Cations M^+ For Which Radiative Decay and Fragmentation Are Both Detected[a]

Fragmentation		M^+	Emission	Reference
C_2HF^+	⟵	$FHC{=}CHF^+$ (*cis*; F, F⁺ above; H, H below)	$\tilde{A}\,^2A_1 \rightarrow \tilde{X}\,^2B_1$	b
$C_3H_3^+$	⟵	$CH_3{-}C{\equiv}C{-}X^+$ X = Cl, Br	$\tilde{A}\,^2E \rightarrow \tilde{X}\,^2E$	c
$C_5H_3^+$	⟵	$CH_3{-}(C{\equiv}C)_2{-}X^+$ X = H, Cl, Br	$\tilde{A}\,^2E \rightarrow \tilde{X}\,^2E$	c
$C_6H_5^+$, $C_6H_4^+$	⟵	$CH_3{-}(C{\equiv}C)_2{-}CH_3^+$	$\tilde{A}\,^2E_u \rightarrow \tilde{X}\,^2E_g$	d
$C_4H_4^+$	⟵	$C_2H_5{-}(C{\equiv}C)_2{-}H^+$	$\tilde{A}\,^2A'' \rightarrow \tilde{X}\,^2A''$	d
$C_7H_7^+$	⟵	$C_2H_5{-}(C{\equiv}C)_2{-}C_2H_5^+$	$\tilde{A}\,^2A'' \rightarrow \tilde{X}\,^2A''$	e
$C_6H_7^+$	⟵	(*trans*-hexatriene)⁺	$\tilde{A}\,^2B_g \rightarrow \tilde{X}\,^2A_u$	f
$C_6H_7^+$	⟵	(*cis*-hexatriene)⁺	$\tilde{A}\,^2A_2 \rightarrow \tilde{X}\,^2B_1$	f
$C_6H_7^+$	⟵	(heptatriene)⁺	$\tilde{A}\,^2A'' \rightarrow \tilde{X}\,^2A''$	f
$C_7H_7^+$	⟵	(octatetraene)⁺	$\tilde{A}\,^2A_u \rightarrow \tilde{X}\,^2B_g$	g

[a] Isotopic derivatives studied and minor fragmentation pathways are not listed.
[b] Maier and Thommen (1981b).
[c] Maier and Thommen (1982c).
[d] Forster *et al.* (1981).
[e] Maier *et al.* (1981b).
[f] Allan *et al.* (1980).
[g] Jones and Maier (1979).

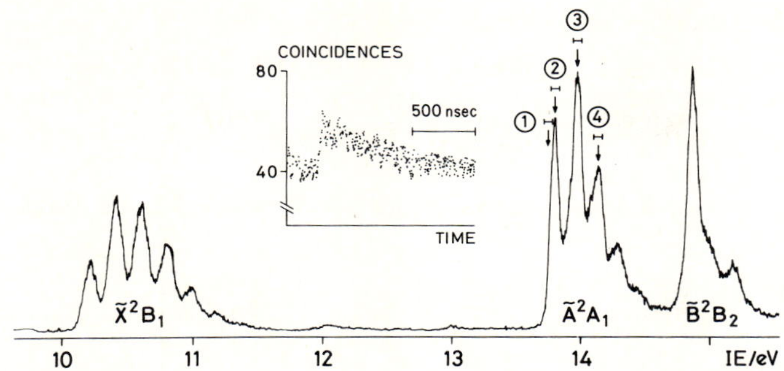

FIG. 12 Photoelectron–photon coincidence curve for *cis*-1,2-difluoroethylene cation accumulated for location 2 in 75 hr, $N_e = 150$ Hz, $N_{h\nu} = 850$ Hz, $N_t = 0.02$ Hz, and the He(Iα) photoelectron spectrum showing the levels studied. [From Maier and Thommen (1981b).]

increasing ionization energy (Stadelmann and Vogt, 1980). Corresponding photoelectron-photon coincidence measurements have also been carried out (Maier and Thommen, 1981b) (cf. Fig. 12). True coincidences were detected for locations ①, ②, and ③; but, for peak ④, only random coincidences were accumulated, indicating that at this energy the fluorescence quantum yield is $<10^{-3}$. The measured data are summarized in Table VI. The values have been corrected for the losses due to the escape from the ionization region.

With increasing internal energy within the $\tilde{A}\ ^2A_1$ state, the branching ratios $BR(C_2H_2F_2^+)$ and the fluorescence quantum yields $\phi_F(v')$ drop off rapidly. For the lowest level of the $\tilde{A}\ ^2A_1$ state, $\phi_F(0^0) = 0.067 \pm 0.013$ is less than the $C_2H_2F_2^+$ branching ratio, BR $(0^0) = 0.25 \pm 0.01$, which has

TABLE VI

Fluorescence Quantum Yields $\phi_F(v')$, Lifetimes $\tau(v')$, and Branching Ratios $BR(C_2H_2F_2^+)$ at Different IE Within the $\tilde{A}\ ^2A_1$ State of the cis-1,2-Difluoroethylene Cation

IE/eV	$\phi_F(v')^c$	$\tau(v')$/nsec	$k_{nr}(v')/\text{sec}^{-1}$	$BR(C_2H_2F_2^+)^e$
13.75 ①[b]	0.067 ± 0.013	320 ± 30^d	$2.9 \pm 0.3 \times 10^6$	0.25 ± 0.01
13.80 ②	0.052 ± 0.010	309 ± 31^c	$3.8 \pm 1.1 \times 10^6$	0.21 ± 0.01
13.97 ③	0.008 ± 0.002		$2.6 \pm 0.9 \times 10^7$	0.006 ± 0.004
14.14 ④	$<10^{-3}$		$>3 \times 10^9$	$<10^{-3}$

[a] The $k_{nr}(v')$ values are calculated using $k_r(0^0) = 2.2 \times 10^5\ \text{sec}^{-1}$.
[b] These numbers correspond to the locations in Fig. 12.
[c] Maier and Thommen (1981b).
[d] Maier *et al.* (1979).
[e] Stadelmann and Vogt (1980).

been measured after an ion flight time of ≈ 35 μsec in the photoelectron-photoion coincidence experiment. This indicates that after this time about two-thirds of the parent ions have not relaxed radiatively but rather by a nonradiative decay, resulting initially in vibrationally highly excited species in the $\tilde{X}\,^2B_1$ state.

The fragmentation rate constant $k_f(v')$ can be evaluated from the relationship

$$BR(v') = [1 - \phi_F(v')] \exp[-k_f(v')t] + \phi_F(v'),$$

and one finds

$$k_f(0^0) = 4.7 \pm 0.2 \times 10^4 \text{ sec}^{-1}.$$

This rate constant is about two orders of magnitude smaller than that for the nonradiative pathway

$$\tilde{A} \rightsquigarrow \tilde{X}^*, \qquad k_{nr}(0^0) = 2.9 \pm 0.3 \times 10^6 \text{ sec}^{-1}$$

and is in accord with the fact that the fragmentations for fluoro- and difluoroethylene cations near the observed onsets for the loss of HF are metastable (Jennings, 1970). The inferred picture of the relaxation pathways for *cis*-1,2-difluoroethylene cation in its $\tilde{A}\,^2A_1 0^0$ state is presented in Fig. 13. The result of the radiationless transition is a highly vibrationally

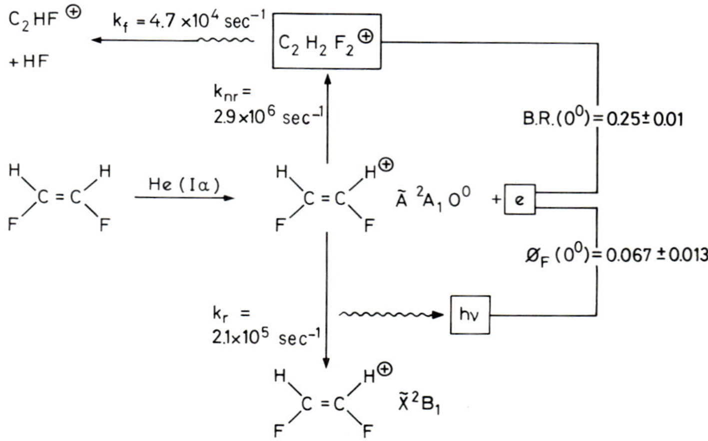

FIG. 13 Summary of the evaluated rate constants for the indicated relaxation pathways of *cis*-1,2-difluoroethylene cation produced initially in the lowest vibrational level of the $\tilde{A}\,^2A_1$ state from the given photoelectron–photon and photoelectron-photoion coincidence data.

excited cation in the ground state, which has enough time to isomerize before the elimination of HF takes place (Jochims *et al.*, 1979).

C. *Nonexponential Decay of Open-Shell Cations*

A few polyatomic linear open-shell cations, i.e., CS_2^+ $\tilde{B}\,^2\Sigma_u^+$, X—C≡N$^+$ $\tilde{B}\,^2\Pi$, and X—C≡C—H$^+$ $\tilde{A}\,^2\Pi$, where X = Cl, Br, I, show a pronounced nonexponential decay behavior. This was first apparent from decay curves recorded using electron beam excitation (Maier, 1979) and recently confirmed by photoelectron–photon coincidence measurements (Maier and Thommen, 1982a). Quantitative measurements of the relaxation behavior of such cations are of considerable interest, since these provide a good opportunity for comparison with the theoretical predictions for the intermediate case model of radiationless transitions (Tramer and Voltz, 1979).

In this section, the observations from emission spectroscopy and photoelectron–photon coincidence measurements on H—C≡C—X$^+$ $\tilde{A}\,^2\Pi$ and X—C≡C—X$^+$ $\tilde{B}\,^2\Pi_u$, where X = Cl, Br, are interpreted with the intermediate case model presented in Section III.

It should be noted that the photoelectron–photon coincidence measurements have also proved that the nonexponential decay curves found with electron impact excitation for CH_3—C≡C—X$^+$ and X—(C≡C)$_n$—X$^+$, where $n = 1$, 2 with X = Cl, Br (Maier *et al.*, 1980), are the result of cascading processes or due to overlapping from fragment emission (Maier and Thommen, 1982c,d).

Haloacetylene Cations

Photoelectron–photon coincidence measurements have been carried out for the haloacetylene cations H—C≡C—X$^+$, where X = Cl, Br, at selected energies within their $\tilde{A}\,^2\Pi$ states. The ions were generated using Ne(I) photoionization so that the effective bandpass of the electrons, which defines the internal energy of the cations, was ≈25 meV. In all the cases, quasi-biexponential decay curves were observed. Four of such curves for the chloroacetylene cation recorded at the locations ① to ④ are displayed in Fig. 14. It can be clearly seen that with increasing vibrational excess energy the short and long lifetime components become shorter and longer, respectively. The data which were evaluated from the coincidence curves are summarized in Table VII. The lifetimes were obtained by fitting two exponentials to the decay part of the curves (see Section II.C). The limit for short lifetime evaluations is set by small fluorescence quantum yields ($\phi_{F,1} < 0.05$) and the time resolution of the electron analyzer (≈10 nsec). Thus, for the bromoacetylene cation, τ_1 was taken from laser excitation studies (Maier and Misev, 1982), and the coincidence data for the long component τ_2 are corrected for the

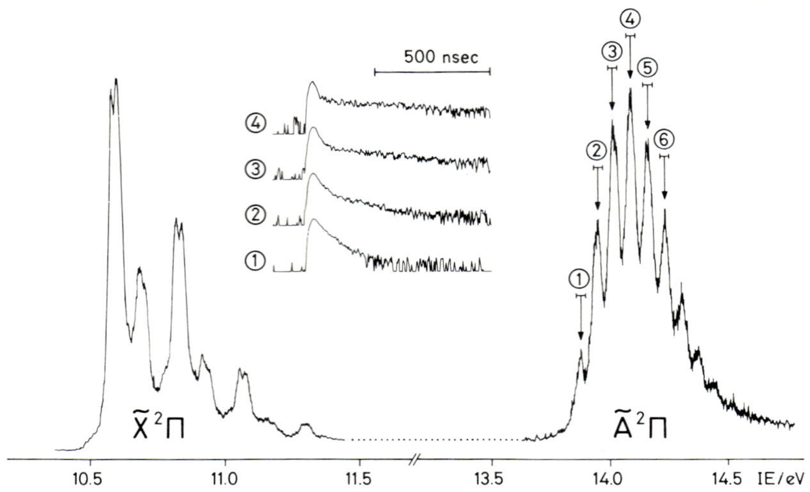

FIG. 14 Ne(I) photoelectron spectrum of chloroacetylene and semilogarithmic plots of the photoelectron–photon coincidence curves accumulated at locations ①–④. [Adapted from Maier and Thommen (1982a).]

thermal escape of the cations (see Section II.C). The uncertainties for the given ϕ_F and τ values are 10–15% and for the amplitude ratios, $A_1/A_2 \approx 20\%$.

As predicted from the intermediate case model of radiationless transitions (Tramer and Voltz, 1979), the short-time behavior is similar to the statistical

TABLE VII

Fluorescence Quantum Yields and Lifetimes for the Prompt (1) and Long (2) Decay Components and Their Relative Amplitude Ratio A_1/A_2, Measured at Selected Energies Within the $\tilde{A}\ ^2\Pi$ State of the Two Haloacetylene Cations

Cation		IE/eV	$\phi_{F,tot}(v')$	$\phi_{F,1}$	τ_1/nsec	$\phi_{F,2}$[a]	τ_2/nsec[a]	A_1/A_2
H—C≡C—Cl$^+$	①[b]	13.87	0.81	0.45	35	0.36	140	5
	②	13.92	0.76	0.32	25	0.44	350	10
	③	13.97	0.72	0.22	18	0.50	480	12
	④	14.02	0.66	0.10	9	0.56	740	15
	⑤	14.07	0.43	0.06	—	0.37	780	20[c]
	⑥	14.12	0.31	0.03	—	0.28	1100	50[c]
H—C≡C—Br$^+$		12.80	0.66	0.08	13	0.58	210	2
		13.00	0.51	0.06	9[d]	0.47	290	4
		13.18	0.40	0.03	5[d]	0.37	430	8

[a] Corrected values. (See text.)

[b] These numbers correspond to the locations in Fig. 14.

[c] Calculated, assuming $k_r(v')$ constant.

[d] From laser excitation spectroscopy [Maier and Misev (1982)].

TABLE VIII

Radiative $k_r(v')$ and Nonradiative Rate Constants $k_{nr}(v')$ and $\Delta/\hbar$ for Different Levels Within the $\tilde{A}\,^2\Pi$ State of the Haloacetylene Cations

Cation	IE/eV	$k_r(v')/10^7$ sec^{-1}	$k_{nr}(v')/10^7$ sec^{-1}	$\Delta/\hbar \cdot 10^7$ sec^{-1}
H—C≡C—Cl$^+$	13.87	1.3	0.5	1.1
	13.92	1.3	1.0	1.7
	13.97	1.2	1.6	2.7
	14.02	1.1	3.8	6.0
	14.07	—	7.4[b]	12.6[b]
	14.12	—	13.2[b]	28.2[b]
H—C≡C—Br$^+$	12.80	0.6	2.6	4.5
	13.00	0.7	5.4	5.0
	13.18	0.6	12.0	7.4

[a] Evaluated from the data given in Table VII.
[b] Calculated, assuming $k_r(v') = 1.3 \times 10^7$ sec^{-1}.

limit. The initially populated level of the $\tilde{A}$ state relaxes radiatively with the rate $k_r(v') = \phi_{F,1}/\tau_1$, which is constant within the error limits (cf. Table VIII), and nonradiatively to the quasi-continuum formed by isoenergetic vibronic levels of the $\tilde{X}$ state. An upper limit for this irreversible nonradiative transition can be estimated by $k_{nr}(v') = [1 - \phi_{F,tot}(v')]/\tau_1$, and its exponential increase with excess energy is typical for this case. $\Delta/\hbar$ is the rate of populating levels with $\tilde{A}$ and $\tilde{X}$ character and also increases exponentially. In other words, the energy spread Δ of $\tilde{A}$ containing resonant levels becomes progressively larger and, therefore, also the number $N \propto A_1/A_2$ of these levels. This effect is responsible for the long-time behavior of the system, which describes the relaxation of resonant levels with $\tilde{A}$ and $\tilde{X}$ character. The uncertainties in $k(\tilde{A})$ and N do not allow us to derive the $k(\tilde{A}, \tilde{X})$ values, but the trend showing the increase of the long component is apparent. An upper limit for the radiationless transition from resonant levels to the dense manifold of $\tilde{X}$ which is part of $k(\tilde{A}, \tilde{X})$ can be estimated from $[1 - \phi_{F,tot}(v')]/\tau_2$ and is of the order of 10^5–10^6 sec^{-1} whereas the collisional deactivation and IR relaxation are probably an order of magnitude smaller.

As a consequence, the time-resolved emission spectra in Fig. 15, which were recorded in different periods after the formation of the cations (Allan *et al.*, 1977a), correspond in the one instance to the prompt decay of the initially populated level of $\tilde{A}$ while the long component spectrum reflects the decay of resonant levels with $\tilde{A}$ and $\tilde{X}$ character. Since the resolution of the time-resolved emission spectrum is only 1.6 nm, perturbation effects due to the resonant coupling, which are expected for the long component spectra, have not yet been observed.

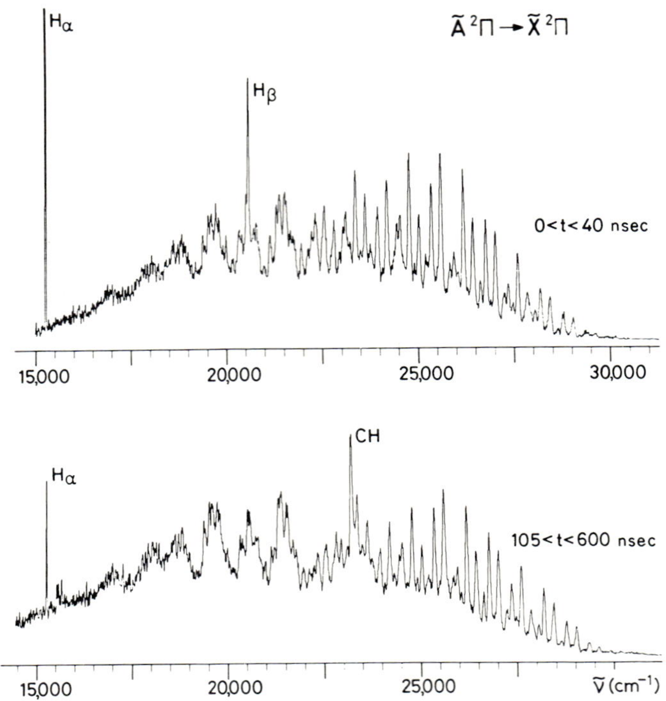

FIG. 15 Time-resolved emission spectra of H—C≡C—Cl^+ recorded within the indicated time periods after the formation of the excited cation by electron impact. [From Allan *et al.* (1977a).]

Dihaloacetylene Cations

In the first studies on dihaloacetylene cations by emission (Allan *et al.*, 1977b), and subsequently by photoion–photon coincidence spectroscopy (Dujardin *et al.*, 1980b), pronounced nonexponential decay curves were found. However, since the lower levels of the B̃ states of the dihaloacetylene cations lie below the fragmentation onsets, population of the Ã state by non-radiative or radiative transition from the B̃ state could not be excluded. Photoelectron–photon coincidence studies clarify the situation because the internal energy of the initially generated cation is well defined.

The measured data for the diacetylene cations, X—C≡C—X^+, where X = Cl, Br in their Ã $^2\Pi_g$ and B̃ $^2\Pi_u$ states, are given in Table IX (Maier *et al.*, 1982). As an example, the He(Iα) photoelectron spectrum of the di-chloroacetylene cation together with coincidence curves for the Ã and B̃ states are shown in Fig. 16.

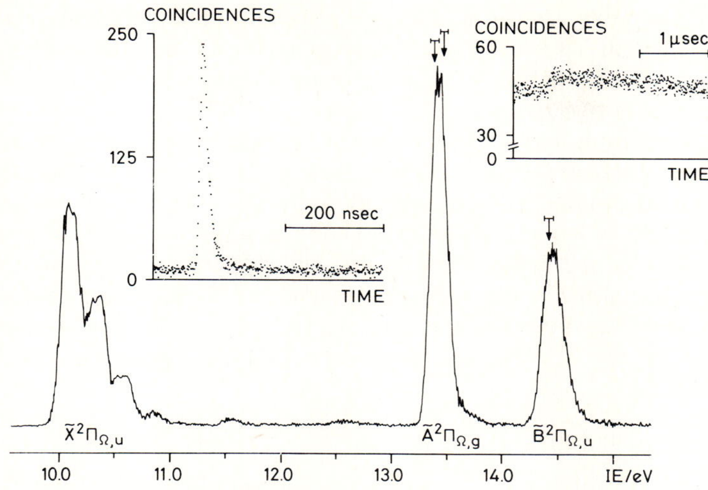

FIG. 16 He(Iα) photoelectron spectrum of Cl—C≡C—Cl showing the levels studied and the photoelectron–photon coincidence curves accumulated for the $\tilde{A}\,^2\Pi_g$ (left) and $\tilde{B}\,^2\Pi_u$ (right) states.

All coincidence decay curves are found to be monoexponential, and the determined lifetimes of the $\tilde{A}$ states agree with the short component values of the quasi-biexponential decay curves obtained from the emission experiment (Allan *et al.*, 1977b). The long component is seen to have its origin in the $\tilde{B}$ state; furthermore, coincidence measurements prove that the wavelength of emitted photons which correspond to the long component lie in the same wavelength region as for the $\tilde{A}\,^2\Pi_g \rightarrow \tilde{X}\,^2\Pi_u$ transition. In addition,

TABLE IX

Fluorescence Quantum Yields $\phi_F(v')$ and Lifetimes $\tau(v')$ at Selected Energies, IE, Within the $\tilde{A}\,^2\Pi_g$ and $\tilde{B}\,^2\Pi_u$ States of the Dihaloacetylene Cations

Cation	State	IE/eV	$\phi_F(v')$	$\tau(v')$/nsec	τ_{em}/nsec[a]
$Cl—C{\equiv}C—Cl^+$	$\tilde{A}\,^2\Pi_g$	13.37	0.38 ± 0.04	13 ± 2	13 ± 2
		13.47	0.36 ± 0.04	13 ± 2	13 ± 2
	$\tilde{B}\,^2\Pi_u$	14.40	≈ 0.13[b]	≈ 2850[b]	≈ 1100
$Br—C{\equiv}C—Br^+$	$\tilde{A}\,^2\Pi_{3/2,g}$	12.12	0.99 ± 0.05	31 ± 3	29 ± 3
	$^2\Pi_{1/2,g}$	12.42	0.67 ± 0.05	25 ± 3	27 ± 3
	$\tilde{B}\,^2\Pi_u$	13.40	≈ 0.05[b]	≈ 1500[b]	≈ 850

[a] This column lists lifetimes evaluated from the biexponential decay curves from Allan *et al.* (1977b).

[b] Not corrected for thermal escape.

the $\tilde{B}\,^2\Pi_u \rightarrow \tilde{X}\,^2\Pi_u$ transition is dipole-forbidden, and the allowed $\tilde{B}\,^2\Pi_u \rightarrow \tilde{A}\,^2\Pi_g$ transition lies outside the detection region of the photomultiplier. In view of these data, the decay of the dihaloacetylene cations in their $\tilde{A}\,^2\Pi_g$ and $\tilde{B}\,^2\Pi_u$ states can be rationalized.

Smaller-than-unity fluorescence quantum yields and monoexponential decay curves obtained for the cations in their $\tilde{A}\,^2\Pi_g$ state are typical of species in the statistical limit. The radiative decay $\tilde{A}\,^2\Pi_g \rightarrow \tilde{X}\,^2\Pi_u$ and internal conversion $\tilde{A}\,^2\Pi_g \rightsquigarrow \tilde{X}^*\,^2\Pi_u$ are the competing processes. The observation that the lifetimes for the higher excited state are larger than for the first excited state is unusual for open-shell organic cations. It can nevertheless be explained with the intermediate case model, whereas a similar decay behavior as in the hexafluorobenzene cation would imply $\tau(\tilde{B}) < \tau(\tilde{A})$ (See Section IV.B).

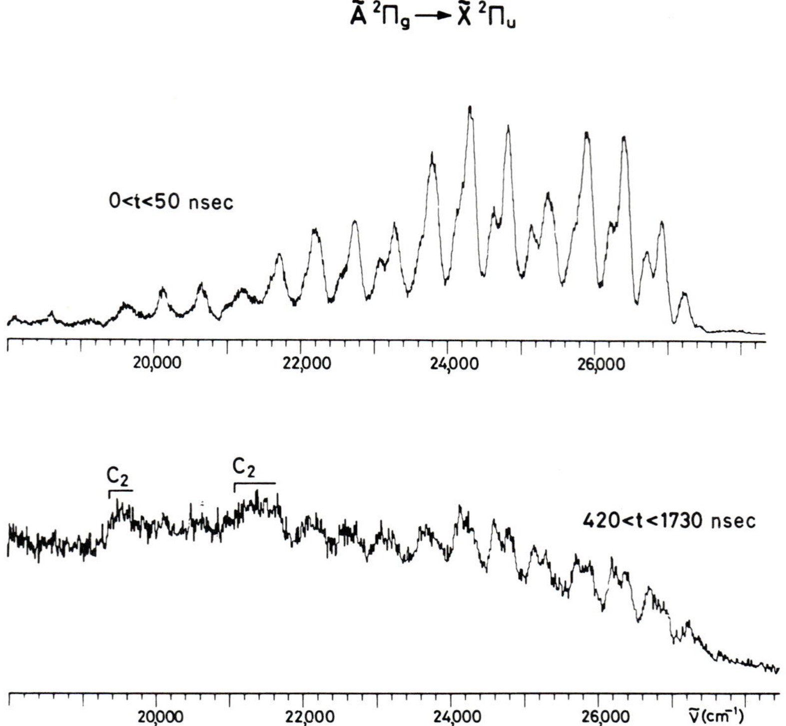

FIG. 17 Time-resolved emission spectra of Cl—C≡C—Cl$^+$ recorded in the indicated time intervals. The cations were formed using ≈ 30-eV electron excitation. [From Allan *et al.* (1977b).]

The density of $\tilde{A}$ levels of the diacetylene cations isoenergetic with the $\tilde{B}$ state is about $10^3/cm^{-1}$. These levels can provide a quasi-continuum and discrete levels for the initially populated $\tilde{B}$ state level. Resonance between these latter levels leads to states with $\tilde{B}$ and $\tilde{A}$ character. Because only the $\tilde{A}$ state carries oscillator strength to the ground-state $\tilde{X}$ and the $\tilde{B}$ character in the resonant state is not irreversibly lost, long lifetimes are observed. In addition, smaller-than-unity fluorescence quantum yields for the $\tilde{B}$ state reflect the irreversible radiationless transition from the $\tilde{B}$ state to the quasi-continuum formed by the $\tilde{A}$ and $\tilde{X}$ states.

The time-resolved emission spectra (Allan *et al.*, 1977b), reproduced in Fig. 17, reflect the different origin of the emitted photons. The short component spectrum corresponds to the radiative transition $\tilde{A} \rightarrow \tilde{X}$, and the long component one is the result of resonant coupling $\tilde{B} \rightsquigarrow \tilde{A}^*$ followed by the radiative transition to highly excited vibronic levels of the ground state.

V. SUMMARY

The application of the photoelectron–photon coincidence technique to determine the fluorescence quantum yields and lifetimes of a variety of open-shell polyatomic cations at selected energies within their excited electronic states has been discussed. Such data yield the radiative and nonradiative rate constants for the decay channels depleting the chosen level, under collision-free conditions, and can be rationalized in terms of the theories of radiationless transitions. These aspects have been illustrated with examples throughout. For the future, the studies of the decay processes of open-shell cations can be further refined in energy and time resolution by combining the coincidence results with lifetime measurements from laser excitation spectroscopy. This will enable their radiationless processes, in many vibrational or rotational levels at high resolution, to be quantitatively followed.

ACKNOWLEDGMENTS

The work summarized in this chapter, as a result of the research projects carried out at Basel, has been financed throughout the years by the "Schweizerischer Nationalfonds zur Förderung der wissenschaftlichen Forschung."

REFERENCES

Allan, M., and Maier, J. P. (1975). *Chem. Phys. Lett.* **34**, 442

Allan, M., Kloster-Jensen, E., and Maier, J. P. (1977a). *J. Chem. Soc. Faraday Trans. 2* **73**, 1406.

Allan, M., Kloster-Jensen, E., and Maier, J. P. (1977b). *J. Chem. Soc. Faraday Trans. 2* **73**, 1417.

Allan, M., Kloster-Jensen, E., Maier, J. P., and Marthaler, O. (1978). *J. Electron Spectrosc. Relat. Phenom.* **14**, 359.

Allan, M., Dannacher, J., and Maier, J. P. (1980). *J. Chem. Phys.* **71**, 3114.

Ames, D. L., Bloch, M., Porter, H. Q., and Turner, D. W. (1976). *In* "Molecular Spectroscopy," pp. 399–421. Institute of Petroleum, London.

Avouris, P., Gelbart, W. M., and El-Sayed, M. A. (1977). *Chem. Rev.* **77**, 793.

Baer, T. (1979). *In* "Gas Phase Ion Chemistry" (M. T. Bowers, ed.), Vol. 1, pp. 153–196. Academic Press, New York.

Beswick, J. A., and Horani, M. (1981). *Chem. Phys. Lett.* **78**, 4.

Bloch, M., and Turner, D. W. (1975). *Chem. Phys. Lett.* **30**, 344.

Brehm, B., Frey, R., Küstler, A., and Eland, J. H. D. (1974). *Int. J. Mass Spectrom. Ion Phys.* **13**, 251.

Carsky, P., and Zahradnik, R. (1973). *Top. Curr. Chem.* **43**, 1.

Dujardin, G., Leach, S., and Taieb, G. (1980a). *Chem. Phys.* **46**, 407.

Dujardin, G., Leach, S., Taieb, G., Maier, J. P., and Gelbart, W. M. (1980b). *J. Chem. Phys.* **73**, 4987.

Eland, J. H. D., Devoret, M., and Leach, S. (1976). *Chem. Phys. Lett.* **43**, 97.

Erman, P., and Larsson, M. (1981). *Physica Scripta* **23**, 1052.

Forster, P., Maier, J. P., and Thommen, F. (1981). *Chem. Phys.* **59**, 85.

Freed, K. F. (1976). *Top. Appl. Phys.* **15**, 23.

Frey, R., Gotchev, B., Peatman, W. B., Pollak, H., and Schlag, E. W. (1978). *Chem. Phys. Lett.* **54**, 411.

Imhof, R. E., and Read, F. H. (1971). *J. Phys. B. Phys.* **4**, 450.

Jennings, K. R. (1970). *Org. Mass Spectrom.* **3**, 85.

Jochims, H. W., Lohr, W., and Baumgärtel, H. (1979). *Nouv. J. Chim.* **3**, 109.

Jones, T. B., and Maier, J. P. (1979). *Int. J. Mass Spectrom. Ion Phys.* **31**, 287.

Jortner, J., and Mukamel, S. (1973). *In* "The World of Quantum Chemistry" (R. Daudel and B. Pullman, eds.), pp. 145–209. Reidel, Holland.

Klapstein, D., and Maier, J. P. (1981). *Chem. Phys. Lett.* **83**, 590.

Klapstein, D., Maier, J. P., Misev, L., Thommen, F., and Zambach, W. (1982). *J. Chem. Soc. Faraday Trans. 2* **78**, 1765.

Lahmani, F., Tramer, A., and Tric, C. (1974). *J. Chem. Phys.* **60**, 4431.

Larzilliere, M., Carre, M., Gaillard, M. L., Rostas, J., Horani, M., and Velghe, M. (1980). *J. Chim. Phys.* **77**, 689.

Leach, S., Devoret, M., and Eland, J. H. D. (1978). *Chem. Phys.* **33**, 113.

Leach, S., Dujardin, G., and Taieb, G. (1980). *J. Chim. Phys.* **77**, 705.

Maier, J. P. (1979). *In* "Kinetics of Ion-Molecule Reactions" (P. Ausloos, ed.), pp. 437–462. Plenum Press, New York.

Maier, J. P. (1980). *Chimia* **34**, 219.

Maier, J. P. (1981). *Angew. Chem. Int. Ed. Engl.* **20**, 638.

Maier, J. P. (1982). *Acc. Chem. Res.* **15**, 18.

Maier, J. P., and Misev, L. (1982). Unpublished results.

Maier, J. P., and Thommen, F. (1980a). *Chem. Phys.* **51**, 319.

Maier, J. P., and Thommen, F. (1980b). *J. Chem. Phys.* **73**, 5616.

Maier, J. P., and Thommen, F. (1981a). *Chem. Phys.* **57**, 319.

Maier, J. P., and Thommen, F. (1981b). *J. Chem. Soc. Faraday Trans. 2* **77**, 845.

Maier, J. P., and Thommen, F. (1982a). *In* "Intramolecular Dynamics" (J. Jortner and B. Pullman, eds.). Reidel, Holland.

Maier, J. P., and Thommen, F. (1982b). *J. Chem. Phys.* **74**, 4427.

Maier, J. P., and Thommen, F. (1982c). *J. Electron Spectrosc. Relat. Phenom.* **28**, 111.

Maier, J. P., and Thommen, F. (1982d). *Chem. Phys.* **70**, 325.

Maier, J. P., Marthaler, O., and Bieri, G. (1979). *Chem. Phys.* **44**, 131.

Maier, J. P., Marthaler, O., and Kloster-Jensen, E. (1980). *J. Electron Spectrosc. Relat. Phenom.* **18**, 251.

Maier, J. P., Marthaler, O., Misev, L., and Thommen, F. (1981a). *Faraday Discuss.* **71**, 181.

Maier, J. P., Misev, L., and Thommen, F. (1981b). *Helv. Chim. Acta* **64**, 1985.

Maier, J. P.. Misev, L., and Thommen, F. (1982). *J. Phys. Chem.* **86**, 514.

Möhlmann, G. R., and de Heer, F. J. (1976). *Chem. Phys. Lett.* **43**, 170.

Schlag, E. W., Frey, R., Gotchev, B., Peatman, W. B., and Pollak, H. (1977). *Chem. Phys. Lett.* **51**, 406.

Smith, A. J., Read, F. H., and Imhof, R. E. (1975). *J. Phys. B* **8**, 2869.

Stadelmann, J.-P., and Vogt, J. (1980). *Int. J. Mass Spectrom. Ion Phys.* **35**, 83.

Tramer, A., and Voltz, R. (1979). *In* "Excited States" (E. C. Lim, ed.), Vol. 4, pp. 281. Academic Press, New York.

Chapter 26

Spectroscopy and structure of the hydrogen molecular ion

ALAN CARRINGTON
RICHARD A. KENNEDY
DEPARTMENT OF CHEMISTRY
UNIVERSITY OF SOUTHAMPTON
SOUTHAMPTON, ENGLAND

I. INTRODUCTION

The hydrogen molecular ion $H_2{}^+$ and its isotopic relations HD^+ and $D_2{}^+$ have occupied an important position in the development of molecular quantum mechanics since the foundations of the subject some 60 years ago.

ISBN 0-12-120803-6

Many different approximations and methods have been tested on H_2^+ as a model system, and it is, of course, the molecule for which the most accurate calculations can be performed, essentially because of the absence of inter-electron interactions. H_2^+ is, however, rather more amenable to theoretical than to experimental investigation. There are two main reasons for this, chemical and structural. The first difficulty is that H_2^+ reacts extremely rapidly with molecular hydrogen to form the H_3^+ ion; electrical discharges in H_2, for example, lead to extensive ionization but unless the gas pressure is extremely low, H_3^+ is the predominant ionic species present. This fact has been used recently to advantage by Oka (1980) and by Carrington *et al.* (1982) in their spectroscopic investigations of H_3^+. As we shall see, however, spectroscopic studies of H_2^+ have necessarily made use of collision-free media. H_2^+ is then the predominant ion, but its low concentration requires special techniques to permit its study by spectroscopy.

The second difficulty is that although H_2^+ has a stable ground electronic state ($D_0^0 = 2.6507$ eV), its first excited state is entirely repulsive and does not, therefore, give rise to a discrete electronic spectrum. Two excited electronic states are predicted to be weakly bound, but they are more than 11 eV above the ground state and not accessible by normal spectroscopic methods.

In Section II we review the development of the theory, from the introduction of the Schrödinger equation through to the present. In Section III we give a comprehensive description of the theory as it is understood at present, and we draw attention to areas which still deserve closer attention. In Section IV we describe photoelectron, photoionization, and photodissociation studies, and in Section V we review investigations of the radiofrequency spectra. Finally, in Section VI we describe recent studies of the vibration–rotation spectra of the HD^+ ion, which have provided accurate data for the vibrational and rotational levels of the ground electronic state. The balance of the discussion in this review is tilted strongly in the direction of the theory; that is but a reflection of the present position with regard to this, the simplest of all molecules.

II. HISTORICAL SURVEY OF THE THEORY

A very large number of papers have been published on the theory of the H_2^+ ion, and in this review we can only give an outline of how the theory has developed. The emphasis will be on work leading to the prediction or explanation of properties which can be measured experimentally. A detailed mathematical treatment is deferred to the next section; our historical survey of the development of the theory will be essentially descriptive.

The hydrogen molecular ion was first detected in the laboratory by J. J. Thomson in 1907. Attempts were made to describe H_2^+, treated as two fixed protons and a moving electron, within the framework of the old quantum theory (Urey, 1926) (phase integral quantization of classical motion). But, by 1926 the deficiencies of the old quantum theory were apparent and Schrödinger proposed the wave equation (Schrödinger, 1926a,b). The equation was soon applied to the problem of two fixed protons and a moving electron. By using elliptical coordinates (R, η, ξ), the three-dimensional wave equation separates into three one-dimensional differential equations (Alexandrow, 1926). The problem can therefore be solved exactly by numerical integration methods, and the first successful solution was obtained by Burrau in 1927, who calculated the electronic energy of the ion in its ground state as a function of internuclear distance. Also in 1927, Born and Oppenheimer gave their now famous justification for treating the nuclei as fixed as a first approximation in molecular quantum mechanics. Approximate methods of obtaining the electronic energy for various states by perturbation theory were investigated by a number of authors (Unsöld, 1927; Morse and Stueckelberg, 1929; Gilbert, 1933; Coulson, 1937), while others applied the variation method (Teller, 1930; Dickinson, 1933). These papers were the first to treat the excited states of H_2^+. Using these results, Condon (1930) was able to predict the behavior of hydrogen under electron-impact ionization by electrons with different energies. Subsequent experimental work in 1930 (Bleakney, 1930; Lozier, 1930) confirmed the predictions, a major success for the quantum treatment of H_2^+.

Methods for solving the separated electronic equations exactly by series expansion techniques were investigated in some detail by Wilson in 1928. These methods were further explored and applied by Hylleraas (1931) and by Jaffé (1934). Specifically, the expansions used by Hylleraas were

$$M(R; \eta) = \sum_{l=1}^{\infty} f_l(R) P_l^{\Lambda}(\eta)$$

$$L(R; \xi) = (\xi^2 - 1)^{\Lambda/2} \exp(-x/2) \sum_{n=\Lambda}^{\infty} \frac{g_n(R)}{n!} \mathscr{L}_n^{\Lambda}(x).$$

$P_l^{\Lambda}(\eta)$ is an associated Legendre function, $\mathscr{L}_n^{\Lambda}(x)$ is an associated Laguerre function in $x = 2p(\xi - 1)$, and the definitions of the various parameters are given in Section III. This expansion is suggested by considering the exact atomic wave functions at the united atom limit (He^+).

The ξ expansion proposed by Jaffé was

$$L(R; \xi) = (\xi^2 - 1)^{\Lambda/2} (\xi + 1)^{\sigma} \exp(-p\xi) \sum_{n=0}^{\infty} g_n(R)[(\xi - 1)/(\xi + 1)]^n.$$

The first three terms reduce the ξ equation to a form which can be solved by a rapidly convergent series in $(\xi - 1)/(\xi + 1)$. They also reproduce the united atom asymptotic behavior at large ξ. An expansion in $(\xi - 1)/(\xi + 1)$ is convenient since the ξ range $+1 \leqslant \xi < \infty$ is mapped onto $0 \leqslant [(\xi - 1)/(\xi + 1)] < 1$.

Both authors proceeded to obtain numerical solutions for the energy of H_2^+ by truncating the ξ and η expansions. They obtained relations between the energy, the separation constant in the differential equations, and the internuclear separation. Hylleraas calculated the energy at a number of internuclear separations by this method. He located the minimum of the internuclear potential energy curve close to 2.0 Bohr radii, with an energy of -1.2053 Rydbergs (relative to $2H^+ + e^-$). In order to obtain the zero-point energy relative to this limit, he fitted his results to a Morse potential function (Morse, 1929), obtaining 16.18 eV. The zero-point dissociation energy of H_2^+ (to $H^+ + H$) is then 2.64 eV, a value subsequently confirmed by experiment (Lozier, 1933).

The results of Jaffé's calculations agreed with those obtained by Hylleraas. The properties of the series expansions were reviewed in 1935 by Baber and Hassé. Further calculations on H_2^+ using Jaffé's method were performed in 1935 by Sandeman, who fitted the potential he obtained to a Dunham (1932) expansion. A comparison between these theoretical results and the available experimental estimates, obtained by the analysis of Rydberg spectra of H_2, can be found in a review by Richardson (1935).

By 1935, the mathematical problems involved in obtaining exact electronic energies for H_2^+ (within the Born–Oppenheimer approximation) had been solved, but it was not until the development of fast computational methods that this work could be explored fully. Extensive tabulations of the results of numerical solutions to the problem appeared during the 1950s and early 1960s (Bates *et al.*, 1953; Wallis and Hulbert, 1954; Wind, 1965a; Peek, 1965). From the electronic wave functions which were obtained, various other physical properties were calculated. Bates (1951) calculated the electronic transition moment between the ground state and first excited state as a function of the internuclear separation. The nuclear hyperfine interactions were considered by Dalgarno *et al.* (1960) and also by Stephen and Auffray (1959). Vibration–rotation energy levels were calculated by Cohen *et al.* (1950) and by Wind (1965b) for H_2^+, and by Dunn (1966) for D_2^+.

During the 1960s the need to take the calculations beyond the Born–Oppenheimer (clamped nuclei) limit became apparent, especially after the important paper on a nonadiabatic theory for diatomic molecules published by Kołos and Wolniewicz in 1963. Many years earlier, Van Vleck (1936) had pointed out that corrections to the clamped nuclei calculations would be needed to obtain the highest accuracy in calculations on diatomic molecules. The diagonal (adiabatic) corrections to the energy due to nuclear motion were calculated at various internuclear separations for H_2^+ as early as 1941 (Johnson, 1941). The application of the Born–Oppenheimer separation to three-particle systems was investigated in detail by Hunter *et al.* (1966), and Hunter and Pritchard (1967a,b). As part of this work, they performed adiabatic, and subsequently the first large-scale nonadiabatic calculations, on the hydrogen molecular ion. The calculations were carried out by evaluating the nonadiabatic coupling between the ground state and the first few excited states of H_2^+, HD^+, and D_2^+; exact (within the Born–Oppenheimer limit) electronic wave functions were used for all states considered.

Much of the recent work on the theory of the hydrogen molecular ion has been devoted to obtaining increasingly accurate vibration–rotation energy levels. Reviews of such work have been given by Kołos (1970), and more recently by Bishop and Cheung (1980). This work has become increasingly important in view of recent measurements of high-resolution infrared spectra of HD^+ (Wing *et al.*, 1976; Carrington and Buttenshaw, 1981), which are described in detail in a later section of this review. During the early 1970s, calculations to the adiabatic limit (inclusion of the diagonal corrections to the internuclear potential due to nuclear motion) of the vibration–rotation levels of H_2^+ and its isotopic modifications have been performed by various authors (Kołos, 1969; Beckel *et al.*, 1970; Hunter *et al.*, 1974; Bishop and Wetmore, 1973; Bishop, 1976; Colbourn, 1976). The paper by Hunter *et al.* (1974) gives a complete tabulation of the vibration–rotation energy levels for H_2^+, HD^+, and D_2^+. Bishop (1976) gives a similar tabulation for HD^+, HT^+, and DT^+, while Bishop and Wetmore (1973) have listed the adiabatic corrections as a function of the internuclear separation. Adiabatic calculations have also been performed for two of the excited states of H_2^+ (Bishop *et al.*, 1975).

More recently, nonadiabatic calculations have been performed. Estimates of the size and effects of nonadiabatic coupling on the ground-state vibration–rotation levels can be made by perturbation theory (Bishop and Wetmore, 1973). Performing complete and accurate nonadiabatic calculations is, however, a daunting task since the electronic and nuclear motions cannot be treated separately. Two methods have been used to perform accurate nonadiabatic calculations. In the first, due to Kołos (1969) and pursued by Bishop (1974) and Bishop and Cheung (1977), a variational solution to the complete nonrelativistic Hamiltonian of the three-particle system is sought. By using a very large expansion over products of functions to describe the electronic and nuclear motions (up to 515 functions were used), energies were obtained for the first few nonrotating vibrational states of H_2^+, D_2^+, HD^+, HT^+, and DT^+ with an estimated accuracy of

± 0.002 cm^{-1}. The calculations were extended to the rotating states by the use of an approximate formula (Colbourn and Bunker, 1976).

Wolniewicz and Poll (1978, 1980) have devised a different method for performing nonadiabatic calculations, which appears to be more elegant and efficient. They treated the nonadiabatic coupling between the ground and excited states as a perturbation to the adiabatic solution. The first-order perturbation to the adiabatic wave function was obtained variationally using expansions of the same form as the exact wave functions for the excited states which can couple directly to the ground state. Using this modified wave function, they then obtained the nonadiabatic correction to the adiabatic vibration–rotation energy as the second-order energy correction. They tabulated the nonadiabatic energies obtained for HD^+ with $v =$ 0 to 21 and $N = 0$ to 5, with an anticipated accuracy of 0.001 cm^{-1} (Wolniewicz and Poll, 1980). These calculations represent the present state of the art and are probably the most accurate calculations yet carried out for any molecule.

Other methods have been proposed for performing nonadiabatic calculations; e.g., the generator coordinate method has received some attention (Lathouwers, 1978), although numerical results have yet to be obtained.

In obtaining the most accurate vibration–rotation energies possible, there are a number of other small effects to be considered, particularly relativistic and radiative corrections. The relativistic corrections allow for the fact that the Schrödinger equation is used rather than the Dirac equation. These corrections have been investigated in 1969 by Luke *et al.* and more recently by Gonsalves and Moss (1979); accurate calculations of the corrections were made by Bishop in 1977. The radiative corrections arise from the interaction of the electron with the zero-point electromagnetic field. The Lamb shift due to this interaction was first calculated for H_2^+ by Gersten (1969), while Bishop and Cheung (1978) have performed further calculations for H_2^+ and, more recently, for HD^+ (Bishop and Cheung, 1980). Both the relativistic and radiative corrections were included in the nonadiabatic calculations of Bishop and Cheung (1977) and of Wolniewicz and Poll (1980).

Other work during the 1970s was concerned with the evaluation of properties of the hydrogen molecular ion, which can be investigated experimentally. These calculations have generally used the exact Born–Oppenheimer electronic wave functions and adiabatic nuclear wave functions (Bishop and Wetmore, 1973). Examples include the study of hyperfine interactions by McEachran *et al.* (1978) for H_2^+, and by Ray and Certain (1977) for H_2^+ and HD^+. Both groups obtained excellent agreement with the available experimental data (Jefferts, 1969; Wing *et al.*, 1976), which will be discussed in detail later. The photodissociation of H_2^+ and its isotopic modifications has also received considerable attention, starting with the work of Dunn (1968; Von Busch and Dunn, 1972). More recently, Tadjeddine and Parlant (1977) studied HD^+, and their results may be compared with the experimental work of van Asselt *et al.* (1975) described in Section IV. Photodissociation by infrared lasers has been considered theoretically by Fournier *et al.* (1979) and observed in the experiments described by Carrington and Buttenshaw (1981). Transition dipole moments for a number of infrared transitions of HD^+, HT^+, and DT^+ have been estimated (Bunker, 1974; Colbourn and Bunker, 1976). Matrix elements of a number of other operators, averaged over vibrational state, have been given by Bishop and Cheung (1978). Rotationally quasibound levels have been considered by Peek *et al.*, (1975), and Peek (1969) has also drawn attention to the importance of long-range forces in H_2^+ and their effects on the potential curve for the first excited electronic state.

Despite the large amount of theoretical work on the hydrogen molecular ion which has been described, important problems still remain to be investigated. As we will show later, the infrared frequencies measured for HD^+ (Carrington and Buttenshaw, 1981) do not all agree with the theory of Wolniewicz and Poll (1980) to the expected level of accuracy, so that further refinement of the calculations is required. Hyperfine constants for the high vibrational levels of HD^+ studied by Carrington and Buttenshaw (1981) have not yet been calculated, even within the adiabatic approximation. Improvements in the range and quality of experimental measurements are likely and could lead to improved values of certain fundamental constants, provided the molecular quantum mechanics is sufficiently accurate.

III. PRESENT STATE OF THE THEORY

In the previous section, we reviewed the historical development of the theory of the hydrogen molecular ion. In this section we will give a mathematical treatment of the theory, leading up to the present best calculations, those of Wolniewicz and Poll (1980) on HD^+. Although most of this material can be found elsewhere, (Kołos and Wolniewicz, 1963; Hunter and Pritchard, 1967; Kołos, 1969, 1970; Bishop and Wetmore, 1973; Bishop, 1974; Wolniewicz and Poll, 1978, 1980) we feel that a complete summary of the theory in one place is now required, and we hope that this section will satisfy this need.

A. The Complete Nonrelativistic Hamiltonian

For a system of point charges interacting electrostatically and moving through field-free space, the complete nonrelativistic Hamiltonian can be written in the form

$$\mathscr{H} = \sum_i \frac{-\hbar^2\nabla_i^{\,2}}{2m_i} + \sum_i \sum_{j>i} \frac{z_i z_j e^2}{4\pi\varepsilon_0 r_{ij}}, \tag{1}$$

where all the symbols have their usual meanings. In order to obtain a convenient form of this Hamiltonian for the hydrogen molecular ion, we separate out the motion of the center of mass of the particles by applying the transformation

$$\begin{pmatrix} \mathbf{r}_g \\ \mathbf{R} \\ \mathbf{R}_{cm} \end{pmatrix} = \begin{pmatrix} -\frac{1}{2} & -\frac{1}{2} & 1 \\ -1 & 1 & 0 \\ m_1/M & m_2/M & m_e/M \end{pmatrix} \begin{pmatrix} \mathbf{r}_1 \\ \mathbf{r}_2 \\ \mathbf{r}_e \end{pmatrix}, \tag{2}$$

where $M = m_1 + m_2 + m_e$. The coordinate systems are illustrated in Fig. 1. $\mathbf{r}_1$, $\mathbf{r}_2$, and $\mathbf{r}_e$ are the position vectors of the three particles relative to an arbitrary space-fixed origin. The new basis vectors are the internuclear vector $\mathbf{R}\ (=\mathbf{r}_2 - \mathbf{r}_1)$, the position of the center of mass of the system relative to the space-fixed origin $\mathbf{R}_{cm}$, and the position of the electron relative to the geometric center of the nuclei $\mathbf{r}_g$. The separation can be achieved by other transformations (Hunter *et al.*, 1966), but that described here is the one most commonly applied to the hydrogen molecular ion. Applying the trans-

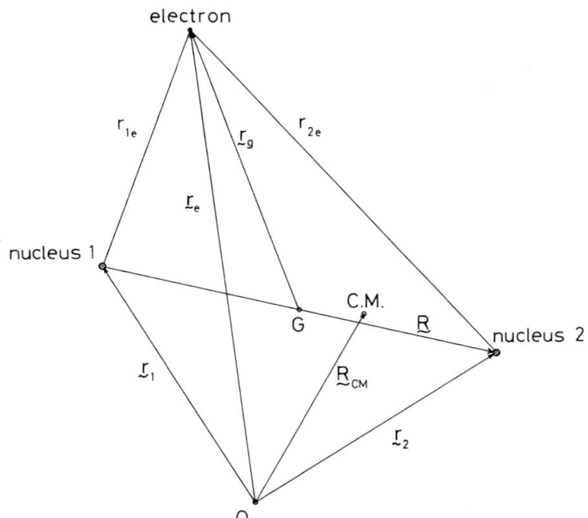

FIG. 1 Coordinate system for the hydrogen molecular ion. O is the arbitrary space-fixed origin, C. M. is the center of mass of the system, and G is the geometric center of the nuclei.

formation (2) to the kinetic energy operators in the Hamiltonian (1), we obtain

$$-\hbar^2 \sum_i \frac{\nabla_i^2}{2m_i} = -\hbar^2 \left(\frac{\nabla_g^2}{2m_e} + \frac{\nabla_R^2}{2\mu} + \frac{\nabla_g^2}{8\mu} + \frac{\nabla_g \cdot \nabla_R}{2\mu_a} + \frac{\nabla_{cm}^2}{2M} \right), \tag{3}$$

where $1/\mu = (1/m_1) + (1/m_2)$ and $1/\mu_a = (1/m_1) - (1/m_2)$. Note that $1/\mu_a = 0$ for the homonuclear case, $m_1 = m_2$. The electrostatic potential V is unchanged by the transformation

$$V = \frac{e^2}{4\pi\varepsilon_0} \left(\frac{1}{R} - \frac{1}{r_{1e}} - \frac{1}{r_{2e}} \right) \tag{4}$$

because it depends on the relative positions of the particles. The translational motion can now be separated out of the Schrödinger equation [Eq. (5)],

$$\left\{ -\hbar^2 \left[\frac{\nabla_g^2}{2m_e} + \frac{\nabla_R^2}{2\mu} + \frac{\nabla_g^2}{8\mu} + \frac{\nabla_g \cdot \nabla_R}{2\mu_a} + \frac{\nabla_{cm}^2}{2M} \right] + V \right\} \Psi_{tot} = E_{tot} \Psi_{tot}. \tag{5}$$

We use the form

$$\Psi_{tot}(\mathbf{R}_{cm}, \mathbf{R}, \mathbf{r}_g) = A_{cm}(\mathbf{R}_{cm}) \psi_{mol}(\mathbf{R}, \mathbf{r}_g)$$

in order to achieve the separation, which is

$$\left\{ -\hbar^2 \left[\frac{\nabla_g^2}{2m_e} + \frac{\nabla_R^2}{2\mu} + \frac{\nabla_g^2}{8\mu} + \frac{\nabla_g \cdot \nabla_R}{2\mu_a} \right] + V \right\} \psi_{mol} = E_{int} \psi_{mol} \tag{6}$$

and

$$(-\hbar^2/2M)\, \nabla_{cm}^2\, A_{cm} = (E_{tot} - E_{int}) A_{cm}. \tag{7}$$

Equation (7) is simply the time-independent Schrödinger equation for a body of mass M freely translating through space with kinetic energy $(E_{tot} - E_{int})$.

In summary, the nonrelativistic Hamiltonian, in atomic units, for the internal motion of the hydrogen molecular ion is

$$\mathscr{H}_{int} = -\frac{\nabla_g^2}{2} - \frac{\nabla_R^2}{2\mu} - \frac{\nabla_g^2}{8\mu} - \frac{\nabla_g \cdot \nabla_R}{2\mu_a} + \frac{1}{R} - \frac{1}{r_{1e}} - \frac{1}{r_{2e}}. \tag{8}$$

B. *Solution of the Schrödinger Equation*

The nonrelativistic Hamiltonian (8) contains terms which couple the electronic and nuclear motions; this makes it impossible to obtain exact eigen functions and eigenvalues. However, by making certain approximations, it is possible to reduce the problem to one which can be solved exactly. In order to see how this may be achieved, we follow the method suggested by Born and Huang (1954), and expand the complete molecular wave function as the series

$$\psi_{\text{mol}}(\mathbf{R}, \mathbf{r}_{\text{g}}) = \sum_t F_t(\mathbf{R})\phi_t(R, \mathbf{r}_{\text{g}}), \tag{9}$$

where the $\phi_t(R, \mathbf{r}_{\text{g}})$ are the exact solutions of what we will call the electronic Born–Oppenheimer equation for the hydrogen molecular ion:

$$\left(-\frac{\nabla_{\text{g}}^2}{2} - \frac{1}{r_{1\text{e}}} - \frac{1}{r_{2\text{e}}} + \frac{1}{R}\right)\phi_t(R, \mathbf{r}_{\text{g}}) = E_t(R)\phi_t(R, \mathbf{r}_{\text{g}}). \tag{10}$$

The methods for solving this equation are well established (Hunter and Pritchard, 1967a), and we will describe them later. Substituting the Born expansion [Eq. (9)] into the complete nonrelativistic Schrödinger equation [Eq. (6)], we obtain a set of coupled differential equations for the functions $F_t(\mathbf{R})$:

$$\mathscr{H}_{\text{int}} \sum_t F_t(\mathbf{R})\phi_t(R, \mathbf{r}_{\text{g}}) = E_{\text{int}} \sum_t F_t(\mathbf{R})\phi_t(R, \mathbf{r}_{\text{g}}). \tag{11}$$

The equation is simplified by premultiplying by $\phi_s{}^*(R, \mathbf{r}_{\text{g}})$ and integrating over the electronic coordinates $\mathbf{r}_{\text{g}}$ to obtain

$$\begin{aligned} E_s(R)F_s(\mathbf{R}) &+ \int \phi_s{}^*(R, \mathbf{r}_{\text{g}})\left[-\frac{\nabla_R{}^2}{2\mu} - \frac{\nabla_{\text{g}}{}^2}{8\mu} - \frac{\boldsymbol{\nabla}_{\text{g}} \cdot \boldsymbol{\nabla}_R}{2\mu_a}\right] F_s(\mathbf{R})\phi_s(R, \mathbf{r}_{\text{g}})\, d\mathbf{r}_{\text{g}} \\ &+ \sum_{t \neq s} \int \phi_s{}^*(R, \mathbf{r}_{\text{g}})\left[-\frac{\nabla_R{}^2}{2\mu} - \frac{\nabla_{\text{g}}{}^2}{8\mu} - \frac{\boldsymbol{\nabla}_{\text{g}} \cdot \boldsymbol{\nabla}_R}{2\mu_a}\right] F_t(\mathbf{R})\phi_t(R, \mathbf{r}_{\text{g}})\, d\mathbf{r}_{\text{g}} = E_{\text{int}}F_s(\mathbf{R}). \end{aligned} \tag{12}$$

Since in the electronic Born–Oppenheimer equation [Eq. (10)], the nuclei are implicitly treated as fixed-point charges, the functions $\phi_t(R, \mathbf{r}_{\text{g}})$ must be either symmetric or antisymmetric with respect to exchange of nuclei and electron inversion through the geometric center of the nuclei. The operator $\boldsymbol{\nabla}_R$ is antisymmetric with respect to nuclear permutation, while $\boldsymbol{\nabla}_{\text{g}}$ is antisymmetric under electron inversion. Consequently, by symmetry, we have

$$\begin{aligned} \int \phi_s{}^*(R, \mathbf{r}_{\text{g}})\boldsymbol{\nabla}_R\phi_s(R, \mathbf{r}_{\text{g}})\, d\mathbf{r}_{\text{g}} &= 0 \\ \int \phi_s{}^*(R, \mathbf{r}_{\text{g}})\boldsymbol{\nabla}_{\text{g}}\phi_s(R, \mathbf{r}_{\text{g}})\, d\mathbf{r}_{\text{g}} &= 0 \\ \int \phi_s{}^*(R, \mathbf{r}_{\text{g}})\boldsymbol{\nabla}_{\text{g}} \cdot \boldsymbol{\nabla}_R\phi_s(R, \mathbf{r}_{\text{g}})\, d\mathbf{r}_{\text{g}} &= 0 \end{aligned} \tag{13}$$

so that Eq. (12) becomes

$$\begin{aligned}
&\left\{E_s(R) - \frac{\nabla_R^2}{2\mu} - \int \phi_s^*(R, \mathbf{r}_g)\left[\frac{\nabla_g^2}{8\mu} + \frac{\nabla_R^2}{2\mu}\right]\phi_s(R, \mathbf{r}_g)\, d\mathbf{r}_g\right\} F_s(\mathbf{R}) \\
&\quad + \sum_{t \neq s}\left\{\int \phi_s^*(R, \mathbf{r}_g)\left[-\frac{\nabla_R^2}{2\mu} - \frac{\nabla_g^2}{8\mu} - \frac{\boldsymbol{\nabla}_g \cdot \boldsymbol{\nabla}_R}{2\mu_a}\right]\phi_t(R, \mathbf{r}_g)\, d\mathbf{r}_g\right. \\
&\quad \left. + \int \phi_s^*(R, \mathbf{r}_g)\left[-\frac{\boldsymbol{\nabla}_R}{\mu} - \frac{\boldsymbol{\nabla}_g}{2\mu_a}\right]\phi_t(R, \mathbf{r}_g)\, d\mathbf{r}_g \cdot \boldsymbol{\nabla}_R\right\} F_t(\mathbf{R}) \\
&\quad = E_{\text{int}} F_s(\mathbf{R}).
\end{aligned} \tag{14}$$

An exact solution of this set of coupled differential equations is not feasible since the couplings between the infinite set of functions $F_t(\mathbf{R})\phi_t(R, \mathbf{r}_g)$ would have to be considered. In practice, it is necessary to neglect some of the couplings in order to make the problem tractable or to seek an alternative method of solving the problem.

C. *The Electronic Born–Oppenheimer Equation*

Before proceeding to describe the approximations that can be made to simplify Eq. (14), we return to the electronic Born–Oppenheimer equation [Eq. (10)]. Historically, this equation was the starting point of the modern quantum theory of the hydrogen molecular ion (Hylleraas, 1931; Jaffé, 1934). The problem is to solve for the motion of a single electron about two fixed nuclear charge centers. By using prolate spheroidal coordinates, the three-dimensional equation [Eq. (10)] separates into three one-dimensional equations. Figure 2 shows the Cartesian coordinate system; the prolate

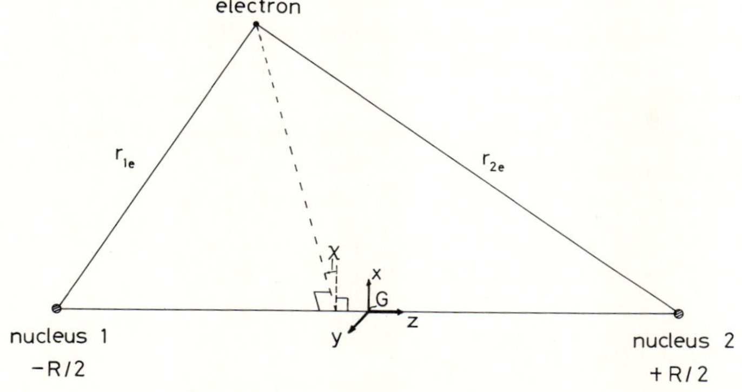

FIG. 2 Cartesian coordinate system.

spheroidal coordinates are

$$\xi = \frac{(r_{1e} + r_{2e})}{R} \quad \text{with} \quad 1 \leqslant \xi < \infty;$$

$$\eta = \frac{(r_{1e} - r_{2e})}{R} \quad \text{with} \quad -1 \leqslant \eta \leqslant 1 \tag{15}$$

and χ, rotation of the electron about the z axis ($0 \leqslant \chi < 2\pi$). In this coordinate system, we have

$$\begin{aligned}\nabla_g &= \frac{\partial}{\partial x}\mathbf{i} + \frac{\partial}{\partial y}\mathbf{j} + \frac{\partial}{\partial z}\mathbf{k} \\ &= \frac{2[(\xi^2 - 1)(1 - \eta^2)]^{1/2}}{R(\xi^2 - \eta^2)} \left\{ \left[\xi \cos\chi \frac{\partial}{\partial \xi} - \eta \cos\chi \frac{\partial}{\partial \eta} - \frac{\sin\chi(\xi^2 - \eta^2)}{(\xi^2 - 1)(1 - \eta^2)} \frac{\partial}{\partial x} \right] \mathbf{i} \right. \\ &\quad + \left[\xi \sin\chi \frac{\partial}{\partial \xi} - \eta \sin\chi \frac{\partial}{\partial \eta} + \frac{\cos\chi(\xi^2 - \eta^2)}{(\xi^2 - 1)(1 - \eta^2)} \frac{\partial}{\partial x} \right] \mathbf{j} \\ &\quad \left. + \left[\frac{\eta(\xi^2 - 1)^{1/2}}{(1 - \eta^2)^{1/2}} \frac{\partial}{\partial \xi} + \frac{\xi(1 - \eta^2)^{1/2}}{(\xi^2 - 1)^{1/2}} \frac{\partial}{\partial \eta} \right] \mathbf{k} \right\}.\end{aligned} \tag{16}$$

The Laplacian operator in the new coordinates is

$$\nabla_g{}^2 = \frac{4}{R^2(\xi^2 - \eta^2)} \left\{ \frac{\partial}{\partial \xi}(\xi^2 - 1)\frac{\partial}{\partial \xi} + \frac{\partial}{\partial \eta}(1 - \eta^2)\frac{\partial}{\partial \eta} + \frac{(\xi^2 - \eta^2)}{(\xi^2 - 1)(1 - \eta^2)} \frac{\partial^2}{\partial \chi^2} \right\} \tag{17}$$

and the volume element is

$$d\mathbf{r}_g = [R^3(\xi^2 - \eta^2)/8]\, d\xi\, d\eta\, d\chi. \tag{18}$$

The electron-nuclear electrostatic attraction operator may also be transformed into the (ξ, η, χ) coordinate system, with the result

$$-1/r_{1e} - 1/r_{2e} = -4\xi/R(\xi^2 - \eta^2). \tag{19}$$

Consequently, in prolate spheroidal coordinates the electronic Born–Oppenheimer equation becomes

$$\begin{aligned}&\left[-\frac{2}{R^2(\xi^2 - \eta^2)} \left\{ \frac{\partial}{\partial \xi}(\xi^2 - 1)\frac{\partial}{\partial \xi} + \frac{\partial}{\partial \eta}(1 - \eta^2)\frac{\partial}{\partial \eta} + \frac{(\xi^2 - \eta^2)}{(\xi^2 - 1)(1 - \eta^2)} \frac{\partial^2}{\partial \chi^2} \right\} \right. \\ &\quad \left. - \frac{4\xi}{R(\xi^2 - \eta^2)} + \frac{1}{R} \right] \phi_t(R, \xi, \eta, \chi) \\ &\quad = E_t(R)\phi_t(R, \xi, \eta, \chi),\end{aligned} \tag{20}$$

which may be written in the more compact form

$$\left[\frac{\partial}{\partial \xi}(\xi^2 - 1)\frac{\partial}{\partial \xi} + \frac{\partial}{\partial \eta}(1 - \eta^2)\frac{\partial}{\partial \eta} + \left\{\frac{1}{(\xi^2 - 1)} + \frac{1}{(1 - \eta^2)}\right\}\frac{\partial^2}{\partial \chi^2} + 2R\xi - p^2(\xi^2 - \eta^2)\right]\phi_t(R, \xi, \eta, \chi) = 0, \tag{21}$$

where $p^2 = -[E_t(R) - (1/R)]R^2/2$. The separation into three one-dimensional equations is now easily achieved by making the substitution

$$\phi_t(R, \xi, \eta, \chi) = L(R, \xi)M(R, \eta)N(\chi). \tag{22}$$

We then obtain the three separated equations:

$$\begin{aligned} \left[\frac{\partial^2}{\partial \chi^2} + \Lambda^2\right] N &= 0 \\ \left[\frac{\partial}{\partial \xi}(\xi^2 - 1)\frac{\partial}{\partial \xi} + A - \frac{\Lambda^2}{(\xi^2 - 1)} + 2R\xi - p^2\xi^2\right] L &= 0 \\ \left[\frac{\partial}{\partial \eta}(1 - \eta^2)\frac{\partial}{\partial \eta} - A - \frac{\Lambda^2}{(1 - \eta^2)} + p^2\eta^2\right] M &= 0, \end{aligned} \tag{23}$$

where Λ^2 and A are the separation constants.

The χ equation can be solved analytically, with the result

$$N(\chi) = \exp(i\Lambda\chi)/(2\pi)^{1/2}, \tag{24}$$

where $\Lambda = 0, \pm 1, \pm 2, \ldots$

Certain aspects of the symmetry of the problem are evident from Eqs. (23). Λ appears as Λ^2 in the ξ and η equations, so that the eigenvalues have a twofold degeneracy for $\Lambda \neq 0$, due to the cylindrical symmetry of the system. The η equation is invariant to the operation $\eta \to -\eta$, so that the equation possesses separate even (g) and odd (u) solutions for each value of A, caused by the nuclear charge permutation symmetry of the system.

In order to obtain a numerical solution of the ξ equation, we will consider the Jaffé expansion

$$L(R, \xi) = (\xi^2 - 1)^{\Lambda/2}(\xi + 1)^{\sigma}\exp(-p\xi)\sum_{n=0}^{\infty} g_n(R)[(\xi - 1)/(\xi + 1)]^n, \tag{25}$$

where $\sigma = (R/p) - \Lambda - 1$. Substitution of this expansion into Eq. (23) leads to a recursion relation between the coefficients $g_n(R)$,

$$\begin{aligned} g_{n+1}[(n + 1)(n + \Lambda + 1)] + g_n[A - p^2 + 2p\sigma + (\sigma + \Lambda)(\Lambda + 1) \\ + 2n(\sigma - 2p) - 2n^2] + g_{n-1}[(n - \sigma - 1)(n - \sigma - \Lambda - 1)] = 0. \end{aligned} \tag{26}$$

We further require $g_n = 0$ for $n < 0$, and for numerical calculations we truncate the series so that $g_n = 0$ for $n > n_{\max}$ (the series converges rapidly). The set of recursion relations (26) may then be written as a matrix eigenvalue equation (Hunter and Pritchard, 1967),

$$\mathbf{G} \cdot \mathbf{g} = -A\mathbf{g} \tag{27}$$

where

$$\mathbf{g} = \begin{pmatrix} g_0 \\ g_1 \\ \vdots \\ g_{n_{\max}} \end{pmatrix},$$

and the nonzero elements of the tridiagonal matrix **G** are

$$\begin{aligned} G_{n,n-1} &= (n - \sigma - 1)(n - \sigma - \Lambda - 1) \\ G_{n,n} &= -p^2 + 2p\sigma + (\sigma + \Lambda)(\Lambda + 1) + 2n(\sigma - 2p) - 2n^2 \\ G_{n,n+1} &= (n + 1)(n + \Lambda + 1). \end{aligned} \tag{28}$$

Other expansions, notably that due to Hylleraas (1931), have been used to solve the ξ equation; they also lead to recursion relations which may be expressed as matrix eigenvalue equations.

The η equation may be solved by using an expansion over associated Legendre functions,

$$M(R, \eta) = \sum_{s=0}^{\infty} f_s(R) P_{\Lambda+s}^{\Lambda}(\eta). \tag{29}$$

Because of the symmetry of the η equation, only odd or even terms will appear in the expansion for a given state. Substituting the expansion (29) into (23), and using the known properties of the functions $P_{\Lambda+s}^{\Lambda}(\eta)$,

$$\left[\frac{d}{d\eta}(1 - \eta^2)\frac{d}{d\eta} - \frac{\Lambda^2}{(1 - \eta^2)} + (\Lambda + s)(\Lambda + s + 1)\right] P_{\Lambda+s}^{\Lambda}(\eta) = 0$$

$$(s + 1)P_{\Lambda+s+1}^{\Lambda}(\eta) - (2\Lambda + 2s + 1)\eta P_{\Lambda+s}^{\Lambda}(\eta) + (2\Lambda + s)P_{\Lambda+s-1}^{\Lambda}(\eta) = 0, \tag{30}$$

we obtain a recursion relation between the coefficients $f_s(R)$,

$$\begin{aligned} & f_{s-2}\left[\frac{-p^2(s - 1)s}{(2\Lambda + 2s - 3)(2\Lambda + 2s - 1)}\right] + f_s\left[(\Lambda + s)(\Lambda + s + 1)\right. \\ & \qquad \left. - \frac{p^2(2\Lambda + 2)s}{(2\Lambda + 2s - 1)(2\Lambda + 2s + 1)} - \frac{p^2(2\Lambda + s + 1)(s + 1)}{(2\Lambda + 2s + 1)(2\Lambda + 2s + 3)} + A\right] \\ & \qquad + f_{s+2}\left[\frac{-p^2(2\Lambda + s + 1)(2\Lambda + s + 2)}{(2\Lambda + 2s + 3)(2\Lambda + 2s + 5)}\right] = 0. \end{aligned} \tag{31}$$

As in the case of the ξ equation, this set of recursion relations may be expressed as a matrix eigenvalue equation (Hunter and Pritchard, 1967),

$$\mathbf{F} \cdot \mathbf{f} = A\mathbf{f}, \tag{32}$$

where

$$\mathbf{f} = \begin{pmatrix} f_0 \\ f_2 \\ \vdots \\ f_{2N} \end{pmatrix} \quad \text{or} \quad \mathbf{f} = \begin{pmatrix} f_1 \\ f_3 \\ \vdots \\ f_{2N+1} \end{pmatrix}$$

for even (g symmetry) or odd (u symmetry) states, respectively. The nonzero elements of the matrix $\mathbf{F}$ are then given by

$$\begin{aligned} F_{s,s-2} &= \frac{p^2(s-1)s}{(2\Lambda + 2s - 3)(2\Lambda + 2s - 1)} \\ F_{s,s} &= -(\Lambda + s)(\Lambda + s + 1) + \frac{p^2}{(2\Lambda + 2s + 1)} \\ &\quad \times \left\{ \frac{(2\Lambda + s)s}{(2\Lambda + 2s - 1)} + \frac{(2\Lambda + s + 1)(s + 1)}{(2\Lambda + 2s + 3)} \right\} \\ F_{s,s+2} &= \frac{p^2(2\Lambda + s + 1)(2\Lambda + s + 2)}{(2\Lambda + 2s + 3)(2\Lambda + 2s + 5)}. \end{aligned} \tag{33}$$

The solution of the electronic Born–Oppenheimer equation [Eq. (10)], determining $f(R)$, $g(R)$, and $E_t(R)$ for a given state at a particular R value, is achieved by requiring that A simultaneously satisfies Eqs. (27) and (32). Numerical methods have been outlined by Hunter and Pritchard (1967); the problem is particularly simple for the hydrogen molecular ion because both $\mathbf{F}$ and $\mathbf{G}$ are tridiagonal matrices. The results of such calculations can be found in many places; see, for example, the paper by Peek (1965).

D. Levels of Approximation

1. The Born–Oppenheimer Approximation

At this level of approximation, all terms coupling the electronic and nuclear motions are neglected. Equation (14) then reduces to the Born–Oppenheimer equation for nuclear motion:

$$\{E_s(R) - (\nabla_R{}^2/2\mu)\}F_s^{\mathrm{BO}}(\mathbf{R}) = E_{\mathrm{int}}^{\mathrm{BO}} F_s^{\mathrm{BO}}(\mathbf{R}). \tag{34}$$

The advantage of this approximation is the great simplification we achieve over Eq. (14). However, this simplification is achieved at the expense of a considerable loss of accuracy in any calculations.

2. *The Adiabatic Approximation*

The adiabatic approximation has been extensively reviewed by Kołos (1970). At this level of approximation, we retain the terms coupling the electronic and nuclear motions which are diagonal in the electronic state; Eq. (14) reduces to

$$\left\{E_s(R) - \int \phi_s^*(R, \mathbf{r}_g) \frac{\nabla_g^2}{8\mu} \phi_s(R, \mathbf{r}_g)\, d\mathbf{r}_g \right.$$
$$\left. - \int \phi_s^*(R, \mathbf{r}_g) \frac{\nabla_R^2}{2\mu} \phi_s(R, \mathbf{r}_g)\, d\mathbf{r}_g - \frac{\nabla_R^2}{2\mu}\right\} F_s^{\mathrm{AD}}(\mathbf{R}) = E_{\mathrm{int}}^{\mathrm{AD}} F_s^{\mathrm{AD}}(\mathbf{R}). \quad (35)$$

The effect of the approximation is that the nuclear motion is now governed by an effective potential

$$E_s(R) - \int \phi_s^*(R, \mathbf{r}_g) \frac{\nabla_g^2}{8\mu} \phi_s(R, \mathbf{r}_g)\, d\mathbf{r}_g - \int \phi_s^*(R, \mathbf{r}_g) \frac{\nabla_R^2}{2\mu} \phi_s(R, \mathbf{r}_g)\, d\mathbf{r}_g \quad (36)$$

obtained by averaging the complete Hamiltonian (8) over the Born–Oppenheimer electronic wave function $\phi_s(R, \mathbf{r}_g)$. The effective adiabatic potential [Eq. (36)] is isotope dependent because of its dependence on μ. The Born–Oppenheimer potential $E_s(R)$ is isotope independent. Working at this level of approximation, we retain the basic simplicity of the Born–Oppenheimer equation for nuclear motion [Eq. (34)] because the electronic and nuclear motions are effectively separated. But because we have included the diagonal corrections, calculations made within the adiabatic approximation will be more accurate. The adiabatic approximation is particularly successful when there are no close-lying states which are nonadiabatically coupled to the state of interest. This is the case for the ground electronic state of H_2^+ and D_2^+, but for the heteronuclear system HD^+, the $\nabla_g \cdot \nabla_R/2\mu_a$ term in Eq. (14) ($1/\mu_a = 0$ for homonuclear systems) couples the ground (Σ_g) and first excited (Σ_u) electronic states. In order to obtain accurate results for HD^+, it will be necessary to perform nonadiabatic calculations, particularly for the high vibrational levels. The problem can be illustrated by considering the behavior of the adiabatic potentials for these states at large R. Within the adiabatic treatment, they are identical at very large R, yet in reality the dissociation limits $H^+ + D(1s)$ and $H(1s) + D^+$ are separated by 29.8 cm^{-1}. This degeneracy is lifted by nonadiabatic coupling.

Hunter *et al.* (1974) have given an extensive tabulation of the results of adiabatic calculations of vibration–rotation levels of the ground electronic state of H_2^+, D_2^+, and HD^+. Bishop and Wetmore (1973) have provided values for the adiabatic corrections to the potential

$$-\int \phi_s^*(R, \mathbf{r}_g)(\nabla_g^2/8\mu)\phi_s(R, \mathbf{r}_g)\, d\mathbf{r}_g \quad \text{and} \quad -\int \phi_s^*(R, \mathbf{r}_g)(\nabla_R^2/2\mu)\phi_s(R, \mathbf{r}_g)\, d\mathbf{r}_g.$$

Figure 3 is a plot of the Born–Oppenheimer potential $E_s(R)$ and the adiabatic corrections, which are both positive, for the ground electronic state of H_2^+.

The adiabatic approximation is widely used for calculating other properties of the hydrogen molecular ion because of the convenient separation it allows of the electronic and nuclear motions through the use of an effective potential. Evaluation of properties for a given vibration–rotation state is achieved by averaging electronic matrix elements obtained from the $\phi_s(R, \mathbf{r}_g)$

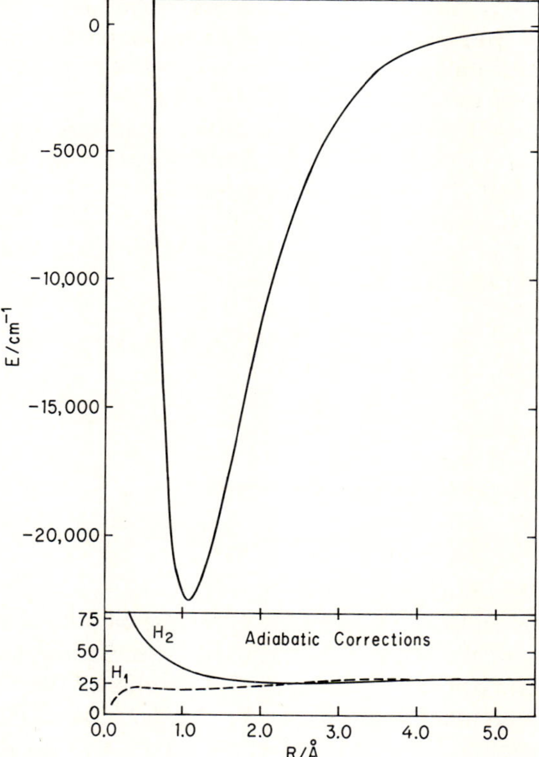

FIG. 3 Born–Oppenheimer potential and adiabatic corrections $H_1 = -\int\phi_s^*(\nabla_R^2/2\mu)\phi_s\, d\mathbf{r}_g$ and $H_2 = -\int\phi_s^*(\nabla_g^2/8\mu)\phi_s\, d\mathbf{r}_g$ from the calculations of Bishop and Wetmore (1973).

over the adiabatic vibration–rotation wave function $F_s^{AD}(\mathbf{R})$. A good example is the calculation of nuclear hyperfine constants (Ray and Certain, 1977).

3. *Nonadiabatic Calculations*

The exact solution of the nonadiabatic problem as expressed by Eq. (14) is impossible, as it requires the solution of an infinite set of coupled differential equations. Following the approximations described above, we could reduce the problem by considering the coupling with only a limited number of other states, and hope that the Born expansion (9) converges rapidly. This approach was followed by Hunter and Pritchard (1967b) for the four lowest vibrational states of H_2^+, HD^+, and D_2^+, by investigating the convergence of the energy levels as progressively higher states $\phi_t(R, \mathbf{r}_g)$ were included in the expansion (9). Initial rapid convergence was found for these low vibrational levels. In order to improve the accuracy of these calculations and extend them to higher vibrational levels, much longer Born expansions will be required. To perform efficient and accurate nonadiabatic calculations, alternative formulations of the problem are desirable, and two techniques have been investigated, which we now describe.

a. Variational Approach. In this method, investigated by Bishop (1974) and by Bishop and Cheung (1977), the eigen energies of the complete Hamiltonian (8) are sought by variational adjustment of a trial wave function. Only a few low-lying nonrotating vibrational levels of the ground electronic state were considered for various isotopic modifications of the hydrogen molecular ion. The trial wave function used was

$$\Psi = \sum_{i=0}^{i_m} \sum_{j=0,2}^{j_m} \sum_{k=0}^{k_m} C_{ijk}\phi_{ijk}(\xi, \eta, R) + \sum_{i=0}^{i_m'} \sum_{j=1,3}^{j_m'} \sum_{k=0}^{k_m'} C_{ijk}\phi_{ijk}(\xi, \eta, R), \tag{37}$$

where the basis functions of the expansion were

$$\phi_{ijk}(\xi, \eta, R) = \exp(-\alpha\xi)\cosh(\beta\eta)\xi^i\eta^j R^{-3/2} \exp[\tfrac{1}{2}(-x^2)]H_k(x). \tag{38}$$

The $H_k(x)$ are Hermite polynomials, $x = \gamma(R - \delta)$, and α, β, γ, δ are adjustable parameters chosen to minimize the energy; with large expansions, the values are not very critical. This basis was selected because the integrals required for a variational solution are easily evaluated (Bishop, 1974). Even and odd states are only mixed by the $\nabla_g \cdot \nabla_R/2\mu_a$ term in the Hamiltonian, so that the second triple summation in the expansion only appears for heteronuclear systems like HD^+. The convergence of the variationally calculated energies with respect to basis set size was investigated very carefully.

The final results, obtained with an expansion of up to 515 terms, should be accurate to $\pm 0.001\ \text{cm}^{-1}$, after the inclusion of radiative and relativistic corrections (Bishop and Cheung, 1977). Energies of the rotating states were also calculated by means of an approximate formula for the change in non-adiabatic energy correction with rotational quantum number (Colbourn and Bunker, 1976). Infrared transition frequencies accurate to $\pm 0.002\ \text{cm}^{-1}$ could then be obtained for HD^+ and compared with the results of Wing *et al.* (1976).

b. Variation–Perturbation Approach. By using the variational approach, the problem of having to consider explicitly the couplings between the adiabatic states, described by Eq. (14), is avoided. However, this method makes no use of the exact adiabatic eigenvalues and wave functions which can be readily obtained for the hydrogen molecular ion. As we have discussed, the adiabatic approximation is usually quite accurate, so that a method which makes use of the adiabatic solutions as a starting point for nonadiabatic calculations would appear to offer greater efficiency and elegance than the fully variational approach.

Wolniewicz and Poll (1978, 1980) have developed such an alternative formulation of the problem, in which the nonadiabatic effects are treated as a perturbation to the adiabatic approximation. They define an adiabatic Hamiltonian $\mathscr{H}^{\text{AD}}$ by the equation

$$\mathscr{H}^{\text{AD}} = \sum_n E_n^{\text{AD}} |\psi_n^{\text{AD}}\rangle \langle \psi_n^{\text{AD}}| \tag{39}$$

such that its eigenvalues are the adiabatic molecular energy levels E_n^{AD}, and its eigen functions are the adiabatic states of the molecule

$$\psi_n^{\text{AD}} = \phi_n^{\text{BO}}(R, \mathbf{r}_g) F_n^{\text{AD}}(\mathbf{R}), \tag{40}$$

which can be obtained exactly from Eqs. (10) and (35). This abstract definition allows the complete nonrelativistic Hamiltonian (8) to be decomposed into an adiabatic part, and a nonadiabatic term,

$$\mathscr{H} = \mathscr{H}^{\text{AD}} + \mathscr{H}'^{\text{N.AD}}. \tag{41}$$

The methods of perturbation theory may now be used to calculate non-adiabatic corrections to the adiabatic energies E_n^{AD},

$$\mathscr{H}|\psi_n^{\text{AD}}\rangle = E_n^{\text{AD}}|\psi_n^{\text{AD}}\rangle + \mathscr{H}'^{\text{N.AD}}|\psi_n^{\text{AD}}\rangle. \tag{42}$$

Since from Eq. (35) we have

$$\langle \psi_n^{\text{AD}}|\mathscr{H}|\psi_n^{\text{AD}}\rangle = E_n^{\text{AD}}, \tag{43}$$

the first-order nonadiabatic energy correction is zero:

$$\langle \psi_n^{\text{AD}}|\mathscr{H}'^{\text{N.AD}}|\psi_n^{\text{AD}}\rangle = 0. \tag{44}$$

The variation–perturbation method of Wolniewicz and Poll is to obtain the second-order nonadiabatic energy corrections from the first-order correction to the wave function,

$$\psi_n = \psi_n^{\mathrm{AD}} + \psi_n'. \tag{45}$$

The first-order correction ψ_n' satisfies

$$(\mathscr{H}^{\mathrm{AD}} - E_n^{\mathrm{AD}})\psi_n' = -\mathscr{H}'^{\mathrm{N.AD}}\psi_n^{\mathrm{AD}}, \tag{46}$$

where we have used Eq. (44). ψ_n' is obtained by solving this equation variationally. The nonadiabatic energy correction to second order in perturbation theory is then

$$E'' = \langle\psi_n'|\mathscr{H}'^{\mathrm{N.AD}}|\psi^{\mathrm{AD}}\rangle. \tag{47}$$

Equation (46), from which the ψ_n' are to be obtained, can be simplified by separating out those parts of the wave functions which describe their angular momentum:

$$\psi^{\mathrm{AD}} = U^{\mathrm{AD}}\Omega^J_{M\Lambda_0} \tag{48}$$

and

$$\psi' = \sum_{\Lambda} U_\Lambda \Omega^J_{M\Lambda}, \tag{49}$$

where the $\Omega^J_{M\Lambda}$ are the eigen functions of a symmetric top, and the quantum numbers J, M, and Λ have their usual meanings. After premultiplying by $\Omega^{J*}_{M\Lambda}$ and integrating over the angular variables, Eq. (46) becomes

$$\{\langle JM\Lambda|\mathscr{H}^{\mathrm{AD}}|JM\Lambda\rangle - E^{\mathrm{AD}}\}U_\Lambda = -\langle JM\Lambda|\mathscr{H}'^{\mathrm{N.AD}}|JM\Lambda_0\rangle U^{\mathrm{AD}}. \tag{50}$$

Appropriate differential forms of the operators $\mathscr{H}^{\mathrm{AD}}$ and $\mathscr{H}'^{\mathrm{N.AD}}$ are obtained as follows. From Eq. (42) we have

$$\mathscr{H}'^{\mathrm{N.AD}} = \mathscr{H} - E^{\mathrm{AD}} \tag{51}$$

while, as indicated in the Born development of the theory, the adiabatic Hamiltonian $\mathscr{H}^{\mathrm{AD}}$, given by

$$\mathscr{H}^{\mathrm{AD}} = -\frac{\nabla_{\mathrm{g}}{}^2}{2} + \frac{1}{R} - \frac{1}{r_{1\mathrm{e}}} - \frac{1}{r_{2\mathrm{e}}} - \frac{\nabla_R{}^2}{2\mu} - \frac{\nabla_{\mathrm{g}}{}^2}{8\mu}, \tag{52}$$

contains all the nonzero diagonal terms. This form is acceptable for calculating the first-order correction ψ'. Equation (46) finally becomes

$$\{\langle JM\Lambda|\mathscr{H}^{\mathrm{AD}}|JM\Lambda\rangle - E^{\mathrm{AD}}\}U_\Lambda = -\langle JM\Lambda|\mathscr{H} - E^{\mathrm{AD}}|JM\Lambda_0\rangle U^{\mathrm{AD}}. \tag{53}$$

Explicit forms of the operators appearing in Eq. (53) are given later. In particular, it is shown that the only nonzero matrix elements of the complete Hamiltonian (8) are between states which satisfy the conditions

$$\Lambda = \Lambda_0, \Lambda_0 \pm 1. \tag{54}$$

For the homonuclear systems there is the further selection rule,

$$\mathrm{u} \leftrightarrow \mathrm{u}, \quad \mathrm{g} \leftrightarrow \mathrm{g}, \quad \mathrm{u} \not\leftrightarrow \mathrm{g}. \tag{55}$$

For the ground electronic state of $HD^+(\Sigma_g)$, on which Wolniewicz and Poll (1980) have performed their most detailed calculations, the only states which are directly coupled nonadiabatically are states of symmetry $\Sigma_g(\Lambda = 0)$, $\Sigma_u(\Lambda = 0)$, $\Pi_g(\Lambda = \pm 1)$, and $\Pi_u(\Lambda = \pm 1)$. Each of these is coupled in by a separate term of the complete Hamiltonian so that the second-order energy correction may be written as a sum:

$$E'' = E''_{0,g} + E''_{0,u} + 2E''_{+1,g} + 2E''_{+1,u}, \tag{56}$$

where we have used the fact that

$$E''_{+1,g} = E''_{-1,g} \qquad \text{and} \qquad E''_{+1,u} = E''_{-1,u}. \tag{57}$$

Since the symmetries of the states which couple are known, it is easy to write an appropriate expansion for the U_Λ by comparison with Eqs. (25) and (29) for use in Eq. (53), as follows:

$$U_\Lambda{}^\alpha = R^{-5/2} \sum_{i,j} g_{ij}(\xi, \eta)\chi_{ij}(R). \tag{58}$$

In this expansion, for Σ_g (j even) and Σ_u (j odd), we have

$$g_{ij}(\xi, \eta) = \exp(-\gamma\xi)(\xi + 1)^\beta[(\xi - 1)/(\xi + 1)]^i P_j{}^0(\eta) \tag{59}$$

while for Π_g (j odd) and Π_u (j even), we have

$$g_{ij}(\xi, \eta) = (\xi^2 - 1)^{1/2} \exp(-\gamma\xi)(\xi + 1)^\beta[(\xi - 1)/(\xi + 1)]^i P_j{}^1(\eta). \tag{60}$$

The $R^{-5/2}$ term serves to simplify the explicit form of Eq. (53). The β and γ are adjustable parameters, determined variationally; with reasonably large expansions (58) the nonadiabatic energy corrections (56) are relatively insensitive to β and γ, and to additional terms in the expansion for $U_\Lambda{}^\alpha$. The numerical implementation of this approach is described fully by Wolniewicz and Poll (1978, 1980). For HD^+ (Wolniewicz and Poll, 1980), they present nonadiabatic calculations of the energy levels for all bound vibrational states ($v = 0$–21) with $N = 0$–5, with an estimated accuracy of the order of $0.001\ \mathrm{cm}^{-1}$. Relativistic and radiative corrections were included in these calculations. The results agree well with the available experimental data (Wing *et al.*, 1976; Carrington and Buttenshaw, 1981), but there do appear to be significant deviations, particularly at higher rotational quantum numbers (see Table I).

A number of refinements of the calculations could be envisaged. In calculating the nonadiabatic corrections, Wolniewicz and Poll neglected the

TABLE I

Comparison of Experimental Transition Frequencies for HD^+ with Adiabatic and Nonadiabatic Calculations

Transition $(v', J')-(v'', J'')$	Experiment (cm^{-1})	Adiabatic[e] (cm^{-1})	Nonadiabatic[f] (cm^{-1})	Experiment − nonadiabatic (cm^{-1})
(1, 0)–(0, 1)	1869.134[a]	1869.25	1869.135	−0.001
(1, 1)–(0, 2)	1823.533[a]	1823.67	1823.533	0.000
(1, 2)–(0, 3)	1776.459[b]	1776.60	1776.461	−0.002
(2, 1)–(1, 0)	1856.778[a]	1856.92	1856.779	−0.001
(3, 1)–(2, 0)	1761.616[a]	1761.74	1761.616	0.000
(3, 2)–(2, 1)	1797.522[a]	1797.64	1797.519	0.003
(3, 3)–(2, 2)	1831.083[b]	1831.21	1831.078	0.005
(3, 1)–(2, 2)	1642.108[a]	1642.23	1642.111	−0.003
(17, 1)–(14, 0)	1813.852[c]	1813.66	1813.852	0.000
(17, 2)–(14, 1)	1820.209[c]	1820.09	1820.201	0.008
(17, 3)–(14, 2)	1820.199[c]	1819.99	1820.187	0.014
(17, 4)–(14, 3)	1813.644[c]	1813.43	1813.627	0.017
(17, 5)–(14, 4)	1800.358[c]	1800.14	1800.333	0.025
(17, 6)–(14, 5)	1780.145[c]	1779.92	1780.115	0.030
(17, 0)–(14, 1)	1782.772[c]	1782.57	1782.776	−0.004
(18, 1)–(16, 0)	926.4895[d]	926.31	926.490	0.000
(18, 2)–(16, 1)	932.2237[d]	932.03	932.220	0.004
(18, 3)–(16, 2)	933.2129[d]	933.02	933.207	0.006
(18, 4)–(16, 3)	929.2471[d]	929.06	929.238	0.009
(18, 5)–(16, 4)	920.1001[d]	919.91	920.089	0.011
(18, 6)–(16, 5)	905.5191[d]	905.33	905.512[g]	0.007
(18, 7)–(16,6)	885.2183[d]	885.03	885.229[g]	−0.011
(18, 0)–(16, 1)	901.5648[d]	901.38	901.571	−0.006
(18, 1)–(16, 2)	882.7312[d]	882.55	882.743	−0.012
(17, 8)–(15, 7)	1078.8532[d]	1078.67	1078.907[g]	−0.054

[a] From Wing *et al.* (1976).

[b] From Spezeski (1977).

[c] From Carrington *et al.* (1983).

[d] From Carrington and Buttenshaw (1981).

[e] Adiabatic calculations from Hunter *et al.* (1974).

[f] Nonadiabatic calculations from Wolniewicz and Poll (1980).

[g] Extrapolated values, obtained by fitting data of Wolniewicz and Poll (1980) to B, D, and H constants.

possible effects of the vibration–rotation interaction. The Σ corrections, which are otherwise N independent, were only calculated for $N = 0$; the Π corrections, which are proportional to $N(N + 1)$, were only calculated for $N = 1$. Corrections for other N values were obtained by using these results. Wolniewicz and Poll encountered some problems in evaluating the Σ_g correction, related to the difficulty of constructing $\psi'(\Sigma_g)$ to be exactly orthogonal to $\psi^{AD}(\Sigma_g)$ (Wolniewicz and Poll, 1980). The use of second-order perturbation theory may be inappropriate for considering the Σ_u correction, particularly for high vibrational levels. This is because, as mentioned earlier, the adiabatic potentials for the ground state and first excited state (Σ_u) become identical at very large separations.

4. *Relativistic and Radiative Corrections*

A detailed treatment of the origin and calculation of these small corrections lies beyond the scope of this review (Luke *et al.*, 1969; Gonsalves and Moss, 1979; Bishop, 1977; Gersten, 1969; Bishop and Cheung, 1978, 1981). However, the corrections are significant and must be included to obtain the most accurate energy levels. The important corrections only involve the motion of the electron and they are diagonal in the adiabatic states, so that they may be introduced as additional terms,

$$\Delta E^r(R) = \langle \phi^{BO} | \mathscr{H}^r(R, \mathbf{r}_g) | \phi^{BO} \rangle \tag{61}$$

to be added to the adiabatic potential (36). The corrections to the adiabatic energy levels are found by obtaining the vibration–rotation eigenvalues for this modified potential. These corrections were included in the nonadiabatic calculations of Bishop and Cheung (1977) and of Wolniewicz and Poll (1980). As we seek to perform still more accurate calculations, it may be necessary to consider other small corrections, such as the effects of finite nuclear size.

In order to give an indication of the relative sizes of the various corrections, we tabulate them for the $v = 1$, $N = 1$ level of HD^+ in Table II.

In concluding this section, we note that the vibration–rotation transition frequencies for HD^+ are sufficiently sensitive to allow a redetermination of the electron/proton mass ratio if the accuracy of both theory and experiment can be improved by an order of magnitude (Wing *et al.*, 1976).

E. *Matrix Elements of the Hamiltonian*

In order to apply the formal Hamiltonian (8) to the hydrogen molecular ion, we require explicit forms for the operators in terms of the internal coordinates of the system. This is achieved by first separating out the angular

TABLE II

Corrections to the Energy of the $v = 1$, $J = 1$ Level of HD^+ (in cm^{-1})

Final nonadiabatic energy (rel. to $H^+ + D$)	Adiabatic correction[a]	Nonadiabatic corrections Σ_g	Σ_u	Π_g	Π_u	Relativistic correction[b]	Radiative correction[c]
-19561.222[c]	$\sim +50$	-0.216	-0.038	-0.002	-0.001	-1.572	0.342

[a]A tabulated value does not seem to be available; this is a crude estimate from the adiabatic potential corrections of Bishop and Wetmore (1973).

[b]Bishop (1977).

[c]Wolniewicz and Poll (1980).

motion of the nuclei and then expressing the resulting matrix elements in terms of the internal coordinates (R, ξ, η). This has been described previously for the two-electron systems by Kołos and Wolniewicz (1963), but an explicit derivation for the one-electron hydrogen molecular ion does not appear to have been given. Attempts to write down the one-electron matrix elements from the two-electron formulas in Kołos and Wolniewicz (1963) appear to have led to some errors in published formulas.

We separate the angular motion of the nuclei by transforming from the space-fixed axis system (X, Y, Z) to a set of rotating molecule-fixed axes. The transformation is defined by two Euler rotations: (i) ϕ about the initial Z axis $(0 \leqslant \phi < 2\pi)$ and (ii) θ about the resultant y axis $(0 \leqslant \theta < \pi)$. Following the usual right-handed conventions, we obtain the results:

$$\begin{pmatrix} x \\ y \\ z \end{pmatrix} = \begin{pmatrix} \cos\phi\cos\theta & \sin\phi\cos\theta & -\sin\theta \\ \sin\phi & \cos\phi & 0 \\ \cos\phi\sin\theta & \sin\phi\sin\theta & \cos\theta \end{pmatrix} \begin{pmatrix} X \\ Y \\ Z \end{pmatrix}. \tag{62}$$

The coordinates R, θ, and ϕ are sufficient to describe the motion of the nuclei.

Applying this transformation, the operator ∇_R becomes:

$$\begin{aligned} \nabla_R = {} & \left[\cos\phi\sin\theta \left.\frac{\partial}{\partial R}\right)_s + \frac{\cos\phi\cos\theta}{R} \left.\frac{\partial}{\partial\theta}\right)_s - \frac{\sin\phi}{R\sin\theta} \left.\frac{\partial}{\partial\phi}\right)_s \right] \mathbf{i} \\ & + \left[\sin\phi\sin\theta \left.\frac{\partial}{\partial R}\right)_s + \frac{\sin\phi\cos\theta}{R} \left.\frac{\partial}{\partial\theta}\right)_s + \frac{\cos\phi}{R\sin\theta} \left.\frac{\partial}{\partial\phi}\right)_s \right] \mathbf{j} \\ & + \left[\cos\theta \left.\frac{\partial}{\partial R}\right)_s - \sin\theta \left.\frac{\partial}{\partial\theta}\right)_s \right] \mathbf{k}, \end{aligned} \tag{63}$$

where **i**, **j**, and **k** are unit vectors along the space-fixed axes X, Y, and Z, respectively. The partial differentials in the above operator are to be performed with the electron coordinates in the space-fixed axis system held constant, denoted by the subscript s. The nuclear Laplacian operator becomes

$$\nabla_R{}^2 = \frac{1}{R^2} \left.\frac{\partial}{\partial R}\right)_s \left\{ R^2 \left.\frac{\partial}{\partial R}\right)_s \right\} + \frac{1}{R^2} \left[\frac{1}{\sin\theta} \left.\frac{\partial}{\partial\theta}\right)_s \left\{ \sin\theta \left.\frac{\partial}{\partial\theta}\right)_s \right\} + \frac{1}{\sin^2\theta} \left.\frac{\partial^2}{\partial\phi^2}\right)_s \right]. \tag{64}$$

Use of this form for $\nabla_R{}^2$ in Eqs. (34) and (35) allows the rotational motion to be separated; the equations may then be solved numerically as shown, for example, by Cooley (1961).

Since the motion of the electron is governed by a molecule-fixed potential,

$$-1/r_{1\mathrm{e}} - 1/r_{2\mathrm{e}}, \tag{65}$$

it is physically more reasonable to transform the operators to a molecule-fixed electron coordinate system, referred to the geometric center of the internuclear vector. Partial differentials in this coordinate system are denoted by the subscript m. Thus,

$$\begin{aligned}
\left.\frac{\partial}{\partial\phi}\right)_{\mathrm{s}} &= \left.\frac{\partial}{\partial\phi}\right)_{\mathrm{m}} + \left.\frac{\partial x}{\partial\phi}\right)_{\mathrm{s}}\left.\frac{\partial}{\partial x}\right)_{\mathrm{m}} + \left.\frac{\partial y}{\partial\phi}\right)_{\mathrm{s}}\left.\frac{\partial}{\partial y}\right)_{\mathrm{m}} + \left.\frac{\partial z}{\partial\phi}\right)_{\mathrm{s}}\left.\frac{\partial}{\partial z}\right)_{\mathrm{m}} \\
&= \left.\frac{\partial}{\partial\phi}\right)_{\mathrm{m}} - i\cos\theta\, L_z + i\sin\theta\, L_x \\
\left.\frac{\partial}{\partial\theta}\right)_{\mathrm{s}} &= \left.\frac{\partial}{\partial\theta}\right)_{\mathrm{m}} - iL_y \\
\left.\frac{\partial}{\partial R}\right)_{\mathrm{s}} &= \left.\frac{\partial}{\partial R}\right)_{\mathrm{m}}.
\end{aligned} \tag{66}$$

The$\nabla_R{}^2$ operator may now be written as

$$\begin{aligned}
\nabla_R{}^2 = {} & \left.\frac{\partial^2}{\partial R^2}\right)_{\mathrm{m}} + \left.\frac{2}{R}\frac{\partial}{\partial R}\right)_{\mathrm{m}} + \frac{1}{R^2}\left\{\left.\frac{\partial^2}{\partial\theta^2}\right)_{\mathrm{m}} + \cot\theta\left.\frac{\partial}{\partial\theta}\right)_{\mathrm{m}} + \left.\frac{1}{\sin^2\theta}\frac{\partial^2}{\partial\phi^2}\right)_{\mathrm{m}}\right\} \\
& + \frac{1}{R^2}\left\{L_z - L^+L^- - \cot^2\theta\, L_z{}^2 - 2i\frac{\cot\theta}{\sin\theta}L_z\left.\frac{\partial}{\partial\phi}\right)_{\mathrm{m}}\right\} \\
& + \frac{L^+}{R^2}\left\{-\left.\frac{\partial}{\partial\theta}\right)_{\mathrm{m}} + \left.\frac{i}{\sin\theta}\frac{\partial}{\partial\phi}\right)_{\mathrm{m}} + \cot\theta\, L_z\right\} \\
& + \frac{L^-}{R^2}\left\{\left.\frac{\partial}{\partial\theta}\right)_{\mathrm{m}} + \left.\frac{i}{\sin\theta}\frac{\partial}{\partial\phi}\right)_{\mathrm{m}} + \cot\theta\, L_z\right\},
\end{aligned} \tag{67}$$

where $L^{\pm} = L_x \pm iL_y$, and L_x, L_y, and L_z are the components of the electron angular momentum in the rotating coordinate system. The operator $\nabla_{\mathrm{g}} \cdot \nabla_R$ is treated similarly:

$$\begin{aligned}
\left.\frac{\partial}{\partial X}\right)_{\mathrm{s}} &= \cos\phi\cos\theta\left.\frac{\partial}{\partial x}\right)_{\mathrm{m}} - \sin\phi\left.\frac{\partial}{\partial y}\right)_{\mathrm{m}} + \cos\phi\sin\theta\left.\frac{\partial}{\partial z}\right)_{\mathrm{m}} \\
\left.\frac{\partial}{\partial Y}\right)_{\mathrm{s}} &= \sin\phi\cos\theta\left.\frac{\partial}{\partial x}\right)_{\mathrm{m}} + \cos\phi\left.\frac{\partial}{\partial y}\right)_{\mathrm{m}} + \sin\phi\sin\theta\left.\frac{\partial}{\partial z}\right)_{\mathrm{m}} \\
\left.\frac{\partial}{\partial Z}\right)_{\mathrm{s}} &= -\sin\theta\left.\frac{\partial}{\partial x}\right)_{\mathrm{m}} + \sin\theta\left.\frac{\partial}{\partial z}\right)_{\mathrm{m}}
\end{aligned} \tag{68}$$

leading to:

$$\begin{aligned}\nabla_g \cdot \nabla_R = {} & iP_z \left.\frac{\partial}{\partial R}\right)_m + \frac{i}{2R}(P^+L^- - P^-L^+) \\ & + \frac{iP^+}{2R}\left\{\left.\frac{\partial}{\partial \theta}\right)_m - \frac{i}{\sin\theta}\left.\frac{\partial}{\partial \phi}\right)_m - \cot\theta\, L_z\right\} \\ & + \frac{iP^-}{2R}\left\{\left.\frac{\partial}{\partial \theta}\right)_m + \frac{i}{\sin\theta}\left.\frac{\partial}{\partial \phi}\right)_m + \cot\theta\, L_z\right\},\end{aligned} \tag{69}$$

where $P^{\pm} = P_x \pm iP_y$, and P_x, P_y, and P_z are the components of the electron's impulse operator $\mathbf{P} = -i\nabla_g$ in the molecule-fixed coordinate system. The operators

$$\begin{aligned}\nabla_g{}^2 &= \left.\frac{\partial^2}{\partial x^2}\right)_m + \left.\frac{\partial^2}{\partial y^2}\right)_m + \left.\frac{\partial^2}{\partial z^2}\right)_m \\ V &= \frac{1}{R} - \frac{1}{r_{1e}} - \frac{1}{r_{2e}}\end{aligned} \tag{70}$$

are both now independent of the Euler angles ϕ and θ.

The forms derived for the operators in the Hamiltonian (8), which can also be found in Kołos and Wolniewicz (1963), allow the rotational motion of the molecule to be separated. This is achieved by expanding the total wave function as

$$\psi = \sum_{\Lambda} \Omega^J_{M_J,\Lambda}(\theta, \phi, \chi) U_\Lambda{}^J(\xi, \eta, R), \tag{71}$$

where the functions $U_\Lambda{}^J$ depend only on the relative positions of the particles, and the $\Omega^J_{M_J,\Lambda}(\theta, \phi, \chi)$ are the normalized symmetric top eigen functions

$$\Omega^J_{M_J,\Lambda} = \frac{1}{2\pi}\exp(iM_J\phi)\exp(i\Lambda\chi) d^J_{M_J,\Lambda}(\theta). \tag{72}$$

J is the total angular momentum in the space-fixed system, M_J is its component along the space-fixed Z axis, and Λ is its component along the molecule-fixed z axis. The motion of the electron about the z axis is described by the angle χ,

$$L_z = -i\frac{\partial}{\partial \chi}. \tag{73}$$

The functions $d^J_{M_J,\Lambda}(\theta)$ satisfy the following relations:

$$\begin{aligned}&\left\{\frac{\partial^2}{\partial\theta^2} + \cot\theta\frac{\partial}{\partial\theta} + 2M_J\Lambda\frac{\cot\theta}{\sin\theta} - \frac{M_J{}^2 + \Lambda^2}{\sin^2\theta} + J(J+1)\right\} d^J_{M_J,\Lambda} = 0 \\ &\left\{\frac{\partial}{\partial\theta} - \Lambda\cot\theta + \frac{M_J}{\sin\theta}\right\} d^J_{M_J,\Lambda} = [(J+\Lambda+1)(J-\Lambda)]^{1/2} d^J_{M_J,\Lambda+1} \\ &\left\{-\frac{\partial}{\partial\theta} - \Lambda\cot\theta + \frac{M_J}{\sin\theta}\right\} d^J_{M_J,\Lambda} = [(J+\Lambda)(J-\Lambda+1)]^{1/2} d^J_{M_J,\Lambda-1}.\end{aligned} \tag{74}$$

Substituting this form for ψ into the Schrödinger equation [Eq. (6)] leads to the elimination of the Euler angles ϕ and θ, and eventually χ also. Because the Hamiltonian is diagonal in J and M_J, we adopt the following designation for its matrix elements:

$$\langle JM_J\Lambda'|\mathscr{H}|JM_J\Lambda\rangle = \langle\Lambda'|\mathscr{H}|\Lambda\rangle. \tag{75}$$

From the expressions derived in Eqs. (67), (69), and (70) for the operators in the Hamiltonian, it is apparent that the only nonzero matrix elements of $\mathscr{H}$ are those which satisfy the Λ selection rule,

$$\Lambda' = \Lambda, \Lambda \pm 1. \tag{76}$$

Specifically, the nonvanishing matrix elements are as follows:

$$\begin{aligned}\langle\Lambda'|\mathscr{H}|\Lambda\rangle = {} & -\frac{1}{2}\langle\Lambda'|\nabla_g{}^2|\Lambda\rangle + \langle\Lambda'|V|\Lambda\rangle - \frac{1}{8\mu}\langle\Lambda'|\nabla_g{}^2|\Lambda\rangle \\ & -\frac{1}{2\mu}\langle\Lambda'|\nabla_R{}^2|\Lambda\rangle - \frac{1}{2\mu_a}\langle\Lambda'|\mathbf{\nabla}_g\cdot\mathbf{\nabla}_R|\Lambda\rangle\end{aligned} \tag{77}$$

$$\langle\Lambda'|\nabla_g{}^2|\Lambda\rangle = 0 \qquad \text{unless } \Lambda' = \Lambda \tag{78}$$

$$\langle\Lambda'|V|\Lambda\rangle = 0 \qquad \text{unless } \Lambda' = \Lambda \tag{79}$$

$$\begin{aligned}\langle\Lambda|\nabla_R{}^2|\Lambda\rangle &= \frac{\partial^2}{\partial R^2} + \frac{2}{R}\frac{\partial}{\partial R} - \frac{1}{R^2}\left\{J(J+1) - \Lambda(\Lambda+1)\right\} \\ &\quad - \frac{1}{R^2}\langle\Lambda|L^+L^-|\Lambda\rangle \\ \langle\Lambda-1|\nabla_R{}^2|\Lambda\rangle &= -\frac{1}{R^2}[(J+\Lambda)(J-\Lambda+1)]^{1/2}\langle\Lambda-1|L^-|\Lambda\rangle \\ \langle\Lambda+1|\nabla_R{}^2|\Lambda\rangle &= -\frac{1}{R^2}[(J+\Lambda+1)(J-\Lambda)]^{1/2}\langle\Lambda+1|L^+|\Lambda\rangle\end{aligned} \tag{80}$$

$$\begin{aligned}\langle\Lambda|\mathbf{\nabla}_g\cdot\mathbf{\nabla}_R|\Lambda\rangle &= i\left\{P_z\frac{\partial}{\partial R} + \frac{1}{2R}\langle\Lambda|P^+L^- - P^-L^+|\Lambda\rangle\right\} \\ \langle\Lambda-1|\mathbf{\nabla}_g\cdot\mathbf{\nabla}_R|\Lambda\rangle &= -\frac{i}{2R}[(J+\Lambda)(J-\Lambda+1)]^{1/2}\langle\Lambda-1|P^-|\Lambda\rangle \\ \langle\Lambda+1|\mathbf{\nabla}_g\cdot\mathbf{\nabla}_R|\Lambda\rangle &= \frac{i}{2R}[(J+\Lambda+1)(J-\Lambda)]^{1/2}\langle\Lambda+1|P^+|\Lambda\rangle\end{aligned} \tag{81}$$

so that we obtain the results:

$$
\begin{aligned}
\langle\Lambda|\mathscr{H}|\Lambda\rangle &= -\frac{1}{2}\langle\Lambda|\nabla_g{}^2|\Lambda\rangle + V - \frac{1}{8\mu}\langle\Lambda|\nabla_g{}^2|\Lambda\rangle \\
&\quad - \frac{1}{2\mu}\left\{\frac{\partial^2}{\partial R^2} + \frac{2}{R}\frac{\partial}{\partial R} - \frac{1}{R^2}[J(J+1) \right. \\
&\quad \left. - \Lambda(\Lambda+1) + \langle\Lambda|L^+L^-|\Lambda\rangle]\right\} \\
&\quad - \frac{i}{2\mu_a}\left\{P_z\frac{\partial}{\partial R} + \frac{1}{2R}\langle\Lambda|P^+L^- - P^-L^+|\Lambda\rangle\right\} \\
\langle\Lambda-1|\mathscr{H}|\Lambda\rangle &= [(J+\Lambda)(J-\Lambda+1)]^{1/2} \\
&\quad \times\left\{\frac{1}{2\mu R^2}\langle\Lambda-1|L^-|\Lambda\rangle + \frac{i}{4\mu_a R}\langle\Lambda-1|P^-|\Lambda\rangle\right\} \\
\langle\Lambda+1|\mathscr{H}|\Lambda\rangle &= [(J+\Lambda+1)(J-\Lambda)]^{1/2} \\
&\quad \times\left\{\frac{1}{2\mu R^2}\langle\Lambda+1|L^+|\Lambda\rangle - \frac{i}{4\mu_a R}\langle\Lambda+1|P^+|\Lambda\rangle\right\}.
\end{aligned} \tag{82}
$$

We have now eliminated the Euler angles ϕ and θ from the Hamiltonian [Eq. (8)] by separating the rotational motion of the nuclei. The final stage is to express the above matrix elements in terms of the internal coordinates (R, ξ, η, χ) of the system. This will lead to the elimination of the coordinate χ. We first define some useful operator abbreviations:

$$
\begin{aligned}
X_\Lambda &= \frac{1}{(\xi^2-\eta^2)}\left\{\frac{\partial}{\partial\xi}(\xi^2-1)\frac{\partial}{\partial\xi} + \frac{\partial}{\partial\eta}(1-\eta^2)\frac{\partial}{\partial\eta}\right\} - \frac{\Lambda^2}{(\xi^2-1)(1-\eta^2)} \\
Y &= \frac{1}{(\xi^2-\eta^2)}\left\{\xi(\xi^2-1)\frac{\partial}{\partial\xi} + \eta(1-\eta^2)\frac{\partial}{\partial\eta}\right\} \\
Z &= \frac{1}{(\xi^2-\eta^2)}\left\{\eta(\xi^2-1)\frac{\partial}{\partial\xi} + \xi(1-\eta^2)\frac{\partial}{\partial\eta}\right\} \\
A &= \frac{[(\xi^2-1)(1-\eta^2)]^{1/2}}{(\xi^2-\eta^2)}\left\{\xi\frac{\partial}{\partial\xi} - \eta\frac{\partial}{\partial\eta}\right\} \\
B &= \frac{[(\xi^2-1)(1-\eta^2)]^{1/2}}{(\xi^2-\eta^2)}\left\{\eta\frac{\partial}{\partial\xi} - \xi\frac{\partial}{\partial\eta}\right\}.
\end{aligned} \tag{83}
$$

The partial derivative $\partial/\partial R$ appearing in the formulas [Eqs. (82)] above is performed with the molecule-fixed Cartesian coordinates (x, y, z) held constant. With the more useful set of coordinates (ξ, η, χ) held constant, we obtain

$$\left.\frac{\partial}{\partial R}\right)_{xyz} = \left.\frac{\partial}{\partial R}\right)_{\xi\eta\chi} - \frac{Y}{R}. \tag{84}$$

Using the results given in the treatment of the Born–Oppenheimer electronic equation, namely, Eqs. (16), (17), and (19), we can write the results:

$$\langle\Lambda|\nabla_g{}^2|\Lambda\rangle = \frac{4}{R^2} X_\Lambda, \tag{85}$$

$$\langle\Lambda|V|\Lambda\rangle = \frac{1}{R} - \frac{4\xi}{R(\xi^2 - \eta^2)}, \tag{86}$$

$$L^+ = \exp(i\chi)\left\{B + \frac{i\xi\eta}{[(\xi^2 - 1)(1 - \eta^2)]^{1/2}}\frac{\partial}{\partial\chi}\right\},$$

$$\langle\Lambda + 1|L^+|\Lambda\rangle = B - \frac{\xi\eta\Lambda}{[(\xi^2 - 1)(1 - \eta^2)]^{1/2}}, \tag{87}$$

$$L^- = \exp(-i\chi)\left\{-B + \frac{i\xi\eta}{[(\xi^2 - 1)(1 - \eta^2)]^{1/2}}\frac{\partial}{\partial\chi}\right\},$$

$$\langle\Lambda - 1|L^-|\Lambda\rangle = -B - \frac{\xi\eta\Lambda}{[(\xi^2 - 1)(1 - \eta^2)]^{1/2}}, \tag{88}$$

$$P^+ = \frac{2\exp(i\chi)}{R}\left\{-iA + \frac{1}{[(\xi^2 - 1)(1 - \eta^2)]^{1/2}}\frac{\partial}{\partial\chi}\right\},$$

$$\langle\Lambda + 1|P^+|\Lambda\rangle = \frac{2i}{R}\left\{-A + \frac{\Lambda}{[(\xi^2 - 1)(1 - \eta^2)]^{1/2}}\right\}, \tag{89}$$

$$P^- = \frac{2\exp(-i\chi)}{R}\left\{-iA - \frac{1}{[(\xi^2 - 1)(1 - \eta^2)]^{1/2}}\frac{\partial}{\partial\chi}\right\},$$

$$\langle\Lambda - 1|P^-|\Lambda\rangle = \frac{2i}{R}\left\{-A - \frac{\Lambda}{[(\xi^2 - 1)(1 - \eta^2)]^{1/2}}\right\}, \tag{90}$$

$$P_z = -(2i/R)Z. \tag{91}$$

With these results we can obtain, after some straightforward manipulations, the matrix elements of the operators appearing in the Hamiltonian (8). They are as follows:

$$\langle\Lambda|-(\nabla_g{}^2/2) + V|\Lambda\rangle = \frac{-2X_\Lambda}{R^2} + \frac{1}{R}\left[1 - \frac{4}{(\xi^2 - \eta^2)}\right], \tag{92}$$

$$\langle\Lambda|-(\nabla_g{}^2/8\mu)|\Lambda\rangle = -X_\Lambda/2\mu R^2, \tag{93}$$

$$\langle\Lambda| - (\nabla_R{}^2/2\mu)|\Lambda\rangle = -\frac{1}{2\mu}\left\{\frac{1}{R^2}\frac{\partial}{\partial R}\left(R^2\frac{\partial}{\partial R}\right) - \frac{2}{R^2}Y\frac{\partial}{\partial R}\times R + \frac{(\xi^2+\eta^2-1)}{R^2}X_\Lambda - \frac{J(J+1)}{R^2} + \frac{2\Lambda^2}{R^2}\right\}, \tag{94}$$

$$\langle\Lambda|-(\mathbf{\nabla}_g\cdot\mathbf{\nabla}_R/2\mu_a)|\Lambda\rangle = -\frac{1}{\mu_a R^2}\left[Z\frac{\partial}{\partial R}R - \xi\eta X_\Lambda\right]. \tag{95}$$

The elements off-diagonal in Λ are:

$$\langle\Lambda-1|-(\nabla_R{}^2/2\mu)|\Lambda\rangle = \frac{[(J+\Lambda)(J-\Lambda+1)]^{1/2}}{2\mu R^2}\left\{-B - \frac{\xi\eta\Lambda}{[(\xi^2-1)(1-\eta^2)]^{1/2}}\right\},$$

$$\langle\Lambda+1|-(\nabla_R{}^2/2\mu)|\Lambda\rangle = \frac{[(J+\Lambda+1)(J-\Lambda)]^{1/2}}{2\mu R^2}\left\{B - \frac{\xi\eta\Lambda}{[(\xi^2-1)(1-\eta^2)]^{1/2}}\right\}, \tag{96}$$

$$\langle\Lambda-1|-(\mathbf{\nabla}_g\cdot\mathbf{\nabla}_R/2\mu_a)|\Lambda\rangle = \frac{[(J+\Lambda)(J-\Lambda+1)]^{1/2}}{2\mu_a R^2}\left\{A + \frac{\Lambda}{[(\xi^2-1)(1-\eta^2)]^{1/2}}\right\},$$

$$\langle\Lambda+1|-(\mathbf{\nabla}_g\cdot\mathbf{\nabla}_R/2\mu_a)|\Lambda\rangle = \frac{[(J+\Lambda+1)(J-\Lambda)]^{1/2}}{2\mu_a R^2}\left\{-A + \frac{\Lambda}{[(\xi^2-1)(1-\eta^2)]^{1/2}}\right\}. \tag{97}$$

There is an additional symmetry selection rule on the above matrix elements. The diagonal operator (94) is even in η and so there is the rule,

$$\mathrm{g}\leftrightarrow\mathrm{g},\qquad \mathrm{u}\leftrightarrow\mathrm{u},\qquad \text{but}\qquad \mathrm{u}\not\leftrightarrow\mathrm{g} \tag{98}$$

while for the odd operator (95) the selection rule is

$$\mathrm{g}\not\leftrightarrow\mathrm{g},\qquad \mathrm{u}\not\leftrightarrow\mathrm{u},\qquad \text{but}\qquad \mathrm{u}\leftrightarrow\mathrm{g}. \tag{99}$$

For the operators off-diagonal in Λ, [Eqs. (96) and (97)], the matrix elements of the even operator (97) satisfy (99), while those of the odd operator (96) obey (98). For the specific case of the ground (Σ_g) state of the hydrogen molecular ion, the matrix element of the adiabatic Hamiltonian, defined by

Eq. (52), is given by the result

$$\langle 0_g|\mathscr{H}^{AD}|0_g\rangle = \frac{-2X_0}{R^2} + \frac{1}{R}\left[1 - \frac{4}{(\xi^2 - \eta^2)}\right] - \frac{1}{2\mu R^2}\left\{\frac{\partial}{\partial R}\left(R^2 \frac{\partial}{\partial R}\right) - 2Y\frac{\partial}{\partial R}R + (\xi^2 + \eta^2)X_0 - J(J+1)\right\}. \tag{100}$$

Finally, the matrix elements of the full Hamiltonian (8) are as follows:

$$\langle 0_g'|\mathscr{H}|0_g\rangle = \frac{-2X_0}{R^2} + \frac{1}{R}\left[1 - \frac{4}{(\xi^2 - \eta^2)}\right] - \frac{1}{2\mu R^2}\left\{\frac{\partial}{\partial R}\left(R^2 \frac{\partial}{\partial R}\right) - 2Y\frac{\partial}{\partial R}R + (\xi^2 + \eta^2)X_0 - J(J+1)\right\}, \tag{101}$$

$$\langle 0_u|\mathscr{H}|0_g\rangle = -\frac{1}{\mu_a R^2}\left\{Z\frac{\partial}{\partial R}R - \xi\eta X_0\right\}, \tag{102}$$

$$\langle \pm 1_g|\mathscr{H}|0_g\rangle = \pm\frac{[J(J+1)]^{1/2}}{2\mu R^2}B, \tag{103}$$

$$\langle \pm 1_u|\mathscr{H}|0_g\rangle = \mp\frac{[J(J+1)]^{1/2}}{2\mu_a R^2}A. \tag{104}$$

Matrix elements [Eqs. (101)–(104)] are those required for performing nonadiabatic calculations, such as those described by Wolniewicz and Poll (1980).

IV. PHOTOELECTRON, PHOTOIONIZATION, AND PHOTODISSOCIATION SPECTROSCOPY

Prior to 1976, most of the experimental data for the hydrogen molecular ion had been obtained by photoelectron, photoionization, or photodissociation spectroscopy. Other important information was obtained from studies of the Rydberg spectra of H_2 by Takezawa (1970a,b), and by Herzberg and Jungen (1972), from numerous studies of the electron-impact dissociation of H_2 [see, for example, Dunn and Van Zyl (1967)], and from investigations of the collision-induced (Gibson and Los, 1967) or unimolecular dissociation (Maas *et al.*, 1975) of H_2^+ and its isotopic modifications. We shall confine our review to experiments involving photon impact; even so, the techniques discussed in this section are not purely spectroscopic in that they do not involve direct measurements of the energies of absorbed or emitted photons.

The most extensive information about the vibrational levels of the hydrogen molecular ion has come from photoelectron spectroscopy. In this method, a target neutral gas (for example, H_2) is subjected to impact by vacuum-ultraviolet photons, which cause ionization to produce the molecular ion (H_2^+) in different electronic and vibrational states. The kinetic energies of the ejected electrons are measured and, combined with knowledge of the initial photon energy, provide relative values of the energies of the internal states of the ion. The photoelectron spectrum of molecular hydrogen has been studied by several authors, notably Åsbrink (1970), but the most complete and best-resolved spectrum has been described recently by Pollard *et al.* (1982). The photon source was a helium Iα resonance lamp with an output at 584 Å, and by using a supersonic molecular beam source of $n - H_2$, $p - H_2$, HD, or D_2, they were able to remove the Doppler broadening. Variation of the molecular beam nozzle stagnation temperature and pressure provided control of the rotational level populations. At a nozzle temperature of 77 K and pressure of 200 torr, the neutral molecules are essentially in the lowest rotational level only; rotational broadening is therefore removed and the vibrational structure of the ions very cleanly resolved, as shown in Fig. 4. The main results are summarized in Table III, in which the observed vibrational intervals $\Delta G(v + \frac{1}{2})$ are listed and compared with the best available theoretical values. For H_2^+ and D_2^+, the comparison is with the adiabatic calculations of Hunter *et al.* (1974), while for HD^+ the comparison is with the nonadiabatic calculations of Wolniewicz and Poll (1980). Agreement between experiment and theory is generally very good, but for the highest vibrational levels of H_2^+ and HD^+ the discrepancy seems to be outside the range of experimental error. It remains an open question, therefore, as to whether the nonadiabatic calculations (Wolniewicz and Poll, 1980), which are excellent for HD^+ in $v = 0$ to 18, remain good for the final levels, $v = 19$, 20, and 21.

The earlier photoionization studies of H_2 have now been superseded by the results described above, but they were important at the time of their publication. Following the initial work of Chupka and Berkowitz (1969), Peatman (1976) described a threshold electron photoionization study of H_2 in which measurements of the rotational energy levels for $v = 0$ to 7 were obtained. In these experiments (which are closely related to photoelectron spectroscopy), the wavelength of the ionizing photons was scanned, but only photoelectrons which had threshold energies of 0–3 meV were detected. Consequently, the appearance of H_2^+ in different vibration–rotation levels was detected as a function of the photon energy.

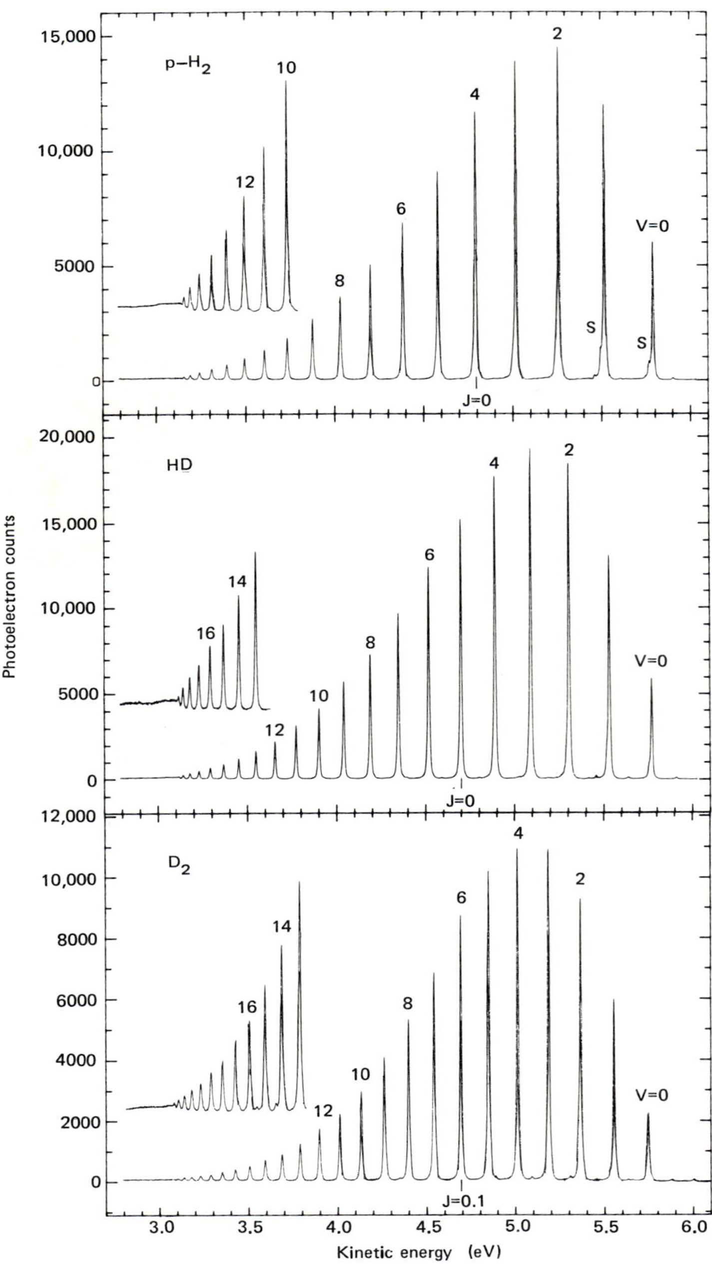

FIG. 4 Photoelectron spectra (584 Å) of para-H_2, HD, and D_2 expanded from 200 torr at 77 K. [From Pollard *et al.* (1982).]

TABLE III

Vibrational Differences, $\Delta G(v+\frac{1}{2})$ for H_2^+, D_2^+, and HD^+, Measured from the Photoelectron Spectra (cm^{-1}) (Pollard et al., 1982)[a]

	H_2^+			D_2^+			HD^+		
$v+\frac{1}{2}$	Experiment	Theory	Difference	Experiment	Theory	Difference	Experiment	Theory	Difference
0.5	2190.9	2191.32	−0.42	1583.5	1577.15	6.35	1913.1	1912.994	0.106
1.5	2065.2	2064.09	1.11	1510.0	1512.47	−2.47	1817.2	1816.862	0.338
2.5	1941.8	1941.08	0.72	1450.3	1449.40	0.90	1722.7	1723.587	−0.887
3.5	1822.1	1821.64	0.46	1388.4	1387.81	0.59	1631.8	1632.799	−0.999
4.5	1706.2	1705.13	1.07	1326.3	1327.51	−1.21	1543.0	1544.134	−1.134
5.5	1591.1	1590.90	0.21	1271.0	1268.36	2.64	1456.8	1457.231	−0.431
6.5	1477.7	1478.32	−0.62	1211.2	1210.18	1.02	1372.0	1371.735	0.265
7.5	1367.0	1366.75	0.25	1157.3	1152.83	4.47	1287.5	1287.287	0.213
8.5	1257.8	1255.51	2.29	1091.7	1096.13	−4.43	1201.7	1203.524	−1.824
9.5	1145.8	1143.88	1.92	1044.7	1039.95	4.75	1119.4	1120.066	−0.666
10.5	1031.9	1031.09	0.81	985.8	984.10	1.70	1035.2	1036.522	−1.322
11.5	913.3	916.31	−3.01	927.0	928.41	−1.41	951.3	952.469	−1.169
12.5	796.6	798.59	−1.99	872.7	872.73	−0.03	866.2	867.459	−1.259
13.5	676.4	676.86	−0.46	814.3	816.86	−2.56	779.8	781.002	−1.202
14.5	548.4	549.97	−1.57	759.3	760.60	−1.30	693.2	692.557	0.643
15.5	415.4	416.62	−1.22	702.4	703.77	−1.37	600.6	601.522	−0.922
16.5	280.3	275.96	4.34	644.7	646.12	−1.42	507.2	507.237	−0.037
17.5	—	131.26	—	584.9	587.43	−2.53	409.0	408.975	0.025
18.5	—	—	—	523.9	527.43	−3.53	306.1	306.048	0.052
19.5	—	—	—	463.7	465.84	−2.14	203.0	198.460	4.540
20.5	—	—	—	399.6	402.35	−2.75	—	90.67	—
21.5	—	—	—	327.3	336.63	−9.33	—	—	—
22.5	—	—	—	271.0	268.41	2.59	—	—	—
23.5	—	—	—	—	197.52	—	—	—	—
24.5	—	—	—	—	124.61	—	—	—	—
25.5	—	—	—	—	54.77	—	—	—	—
26.5	—	—	—	—	11.07	—	—	—	—

[a] Theoretical values for H_2^+ and D_2^+ from Hunter *et al.* (1974), and for HD^+ from Wolniewicz and Poll (1980).

Photodissociation studies of H_2^+, D_2^+, and HD^+ have been described by many authors. Following the initial theoretical and experimental work by Dunn (1968), and by von Busch and Dunn (1972), most investigators have used an accelerated beam of H_2^+ (HD^+, D_2^+) ions, crossed the ion beam with a suitable photon beam, and used a mass filter to separate and collect the photofragment H^+ or D^+ ions. The molecular ions are formed by electron impact on the neutral gas. Because the potential energy curve for the ion is both shallow and displaced to larger internuclear separation, relative to that of the neutral molecule, electron-impact ionization of H_2 in its $v = 0$ level leads to production of H_2^+ in all possible vibrational levels. Franck–Condon factors for the ionizing transitions from the ground states of H_2 and D_2 to all vibrational levels of the ground electronic states of the respective molecular ions have been calculated by a number of workers (Dunn, 1966; Villarejo, 1968; Rothenberg and Davidson, 1967). Similar calculations for the production of HD^+ from HD have been described by Tadjeddine and Parlant (1977) and, by way of illustration, Table IV lists the Franck–Condon factors (Tadjeddine and Parlant, 1977) for the production of HD^+ in the lowest rotational level. The vibrational populations should be proportional to these Franck–Condon factors, and the use of a collision-free environment, such as an accelerated ion beam, enables the vibrational populations to be conserved. Von Busch and Dunn (1972) measured the total photodissociation cross sections of H_2^+ and D_2^+,

$$H_2^+(1s\sigma_g) + h\nu \rightarrow H_2^+(2p\sigma_u) \rightarrow H(1s) + H^+$$

as a function of photon wavelength, using a white light source and a monochromator. In more recent experiments the white light source has been replaced by an argon ion laser operating in the visible. Van Asselt *et al.* (1975) have described experiments on H_2^+ and HD^+ (Van Asselt *et al.*, 1974) in which the momentum spectrum of the photofragment H^+ or D^+ ions is recorded. The principles of the experiment are illustrated in Fig. 5. The

TABLE IV

Franck–Condon Factors for the Ionization of HD to HD^+, with $N = 0$[a]

v	f	v	f	v	f	v	f
0	0.06320	1	0.12862	2	0.15732	3	0.15186
4	0.12845	5	0.10040	6	0.07480	7	0.05416
8	0.03859	9	0.02730	10	0.01928	11	0.01366
12	0.00972	13	0.00696	14	0.00501	15	0.00362
16	0.00261	17	0.00187	18	0.00129	19	0.00083
20	0.00045	21	0.00013				

[a] Taken from Tadjeddine and Parlant (1977).

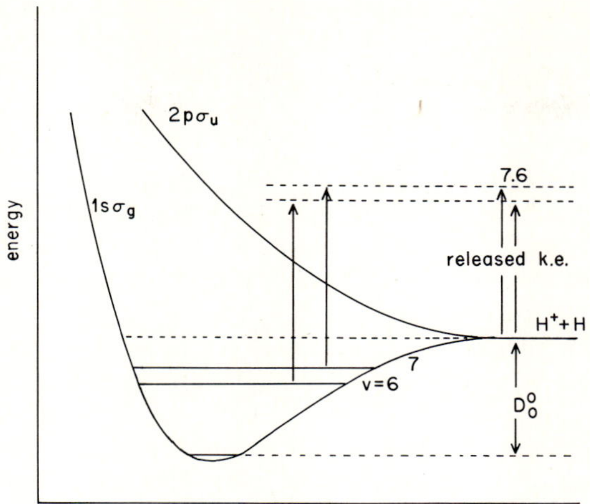

FIG. 5 Principles of measurement of the momentum of the H^+ ion in the laser photofragmentation of $H_2{}^+$.

photofragment ions possess excess center-of-mass kinetic energy, the magnitude of which depends upon the photon energy and the energy of the vibrational level of the molecular ion, relative to the dissociation energy, from which excitation occurs. By using a magnetic analyzer to separate the fragment ions from the parent ions, the momentum spectrum of the fragments is obtained. This spectrum yields the relative energies of several different vibrational levels, although the accuracy is rather poor. Of more interest, however, are the fragment ion intensities which depend upon the Franck–Condon factors for ionization, as discussed above, and also the photodissociation cross sections for different vibrational levels of the molecular ion.

V. RADIO-FREQUENCY SPECTROSCOPY

A most ingenious method for obtaining the radio-frequency spectrum of $H_2{}^+$ has been described by Dehmelt and Jefferts (1962), Richardson *et al.* (1968), and Jefferts (1968, 1969). The method is, in principle, applicable to other ions but it has not yet been extended beyond $H_2{}^+$. Before considering the principles and techniques, it is helpful to examine qualitatively the hyperfine structure of $H_2{}^+$. Figure 6 illustrates the main features. For ortho-$H_2{}^+$, the $N = 1$ level, for example, is split into a doublet by the spin–rotation interaction; each level is then further split by nuclear hyperfine interaction as shown. The quantum number J is obtained by adding **N** and **S**,

$$\mathbf{J} = \mathbf{N} + \mathbf{S}, \qquad J = N + S, N + S - 1, \ldots, |N - S| = \tfrac{3}{2} \text{ or } \tfrac{1}{2}$$

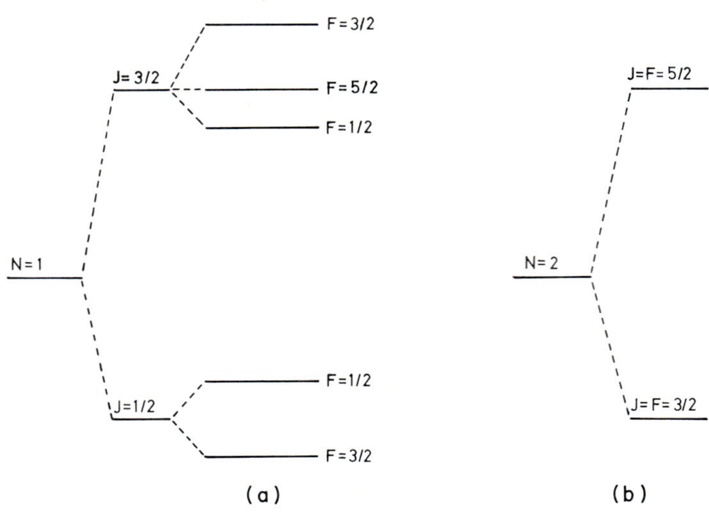

FIG. 6 Hyperfine structure of (a) the $N = 1$ rotational level of ortho-H_2^+ and (b) the $N = 2$ rotational level of para-H_2^+.

and the total angular momentum **F** is formed by addition of the total nuclear spin **I**,

$$\mathbf{F} = \mathbf{J} + \mathbf{I}, \qquad F = J + I, J + I - 1, \ldots, |J - I| = \tfrac{5}{2}, \tfrac{3}{2}, \tfrac{1}{2}.$$

For para-H_2^+ the net nuclear spin **I** is zero so that the substructure for $N = 2$ is simpler, as shown. Figure 6 is not necessarily to scale, and we do not wish to presuppose the relative magnitudes of the spin–rotation and nuclear hyperfine interactions.

Dehmelt and Jefferts' experiment is based upon the fact that the photodissociation rates of ions in the ground electronic state are dependent upon F and M_F when the sample is illuminated by linearly polarized light. This is because the photodissociating transition obeys the normal electric dipole selection rules. Consequently, a sample of H_2^+ ions held in a collision-free environment and subjected to polarized white light irradiation will develop net spatial alignment because the photodissociation rate is proportional to M_F^2. If, however, transitions between two states, F, M_F and F', $M_{F'}$ are saturated with radiation at the appropriate resonance frequency (which, typically, will be in the radio-frequency region), the populations become equalized, thus opposing the tendency toward spatial alignment induced by the photodissociation process. Consequently, the radio-frequency transitions can be detected by monitoring the net photodissociation rate as a function of the radio frequency.

In the experiments of Dehmelt and Jefferts, the H_2^+ ions are formed in an rf quadrupole ion trap by pulsed electron impact on H_2 at a pressure of

10^{-8} torr. Photodissociation is induced by an intense linearly polarized white light beam providing an average dissociation lifetime of about 30 msec for those ions which can be dissociated by the light. The characteristics of the ion trap allow simultaneous trapping of the H_2^+ and photofragment H^+ ions, and at the end of a suitable irradiation period (e.g., 100 msec) the H_2^+ and H^+ ions are extracted and counted, and the ratio H^+/H_2^+ is stored in a multichannel digital memory. The frequency of a radio-frequency magnetic field is slowly swept and the transitions detected as changes in the H^+/H_2^+ ratio. The spectroscopic resolution is extremely high because the Doppler width at radio frequencies is small, and the residence time of the ions in the radiation field is long; linewidths in the range 200–750 Hz are observed. Figure 7 shows a block diagram of the apparatus used by Jefferts (1968); a small static magnetic field (10–30 mG) is used to remove the M_F spatial degeneracy.

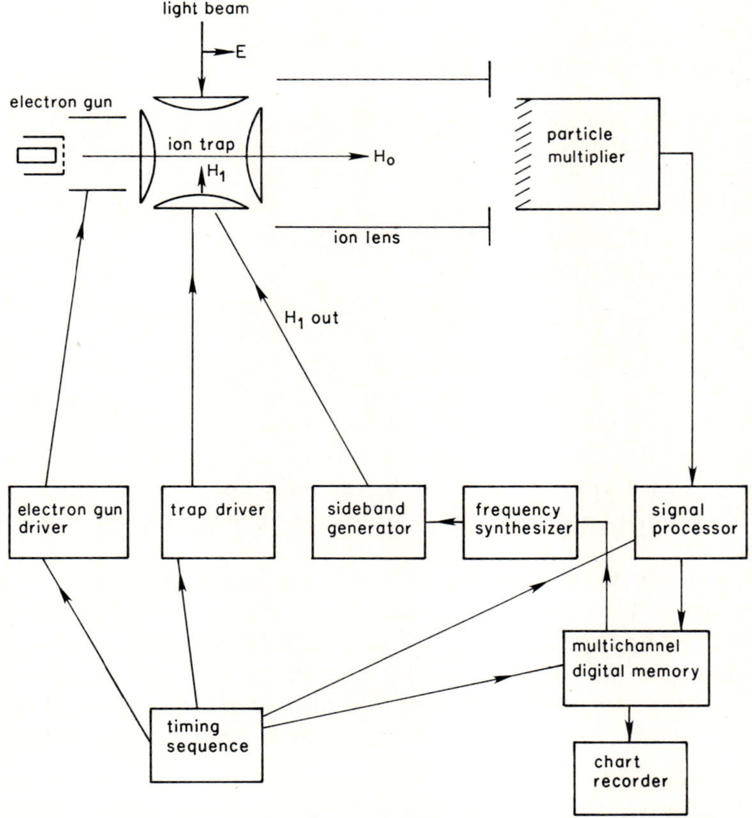

FIG. 7 Block diagram of the quadrupole trap apparatus used by Jefferts (1968).

The effective Hamiltonian used by Jefferts to interpret his observations was,

$$\mathscr{H}_{\text{eff}} = b\mathbf{I}\cdot\mathbf{S} + cI_zS_z + \gamma\mathbf{S}\cdot\mathbf{N} + f\mathbf{I}\cdot\mathbf{N}.$$

We shall discuss the form of this Hamiltonian in more detail in the next section. For ions in the $v = 4$ level, for example, the values of the constants were found to be

$$b = 804.065 \text{ MHz}, \quad c = 98.034, \quad \gamma = 32.636, \quad f = 0.038.$$

Measurements have been made for levels of H_2^+ with $v = 4$–8 and interpreted by Ray and Certain (1977) using the adiabatic potential of Bishop and Wetmore (1973); agreement between theory and experiment is generally very good. Experiments on higher vibrational levels would be desirable, since the nuclear hyperfine and spin–rotation parameters provide excellent tests of the quality of the calculated wave functions.

VI. ION BEAM SPECTROSCOPY

The first infrared spectrum of a gaseous molecular ion was reported in 1976 by Wing *et al.*, who used the interaction of an infrared CO laser with a beam of HD^+ ions. Subsequently, further lines of HD^+ have been observed by the same group (Wing, 1981). In a related series of experiments, Carrington and Buttenshaw (1981) and Carrington *et al.* (1983) have also observed infrared spectra of HD^+. The transitions observed by Wing *et al.* involve the low vibrational levels, while Carrington *et al.* have observed levels close to the dissociation limits; the experimental methods used by the two groups are different but complementary. The general principle of both methods is that a high-velocity, well-collimated beam of ions interacts collinearly with one or two infrared laser beams. Changes in the accelerating potential of the ion beam enable vibration–rotation transitions of the ion to be tuned into resonance with a suitable laser line by means of the Doppler shift. Indirect methods are used to detect population transfer at resonance.

The ions are generally produced by electron-impact ionization of neutral HD, although plasma discharge sources may also be used. The HD^+ ions are extracted by applying a 1–10 kV positive potential to the source and are collimated into a narrow beam by a series of electrostatic lenses. The resulting beam possesses a number of properties offering considerable advantages for spectroscopic study, which we now discuss (Carrington, 1979).

(a) Ions present in the beam enjoy a collision-free environment; hence, the internal energy distribution of the ions in the beam will reflect directly the energetics of the formation processes in the source. The Franck–Condon factors for ionization of HD^+, listed in Table IV, show that ions with high vibrational excitation are present in the beam, and this aspect is central to the techniques used by Carrington and his colleagues.

(b) A high degree of spatial control of the ion beam is possible using collimating and deflecting lenses, enabling very efficient interaction with the laser beam to occur.

(c) Unambiguous identification of the ions in the beam can be made through use of the now standard techniques of mass spectrometry. In addition, ions of different charge-to-mass ratio can be separated from each other, ion beam intensity measurements are straightforward, and even very weak beams of only a few ions per second can be detected by means of high-gain electron multipliers.

(d) The individual ions in the beam travel at velocities which are typically in the range 10^5–10^6 msec^{-1}, depending upon the beam potential and mass of the ion. The effective laser frequency observed by an ion moving at velocity v parallel or antiparallel to the laser beam is given by the relativistic Doppler-shift formula,

$$\nu_{\text{eff}} = \nu_{\text{laser}} \times \left\{\frac{1 \pm (v/c)}{1 \mp (v/c)}\right\}^{1/2},$$

where the upper and lower signs refer to antiparallel and parallel orientations, respectively, and c is the velocity of light. By sweeping the source potential, frequency scanning may be achieved with a fixed-frequency laser line.

(e) Sub-Doppler resolution is obtainable as a result of an effect known as kinematic compression, or velocity bunching. This effect may be understood (Kaufman, 1976) by considering two different ions, a and b, of the same mass and which have velocity components in the beam propagation direction,

$$v_{\mathrm{a}} = 0, \qquad v_{\mathrm{b}} = (2kT/m)^{1/2}$$

After acceleration through a potential difference V, the ions have final velocity components:

$$v_{\mathrm{a}}' = (2eV/m)^{1/2},$$
$$v_{\mathrm{b}}' = [v_{\mathrm{b}}^2 + (2eV/m)]^{1/2} = (v_{\mathrm{b}}^2 + v_{\mathrm{a}}'^2)^{1/2} = v_{\mathrm{a}}' + v_{\mathrm{b}}^2/2v_{\mathrm{a}}'.$$

Hence,

$$(v_{\mathrm{b}}' - v_{\mathrm{a}}')/(v_{\mathrm{b}} - v_{\mathrm{a}}) = \tfrac{1}{2}(kT/eV)^{1/2} = R.$$

For $T = 2000$ K and $V = 10$ kV, the value of R is 2.1×10^{-3}; consequently, the velocity difference and, therefore, Doppler shift between the different ions is greatly reduced. In practice, the resolution is improved by factors of 10–100 over normal Doppler-limited spectra as a result of this velocity bunching effect.

Instantaneous ion densities in typical molecular ion beams are estimated to be in the range 10^4–10^6 ions/cm^3, which is too low for infrared spectroscopy to be carried out with conventional light sources. Even with the use of high-powered continuous wave (cw) lasers, it is necessary to use indirect methods to detect the absorption of radiation. Wing and his colleagues have used charge transfer reactions, allowing the HD^+ beam to be partly neutralized with an appropriate target gas after interaction with the laser beam has occurred. The cross section for charge exchange is dependent upon the vibrational state of the HD^+ ion and, therefore, the population transfer between vibrational states at resonance with the laser can be detected by a corresponding change in the degree of charge neutralization and, hence, HD^+ beam intensity.

Carrington and his colleagues have used photodissociation methods, allowing the HD^+ ions to interact with two infrared photons. The first photon is resonant with a vibration–rotation transition through the Doppler effect, while the second has sufficient energy to photodissociate the upper level, but insufficient energy to photodissociate directly the lower level. Resonant population transfer is thus detected as an increase in the number of photofragment H^+ or D^+ ions. It is noteworthy that the resonant signal will be comparable with the background signal in the photodissociation method, whereas in the charge exchange method the change in parent ion beam current after partial neutralization is relatively small at resonance. The photodissociation method is, therefore, much the more sensitive, but can only be used to study levels near the dissociation limit.

The apparatus used by Wing and his group is illustrated in Fig. 8. The electron-bombardment type source is found to be superior to a plasma discharge source in that it forms ions with less vibrational excitation. Consequently, the off-resonance population differences between the initial and final states are greater, enhancing the signal-to-noise ratio. Although the vibrational populations for HD^+ are determined by the Franck–Condon factors for ionization (Table IV), the rotational distribution is essentially Boltzmann at 320 K. HD^+ ion beam currents of 3×10^{-7} A are produced by admitting a mixture of H_2 and D_2 through a hot palladium leak into the source. After focusing, the ion beam enters a constant potential interaction region, maintained at a pressure of 10^{-6} torr, in which it is crossed at an angle of 11 mrad by an infrared laser beam. The continuous wave CO laser employed is capable of producing 150 lines in the 5.2–6.2 μm region. Single-line selection is achieved using a diffraction grating, and powers of a few

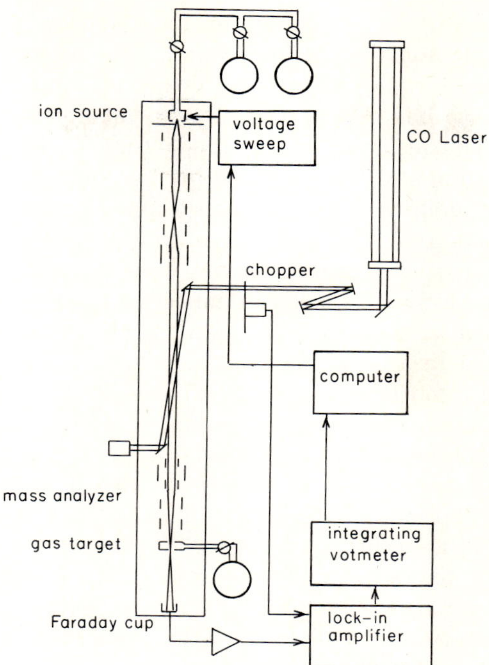

FIG. 8 Ion beam/laser beam apparatus using charge-exchange detection. [From Wing *et al.* (1976).]

milliwatts to 1 W are obtained per line. The source potential is swept by a microcomputer, usually in 0.1-V steps with an integration time of 8 sec per step, in order to Doppler-tune transitions into coincidence with a single selected laser line. After the interaction region, crossed electric and magnetic fields are used to mass-select the HD^+ ions; subsequently, the ion current is attenuated by passing the beam through a gas target such as N_2, Ar, or H_2, and the remaining ion current is measured at a Faraday cup. The current alters by a few parts per million at resonance; hence, the laser beam is chopped at 1 kHz and the alternating-current Faraday cup signal synchronously detected.

Twenty-five vibration–rotation transitions of HD^+ involving $v = 0–5$ and $N = 0–5$ have now been observed. All the transitions obey the harmonic oscillator selection rule $\Delta v = \pm 1$, and the measured frequencies are accurate to $\pm 0.0007\ \text{cm}^{-1}$, the main limitations being voltage calibration and uncertainty in the laser frequencies. The observed linewidths are in the region 7–25 MHz and become smaller at higher source potentials because of increased kinematic compression effects. The experimental results are listed in

Table I; we have already commented on the excellent agreement between experiment and the nonadiabatic calculations (Wolniewicz and Poll, 1980) for the low vibration–rotation levels. Hyperfine and spin–rotation structure is observed for each line with two or three strong components and one to four weaker components; the splittings range from 12–45 MHz, and a satisfactory interpretation has been provided by Ray and Certain (1977).

The apparatus employed by Carrington and his colleagues for photodissociation detection of infrared transitions in HD^+ is shown in Fig. 9. In essence, a tandem mass spectrometer has been adapted for spectroscopic purposes. Ions are formed by electron impact or plasma discharge, and the extracted ion beam is mass-analyzed by a 55° electromagnetic sector. The beam is focused at the intermediate slit, and the ion current may be measured immediately after the slit using an off-axis electron multiplier. The ion beam then passes into an electrostatic analyzer (ESA), which selects a particular fragment ion formed from the parent ion between the two analyzers. The fragment ion current may be measured either by means of a Faraday cup or with a second off-axis electron multiplier. Prior to the intermediate slit, the ion beam passes through a tube of 40-cm length, known as the drift tube. Potentials in the range -500 to $+500$ V may be applied to this tube, and the ESA is able to separate fragment ions formed in the drift tube from those formed elsewhere at earth potential due to their different kinetic energies. The path of the ion beam is evacuated to 7×10^{-8} torr.

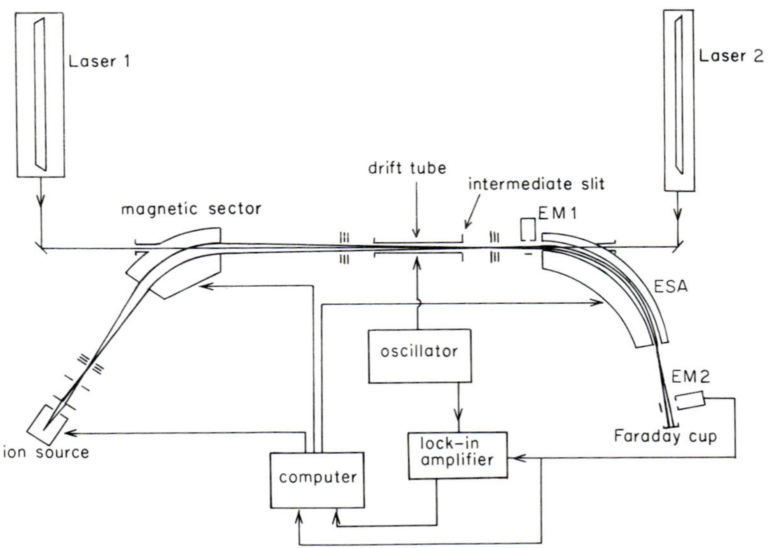

FIG. 9 Tandem ion beam/laser beam apparatus using photodissociation detection. [From Carrington and Buttenshaw (1981).]

Two lasers are available for spectroscopic studies. Laser 1 is of sealed tube design and may be operated with CO as the lasing gas to give maximum powers of 3 W in a single line, or with CO_2 giving powers of up to 25 W per line. The laser beam is reflected into parallel or antiparallel alignment with the ion beam and is focused at the intermediate slit. Laser 2 may be operated in a sealed or flowing gas mode with CO_2 as the lasing gas, and has a maximum output power of 60-W in a single line. The laser beam is aligned antiparallel with the ion beam and focused into the drift tube, although when both lasers are operated simultaneously a small angle is created between the two laser beams to avoid interference.

In experiments using the photodissociation method, both lasers may be used simultaneously. In studies of the $v = 18-16$ band, the vibration–rotation transitions are Doppler-tuned into resonance with one of the lasers (using CO_2), operated at low power to minimize power broadening; the transitions are detected by dissociation of the upper state using the other laser (also with CO_2) at high power to maximize the sensitivity. For the $v = 17-14$ band, however, only laser 1 operating with CO is used. Photodissociation occurs mainly from $v = 17$ with the CO laser, and mainly from $v = 18$ with the CO_2 laser.

Several different Doppler-tuning and signal detection modes are possible, but for the HD^+ studies the instrument is usually operated in the following manner. A PDP-11 minicomputer is used to scan the source potential over the range 2–10 kV, typically in 0.1–1 V steps. The magnetic field is simultaneously scanned by the computer to transmit the HD^+ beam at constant intensity. The ESA potential is linked automatically to the source potential, and is set to transmit photoproduct D^+ ions; the computer repeatedly reads the D^+ ion current and adjusts the magnetic sector for maximum intensity. A small constant offset voltage is applied to the drift tube, so that D^+ ions formed in the drift tube may be separated by the ESA from those formed elsewhere; in addition, a square-wave modulation of frequency up to 10 kHz and amplitude up to 10 V is also applied to the drift tube. Voltage modulation causes velocity modulation of the HD^+ beam and is, through the Doppler effect, equivalent to frequency modulation. Hence, the ESA voltage is synchronously modulated and a phase-sensitive detector used to demodulate the fragment ion signal; the direct-current signal output is read by the computer for each step of the source potential scan.

The experimental results are listed in Table I. A total of nine rotational components of the $v = 18-16$ band, seven components of the $v = 17-14$ band, and one component of the $v = 17-15$ band have now been observed. In many cases, it is possible to obtain two or three independent measurements of the same transition by using different laser lines, parallel or antiparallel, at different Doppler-tuning potentials. Consequently, the transition frequencies are determined to $\pm 0.0005\ \mathrm{cm}^{-1}$ and, as we have seen in Table I, the results provide a severe test of the nonadiabatic calculations of Wolniewicz and Poll (1980).

Linewidths down to 7 MHz are obtained, and each transition shows a doublet splitting, the two lines showing an intensity ratio of about 3:1, and separations of from 15–25 MHz. This splitting arises from nuclear hyperfine interaction, but it is not immediately obvious why a doublet separation is consistently observed, or what determines the size of the splitting. We must, therefore, consider the effects of the nuclear hyperfine and spin–rotation interactions in more detail. The effective Hamiltonian may be written in the form

$$\mathscr{H}_{\mathrm{eff}} = b_1 \mathbf{I}_1 \cdot \mathbf{S} + b_2 \mathbf{I}_2 \cdot \mathbf{S} + c_1 I_{1z} S_z + c_2 I_{2z} S_z + \gamma \mathbf{S} \cdot \mathbf{N}.$$

The first two terms represent the Fermi contact interaction between the electron spin ($\mathbf{S}$) and the proton ($\mathbf{I}_1$) and deuteron ($\mathbf{I}_2$) nuclear spins. The third and fourth terms represent the axial components of the dipolar hyperfine couplings, where z is the internuclear axis, while the fifth term represents the electron spin–rotation interaction. The simplified form of the dipolar terms implies neglect of nonaxial components which have matrix elements connecting different electronic states. We have also neglected the nuclear spin–rotation interaction and terms describing the deuterium nuclear quadrupole interaction.

Within the restrictions of our Hamiltonian, the hyperfine and fine structure of each vibration–rotation level will depend upon the values of the five constants b_1, b_2, c_1, c_2, and γ. As we saw in Section V, information about these constants in H_2^+ has been obtained by Jefferts (1968, 1969) for the levels $v = 4$–8. Although extrapolation of these results to the $v = 16$ and 18 levels of HD^+ is undesirably long, they do suggest that the Fermi contact interaction constant (b_1) for the proton is by far the most important, with a value of approximately 720 MHz. The deuteron constant b_2 would then be expected to be smaller in proportion to the difference in magnetogyric ratios for the proton and deuteron (i.e., by a factor of 0.1535). The dipolar constants c_1 and c_2 are fairly small for the lower vibrational levels and are likely to be even smaller for $v = 16$ and 18. Finally, the spin–rotation constant γ is only 32 MHz for $v = 0$ and becomes progressively smaller the higher the vibrational level; we anticipate γ values of only a few megahertz for $v = 16$ and higher. These considerations suggest that the most appropriate coupling scheme for the spin and rotational angular momenta is the following:

$$\mathbf{S} + \mathbf{I}_1 = \mathbf{G}_1, \qquad G_1 = 1; 0,$$

$$\mathbf{G}_1 + \mathbf{I}_2 = \mathbf{G}_2, \qquad G_2 = 2, 1, 0; 1,$$

$$\mathbf{G}_2 + \mathbf{N} = \mathbf{F}, \qquad F = N + G_2, \ldots, |N - G_2|.$$

Consequently, the hyperfine levels may be characterized by values of the quantum numbers G_1, G_2, N, F, and an appropriate energy level diagram for the transition 18,3–16,2 is shown in Fig. 10.

The selection rules for the electric-dipole allowed vibration–rotation transitions are $\Delta N = \pm 1$, $\Delta G_1 = \Delta G_2 = 0$, $\Delta F = \pm 1, 0$. The strongest lines, therefore, arise from the four transitions shown in Fig. 10, and any splitting between the lines will arise only from differences in the values of the constants for the two vibrational levels involved. The doublet splittings observed in the HD^+ lines, therefore, arise primarily from the difference in b_1 for $v = 16$ and 18. The weaker component involves the levels with $G_1 = 0$, and the stronger component has $G_1 = 1$. Consequently, the stronger component contains unresolved deuterium structure, and each component is probably broadened by unresolved spin–rotation structure. Absolute values of the hyperfine constants could only be obtained through the observation of transitions for which $\Delta G_1 \neq 0$, $\Delta G_2 \neq 0$. Unfortunately, these transitions have

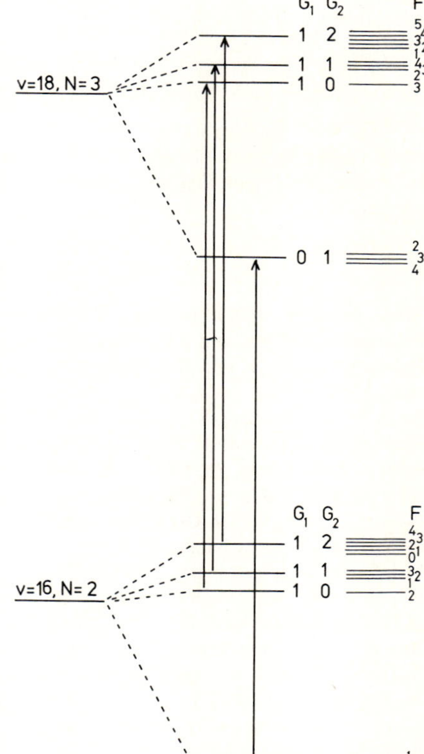

FIG. 10 Hyperfine and spin–rotation splitting for the 18,3–16,2 transition in HD^+.

very little electric dipole intensity, essentially because the electron spin is only very weakly coupled to the internuclear axis.

The present position for HD^+, therefore, is that experiment and theory agree to within a few thousandths of a wave number for levels of moderate v and N value. It may be possible to improve the accuracy of the experiments by an order of magnitude and, thus, to probe certain more subtle aspects of the molecular physics. The agreement between theory and experiment for the higher vibration–rotation levels is less satisfactory. A particularly challenging experimental problem is to detect transitions involving the highest vibrational levels, $v = 20$ and 21. Searches for such transitions have been made (Carrington and Kennedy, 1982), so far without success. The nuclear hyperfine structure for a level which is close to the lower dissociation limit (i.e., within 1 or 2 cm^{-1}) might be most unusual.

Because of the lack of an electric dipole moment in the homonuclear species H_2^+ and D_2^+, vibration–rotation spectra of the type measured for HD^+ are not accessible. It is possible that molecular quadrupole-induced transitions could be detected using the much larger infrared powers available from pulsed sources.

ACKNOWLEDGMENTS

A.C. thanks the Royal Society for a Research Professorship, and R.A.K. thanks the Science and Engineering Research Council, U.K. for a Post-Graduate Research Studentship. The ion beam research at Southampton has been supported generously by the SERC, for which we are also grateful.

REFERENCES

Alexandrow, W. (1926). *Ann. Phys.* **81**, 603.
Åsbrink, L. (1970). *Chem. Phys. Lett.* **7**, 549.
Baber, W. G. and Hassé, H. R. (1935). *Proc. Cambridge Phil. Soc.* **31**, 564.
Bates, D. R. (1951). *J. Chem. Phys.* **19**, 1122.
Bates, D. R., Ledsham, K., and Stewart, A. L. (1953). *Phil. Trans. R. Soc.* **A246**, 215.
Beckel, C. L., Hansen, B. D., III, and Peek, J. M. (1970). *J. Chem. Phys.* **53**, 3681.
Bishop, D. M. (1974). *Mol. Phys.* **28**, 1397.
Bishop, D. M. (1976). *At. Data Nucl. Data Tables* **18**, 521.
Bishop, D. M. (1977). *J. Chem. Phys.* **66**, 3842.
Bishop, D. M., and Cheung, L. M. (1977). *Phys. Rev. A* **16**, 640.
Bishop, D. M., and Cheung, L. M. (1978a). *J. Phys. B* **11**, 3133.
Bishop, D. M., and Cheung, L. M. (1978b). *Mol. Phys.* **36**, 501.
Bishop, D. M., and Cheung, L. M. (1980). *Adv. Quant. Chem.* **12**, 1.
Bishop, D. M., and Cheung, L. M. (1981). *J. Chem. Phys.* **75**, 3155.
Bishop, D. M., and Wetmore, R. W. (1973). *Mol. Phys.* **26**, 145.
Bishop, D. M., Shih, S., Beckel, C. L., Wu, F., and Peek, J. M., (1975). *J. Chem. Phys.* **63**, 4836.
Bleakney, W. (1930). *Phys. Rev.* **35**, 1180.
Born, M., and Huang, K. (1954). "Dynamical Theory of Crystal Lattices," Appendix VIII. Oxford Univ. Press, London and New York.

Born, M., and Oppenheimer, J. R. (1927). *Ann. Phys.* **84**, 457.
Bunker, P. R. (1974). *Chem. Phys. Lett.* **27**, 322.
Burrau, Ø. (1927). *Kgl. Danske Vidensk. Selskal Math. Fys. Med.* **7**, 14.
Carrington, A. (1979). *Proc. R. Soc. London Ser. A* **367**, 433.
Carrington, A., and Buttenshaw, J. (1981). *Mol. Phys.* **44**, 267.
Carrington, A., and Kennedy, R. A. (1982). Unpublished work.
Carrington, A., Buttenshaw, J., and Kennedy, R. A. (1982). *Mol. Phys.* **45**, 753.
Carrington, A., Buttenshaw, J., and Kennedy, R. A. (1983). *Mol. Phys.* **48**, 775.
Chupka, W. A., and Berkowitz, J. (1969). *J. Chem. Phys.* **51**, 4244.
Cohen, S., Hiskes, J. R., and Riddell, R. J., Jr. (1960). *Phys. Rev.* **119**, 1025.
Colbourn, E. A. (1976). *Chem. Phys. Lett.* **44**, 374.
Colbourn, E. A., and Bunker, P. R. (1976). *J. Mol. Spectrosc.* **63**, 155.
Cooley, J. W. (1961). *Math. Comp.* **15**, 363.
Condon, E. U. (1930). *Phys. Rev.* **35**, 658.
Coulson, C. A. (1937). *Trans. Faraday Soc.* **33**, 1479.
Dalgarno, A., Patterson, T. N. L., and Somerville, W. B. (1960). *Proc. R. Soc. London Ser. A* **259**, 100.
Dehmelt, H. G., and Jefferts, K. B. (1962). *Phys. Rev.* **125**, 1318.
Dickinson, B. N. (1933). *J. Chem. Phys.* **1**, 317.
Dunham, J. L. (1932). *Phys. Rev.* **41**, 713, 721.
Dunn, G. H. (1966). *J. Chem. Phys.* **44**, 2592.
Dunn, G. H. (1968). *Phys. Rev.* **172**, 1.
Dunn, G. H., and Van Zyl, B. (1967). *Phys. Rev.* **154**, 40.
Fournier, P., Lassier-Govers, B., and Comtet, G. (1979). *In* "Laser-Induced Processes in Molecules" (K. L. Kompa and S. D. Smith, eds.), p. 247. Springer-Verlag, Berlin and New York.
Gersten, J. I. (1969). *J. Chem. Phys.* **51**, 3181.
Gibson, D. K., and Los, J. (1967). *Physica* **35**, 258.
Gilbert, C. (1933). *Phil. Mag.* **16**, 929.
Gonsalves, J. W., and Moss, R. E. (1979). *Chem. Phys. Lett.* **62**, 534.
Herzberg, G., and Jungen, Ch. (1972). *J. Mol. Spectrosc.* **41**, 425.
Hunter, G., and Pritchard, H. O. (1967a). *J. Chem. Phys.* **46**, 2146.
Hunter, G., and Pritchard, H. O. (1967b). *J. Chem. Phys.* **46**, 2153.
Hunter, G., Gray, B. F., and Pritchard, H. O. (1966). *J. Chem. Phys.* **45**, 3806.
Hunter, G., Yau, A. W., and Pritchard, H. O. (1974). *At. Data Nucl. Data Tables* **14**, 11.
Hylleraas, E. A. (1931). *Z. Phys.* **71**, 739.
Jaffé, G. (1934). *Z. Phys.* **87**, 535.
Jefferts, K. B. (1968). *Phys. Rev. Lett.* **20**, 39.
Jefferts, K. B. (1969). *Phys. Rev. Lett.* **23**, 1476.
Johnson, V. A. (1941). *Phys. Rev.* **60**, 373.
Kaufman, S. L. (1976). *Opt. Commun.* **17**, 309.
Kołos, W. (1969). *Acta Phys. Acad. Sci. Hung.* **27**, 241.
Kołos, W. (1970). *Adv. Quant. Chem.* **5**, 99.
Kołos, W., and Wolniewicz, L. (1963). *Rev. Mod. Phys.* **35**, 473.
Lathouwers, L. (1978). *Phys. Rev. A* **18**, 2150.
Lozier, W. W. (1930). *Phys. Rev.* **36**, 1285.
Lozier, W. W. (1933). *Phys. Rev.* **44**, 575.
Luke, S. K., Hunter, G., McEachran, R. P., and Cohen, M. (1969). *J. Chem. Phys.* **50**, 1644.
Maas, J. G., van Asselt, N. P. F. B. and Los, J. (1975). *Chem. Phys.* **8**, 37.
McEachran, R. P., Veenstra, C. J., and Cohen, M. (1978). *Chem. Phys. Lett.* **59**, 275.
Morse, P. M. (1929). *Phys. Rev.* **34**, 57.
Morse, P. M., and Stueckelberg, E. C. G. (1929). *Phys. Rev.* **33**, 932.

Oka, T. (1980) *Phys. Rev. Lett.* **45**, 531.
Peatman, W. B. (1976). *J. Chem. Phys.* **64**, 4093.
Peek, J. M. (1965). *J. Chem. Phys.* **43**, 3004.
Peek, J. M. (1969). *J. Chem. Phys.* **50**, 4595.
Peek, J. M., Maas, J. G., and Los, J. (1975). *Chem. Phys.* **8**, 46.
Pollard, J. E., Trevor, D. J., Reutt, J. E., Lee, Y. T., and Shirley, D. A. (1982). *J. Chem. Phys.* **77**, 34.
Ray, R. D., and Certain P. R. (1977). *Phys. Rev. Lett.* **38**, 824.
Richardson, O. W. (1935). *Proc. R. Soc. London Ser. A* **152**, 503.
Richardson, C. B., Jefferts, K. B., and Dehmelt, H. G. (1968). *Phys. Rev.* **165**, 80.
Rothenberg, S., and Davidson, E. R. (1967). *J. Mol. Spectrosc.* **22**, 1.
Sandeman, I. (1935). *Proc. R. Soc. Edinburg.* **55**, 72.
Schrödinger, E. (1926a). *Ann. Phys.* **79**, 361, 489.
Schrödinger, E. (1926b). *Ann. Phys.* **81**, 109.
Spezeski, J. J. (1977). Ph.D thesis, Yale University, New Haven, Connecticut.
Stephen, M. J., and Auffray, J. P. (1959). *J. Chem Phys.* **31**, 1329.
Tadjeddine, M., and Parlant, G. (1977). *Mol. Phys.* **33**, 1797.
Takezawa, S. (1970a). *J. Chem. Phys.* **52**, 2575.
Takezawa, S. (1970b). *J. Chem. Phys.* **52**, 5793.
Teller, E. (1930). *Z. Phys.* **61**, 458.
Thomson, J. J. (1907). *Phil. Mag.* **13**, 561.
Unsöld, A. (1927). *Z. Phys.* **43**, 563.
Urey, H. C. (1926). *Phys. Rev.* **27**, 216 and references therein.
van Asselt, N. P. F. B., Maas, J. G., and Los, J. (1974). *Chem. Phys. Lett.* **24**, 555.
van Asselt, N. P. F. B., Maas, J. G., and Los, J. (1975). *Chem. Phys.* **11**, 253.
Van Vleck, J. H. (1936). *J. Chem. Phys.* **4**, 327.
von Busch, F., and Dunn, G. H. (1972). *Phys. Rev. A* **5**, 1726.
Villarejo, D. (1968). *J. Chem. Phys.* **49**, 2523.
Wallis, R. F., and Hulbert, H. M. (1954). *J. Chem. Phys.* **22**, 774.
Wilson, A. H. (1928). *Proc. R. Soc. London Ser. A* **118**, 617, 635.
Wind, H. (1965a). *J. Chem. Phys.* **42**, 2371.
Wind, H. (1965b). *J. Chem. Phys.* **43**, 2956.
Wing, W. H. (1981). Private communication.
Wing, W. H., Ruff, G. A., Lamb, W. E., Jr., and Spezeski, J. J. (1976). *Phys. Rev. Lett.* **36**, 1488.
Wolniewicz, L., and Poll, J. D. (1978). *J. Mol. Spectrosc.* **72**, 264.
Wolniewicz, L., and Poll, J. D. (1980). *J. Chem. Phys.* **73**, 6225.

Index

D

E

L

M

N

O

P

Q

R

S

T

U

V

X